Natural Biomaterials Engineering

Natural Biomaterials for Tissue Engineering

Edited by

Naveen Kumar
Division of Surgery, ICAR-Indian Veterinary Research Institute, Izatnagar, Uttar Pradesh, India

Sonal Saxena
Division of Veterinary Biotechnology, ICAR-Indian Veterinary Research Institute, Izatnagar, Uttar Pradesh, India

Vineet Kumar
Department of Veterinary Surgery and Radiology, College of Veterinary and Animal Sciences, Bihar Animal Sciences University, Kishanganj, Bihar, India

Anil Kumar Gangwar
Department of Veterinary Surgery & Radiology, College of Veterinary Science & Animal Husbandry, Acharya Narendra Deva University of Agriculture and Technology, Ayodhya, Uttar Pradesh, India

Dayamon David Mathew
Department of Veterinary Surgery & Radiology, Faculty of Veterinary and Animal Sciences, Banaras Hindu University, Rajiv Gandhi South Campus, Barkachha, Uttar Pradesh, India

Sameer Shrivastava
Division of Veterinary Biotechnology, ICAR-Indian Veterinary Research Institute, Izatnagar, Uttar Pradesh, India

Naresh Kumar Singh
Department of Veterinary Surgery & Radiology, Faculty of Veterinary and Animal Sciences, Banaras Hindu University, Rajiv Gandhi South Campus, Barkachha, Uttar Pradesh, India

ACADEMIC PRESS
An imprint of Elsevier

Academic Press is an imprint of Elsevier
125 London Wall, London EC2Y 5AS, United Kingdom
525 B Street, Suite 1650, San Diego, CA 92101, United States
50 Hampshire Street, 5th Floor, Cambridge, MA 02139, United States

Copyright © 2025 Elsevier Inc. All rights are reserved, including those for text and data mining, AI training, and similar technologies.

Publisher's note: Elsevier takes a neutral position with respect to territorial disputes or jurisdictional claims in its published content, including in maps and institutional affiliations.

No part of this publication may be reproduced or transmitted in any form or by any means, electronic or mechanical, including photocopying, recording, or any information storage and retrieval system, without permission in writing from the publisher. Details on how to seek permission, further information about the Publisher's permissions policies and our arrangements with organizations such as the Copyright Clearance Center and the Copyright Licensing Agency, can be found at our website: www.elsevier.com/permissions.

This book and the individual contributions contained in it are protected under copyright by the Publisher (other than as may be noted herein).

Notices
Knowledge and best practice in this field are constantly changing. As new research and experience broaden our understanding, changes in research methods, professional practices, or medical treatment may become necessary.

Practitioners and researchers must always rely on their own experience and knowledge in evaluating and using any information, methods, compounds, or experiments described herein. In using such information or methods they should be mindful of their own safety and the safety of others, including parties for whom they have a professional responsibility.

To the fullest extent of the law, neither the Publisher nor the authors, contributors, or editors, assume any liability for any injury and/or damage to persons or property as a matter of products liability, negligence or otherwise, or from any use or operation of any methods, products, instructions, or ideas contained in the material herein.

ISBN: 978-0-443-26470-2

For Information on all Academic Press publications
visit our website at https://www.elsevier.com/books-and-journals

Publisher: Stacy Masucci
Acquisitions Editor: Elizabeth Brown
Editorial Project Manager: Billie Jean Fernandez
Production Project Manager: Manju Paramasivam
Cover Designer: Greg Harris

Typeset by MPS Limited, Chennai, India

Contents

List of contributors	xi
Preface	xv

1   An introduction to biomaterials   1
Naveen Kumar, Vineet Kumar, Anil Kumar Gangwar, Sameer Shrivastava, Sonal Saxena, Sangeeta Devi Khangembam, Swapan Kumar Maiti, Rahul Kumar Udehiya, Mamta Mishra, Pawan Diwan Singh Raghuvanshi and Naresh Kumar Singh

1.1	Introduction	1
1.2	Definitions of biomaterials	3
1.3	Basic features required for the biomaterial	6
1.4	Classification of biomaterials	7
1.5	Biocompatibility	8
1.6	Summary	14
	References	15

2   Decellularization and characterization methods   17
Naveen Kumar, Vineet Kumar, Anil Kumar Gangwar, Sangeeta Devi Khangembam, Naresh Kumar Singh, Pawan Diwan Singh Raghuvanshi, Sameer Shrivastava, Sonal Saxena, Sangeetha P., Rahul Kumar Udehiya and Dayamon David Mathew

2.1	Introduction	17
2.2	Decellularization	18
2.3	Decellularized biomaterials	22
2.4	Characterization methods of decellularized tissues	26
2.5	Development of decellularized naturally derived biomaterials	34
2.6	In vitro evaluation of decellularized naturally derived biomaterials	38
	References	41

3   Crosslinking of biomaterials   47
Naveen Kumar, Anil Kumar Gangwar, Vineet Kumar, Dayamon David Mathew, Pawan Diwan Singh Raghuvanshi, Rahul Kumar Udehiya, Naresh Kumar Singh, Sangeeta Devi Khangembam, Sameer Shrivastava, Sonal Saxena and Rukmani Dewangan

3.1	Introduction	47
3.2	Advantages of crosslinking biomaterials	48
3.3	Disadvantages of crosslinking biomaterials	49

	3.4	Types of crosslinking	49
	3.5	Crosslinking agents	55
	References	68

4 Rumen-derived extracellular matrix scaffolds and clinical application   75

Ajit Kumar Singh, Naveen Kumar, Pawan Diwan Singh Raghuvanshi, Harendra Rathore, Anil Kumar Gangwar, Sameer Shrivastava, Sonal Saxena, Mohar Singh, Dayamon David Mathew and Karam Pal Singh

	4.1	Introduction	75
	4.2	Anatomy of ruminant stomach	77
	4.3	Decellularization of forestomach matrix	78
	4.4	Preparation of acellular matrices from bubaline rumen	79
	4.5	Development of 3-D bioengineered scaffolds from bubaline rumen	89
	4.6	Testing the efficacy of 3-D bioengineered scaffolds in a diabetic rat model	89
	4.7	Evaluation of bubaline rumen matrix in clinical cases	93
	4.8	Preparation of acellular matrices from caprine rumen	95
	4.9	Evaluation of caprine rumen matrix in clinical cases	100
	4.10	Conclusions	102
	References	103

5 Reticulum-derived extracellular matrix scaffolds   105

Naveen Kumar, Pawan Diwan Singh Raghuvanshi, Mohar Singh, Anwarul Hasan, Aswathy Gopinathan, Kiranjeet Singh, Ashok Kumar Sharma, Remya Vellachi, Sameer Shrivastava, Sonal Saxena, Swapan Kumar Maiti and Karam Pal Singh

	5.1	Introduction	105
	5.2	Preparation of acellular bovine reticulum extracellular matrix	107
	5.3	Preparation of acellular caprine reticulum extracellular matrix	114
	5.4	Evaluation of bovine reticulum extracellular matrix in rat model	117
	5.5	Conclusion	127
	References	128

6 Omasum-derived extracellular matrix scaffolds   131

Ashok Kumar Sharma, Naveen Kumar, Priya Singh, Pawan Diwan Singh Raghuvanshi, Mohar Singh, Sangeetha P., Sameer Shrivastava, Sonal Saxena, Swapan Kumar Maiti and Karam Pal Singh

	6.1	Introduction	131
	6.2	Preparation of acellular matrix from buffalo omasum	132
	6.3	Preparation of acellular matrix from goat omasum	138
	6.4	Experimental evaluation of acellular buffalo omasum in wound healing in rat model	141
	6.5	Conclusion	154
	References	155

7	**Gall bladder-derived extracellular matrix scaffolds**	**159**

Naveen Kumar, Anil Kumar Gangwar, Sangeeta Devi Khangembam,
Poonam Shakya, Ashok Kumar Sharma, Amit Kumar Sachan,
Ravi Prakash Goyal, Parvez Ahmed, Kiranjeet Singh, Aswathy Gopinathan,
Sonal Saxena, Sameer Shrivastava, Remya Vellachi, Dayamon David Mathew,
Swapan Kumar Maiti and Karam Pal Singh

7.1	Introduction	159
7.2	Anatomy of gallbladder (cholecyst)	160
7.3	Cholecyst-derived extracellular matrix	160
7.4	Preparation of acellular buffalo cholecyst-derived extracellular matrix	163
7.5	Preparation of acellular pig cholecyst-derived extracellular matrix	168
7.6	Preparation of acellular goat cholecyst-derived extracellular matrix	172
7.7	Experimental evaluation of buffalo cholecyst-derived extracellular matrix in a rat model	178
7.8	Experimental evaluation of pig cholecyst-derived extracellular matrix in a rat model	190
7.9	Conclusion	199
	References	199

8	**Aorta-derived extracellular matrix scaffolds and clinical application**	**203**

Jetty Devarathnam, Ashok Kumar Sharma, Naveen Kumar, Vineet Kumar,
Shruti Vora, Kaarthick D.T., Anil Kumar Gangwar, Rukmani Dewangan,
Himani Singh, Sameer Shrivastava, Sonal Saxena,
Kalaiselvan E., Shivaraju S. and Swapan Kumar Maiti

8.1	Introduction	203
8.2	Optimization of protocols for decellularization of buffalo aorta	206
8.3	Preparation and characterization of the buffalo aortic matrix	215
8.4	In vivo biocompatibility determination of acellular aortic matrix	220
8.5	Clinical applications in different species of animals	232
8.6	Conclusion	236
	References	237

9	**Pericardium-derived extracellular matrix scaffolds**	**241**

Naveen Kumar, Honjon Perme, Ashok Kumar Sharma, Himani Singh,
Rukmani Dewangan and Swapan Kumar Maiti

9.1	Introduction	241
9.2	Preparation of acellular goat pericardium matrix	243
9.3	Crosslinking of native and acellular goat pericardium matrix	243
9.4	Preparation of acellular buffalo pericardium matrix	244
9.5	Crosslinking of native and acellular buffalo pericardium matrix	245
9.6	In vivo evaluation in a rabbit model	256
9.7	Conclusion	264
	References	264

10	**Diaphragm-derived extracellular matrix scaffolds and clinical application**	**269**

Vineet Kumar, Naveen Kumar, Anil Kumar Gangwar, Kaarthick D.T., Harendra Rathore, Swapan Kumar Maiti, Ashok Kumar Sharma, Dayamon David Mathew, Jetty Devarathnam, Sameer Shrivastava, Sonal Saxena, Apra Shahi, Himani Singh and Karam Pal Singh

10.1	Introduction	269
10.2	Preparation and characterization of buffalo diaphragm matrix	271
10.3	Preparation of acellular diaphragmatic scaffold of buffalo origin	278
10.4	Biocompatibility evaluation of pig diaphragm	283
10.5	Biochemical changes in rabbit organs after subcutaneous implantation of bovine diaphragm	285
10.6	Dermal wound healing using primary mouse embryonic fibroblasts seeded buffalo acellular diaphragm matrix in a rat model	289
10.7	Clinical applications	298
10.8	Conclusion	300
	References	301

11	**Fish swim bladder–derived tissue scaffolds**	**307**

Remya Vellachi, Naveen Kumar, Ashok Kumar Sharma, Sonal Saxena, Swapan Kumar Maiti, Vineet Kumar, Dayamon David Mathew and Sameer Shrivastava

11.1	Introduction	307
11.2	Preparation of acellular matrix from fish swim bladder	309
11.3	Collection and isolation of bone marrow–derived mesenchymal stem cells	312
11.4	Seeding of stem cells on acellular fish swim bladder matrix	313
11.5	Attachment and growth of mesenchymal stem cells	314
11.6	Evaluation of bioengineered fish swim bladder matrices for skin wound healing in a rat model	316
11.7	Evaluation of acellular and crosslinked acellular swim bladder matrix for skin wound healing in rabbit model	327
11.8	Conclusion	340
	References	340

12	**Stem cell loading multiwalled carbon nanotubes-based bioactive scaffold for peripheral nerve regeneration in a rat model**	**345**

Mamta Mishra, Merlin Mamachan, Manish Arya, Swapan Kumar Maiti and Naveen Kumar

12.1	Introduction	345
12.2	Isolation culture and expansion of bone marrow-derived mesenchymal stem cell	346
12.3	Stem cell characterization	349
12.4	Cell viability assay	349
12.5	Colony forming assays	350

12.6	Tri-lineage staining characterization	350
12.7	Scaffold preparation and characterization	351
12.8	In vitro cytotoxicity testing	353
12.9	Preparation of stem cell loaded MWCNT-based bioactive nanoneural construct and assessment	353
12.10	Preparation of sciatic nerve injury model	355
12.11	Conclusion	362
	References	363

13 Cellular architects: mesenchymal stem cells crafting the future of regenerative medicine — 365
Rahul Kumar Udehiya and Sarita Kankoriya

13.1	Introduction	365
13.2	History of mesenchymal stem cells	366
13.3	Isolation and culture of mesenchymal stem cells	366
13.4	Characterization of mesenchymal stem cells: key aspects	370
13.5	Mesenchymal stem cells in disease treatment: a multifaceted approach	377
13.6	Summary and conclusions	378
	References	379

14 Polymeric nanoencapsulation for ameliorative application in rodent hepatic regeneration — 387
Deba Brata Mondal, Jithin Mullakkalparambil Velayudhan, Aishwarya Lekshman, Ravi Shankar Kumar Mandal, Raguvaran Raja and Naveen Kumar

14.1	Introduction	387
14.2	Catechin: A potent antioxidant and hepatoprotectant	390
14.3	Excipients	393
14.4	Novel drug delivery systems	393
14.5	Polymers: biomedical excipients for nanostructurization	397
14.6	Encapsulation of nanopolymer	403
14.7	Ionotropic gelation (polyelectrolyte complexation)	403
14.8	Catechin-loaded alginate polymeric nanoparticles (CA-AL-NP)	405
14.9	Catechin-loaded pectin polymeric nanoparticles (CA-PC-NP)	406
14.10	Catechin-loaded chitosan polymeric nanoparticles (CA-CH-NP)	407
14.11	Characterization and evaluation of the polymeric nanoparticles	408
14.12	Evaluation of in vitro antioxidant property of nanopolymer-coated catechin	410
14.13	Ameliorative application of nanopolymer-coated catechin for rodent hepatic regeneration	412
14.14	Conclusion	417
14.15	Notes	417
	References	419

Index — 427

List of contributors

Parvez Ahmed Department of Veterinary Surgery and Radiology, Acharya Narendra Deva University of Agriculture and Technology, Ayodhya, Uttar Pradesh, India

Manish Arya Department of Veterinary Surgery and Radiology, College of Veterinary and Animal Sciences, Bihar Animal Sciences University, Kishanganj, Bihar, India

Kaarthick D.T. Veterinary Clinical Complex, Veterinary College and Research Institute, Thanjavur, Tamil Nadu, India

Jetty Devarathnam Department of Surgery and Radiology, College of Veterinary Science, Sri Venkateswara Veterinary University, Proddatur, Andhra Pradesh, India

Rukmani Dewangan Department of Veterinary Surgery and Radiology, College of Veterinary Science and Animal Husbandry, Dau Shri Vasudev Chandrakar Kamdhenu Vishwavidyalaya, Durg, Chhattisgarh, India

Anil Kumar Gangwar Department of Veterinary Surgery & Radiology, College of Veterinary Science & Animal Husbandry, Acharya Narendra Deva University of Agriculture and Technology, Ayodhya, Uttar Pradesh, India

Aswathy Gopinathan Division of Surgery, ICAR-Indian Veterinary Research Institute, Izatnagar, Uttar Pradesh, India

Ravi Prakash Goyal Department of Veterinary Surgery and Radiology, Acharya Narendra Deva University of Agriculture and Technology, Ayodhya, Uttar Pradesh, India

Anwarul Hasan Division of Surgery, ICAR-Indian Veterinary Research Institute, Izatnagar, Uttar Pradesh, India

Kaarthick D.T. Veterinary Clinical Complex, Veterinary College and Research Institute, Thanjavur, Tamil Nadu, India

Kalaiselvan E. Division of Surgery, ICAR-Indian Veterinary Research Institute, Izatnagar, Uttar Pradesh, India

Sarita Kankoriya Veterinary Hospital, Shivpur, Department of Animal Husbandry, Mirzapur, Uttar Pradesh, India

Sangeeta Devi Khangembam Department of Veterinary Surgery and Radiology, Acharya Narendra Deva University of Agriculture and Technology, Ayodhya, Uttar Pradesh, India

Naveen Kumar Division of Surgery, ICAR-Indian Veterinary Research Institute, Izatnagar, Uttar Pradesh, India

Vineet Kumar Department of Veterinary Surgery and Radiology, College of Veterinary and Animal Sciences, Bihar Animal Sciences University, Kishanganj, Bihar, India

Aishwarya Lekshman Division of Medicine, ICAR-Indian Veterinary Research Institute, Izatnagar, Uttar Pradesh, India

Swapan Kumar Maiti Division of Surgery, ICAR-Indian Veterinary Research Institute, Izatnagar, Uttar Pradesh, India

Merlin Mamachan Division of Surgery, ICAR-Indian Veterinary Research Institute, Izatnagar, Uttar Pradesh, India

Ravi Shankar Kumar Mandal Department of Veterinary Medicine, College of Veterinary and Animal Sciences, Patna, Bihar, India

Dayamon David Mathew Department of Veterinary Surgery & Radiology, Faculty of Veterinary and Animal Sciences, Banaras Hindu University, Rajiv Gandhi South Campus, Barkachha, Uttar Pradesh, India

Mamta Mishra Department of Veterinary Surgery and Radiology, College of Veterinary and Animal Sciences, Bihar Animal Sciences University, Kishanganj, Bihar, India

Deba Brata Mondal Division of Medicine, ICAR-Indian Veterinary Research Institute, Izatnagar, Uttar Pradesh, India

Sangeetha P. Division of Surgery, ICAR-Indian Veterinary Research Institute, Izatnagar, Uttar Pradesh, India

Honjon Perme Division of Surgery, ICAR-Indian Veterinary Research Institute, Izatnagar, Uttar Pradesh, India

Pawan Diwan Singh Raghuvanshi Department of Veterinary Surgery and Radiology, College of Veterinary Science and Animal Husbandry, Nanaji Deshmukh Veterinary Science University, Mhow, Indore, Madya Pradesh, India

List of contributors

Raguvaran Raja Division of Medicine, ICAR-Indian Veterinary Research Institute, Izatnagar, Uttar Pradesh, India

Harendra Rathore Division of Surgery, ICAR-Indian Veterinary Research Institute, Izatnagar, Uttar Pradesh, India; Department of Veterinary Surgery and Radiology, College of Veterinary Science and Animal Husbandry, Nanaji Deshmukh Veterinary Science University, Mhow, Indore, Madya Pradesh, India

Amit Kumar Sachan Department of Surgery and Radiology, Acharya Narendra Deva University of Agriculture and Technology, Ayodhya, Uttar Pradesh, India

Sonal Saxena Division of Veterinary Biotechnology, ICAR-Indian Veterinary Research Institute, Izatnagar, Uttar Pradesh, India

Apra Shahi Department of Surgery and Radiology, College of Veterinary Science and Animal Husbandry, Nanaji Deshmukh Veterinary Science University, Jabalpur, Madhya Pradesh, India

Poonam Shakya Department of Surgery and Radiology, Acharya Narendra Deva University of Agriculture and Technology, Ayodhya, Uttar Pradesh, India

Ashok Kumar Sharma Division of Surgery, ICAR-Indian Veterinary Research Institute, Izatnagar, Uttar Pradesh, India

Shivaraju S. Division of Surgery, ICAR-Indian Veterinary Research Institute, Izatnagar, Uttar Pradesh, India

Sameer Shrivastava Division of Veterinary Biotechnology, ICAR-Indian Veterinary Research Institute, Izatnagar, Uttar Pradesh, India

Ajit Kumar Singh Department of Veterinary Surgery and Radiology, College of Veterinary Science and Animal Husbandry, Nanaji Deshmukh Veterinary Science University, Mhow, Indore, Madya Pradesh, India

Himani Singh Division of Surgery, ICAR-Indian Veterinary Research Institute, Izatnagar, Uttar Pradesh, India

Karam Pal Singh CADRAD, ICAR-Indian Veterinary Research Institute, Izatnagar, Uttar Pradesh, India; Department of Veterinary Surgery and Radiology, College of Veterinary Science and Animal Husbandry, Nanaji Deshmukh Veterinary Science University, Mhow, Indore, Madya Pradesh, India

Kiranjeet Singh Division of Surgery, ICAR-Indian Veterinary Research Institute, Izatnagar, Uttar Pradesh, India

Mohar Singh Division of Surgery, ICAR-Indian Veterinary Research Institute, Izatnagar, Uttar Pradesh, India; Department of Veterinary Surgery and Radiology, College of Veterinary Science and Animal Husbandry, Nanaji Deshmukh Veterinary Science University, Mhow, Indore, Madya Pradesh, India

Naresh Kumar Singh Department of Veterinary Surgery & Radiology, Faculty of Veterinary and Animal Sciences, Banaras Hindu University, Rajiv Gandhi South Campus, Barkachha, Uttar Pradesh, India

Priya Singh Department of Veterinary Surgery and Radiology, College of Veterinary Science and Animal Husbandry, Nanaji Deshmukh Veterinary Science University, Jabalpur, Madhya Pradesh, India

Rahul Kumar Udehiya Department of Veterinary Surgery and Radiology, Faculty of Veterinary and Animal Sciences, Institute of Agricultural Sciences, Rajiv Gandhi South Campus, Banaras Hindu University, Barkachha, Uttar Pradesh, India

Jithin Mullakkalparambil Velayudhan Department of Veterinary Medicine, College of Veterinary and Animal Sciences, Sardar Vallabhbhai Patel University of Agriculture and Technology, Meerut, Uttar Pradesh, India

Remya Vellachi Department of Veterinary Surgery and Radiology, College of Veterinary and Animal Sciences, Wayanad, Kerala, India

Shruti Vora Department of Veterinary Surgery and Radiology, College of Veterinary Science and Animal Husbandry, Junagadh Agricultural University, Junagadh, Gujarat, India

Preface

Tissue or organ defects, either congenital or acquired, pose challenging medical problems and more often require grafting for restoration. Autograft use is restricted by donor site morbidity and limited availability. Allografts and xenografts contain antigens that elicit host immune responses, which in turn lead to graft rejection. These antigens can be removed by tissue processing to reduce their immunogenicity. A lot of work has been done on the preparation of decellularized tissues and tissue-engineered scaffolds by various researchers.

This edition of *Natural Biomaterials for Tissue Engineering* is a comprehensive reference book that provides in-depth principles for supporting and enabling knowledge during tissue processing and scaffold preparation, as well as focusing on the different cell systems. The tissue fabrication process is illustrated with specific examples for over 30 tissues, which may soon lead to new tissue engineering therapies. Section coverage includes an overall introduction, decellularization protocols specific to each tissue, characterization, materials and methods, cell seeding process, preclinical evaluation in laboratory animals, clinical applications, limitations, conclusion, and future challenges. Readers may turn to this up-to-date coverage for a widespread understanding of regenerative medicine, which will be useful to students and experts alike.

Naveen Kumar
Sonal Saxena
Vineet Kumar
Anil Kumar Gangwar
Dayamon David Mathew
Sameer Shrivastava
Naresh Kumar Singh

An introduction to biomaterials

Naveen Kumar[1,], Vineet Kumar[2], Anil Kumar Gangwar[3], Sameer Shrivastava[4], Sonal Saxena[4], Sangeeta Devi Khangembam[5], Swapan Kumar Maiti[1], Rahul Kumar Udehiya[6], Mamta Mishra[2], Pawan Diwan Singh Raghuvanshi[7] and Naresh Kumar Singh[8]*

[1]Division of Surgery, ICAR-Indian Veterinary Research Institute, Izatnagar, Uttar Pradesh, India, [2]Department of Veterinary Surgery and Radiology, College of Veterinary and Animal Sciences, Bihar Animal Sciences University, Kishanganj, Bihar, India, [3]Department of Veterinary Surgery & Radiology, College of Veterinary Science & Animal Husbandry, Acharya Narendra Deva University of Agriculture and Technology, Ayodhya, Uttar Pradesh, India, [4]Division of Veterinary Biotechnology, ICAR-Indian Veterinary Research Institute, Izatnagar, Uttar Pradesh, India, [5]Department of Veterinary Surgery and Radiology, Acharya Narendra Deva University of Agriculture and Technology, Ayodhya, Uttar Pradesh, India, [6]Department of Veterinary Surgery and Radiology, Faculty of Veterinary and Animal Sciences, Institute of Agricultural Sciences, Rajiv Gandhi South Campus, Banaras Hindu University, Barkachha, Uttar Pradesh, India, [7]Department of Veterinary Surgery and Radiology, College of Veterinary Science and Animal Husbandry, Nanaji Deshmukh Veterinary Science University, Mhow, Indore, Madya Pradesh, India, [8]Department of Veterinary Surgery & Radiology, Faculty of Veterinary and Animal Sciences, Banaras Hindu University, Rajiv Gandhi South Campus, Barkachha, Uttar Pradesh, India

1.1 Introduction

A biomaterial can be defined as any material used to make devices to replace a part or a function of the body in a safe, reliable, economic, and physiologically acceptable manner. Some people refer to materials of biological origin, such as wood and bone, as biomaterials, but we refer to such materials as "biological materials." A variety of devices and materials are used in the treatment of disease or injury. Commonplace examples include sutures, tooth fillings, needles, catheters, bone plates, etc. A biomaterial may be a synthetic material used to replace part of a living system or to function in intimate contact with living tissue.

The definitions of a biomaterial cover an equally broad spectrum. In a very general sense, biomaterials are materials (synthetic and natural; solid and sometimes liquid) that are used in medical devices or in contact with biological systems [1].

---

*Present affiliation: Veterinary Clinical Complex, Apollo College of Veterinary Medicine, Jaipur, Rajasthan, India.

A biomaterial is any substance that has been engineered to interact with biological systems for a medical purpose—either a therapeutic (treat, augment, repair, or replace a tissue function of the body) or a diagnostic one. Earlier, these materials were only used in medical devices to treat or replace any tissue or improve the functions of organs. But later, it was found that the term nonviable given to them was inappropriate, as biomaterials have more applications than just implanted devices. Biomaterials are a major part of our routine practice in the diagnosis and treatment of several human diseases. Biomaterials are basically any materials (natural or synthetic) that are biologically compatible with the human body, are used to support, enhance, restore, or replace the biological function of damaged tissues, and are continuously in contact with body fluids. The use of the word "biomaterials" had been largely anticipated by the practical use of materials as biomaterials. Indeed, the presence of exogenous materials in the human body can be dated back to prehistory. In South Africa and India, the heads of large, biting ants were exploited to clamp wound edges together [2]. Over the centuries, other metals have been exploited: lead and silver, among others, with and without evidence of adverse reactions. Moreover, 4000 years ago, the Chinese carved bamboo sticks in the form of natural teeth to be inserted into jaws, just like current dental implants. Egyptians used precious metals for dental implants [3].

As a science, biomaterials are about 50 years old. The study of biomaterials is called biomaterials science or biomaterials engineering. Biomaterials science encompasses elements of medicine, biology, chemistry, tissue engineering, and materials science. The biocompatibility of a particular biomaterial is application-specific. Note that a biomaterial is different from a biological material, such as bone, that is produced by a biological system. A biomaterial that is biocompatible or suitable for one application may not be biocompatible with another [1]. Additionally, the purpose of replacing diseased or damaged parts of the human body has been pursued for centuries. During the 16th century, Gaspare Tagliacozzi and other pioneering plastic surgeons successfully used autogenous skin flaps to replace missing noses [4]. All these original surgical procedures had been performed without any awareness of the problems and limitations related to material science and biological phenomena; moreover, no knowledge of sterilization, immunological reactions, inflammation, or biodegradation was available at those times [2]. However, their "unconscious" success clearly demonstrates that the human body has an impressive ability to adapt itself and accommodate foreign materials. This allowed for traveling on the road to biomaterial evolution before taking into account the fundamental interactions between the body and the implanted materials; the systematic examination thereof only began about 150 years ago, when scientists and physicians started to scientifically evaluate how the body reacts to the presence of exogenous materials. The practical exploitation of materials as biomaterials then began to face the issue of biocompatibility.

The choice of biomaterial depends on the type of procedure being performed, the severity of the patient's condition, and the surgeon's preference. To be successful, the implant should effectively repair the defect it covers without eliciting an adverse tissue reaction while maintaining mechanical and biological integrity for a desired amount of time, from a few weeks to several years. Prostheses made from naturally derived biomaterials are frequently the decellularized extracellular matrix (ECM) of an animal (xenograft) or human (allograft). There are several advantages to using ECM biomaterials. First, all the molecules in an ECM can be broken down by normal enzymatic processes. Second, the three-dimensional structure and morphology of the ECM resemble the structure and morphology of the native tissue that is being replaced. Lastly, because of the nature of the biomaterial, researchers can design a prosthesis that works not only on a macroscopic level but also on the cellular level. Along with the advantages, there are certainly some disadvantages as well. ECMs are frequently immunogenic, causing harsh reactions in the host. There are also many ancillary molecules that change the way the prosthetic will interact when placed in vivo. Some molecules can enhance the regenerative capabilities of the surrounding tissue, while others provoke an immune response [5].

The success of any implant depends so much on the biomaterial used. Naturally derived biomaterials have been demonstrated to show several advantages compared to synthetic biomaterials. These are biocompatibility, biodegradability, and remodeling. Biomaterials can be classified into two main groups: synthetic and natural biomaterials. Synthetic biomaterials are classified as metals, ceramics, nonbiodegradable polymers, and biodegradable polymers. Some synthetic biomaterials are commercialized and applied in clinical treatment, such as metal hips, Dacron, and plastic intraocular lenses. However, synthetic biomaterials have some disadvantages, including that their structure and composition are not similar to those of native tissues or organs, and their biocompatibility and ability to induce tissue remodeling are low. Thus, other biomaterials have been developed that can overcome the disadvantages of synthetic biomaterials. Today, naturally derived biomaterials have been attracting scientists' interest all over the world.

Decellularized biomaterials are created by eliminating all cells from native tissues and organs. Physical, chemical, and enzymatic approaches are combined to make an effective decellularization protocol. Because of their advantages, naturally derived biomaterials are usually applied to replace or restore the structure and function of damaged tissues and organs. They have the ability to adequately support cell adhesion, migration, proliferation, and differentiation. In particular, when implanted into a defective area, naturally derived biomaterials can enhance the attachment and migration of cells from the surrounding environment and, therefore, induce extracellular matrix formation and promote tissue repair.

1.2 Definitions of biomaterials

As was described in the previous paragraph, biomaterials are characterized by a wide range of chemical compositions and properties, and they can be exploited in very many applications. Therefore, it is quite difficult to define them unambiguously Marin et al. [6] ascribed to Jonathan Cohen one of the earliest definitions of biomaterials, which dates to 1967. Dr. Cohen was an orthopedic surgeon, and exogenous materials had been used in orthopedic surgery for many years. He simply defined "biomaterials" as all materials that are used as implants, with the exception of drugs and soft biological tissues [7]. Indeed, this definition comes from the practical use of biomaterials in surgery, focusing on "hard" materials that are typically applied in orthopedics.

In April 1974, the Society for Biomaterials (SFB) was formally established and organized its inaugural annual symposium at Clemson University (SC, USA) [Society for Biomaterials (SFB), 2021]. In this symposium, a new definition of biomaterial was coined: "A biomaterial is a systematically, pharmacologically inert substance designed for implantation within or incorporation with a living system" [8]. These descriptions add to the many ways of looking at the same concept but expressing it in different ways. By contrast, a biological material is a material such as bone, skin, or an artery produced by a biological system. Artificial materials that are simply in contact with the skin, such as hearing aids and wearable artificial limbs, are not included in our definition of biomaterials since the skin acts as a barrier with the external world. Because the ultimate goal of using biomaterials is to improve human health by restoring the function of natural living tissues and organs in the body, it is essential to understand relationships among the properties, functions, and structures of biological materials. Thus, three aspects of study on the subject of biomaterials can be envisioned: biological materials, implant materials, and interaction between the two in the body. The inert nature of biomaterials was then confirmed, as was their difference from a drug; moreover, the role of biomaterials for implantation was stressed again. Professor Hench, who was a member of the SFB board, accepted the definition, even though he had already discovered bioactive glasses. Apparently, at that time, the absence of adverse responses to the presence of biomaterials in vivo was prioritized with respect to their bioactive effects [9].

A broader definition was formulated in 1982 during the "National Institutes of Health Consensus Development Conference Statement on the Clinical Applications of Biomaterials" (Bethesda, MD, USA): Biomaterial is "a substance (other than a drug) or combination of substances, synthetic or natural in origin, that can be used for any period of time, as a whole or as a part of a system that treats, augments, or replaces any tissue, organ, or function of the body" [10]. The difference from a drug is maintained, but now the definition includes materials of "natural" origin and specifies what biomaterials are intended for: they are part of a system that is conceived not only to replace but also to potentially treat and augment each tissue, each organ, and each function of the body. This definition marked a step forward to the most recent exploitation of biomaterials: they are not only simply "spare parts" of the body, but they can also play an active role, whatever their nature is. As an immediate consequence, possible applications increase as much as the availability of biomaterials increases.

As an immediate consequence, possible applications increase as much as the availability of biomaterials increases. In this aspect, the definition given by Prof. D. F. Williams, "a biomaterial is a nonviable material used in a medical device, intended to interact with biological systems," seems to be more appropriate [11]. Besides the current use of biological tissues from human cadavers (tissue banks) and from animals (after chemical treatments), tissue engineering techniques appear as extremely promising approaches to create viable tissues (and organs) by combining cells, scaffolds (biomaterials), and biochemical signals. During the European Society for Biomaterials 9th European Conference (Chester, UK) in 1991, the definition, approved in 1982, was improved, including "in order to maintain or improve the quality of life of the individual" (The European Society for Biomaterials, 1991). It clearly affirms that the aim of any biomaterial is not only the "survival" of the patient but also the maintenance or improvement of their quality of life. The huge impact of this statement can be easily understood considering that the WHO foresees that the proportion of the world's population over 60 years will nearly double (from 12% to 22%) between 2015 and 2050. Given the faster and faster advances in scientific research and technological applications, especially in biomedical field and in the clinical practice as well, we deem it opportune to apply the most inclusive definition of biomaterials.

A biomaterial is a nonviable substance or combination of substances that is used to interact with biological systems for a medical purpose. Biomaterials can be derived from natural sources (such as bone, collagen, and chitosan) or synthesized in the laboratory (such as metals, polymers, and ceramics). They are used in a wide range of medical applications, including implants, tissue engineering, drug delivery, and diagnostics. Biomaterials must be designed to be compatible with the body, meaning that they should not cause harm to the patient. They should also be able to perform their intended function, such as supporting or replacing damaged tissue or delivering drugs in a controlled manner. Some examples of biomaterials include:

- Metals: stainless steel, titanium, and cobalt-chromium alloys
- Polymers: silicone rubber, polyurethane, and polyethylene terephthalate (PET)
- Ceramics: hydroxyapatite, zirconia, and alumina
- Composite materials: carbon fiber-reinforced polymers, metal-ceramic composites
- Natural materials: collagen, chitosan, and hyaluronic acid

Biomaterials are essential for many modern medical treatments. They have improved the quality of life for millions of people around the world. As research in biomaterials science continues to advance, we can expect to see even more innovative and life-saving applications in the years to come. Here are some specific examples of how biomaterials are used today:

- Artificial heart valves: Biomaterials such as titanium and silicone are used to create artificial heart valves that can replace diseased or damaged valves.
- Joint replacements: Biomaterials such as titanium and cobalt-chromium alloys are used to create artificial hip and knee joints that can replace worn or damaged joints.
- Dental implants: Biomaterials such as titanium and zirconia are used to create dental implants that can replace missing teeth.
- Contact lenses: Biomaterials such as silicone hydrogel and polymethyl methacrylate (PMMA) are used to create contact lenses that can correct vision problems.
- Drug delivery systems: Biomaterials such as biodegradable polymers and nanoparticles are used to create drug delivery systems that can deliver drugs to the body in a controlled manner.
- Biomaterials are also being used to develop new and innovative medical treatments, such as:

- Tissue engineering: Biomaterials are being used to create scaffolds and other materials that can be used to grow new tissue or organs. This could potentially lead to treatments for a variety of diseases and conditions, such as spinal cord injuries, burns, and cancer.
- Regenerative medicine: Biomaterials are being used to develop new treatments that can help the body's own cells repair and regenerate damaged tissues and organs. This could potentially lead to treatments for a variety of diseases and conditions, such as heart disease, stroke, and Alzheimer's disease.

1.3 Basic features required for the biomaterial

Since the biomaterials are in direct contact with the body tissues and body fluid, there are some basic features required for the biomaterials, such as biocompatibility, inertness, safety, stability, cost-effectiveness, and ease of fabrication, as shown in Fig. 1.1.

1.3.1 Characteristics of biomaterials

The requirements for designing and selecting biomaterials depend on the type of medical application. The biomaterial must have some unique characteristics that can have potent applications in the biomedical field for a longer duration without immune rejection (Fig. 1.1). Some of these characteristics are described here [12,13].

1. Outstanding biocompatibility

Figure 1.1 Diagrammatic demonstration of substance design requirements for biomaterials.

2. Sufficient mechanical properties
3. High-quality physical and chemical properties
4. Enough resistance to wear
5. Enough resistance to rust
6. Osseo-integration (for bone implants)

1.3.2 Requirements for scaffolds used in tissue engineering

A scaffold is an artificial three-dimensional frame structure that serves as a mimic of the extracellular matrix for cellular adhesion, migration, proliferation, and tissue regeneration in three dimensions. Its architecture and microstructure define the ultimate shape and structure of the regenerated tissue and organs. An ideal scaffold for tissue engineering should possess the following characteristics [9,14,15].

1. It is highly biocompatible and does not elicit an immunological or clinically detectable foreign body reaction.
2. It is three-dimensional and capable of regenerating tissue and organs in their normal physiological shape.
3. It is highly porous, with an interconnected pore network available for cell growth and nutrient and metabolic waste transport.
4. It has a suitable surface chemistry, allowing for cell attachment, migration, proliferation, and differentiation.
5. It has controllable degradation and resorption rates that match the rate of tissue growth in vitro, ex vivo, and in vivo for biodegradable or resorbable materials.
6. It possesses the appropriate mechanical properties that match those of normal tissue and organs.
7. It has a bioactive surface to encourage faster regeneration of the tissue.

1.4 Classification of biomaterials

Biomaterials can be broadly classified on the basis of their source, such as natural and synthetic biomaterials, which can also be further subclassified as shown in Table 1.1.

Start with understanding of the properties of materials that can meet the requirements of biomaterial implantation. Table 1.2 illustrates some of the advantages, disadvantages, and applications of four groups of synthetic (manmade) materials used for implantation. Reconstituted (natural) materials such as collagen have been used for replacements (e.g., arterial wall, heart valve, and skin).

The materials to be used in vivo have to be approved by the FDA (United States Food and Drug Administration). If a proposed material is substantially equivalent to one used before the FDA legislation of 1976, then the FDA may approve its use on a premarket approval (PMA) basis. This process, justified by experience with a similar material, reduces the time and expense associated with the use of the proposed material. Otherwise, the material has to go through a series of "biocompatibility" tests.

Table 1.1 Classification based on the occurrence of biomaterials.

Biomaterials	Example
Naturally extracted biomaterials	
Protein-based biomaterials	Collagen, fibrin, and silk
Polysaccharide-based biomaterials	Chitosan (CS), alginate, and hyaluronan
Gum-based biomaterials	Pectin, xanthum gum, dextran
Biologically derived materials	porcine/bovine pericardium
Synthetically derived biomaterials	
Polymer-based biomaterials	Polymethylmethacrilate (PMMA), ultrahigh molecular weight polyethylene (UHMWPE), polylactic acid (PLA), polytetrafluoroethylene (PTFE), nylon, polyethylene, polyurethane, celluloid, cellophane, polycaprolactone (PCL), polyglycolic acid (PGA), polylactic acid (PLA), poly-lactic-co-glycolic acid (PLGA), poly(ethers) including polyethylene glycol (PEG), polyvinyl alcohol (PVA) and polyurethanes (PUs)
Peptide-based biomaterials	Short amino acids and self-assembling peptides
Ceramic-based biomaterials	Bioactive glass, alumina, zirconia, hydroxyapatite, beta tricalcium phosphate, pyrolytic carbon
Metal-based biomaterials	Stainless steel, CoCrMo, titanium, Ti6Al4V, nitinol, nickel, platinum, tantalum
Biocomposites or composites biomaterials	Polysaccharides, proteins, sugars, lignins, synthetic polymers

1.5 Biocompatibility

Biocompatibility is the ability of a material to perform with an appropriate host response in a specific application. It is a critical factor in the design and development of medical devices and implants, as well as in the selection of materials for other biomedical applications, such as tissue engineering and drug delivery systems. This means that the material should not cause any adverse effects on the living system it is in contact with, such as inflammation, infection, or toxicity. Biocompatibility is a complex topic, and there is still much to learn about how to develop and select biocompatible materials for specific applications. However, advances in biomaterials science have led to the development of a wide range of biocompatible materials that are used to improve the health and well-being of patients around the world.

A biocompatible material is one that does not cause any harmful or adverse effects in the body when it is implanted or used in a medical device. It should not be toxic, carcinogenic, or mutagenic. It should also not cause inflammation,

Table 1.2 Class of materials used in the body.

Materials	Advantages	Disadvantages	Examples
Polymers (nylon, silicone rubber, polyester, polytetrafluoroethylene, etc.)	Resilient, easy to fabricate	Not strong, deforms with time, may degrade	Sutures, blood vessels other soft tissues, sutures, hip socket, ear, nose
Metals (Ti and its alloys, Co-Cr alloys, Au, Ag stainless steels, etc.)	Strong, tough, ductile	May corrode, dense, difficult to make	Joint replacements, dental root implants, pacer and suture wires, bone plates and screws
Ceramics (alumina zirconia, calcium phosphates including hydroxyapatite, carbon)	Very biocompatible	Brittle, not resilient, weak in tension	Dental and orthopedic implants
Composites (carbon-carbon, wire- or fiber-reinforced)	Strong, tailor-made	Difficult to make	Bone cement, dental resin

irritation, or other tissue damage. Biocompatibility testing is a critical part of the regulatory approval process for medical devices and implants. Manufacturers must demonstrate that their products are biocompatible before they can be marketed and sold. Biocompatibility testing is also used in the development of new materials and technologies for biomedical applications. Biocompatibility is important for a wide range of medical devices and implants, including:

- Contact lenses
- Dentures
- Artificial joints
- Pacemakers
- Stents
- Heart valves
- Tissue scaffolds for engineering new tissues
- Drug delivery systems

There are a number of factors that can influence the biocompatibility of a material, including its chemical composition, physical properties, surface characteristics, and how it is processed and sterilized. Biocompatibility is also affected by the type of tissue or organ that the material is in contact with.

Some common biocompatible materials include:

- Metals such as titanium and stainless steel
- Polymers such as silicone and polyurethane
- Ceramics such as alumina and zirconia
- Natural materials such as collagen and bone

Biocompatibility is an important consideration in a wide range of biomedical applications, including:

- Medical implants, such as hip replacements and heart valves
- Drug delivery systems, such as transdermal patches and stents
- Tissue engineering scaffolds
- Bioartificial organs
- Medical devices, such as catheters and surgical instruments

Biocompatibility testing is a critical part of the regulatory approval process for medical devices. Manufacturers must demonstrate that their devices are biocompatible before they can be marketed. Biocompatibility testing is also used to develop new materials and devices for medical use. There are a number of different factors that can affect the biocompatibility of a material, including:

- The type of material
- The surface properties of the material
- The purity of the material
- The duration of contact with the living system
- The location of the material in the body
- The individual's immune response

Biocompatibility testing typically involves a combination of in vitro and in vivo tests. In vitro tests are performed on cells or tissues in a laboratory setting. In vivo tests are performed on animals. Biocompatibility testing can be complex and expensive, but it is essential for ensuring the safety and efficacy of medical devices and implants.

Why is biocompatibility important?

Biocompatibility is important because it can affect the success of medical devices and implants. If a material is not biocompatible, it can cause the body to reject the device or implant. This can lead to complications such as infection, inflammation, and pain. In some cases, it can also lead to serious health problems, such as organ failure. Biocompatibility is also important for the development of new medical technologies. For example, scientists are working on developing new biomaterials that can be used to engineer new tissues and organs. These biomaterials need to be highly biocompatible in order to be safe and effective.

How is biocompatibility tested?

Biocompatibility testing is a complex process that involves a variety of different tests. The specific tests that are performed will depend on the type of material being tested and the intended use of the material. Biocompatibility testing is typically performed in a laboratory setting and involves a combination of in vitro and in vivo tests.

In vitro tests are performed on cells and tissues outside of the body, while in vivo tests are performed on animals. Some common biocompatibility tests include:

- Cytotoxicity testing: This test evaluates the potential of a device material to damage or kill cells.
- Sensitization testing: This test evaluates the potential of a device material to cause an allergic reaction.
- Irritation testing: This test evaluates the potential of a device material to cause inflammation or irritation.
- Pyrogenicity testing: This test evaluates the potential of a device material to cause fever.
- Genotoxicity testing: This test evaluates the potential of a device material to damage DNA.
- Systemic toxicity testing: This test evaluates the potential of a device material to cause harmful effects to the body's organs and systems.
- Hemocompatibility testing: This test evaluates the potential of a device material to interact with blood in a harmful way.

The specific biocompatibility tests that are required for a given medical device will depend on the type of device, the materials it is made of, and how it will be used.

Who performs biocompatibility testing?

Biocompatibility testing can be performed by a variety of different organizations, including medical device manufacturers, contract research organizations, and government agencies. Medical device manufacturers are typically responsible for performing the biocompatibility testing required for their products, but they may outsource some of this testing to contract research organizations. Government agencies, such as the US Food and Drug Administration (FDA), may also perform biocompatibility testing on medical devices as part of the regulatory approval process.

How long does biocompatibility testing take?

The amount of time it takes to perform biocompatibility testing varies depending on the type of tests that are required and the complexity of the device. However, it is important to note that biocompatibility testing can be a lengthy and expensive process. This is one of the reasons why it is important for medical device manufacturers to carefully consider the materials they use in their products and to design their products in a way that minimizes the risk of adverse reactions.

Biocompatibility assessment is a complex procedure aimed at verifying the capacity of a given material to avoid adverse reactions and also to correctly perform the intended function when in contact with (or inserted into) the biological environment. The fundamental concepts related to biocompatibility are available at the FDA (https://www.fda.gov). The FDA assesses the biocompatibility of the whole device, not just the component materials. The ISO 10993-1 establishes criteria for the biological evaluation of medical devices, again confirming that biological tests have to be "performed on the final medical device, or representative samples from the final device or materials processed in the same manner as the final medical device (including sterilization, if needed)" [16]. Thus, the term "biocompatibility" has to include not only what is commonly meant as "biological compatibility" but

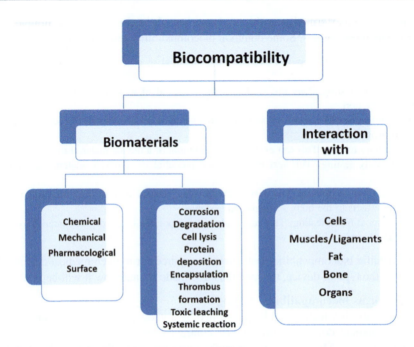

Figure 1.2 Schematic illustration of biocompatibility.

also a functional evaluation of the entire implantable system. For sure, several aspects determine the biocompatibility of a given material, also considering the duration of contact with its biological counterparts: chemical composition, mechanical behavior, and physical shape.

The success of a biomaterial or an implant is highly dependent on three major factors: the properties and biocompatibility of the implant (Fig. 1.2), the health condition of the recipient, and the competency of the surgeon who implants and monitors its progress. It is easy to understand the requirements for an implant by examining the characteristics that a bone plate must satisfy for stabilizing a fractured femur after an accident. These are:

1. Acceptance of the plate to the tissue surface, that is, biocompatibility (this is a broad term and includes points 2 and 3)
2. Pharmacological acceptability (nontoxic, nonallergenic, nonimmunogenic, noncarcinogenic, etc.)
3. Chemically inert and stable (no time-dependent degradation)
4. Adequate mechanical strength
5. Adequate fatigue life
6. Sound engineering design
7. Proper weight and density
8. Relatively inexpensive, reproducible, and easy to fabricate and process for large-scale production.

There are many cases of biologically compatible materials that do not pass the functional check. A clear example is represented by Teflon (polytetrafluoroethylene): it does not evoke any particular biological reaction, and therefore it can be considered "biologically" compatible. When Teflon was used for replacing the temporomandibular joint, it resulted in substantial fragmentation and caused huge foreign body giant cell responses that progressively eroded adjacent structures [17]. So, Teflon is not "functionally" compatible with respect to the foreseen application. Several aspects determine the biocompatibility of a given material, also considering the duration of contact with its biological counterparts: chemical composition, mechanical behavior, and physical shape.

The reactions induced by metals were studied early in the 19th century, since metals were exploited at that time. Gold, silver, lead, nickel, and platinum were studied in animals, and platinum was found to be well tolerated [2]. Other metals resulted in fast corrosion (iron and steel); others in tissue discoloration (copper, magnesium, aluminum alloy, zinc, and nickel); others exhibited inadequate mechanical features (gold, silver, lead, and aluminum). CoCrMo alloys, titanium, and its alloys were then proposed as promising candidates for biomedical applications due to their biocompatibility, which is combined with good mechanical properties. Indeed, the major limitation of metals in contact with biological fluids is corrosion, which is the sum of (electro)chemical phenomena that commonly take place in the presence of water and oxygen. After surgical implantation, all metallic devices (i.e., articular prostheses, plates, and screws) are exposed to the attack of the body's structures that act as a defense system. Under these circumstances, some metals are oxidized; they release ions that can be toxic both locally and systemically. Consequently, the device degrades and is no longer able to appropriately perform the intended function.

Polymeric materials entered the field of biomaterials quite recently: the first plastic material (celluloid) was developed in the 1860s, and others followed thereafter [18]. Ratner and Zhang dated the first use of a polymer (cellophane) as implantable material in 1939: it was applied to wrap blood vessels, inducing a fibrotic reaction to limit the further expansion of the aneurysm. Interestingly, Nobel Prize recipient Albert Einstein was diagnosed with an abdominal aortic aneurysm and treated with cellophane wrapping [19]. Two years later, nylon and poly(methyl methacrylate) were tested in vivo, then polyethylene followed [2]. The world of polymeric materials is continuously growing and now includes a huge number of different substances that can be produced in a variety of physical shapes with a variety of physical features: from solids to fibers, from thin sheets to thick plates, from hard to soft components, from inert to bioactive products. Manufacturing techniques were recently boosted by the introduction of electrospinning apparatuses and 3D printers; they both help to customize polymers with respect to an increasing number of applications. The major aspect limiting polymers' biocompatibility is due to their chemical formulation: they always include additives that can be released in vivo, resulting in adverse reactions. They are plasticizers, pigments, antioxidants, radiopaque agents, polymerization inhibitors and initiators, and, of course, monomers. Their undesired effects were observed as early as the mid-1900s [20]. Ceramics can be defined as inorganic, nonmetallic materials, and they are widely used as biomaterials, especially for dental restoration and bone-contacting applications.

Indeed, ceramics were suggested as an alternative to metals and polymers with the purpose of enhancing bone fixation and integration. Generally, they are biologically compatible in the sense that they are inert or bioactive but do not elicit adverse reactions. Their marked limitation is due to stiffness and brittleness, which both represent severe drawbacks in many practical applications [21]. Only alumina and zirconia have been used for the production of hip prostheses. Interestingly, some ceramics (e.g., bioactive glasses) are able to form a direct bond with living tissues [22,23]. The multistage process taking place on bioactive glass surfaces in vivo results in the formation of a strong interface between bone and a dense layer of hydroxyapatite and carbonate-apatite [24]. Thus, a stable bond to bone is established, which promotes osteoconduction and also osteoinduction [25].

1.6 Summary

Biomaterials are intended for clinical applications. Therefore, rigorous, systematic, and standardized testing protocols must be adhered to that ensure the biomaterial will function effectively for the intended application and that the outcome of that application does not harm the patient. To do no harm and therefore ensure safety must be the hallmark characteristic of any biomaterial that is implanted into a patient. To achieve that goal, standardized testing procedures must carefully transition between in vitro and in vivo methodologies and must provide accurate and reproducible outcome data. Moreover, standardized testing procedures must duplicate as closely as possible the actual intended clinical application. The emphasis is that the outcome of implantation of a biomaterial in a patient, that is, the consequence of the interface with the biological system, should be predictable, as intended, and beneficial, and that the resultant biomaterial interface, biomaterial as a whole, and degradation products of the implanted biomaterial provide an improvement in the quality of life for the patient. The overarching theme for this fundamental is that the biomaterial will be biocompatible. Finally, regardless of whether you are a healthcare provider responsible for the direct, hands-on care of a patient or if you are a bench scientist, the patient receiving the biomaterial you designed and developed trusts you. That trust resides in the belief that the biomaterial being implanted will be helpful, not harmful. Consequently, certainty about biocompatibility must be absolute. In line with this fundamental is the requirement for standards. Standards are the alphabet scripting the common language that permits rational communication within a laboratory and among laboratories. Standards enable confirmation of outcomes among laboratories through reproducible data. Confirmation is validation that an outcome is not serendipitous but due to scientific design. When outcome data are reproducible and based on standardized procedures, decisions can be made about whether to modify a biomaterial, accept it as is for the intended clinical use, or abandon it completely because it is unsafe and/or not efficacious. The intent of adopting standards is to ensure that laboratories have an operational consensus and a clear definition of acceptable property thresholds for biomaterials. For example, if a biomaterial property must comply with a standard that is safe, what is the safety threshold? What is meant by safe? How do we determine safety? What are the criteria that can be measured that will validate safety?

References

[1] Ratner BD. Biomaterials science: An interdisciplinary endeavor. In: Ratner BD, Hoffman AS, Schoen FJ, Lemons JE, editors. Biomaterials Science. Academic Press; 2006.
[2] Ratner BD, Zhang G. A history of biomaterials. In: Ratner BD, Hoffman AS, Schoen FJ, Lemons JE, editors. Biomaterials science: An introduction to materials in medicine. 4th edition London: Elsevier Academic Press; 2020.
[3] Ali F. History of dental implants. In memoriam: Dr. Per-Ingvar Branemark, the man who made people smile. Int J Adv Res Ideas Innov Technol. 2019;5:123–4.
[4] Barker CF, Markmann JF. Historical overview of transplantation. Cold Spring Harb Perspect Med. 2013;3:a014977. Available from: https://doi.org/10.1101/cshperspect. a0149779 –a014977.
[5] Ratner BD, Hoffman AS, Schoen FJ, Lemons JE. An introduction to materials in medicine. Endorsed by society of biomaterials. Elsevier; 1996.
[6] Marin E, Boschetto F, Pezzotti G. Biomaterials and biocompatibility: an historical overview. J Biomed Mater Res Part A. 2020;108:1617–33. Available from: https://doi.org/10.1002/jbm.a.36930.
[7] Cohen J. Biomaterials in orthopedic surgery. Am J Surg. 1967;114:31–41. Available from: https://doi.org/10.1016/0002-9610(67)90037-2.
[8] Park JB, editor. Introduction. In: Biomaterials science and engineering. Boston, MA, USA: Springer; 1984. ISBN 978-1-4612-9710-9.
[9] Hubbell JA. Biomaterials in tissue engineering. Biotechnology. 1995;13(6):565–76.
[10] Bergmann CP, Aisha S. Biomaterials. In Bergmann CP, editor. Dental Ceramics. Berlin, Germany: Springer; 2013, p. 9–13; ISBN 978-3-642–38224-6.
[11] Williams DF, David F. European society for biomaterials definitions in biomaterials. Proceedings of the consensus conference of the European society for biomaterials, 4. Amsterdam, The Netherlands; Chester, UK: Elsevier; 1986. p. 3–5.
[12] Kadam AG, Pawar SA, Abhang SA. A Review on finite element analysis of different biomaterials used in orthopedic implantation. Inter Res J Eng Tech 2017;4:2192–5.
[13] Vesque JL, Hermawan H, Dube D, Mantovani D. Design of a pseudo-physiological test bench specific to the development of biodegradable metallic biomaterials. Acta Biomaterialia 2008;4:284–95.
[14] Cima LG, Vacanti JP, Vacanti C, Ingber D, Mooney D, Langer R, et al. Tissue engineering by cell transplantation using degradable polymer substrates. J Biomech Eng 1991;113(2):143–51.
[15] Peter SJ, Miller MJ, Yasko AW, Yaszemski MJ, Mikos AG, et al. Polymer concepts in tissue engineering. J Biomed Mater Res 1998;43(4):422–7.
[16] ISO 10993-1. Biological evaluation of medical devices – Part 1. Guidance Document; 2009.
[17] Lee KC, Eisig SB, Perrino MA. Foreign body giant cell reaction to a Proplast/Teflon interpositional implant: a case report and literature review. J Oral Maxillofac Surg. 2018;76:1719–24. Available from: https://doi.org/10.1016/j.joms.2018.03.002.

[18] Feldman D. Polymer history. Des Monomers Polym. 2008;11:1−15. Available from: https://doi.org/10.1163/156855508x292383.
[19] Castro JC. Albert Einstein and his abdominal aortic aneurysm. Gac Med Mex. 2011;147:74−6.
[20] LeVeen HH, Rarberio JR. Tissue reaction to plastics used in surgery with special reference to Teflon. Ann Surg. 1949;129:74−84. Available from: https://doi.org/10.1097/00000658-194901000-00008.
[21] Vallet-Regí M. Ceramics for medical applications. J Chem Soc Dalton Trans. 2001;97−108. Available from: https://doi.org/10.1039/b007852m.
[22] Hench LL. Bioactive ceramics. Ann N Y Acad Sci 1988;523:54−71. Available from: https://doi.org/10.1111/j.1749-6632.1988.
[23] Hench LL, Jones JR. Bioactive glasses: frontiers and challenges. Front Bioeng Biotechnol. 2015;3:194. Available from: https://doi.org/10.3389/fbioe.2015.00194.
[24] Nordstrom EG, Munoz OLS. Physics of bone bonding mechanism of different surface bioactive ceramic materials in vitro and in vivo. Bio-Medical Mater Eng 2001;11:221−31.
[25] Baino F, Hamzehlou S, Kargozar S. Bioactive glasses: where are we and where are we going? J Funct Biomater. 2018;9:25. Available from: https://doi.org/10.3390/jfb9010025.

Decellularization and characterization methods

Naveen Kumar[1],*, Vineet Kumar[2], Anil Kumar Gangwar[3], Sangeeta Devi Khangembam[4], Naresh Kumar Singh[5], Pawan Diwan Singh Raghuvanshi[6], Sameer Shrivastava[7], Sonal Saxena[7], Sangeetha P.[1], Rahul Kumar Udehiya[8] and Dayamon David Mathew[5]

[1]Division of Surgery, ICAR-Indian Veterinary Research Institute, Izatnagar, Uttar Pradesh, India, [2]Department of Veterinary Surgery and Radiology, College of Veterinary and Animal Sciences, Bihar Animal Sciences University, Kishanganj, Bihar, India, [3]Department of Veterinary Surgery & Radiology, College of Veterinary Science & Animal Husbandry, Acharya Narendra Deva University of Agriculture and Technology, Ayodhya, Uttar Pradesh, India, [4]Department of Veterinary Surgery and Radiology, Acharya Narendra Deva University of Agriculture and Technology, Ayodhya, Uttar Pradesh, India, [5]Department of Veterinary Surgery & Radiology, Faculty of Veterinary and Animal Sciences, Banaras Hindu University, Rajiv Gandhi South Campus, Barkachha, Uttar Pradesh, India, [6]Department of Veterinary Surgery and Radiology, College of Veterinary Science and Animal Husbandry, Nanaji Deshmukh Veterinary Science University, Mhow, Indore, Madya Pradesh, India, [7]Division of Veterinary Biotechnology, ICAR-Indian Veterinary Research Institute, Izatnagar, Uttar Pradesh, India, [8]Department of Veterinary Surgery and Radiology, Faculty of Veterinary and Animal Sciences, Institute of Agricultural Sciences, Rajiv Gandhi South Campus, Banaras Hindu University, Barkachha, Uttar Pradesh, India

2.1 Introduction

An ideal biomaterial should initiate the minimal immune response possible and allow cellular infiltration while maintaining its structure and performing its intended function. Eventually, it will degrade and promote healthy tissue regeneration rather than fibrous scarring. The physiological similarity of the biomaterials is probably the most important factor governing their ability to obtain approval for use. Native extracellular matrices (ECMs) provoke a more natural healing response than synthetic materials [1], promoting cellular infiltration, proliferation, and differentiation into structures very similar to those of the uninjured host tissue. As previously discussed, many of the first ECM biomaterials were used in prostheses,

---

*Present affiliation: Veterinary Clinical Complex, Apollo College of Veterinary Medicine, Jaipur, Rajasthan, India.

providing structural support or mechanical functionality [2–4]. Consequently, preservation of the original structure and strength while reducing immunogenicity was paramount. The most abundant protein in ECMs is collagen, a fibrous protein that is remarkably preserved across species [5] and, therefore, invokes one of the weakest immune responses of all the proteins. This is, in fact, one reason natural collagen sutures implanted for thousands of years were so effective. Bovine collagen is still one of the most widely used and abundantly available xenogeneic materials used in biomedical applications [6]. Even though it is so well preserved, xenogeneic collagen can still provoke immune reactions in humans who are hypersensitive to it or in extenuating circumstances [7–9]. Typically, proper cleaning with detergents and terminal sterilization by gamma irradiation or ethylene oxide gas is enough to reduce the immune response to a very minimal level, lower even than synthetic meshes [10–13].

2.2 Decellularization

Decellularization is the process of removing cells from a tissue or organ while preserving the ECM. The ECM is a three-dimensional network of proteins and other molecules that provides support and structure to cells. It also contains important signaling molecules that guide cell growth and development. Decellularized ECM can be used as a scaffold for tissue engineering and regenerative medicine applications. When recellularized with new cells, decellularized ECM can help to create new tissues and organs that are more likely to be accepted by the body and function properly. Decellularized ECM has a number of advantages as a scaffold for tissue engineering and regenerative medicine applications. It is biocompatible, biodegradable, and has a natural three-dimensional structure. Decellularized ECM can also be modified to incorporate specific growth factors or other molecules to promote cell growth and differentiation. Decellularized ECM has been used to engineer a variety of tissues and organs, including blood vessels, heart valves, skin, corneas, and livers. It is also being investigated for use in the development of bioartificial organs.

2.2.1 Cellular matrix acellular matrix

Decellularization is a complex process, and there are a variety of different methods that can be used. The specific method chosen will depend on the type of tissue being decellularized and the desired properties of the final product. There are a variety of decellularization methods, which can be classified into three main categories:

- Chemical methods: Chemical decellularization methods use detergents, solvents, and other chemicals to lyse cells and remove cellular debris. Detergents can be used to dissolve the cell membranes, releasing the cells from the ECM.

- Enzymatic methods: Enzymatic decellularization methods use enzymes to degrade cellular components. Enzymes can be used to break down specific proteins and other molecules in the cells without damaging the ECM.
- Physical methods: Physical decellularization methods use mechanical forces, such as sonication, to disrupt and remove cells. Physical methods such as sonication (high-frequency sound waves) and osmotic shock can also be used to disrupt the cells and release them from the ECM.

The best decellularization method for a particular tissue or organ will depend on the specific characteristics of the ECM and the desired outcome. Once the cells have been removed, the decellularized ECM must be thoroughly washed and sterilized to remove any residual cellular material. The decellularized ECM can then be used for a variety of applications, including:

- Regenerative medicine: Decellularized ECM can be used to repair or replace damaged or diseased tissues and organs such as heart valves, blood vessels, and skin grafts.
- Tissue engineering: Decellularized ECM can be used to create new tissues and organs for implantation, such as heart valves, blood vessels, and skin grafts.
- Medical devices: Decellularized ECM can be used to create medical devices, such as stents and patches.
- Disease modeling: Decellularized tissues and organs can be used to model diseases and develop new treatments.
- Drug discovery: Decellularized tissues can be used to test the toxicity and efficacy of new drugs. Decellularized ECM can be used to deliver drugs to the body in a targeted and controlled manner.
- Food production: Decellularized plant and animal tissues could be used to produce new types of food products.

Decellularized tissues and organs have a number of advantages over synthetic scaffolds for tissue engineering and regenerative medicine applications. Decellularized tissues and organs are:

- Biocompatible: They are less likely to trigger an immune response in the recipient.
- Biodegradable: They can be broken down and replaced by the body's own tissues over time.
- Angiogenic: They promote the growth of new blood vessels, which is essential for tissue regeneration.
- Organotypic: They retain the original structure and function of the tissue or organ from which they were derived.

Decellularization is a rapidly developing field with the potential to revolutionize the way we treat and prevent diseases.

Here are some examples of how decellularization is being used today:

- Researchers are using decellularized heart valves to treat patients with heart valve disease.
- Decellularized blood vessels are being used to bypass clogged arteries in patients with heart disease.
- Decellularized skin grafts are being used to treat patients with severe burns and other skin injuries.

- Decellularized ECM is being used to develop new drug delivery systems that can target specific tissues and organs.
- Decellularized ECM is also being used to create new medical devices, such as stents and patches.

Decellularization is a promising new technology with the potential to make a significant impact on human health. Decellularization is a rapidly developing field with a wide range of potential applications. As the technology continues to improve, we can expect to see even more innovative and beneficial uses for decellularized ECM in the future.

Decellularized tissues and organs have been successfully used in a variety of tissue engineering and regenerative medicine applications, and the decellularization methods used vary as widely as the tissues and organs of interest. The efficiency of cell removal from a tissue is dependent on the origin of the tissue and the specific physical, chemical, and enzymatic methods that are used. Each of these treatments affects the biochemical composition, tissue ultrastructure, and mechanical behavior of the remaining ECM scaffold, which in turn affects the host response to the material. The goal of a decellularization protocol is to efficiently remove all cellular and nuclear material while minimizing any adverse effects on the composition, biological activity, and mechanical integrity of the remaining ECM [14].

Decellularization can be brought about by chemical, enzymatic, or mechanical methods which leave a material composed of ECM components. These acellular tissues retained their natural mechanical properties and promoted remodeling of the prosthesis by neovascularization and recellularization by the host [15]. The cells in ECMs have class I and II histocompatibility antigens capable of eliciting rejection reactions. Also, the cells have glycoproteins recognized by the immune system of hosts, which elicit rejection reactions. Therefore, if these substances are eliminated from ECMs, rejection reactions can be prevented. However, complete elimination of all antigens is considerably difficult to perform and verify [16].

To address the decellularization technique, various chemical and enzymatic steps have been employed to remove the cellular components, yet residual cellular components and lipids cannot be removed. That might have promoted undesired effects such as calcification and host immune responses, resulting in an inflammatory reaction in the recipient. It was also mentioned that, even after the removal of cells and debris from the biomaterials, the ECM of the acellular tissue itself might elicit some amount of immune response [17].

Autologous grafts are the "gold standard" for implantation. However, the greatest disadvantage of autologous is quantity. The number of autografts does not meet the needs of patients. Homograft is greater than autograft, but it cannot satisfy the needs of patients. Many patients must wait a long time to take a homogenous organ. Xenografts are the best, but they can evoke serious immune reactions. So, one method developed to process homografts and xenografts is decellularization. Every tissue or organ has cells and an ECM. Cells are the structure and functional units of tissue or organ, but cells are major antigens of tissue or organ. The ECM contains many proteins, polysaccharides, and proteoglycans released by cells. ECM plays an important role in mechanical support, signal transportation, and the adhesion of tissue and organs. Decellularization is a multistep process to remove all cell components from tissue or organs and leave an intact ECM. Many decellularization agents were researched, such as physical methods, chemical methods, and enzyme methods. Every decellularization agent has specific affections for the cell and ECM. So, these agents are combined to make an effective decellularization process that removes all cell components and reverses maximum ECM. Decellularization effectiveness depends on the type of tissue or organ. One agent can be a good detergent for decellularizing one tissue but not for another [18]. Moreover, cell-derived ECM can be used as a matrix for cell culture.

A decellularization protocol generally begins with lysis of the cell membrane using physical treatments or ionic solutions, followed by separation of cellular components from the ECM using enzymatic treatments, solubilization of cytoplasmic and nuclear cellular components using detergents, and finally removal of cellular debris from the tissue. These steps can be coupled with mechanical agitation to increase their effectiveness [14]. Alkaline and acid treatments are used in decellularization protocols to solubilize the cytoplasmic component of the cells as well as remove nucleic acids such as RNA and DNA.

Conventionally, two strategies have been used to alleviate rejection of these substances. One of the two strategies was to reduce the immune reaction of hosts, and the other was to reduce the antigenicity of allografts or xenografts mainly by crosslinking [19]. Yannas et al. [20] demonstrated that the immunogenicity of the collagen was measurable, but the actual significance of such immunogenicity was very low due to the small species differences among different types of collagen. Both residual cellular debris and ECM protein in glutaraldehyde-treated tissue were thought to contribute to such an immune response [17]. The acellular grafts were less immunogenic, had better tolerance by allogenic hosts, and were equally effective as isografts [21].

The decellularization specifically removes cellular components that give rise to a residual immunological response. The decellularized tissues were expected to closely mimic the complex three-dimensional structure and mechanical properties of the native tissues from where they originated [14]. Thus, there is always a need to develop decellularization protocols that can effectively remove cellular components and also maintain the integrity of the ECM. One of the major goals of using natural biodegradable materials is to induce the host to replace the implanted construct with native tissue [22]. Xenographically derived natural materials have been investigated for their apparent ability to elicit a regenerative response from host tissue rather than a normal healing response [6,23].

Physical methods of decellularization include snap freezing [24], mechanical agitation and sonication, and applying direct pressure [25]. The chemical method of decellularization includes the use of acid and alkaline agents, nonionic detergents like triton X-100, ionic detergents like sodium deoxycholate, sodium dodecyl sulfate, triton X-200, and zwitter ionic detergents like 3-[(3-cholamidopropyl) dimethylammonio]-1-propanesulfonate (CHAPS), sulfobetaine-10 and -16 (SB10, SB-16), Tri(n-butyl) phosphate, hypotonic, and hypertonic solutions. The enzymatic method of decellularization includes the use of trypsin, endonucleases, and exonucleases [14]. Ionic detergents are effective for solubilizing both cytoplasmic and nuclear cellular membranes but tend to denature proteins by disrupting protein-–protein interactions. Sodium deoxycholate is very effective for removing cellular remnants but tends to cause greater disruption to the native tissue architecture when compared to sodium dodecyl sulfate [14]. The osmotic shock with a hypotonic solution such as deionized water is used to lyse the cells within tissues and organs [26]. Porcine small intestinal submucosa has been decellularized using peracetic acid (PAA) at concentrations of approximately 0.10%–0.15% (w/v). This treatment was highly efficient at removing cellular material from these thin ECM structures and simultaneously disinfecting the material by entering microorganisms and oxidizing microbial enzymes [27]. Every decellularization agent has specific affections for the cell and ECM. So, these agents are combined to make an effective decellularization process that removes all cell components and reverses maximum ECM. Decellularization effectiveness depends on the type of tissue or organ. One agent can be a good detergent for decellularizing one tissue but not for another. Moreover, cell-derived ECM can be used as a matrix for cell culture (Table 2.1).

2.3 Decellularized biomaterials

Among all the protein biomaterials, collagen is a potentially useful biomaterial as it is a major constituent of connective tissue. This protein has been well characterized and ubiquitous across both kingdoms, namely, animal and plant [5]. Its characteristics offer several advantages, such as being biocompatible and nontoxic in most tissues, cellular mobility, growth, and porous nature. The collagen from bovine tendons was found to contain monomeric, diameric, trimeric, and higher polymeric forms.

An ideal acellular matrix should be able to provide the right biological and physiological environment to ensure homologous cell and ECM distribution. It should also provide the right size and morphology of the neotissue required. Collagen ranks as the most widely used matrix material for the purpose of tissue engineering in the dermis [28]. It is the major constituent of the native dermal ECM (collagen types I and II), and therefore, it is a natural choice for reconstructive surgery and tissue engineering. Burke et al. [29] used a bovine collagen-based material to develop a dermal replacement that provides mechanical strength and acts as a barrier to control the loss of moisture and prevent infection. Due to their advantageous biological properties, collagen-based scaffolds are under extensive investigation and used as matrices for tissue engineering. The advantages are that natural matrix materials mimic natural ECM structure and composition, emulate the native stimulating effects of ECM on cells, and allow the incorporation of growth factors and other matrix proteins to further enhance cell functions.

The use of xenogenic acellular scaffolds resulting after the removal of cells as skin substitutes is an acceptable modality for treating dermal wounds, but there are several drawbacks to such graft-assisted healing strategies. In modern medicine, natural and synthetic biomaterials play an increasingly important role in the

Table 2.1 Decellularization methods, mode of action, and effects on ECM.

Method	Mode of action	Effects on ECM
Physical Mechanical agitation	Can cause cell lysis, but more commonly used to facilitate chemical exposure and cellular material removal	Aggressive agitation or sonication can disrupt ECM as the cellular material is removed
Nonionic detergent Triton X-100	Disrupts lipid−lipid and lipid−protein interactions, while leaving protein−protein interactions intact	Mixed results; efficiency dependent on tissue; removes GAGs
Ionic detergent Sodium dodecyl sulfate (SDS)	Solubilize cytoplasmic and nuclear cellular membranes; tend to denature proteins	Removes nuclear remnants and cytoplasmic proteins; tends to disrupt native tissue structure; removes GAGs and damage collagen
Ionic detergent Sodium deoxycholate		More disruptive to tissue structure than SDS
Zwitterionic detergent Tri(n-butyl) phosphate	Organic solvent that disrupts protein−protein interactions	Variable cell removal; loss of collagen content, although effect on mechanical properties was minimal
Hypotonic and hypertonic solutions	Cell lysis by osmotic shock	Efficient for cell lysis but does not effectively remove the cellular remnants
EDTA	Chelating agents that bind divalent metallic ions, thereby disrupting cell adhesion to ECM	No isolated exposure, typically used with enzymatic methods (e.g., trypsin)
Enzymes Trypsin	Cleaves peptide bonds on the C-side of Arg and Lys	Prolonged exposure can disrupt ECM structure, removes laminin, fibronectin, elastin, and GAGs
Enzymes Endonucleases	Catalyze the hydrolysis of the interior bonds of ribonucleotide and deoxyribonucleotide chains	Difficult to remove from the tissue and could invoke an immune response

Modified from Gilbert TW, Sellaroa TL, Badylak SF. Decellularization of tissues and organs. *Biomaterials* 2006; 27:3675−83.

treatment of diseases and the improvement of health care. The development of novel "smart" biomaterials with optimized characteristics for very specific applications has become a main research focus. For tissue engineering applications, biomaterials often serve as scaffolds for a specific cell type [30].

Decellularization has been done by various chemical and enzymatic techniques to remove the cellular components. Following decellularization, all residual chemicals must be removed to avoid an adverse host tissue response to the chemical [14]. The removal of cells from the tissue leaves the complex mixture of structural and functional proteins that constitute the ECM. The tissue from which ECM is harvested; species of origin, decellularization protocol, and method of sterilization affects the composition and ultarstructure of ECM and accordingly affects the host tissue response to the ECM scaffold following implantation.

Triton-X-100 is the most widely used nonionic detergent for the decellularization of tissue. It led to a nearly complete loss of GAGs and a decrease in the laminin and fibronectin content of the tissue [31]. Sodium deoxycholate is also very effective for removing cellular remnants. Treatment with a hypotonic solution followed by a hypertonic solution can cause lysis but not generally remove the resultant cellular remnants from the tissues [32]. Trypsin is the most common proteolytic enzyme in decellularization protocols. Trypsin and EDTA reduced the laminin and fibronectin content of the ECM. The prolonged exposure reduces GAGs. The remaining ECM still supports endothelial cell growth in vitro despite the removal of ECM components [31]. The residual chemicals (particularly SDS) must be flushed from the ECM after decellularization. The decellularized tissue requires multiple (more than six) agitated washes to completely remove the detergents [33]. Takami et al. [34] reported that the treatment of rat skin with dispase, followed by Triton X-100, completely removed the cellular components from the dermis. Subcutaneously implanted acellular dermal matrix (ADM) in rats evoked no immunological reaction even after 20 weeks of implantation.

Yannas [35] demonstrated that the immunogenicity of the collagen was measurable, but the actual significance of such immunogenicity was very low due to the small species differences among different types of collagen. Acellular grafts were less immunogenic, had better tolerance by allogenic hosts, and were equally effective as isografts [21]. Allaire et al. [36] opined that interspecies rather than intraspecies immunogenicity of the arterial ECM leads to inflammation, biodegradation, dilation, and ultimately chronic rejection of the graft. Cartmell and Dunn [37] reported the efficacy of the acellular matrix in the treatment of abdominal wall defects depended upon its low antigenicity, capacity for rapid vascularization, and stability as abdominal wall tissue, which would be determined largely by the final composition of the acellular matrix. Removal of cells from the graft decreased the antigenicity. It was suggested that an ideal removal method should not compromise graft structure and mechanical properties.

Vezzoni et al. [38] opined that heterologous collagen favored the regeneration of connective tissue in soft tissue defects. It acted as a framework for new tissue formation, which was well tolerated by the host and replaced by new tissue ingrowths. Revascularization of acellular human dermis for soft tissue augmentation was assessed in a rabbit model. Acellular human dermis was found capable of significant revascularization of its compact collagen composition in the early postoperative period. However, the rate and completeness of vessel ingrowths were slower. Jiang et al. [39] reported that the immunogenicity in the xenogenic acellular dermal matrix was stronger than that in the allogenic acellular dermal matrix, which could induce the host to develop immune reactions restricted by IgG. Large sheets of degenerated ADM implants could lower down the antigen-antibody reaction and ameliorate the structural destruction and degeneration absorption of ADM induced by inflammatory immune reactions. Liang et al. [40] reported that the acellular biological tissues provided a natural microenvironment for host cell migration and therefore, could be used as a scaffold for tissue regeneration.

The sodium dodecyl-sulfate treatment presented chronic rejection of arterial allografts and led to the proliferation of elastin-rich intima. The acellular matrix vascular allograft that was produced by detergent and enzymatic extraction of natural canine arteries showed no inflammation when implanted into different breeds of dogs [41]. Courtman et al. [42] used detergent and enzymatic extraction of natural arteries to produce an acellular matrix vascular prosthesis. The allograft was implanted in a canine model, which showed excellent handling characteristics, low thrombo-reactivity, no evidence of aneurysm, and exceptional graft patency in the peripheral vasculature.

The dermis, diaphragm, and urinary bladder were made acellular by placing in distilled water and agitated for 1 hour to lyse the cells and to release the intracellular contents. Tissues were then suspended in sodium deoxycholate for 3 hours followed by the treatment with deoxyribonuclease-1 (2000 Kunitz units) suspended in 1 M sodium chloride solution. This process was repeated 3 times to extract all cells from the tissue [43]. Marzaro et al. [44] isolated satellite cells from rat dorsal muscle and cultured in vitro on homologous acellular matrix obtained by detergent-enzymatic treatment of abdominal muscle fragments. Thus, it was concluded that the presence of autologous satellite cells is an important factor to preserve the structural integrity and improved in vivo biocompatibility of homologous muscular acellular matrix implants. Removal of epidermis using treatment with 0.25% trypsin for 18 hours and 0.1% sodium dodecyl sulfate (SDS) for 12 hours at room temperature was beneficial for the subsequent treatment to remove cells in the dermal structure. Lengthy incubation in 0.25% trypsin (12 hours) and then dispase (12 hours) at 25°C of small pieces of porcine skin from which the epidermis had been removed efficiently removed cells and cellular components from the skin [45].

Bovine pericardium treated with hypotonic lysis and treatment with DNAse/RNAse resulted in only partial removal of histological cellularity and persistence of alpha-gal, MHC I and alpha-actin. Adding standard treatment resulted in apparent acellularity but persistence of xenogeneic antigens. It was found that only after the addition of sodium dodecyl sulfate resulted in complete histological acellularity and removal of xenogeneic antigens [46].

Meyer et al. [47] compared three different techniques, namely, detergent, enzymatic, and osmotic methods for decellularization of aortic valve allograft. Detergent method completely decellularize both the leaflets and aortic wall in addition to preservation of the ECM. Enzymatic decellularization resulted in complete decellularization but extensive degeneration and fragmentation of the ECM. Homologous acellular dermal matrix grafts were used for the reconstruction of abdominal wall defects in thirty weanling Wistar rats derived from rat skin prepared through a detergent-enzymatic method. Histological evaluation demonstrated the migration of fibroblasts and neovascularization within the homologous acellular dermal matrix graft [48].

2.4 Characterization methods of decellularized tissues

Decellularized ECMs (dECM) provide a possible supply of substances to generate different scaffolds. To date, there are no absolute criteria precisely to confirm dECM. Effective decellularization methodology is govern by various factors such as tissue density and structure, geometric and biologic characteristics needed for the targeted clinical purpose, as well as the specific characteristics of the tissue of the origin. Each tissue demands its specific characterization methods. Indeed, efficient cell and genetic elements elimination are crucial in preventing immune rejection of the construct to seeded cells. Quantitative metrics have not been described for the term decellularization yet. To evaluate the quality of decellularized tissue and its ECM, multiple aspects should be examined. Based on current literature and experiments in which in vivo constructive response was established, and immune rejection of the host did not occur, several criteria have been suggested. These minimal criteria are suggested to be exercised to assess the decellularization process and its efficacy. The basis of the suggested criteria lies in the amount and quality of genetic material that is remained in the ECM. First, the amount of dsDNA should be less than 50 nanograms per milligram of the dry weight of the ECM. Second, DNA fragments detected in the ECM should not be more than 200 base pairs. And lastly, histological evaluation should not be able to identify genetic material with hematoxylin and eosin staining (H&E) or 4',6-diamidino-2-phenylindole (DAPI) [49]

The first and second principles are considered a quantifiable approach to assessing the decellularized ECM, whereas they are qualitatively verified by histological staining such as H&E or DAPI. Quantitative assessment is readily accomplished by available dsDNA intercalators. For the second criterion, endonucleases such as DNase and RNase are applied to break down nucleic acid base pair fragments. These enzymes, fortunately, decrease the length of fragments, but they do little to part the fragments of the ECM. In pursuit of a decreasing immune response, the intracellular membrane compartment (e.g., phospholipids) must be noticed via enzyme-based measurement. The ECM is a vital component during development, influencing cell differentiation, proliferation, and migration, and its prominent role in providing structural stability and support for cells and tissue is indispensable [49]. The ECM, a noncellular element of the tissue microenvironment, comprises proteins such as collagens, laminins, fibronectin, and polysaccharide glycosaminoglycans (GAGs) [50]. Collagens are the most abundant component in ECM; so far, they are the main aim to modify. Several methods have been applied to determine the amount and quality of ECM components. For example, to maintain the integrity of collagen, histological collagen stains are used, whereas scanning electron microscope (SEM) provides more information about the structure and architecture. Additionally, second harmonic generation (SHG) detects structural changes in collagen fibers through the loss of signals.

Proteins that help with structural abilities as well as mechanical properties of the tissue should match the original tissue to a reasonable extent so that the process would have a better chance of being successful. It is assumed some decellularization agents and protocols destroy basic ECM elements; for example, detergents may disrupt collagen structure, causing the mechanical strength of ECM to change, or most detergents eliminate the amount of GAG, thus decreasing the viscoelasticity feature of ECV [51,52]. Consequently, mechanical properties such as elastic modulus, viscous modulus, tensile strength, and yield strength should be assessed. However, all in all, the characteristic is primarily dependent on the type of tissue or organ sought to function. The process of decellularization via a cell removal agent will alter the ECM composition and structure. The goal is to try and minimize these alterations to have a robust ECM.

2.4.1 Hematoxylin and eosin

H&E is a well-known stain used to assess the overall histologic appearance of samples, showing cells, cytoplasm, nuclei, and ECM constitution. H&E stain is one of the principal tissue stains used in histology. It is the most widely used stain in medical diagnosis and is often the gold standard. For example, when a pathologist looks at a biopsy of a suspected cancer, the histological section is likely to be stained with H&E. Nucleoli stain with eosin; if abundant polyribosomes are present, the cytoplasm will have a distinct blue cast. This stain reveals sufficient structural data with specific functional implications [53]. H&E is the combination of two histological stains: H&E. H&E are both natural dyes. Hematoxylin is extracted from the logwood tree, *Haematoxylum campechianum*, and eosin is extracted from the eosinophil worm, *Haliclystus octoradiatus*. Hematoxylin stains cell nuclei a purplish blue, and eosin stains the ECM and cytoplasm pink, with other structures taking on different shades, hues, and combinations of these colors. Hematoxylin is a basic dye that stains nucleic acids, including DNA and RNA. The exact mechanism of staining is not fully understood, but it is thought to involve the formation of a complex between the dye and the phosphate groups in the nucleic acids. Eosin is an acidic dye that stains proteins and other acidic substances. It is thought to bind to the amino acids in proteins, particularly arginine and lysine. When a tissue section is stained with H&E, the nuclei of the cells are stained blue-purple, while the cytoplasm and ECM are stained pink. This allows the pathologist to easily identify different cell types and tissues and assess for any abnormalities.

H&E staining is a relatively inexpensive and simple procedure, but it requires careful attention to detail to produce high-quality results. The process begins with the fixation of the tissue sample in a chemical solution, such as formalin. This preserves the tissue's structure and makes it easier to section. Once the tissue is fixed, it is embedded in paraffin wax and then sectioned into thin slices using a microtome. The sections are then mounted on glass slides and deparaffinized. The staining process then begins with the hematoxylin stain. The slides are placed in a hematoxylin solution for a specific period of time, depending on the desired intensity of the stain. The slides are then washed and rinsed to remove any excess hematoxylin. Next, the slides are placed in an eosin solution for a specific period of time. The eosin stain is typically less concentrated than the hematoxylin stain. After the eosin stain, the slides are washed and rinsed again. Finally, the slides are dehydrated and cleared using a series of alcohol and xylene baths. The slides are then mounted with a coverslip and are ready for viewing under a microscope.

H&E staining is a relatively simple and inexpensive procedure, and it can be performed on a variety of different tissue types. H&E staining provides a wealth of information about the structure and cellular composition of tissues. It can be used to identify different types of cells as well as diagnose a wide range of diseases and conditions. Decellularized tissue can show a decrease or lack of hematoxylin-stained nuclei, indicating adequate cell elimination. Such as seen, in many studies that by use of H&E stain, the first and second criteria for decellularized tissue can be established. The H&E stain of the decellularized tissue is altering the color to pink with considerably fewer blue-purple portions, indicating the presence of the nucleus in the native tissue stain.

2.4.2 4′,6-Diamidino-2-phenylindole (DAPI)

4′,6-Diamidino-2-phenylindole (DAPI) is a fluorescent stain that binds to DNA. It is commonly used in fluorescence microscopy to visualize the nuclei of cells. DAPI is a blue-fluorescent stain, and it emits light at a wavelength of 461 nm when excited by light at a wavelength of 358 nm. DAPI is a relatively simple molecule, but it is very effective at binding to DNA. DAPI binds to the minor groove of double-stranded DNA, and it is particularly attracted to AT-rich regions. This makes DAPI ideal for visualizing the nuclei of cells, which are rich in AT-rich DNA. DAPI is a very versatile stain, and it can be used to stain both fixed and live cells. It is also relatively nontoxic, so it can be used to stain cells in culture. DAPI is also used in a variety of other applications, such as flow cytometry, chromosome staining, and karyotyping. For visualization of intact and undamaged nuclei, DAPI staining is used. The tissue sections are prepared as mentioned for H&E staining. Immunofluorescence staining will be performed, and a fluorescent microscope with Slide Book software is utilized to get a photo of the slide. If intact nuclei are present, the fluorescence dye will manifest in the camera. In the majority of cases with successful decellularization, the DAPI image would show a much more limited fluorescent dye than in the native tissue [54].

DAPI is used in a wide variety of applications, including:

- Counting cells: DAPI can be used to count cells in a sample by counting the number of nuclei stained with DAPI. This is a useful technique for cell culture and other applications where it is important to know the number of cells in a sample. This is useful for a variety of applications, such as determining cell growth rates and assessing the effects of drugs or other treatments on cells.
- Visualizing nuclei and chromosomes in fluorescence microscopy: DAPI is commonly used to visualize nuclei and chromosomes in fluorescence microscopy. This is useful for a variety of applications, such as studying cell division, identifying different cell types, and diagnosing diseases. This allows scientists to visualize the nucleus of a cell and distinguish it from other cellular components.
- Measuring apoptosis: Apoptosis is a type of programmed cell death. DAPI can be used to measure apoptosis by staining apoptotic cells with DAPI and counting the number of cells

with condensed or fragmented nuclei. DAPI can be used to visualize this fragmentation, which can be used to diagnose and study diseases that involve apoptosis.
- Karyotyping: DAPI is used to stain chromosomes in karyotyping, which is a technique used to identify and classify chromosomal abnormalities.
- Detecting amyloid plaques: Amyloid plaques are a characteristic feature of Alzheimer's disease. DAPI can be used to detect amyloid plaques in tissue samples by staining them with DAPI and looking for areas of increased fluorescence.

DAPI is a safe and nontoxic dye, and it is relatively inexpensive. This makes it a widely used and versatile tool in biology research and clinical diagnostics.

Here are some specific examples of how DAPI is used:

- In cancer research, DAPI is used to study the effects of chemotherapy drugs on tumor cells.
- In infectious disease research, DAPI is used to study the replication and spread of viruses and bacteria.
- In developmental biology, DAPI is used to study the formation and development of organs and tissues.
- In clinical diagnostics, DAPI is used to diagnose a variety of diseases, including cancer, genetic disorders, and infectious diseases.

2.4.3 The MTT cell proliferation assay

MTT (3 4 5 dimethylthiazol 2 y 2 5 diphenyl tetrazolium) is a yellow water-soluble tetrazolium that is reduced by the mitochondria of live cells. After reduction, it turns into a water-insoluble purple or blue formazan product. Classically, 10,000 cells suspended in 100 µL of media are incubated with 10 microliters of MTT reagent. After 3 hours, detergents should be added to the media to lyse the cells and dissolve the colored crystal. Ethanol or propanol, acid-isopropanol, acid-isopropanol plus 10% Triton X 100, mineral oil, and dimethyl sulfoxide are all suggested as solubilized agents. The amount of purple or blue formazan production is then detected by spectrophotometry at 570 nm and is directly proportional to the number of viable cells. MTT is a sensitive and quantitative assay for cell proliferation and determining the absence of viable cells [55]. The strength of using the MTT assay to confirm decellularized ECM is that even small changes in metabolic activity can provoke an alteration in MTT.

2.4.4 Second harmonic generation (SHG)

Second harmonic generation (SHG) is a nonlinear optical process in which two photons of the same frequency interact with a material and generate a single photon with twice the frequency. This process is also known as frequency doubling. SHG can only occur in materials that lack inversion symmetry. This means that the material has different properties when it is flipped upside down. Some examples of materials that lack inversion symmetry include quartz, lithium niobate, and potassium dihydrogen phosphate (KDP). SHG is a very efficient process, and it can be used to generate high-power laser light at a variety of wavelengths. SHG is also used in a variety of other applications, such as nonlinear microscopy and surface science.

When an electromagnetic wave interacts with a material, it induces polarization in the material. Polarization is a measure of how the electric field of the wave distorts the electron cloud of the atoms in the material. In a linear material, the polarization is proportional to the electric field of the wave. However, in a nonlinear material, the polarization is also proportional to the higher powers of the electric field. This means that the polarization in a nonlinear material will contain terms that oscillate at twice the frequency of the wave. These second-harmonic polarization terms will radiate electromagnetic waves at twice the frequency of the original wave. This is how SHG occurs.

SHG imaging is used for noncentrosymmetric spatial arrangements such as collagen-based structures or birefringent crystals, membranes, and proteins. The strength of SHG is that stationing procedures are not required, and it also preserves the molecular architecture's polarization dependence. SHG intensity depends on both the size and organization of the collagen fibers. This imaging tool represents the integrity and uniformity of collagen fibers in the context of scaffolding, and any disorganization in the stroma of the tissue can be demonstrated with no need for any specific pretreatment or protocols, so it can be conducted in the fresh tissue structure [56,57].

Here are some examples of how SHG is used:

- To study the structure and function of proteins: SHG can be used to image the structure and dynamics of proteins at high resolution. This is because the SHG signal is generated by the nonlinear optical properties of the protein's amino acids.
- To detect cancer cells: SHG can be used to detect cancer cells in biological samples. This is because cancer cells often have a different nonlinear optical response than healthy cells.
- To develop new materials: SHG can be used to study the nonlinear optical properties of new materials. This information can be used to develop new materials for applications such as optics and lasers.
- To image collagen fibers in the skin: SHG can be used to image collagen fibers in the skin, which can be used to diagnose and monitor skin conditions such as scleroderma and psoriasis.
- To generate ultraviolet (UV) light from infrared (IR) light: SHG can be used to generate UV light from IR light, which is useful for a variety of applications, including sterilization, lithography, and laser surgery.

SHG is a rapidly growing field of research, and new applications for SHG are being discovered all the time. It is a powerful tool that has the potential to revolutionize a wide range of fields, including biology, medicine, and materials science.

2.4.5 Electron microscopy

Electron microscopy (EM) is a microscopy technique in which a beam of electrons is transmitted through a specimen to form an image. The specimen is usually very thin and is often stained or coated with metal to increase its contrast. The electrons interact with the specimen in a variety of ways, including scattering, absorption, and diffraction. These interactions produce an image that can be magnified up to a million times. Electron microscopes are used in a wide variety of fields, including biology, materials science, and semiconductor manufacturing. They are used to study the structure and composition of materials at the atomic and molecular levels.

There are two main types of electron microscopes: transmission electron microscopes (TEMs) and scanning electron microscopes (SEMs). The critical difference between these is in the optics.

2.4.5.1 Transmission electron microscopy (TEM)

TEMs are used to produce images of the internal structure of materials. The electron beam is transmitted through the specimen, and the image is formed by the electrons that are scattered or absorbed by the specimen. TEMs can produce images with resolutions of up to 0.1 nm, which is equivalent to the size of a single atom. TEMs work by passing a beam of electrons through a thin sample. The electrons interact with the sample in different ways, depending on the composition of the sample. For example, electrons are more likely to be scattered by heavy atoms than by light atoms. This scattering creates an image of the sample on a detector. TEM is appropriate for imaging microscopic particles such as viruses and organelles inside the cells.

2.4.5.2 Scanning electron microscopy (SEM)

SEMs are used to produce images of the surface of materials. The electron beam is scanned across the surface of the specimen, and the image is formed by the secondary electrons that are emitted from the surface. SEMs can produce images with resolutions of up to 1 nm and can also be used to produce three-dimensional images of surfaces. SEMs work by scanning a beam of electrons across the surface of a sample. The electrons interact with the surface of the sample, and the resulting secondary electrons are collected by a detector. The detector creates an image of the sample based on the number of secondary electrons collected from each point on the sample. The SEM is suitable for imaging surfaces such as tissues, bacteria, cells, and organisms. In order to image the surface of the sample, the SEM utilizes a beam of electrons in a raster pattern through the sample and provides information about topography, composition, and directionality [58].

In order to get accurate images, the sample must first be fixed in a serial solution such as glutaraldehyde or paraformaldehyde in either phosphate or cacodylate buffer. For biological samples, distilled water is the right choice. Afterward, the sample must be entirely dried prior to being placed in a high vacuum environment. These two tests confirm the preservation of the ECM, the lack of distortion in the scaffold, and more specifically, the removal of organelle residues in the decellular scaffold, which is vital for implantation purposes [59]. Electron microscopy is a powerful tool that has revolutionized our understanding of the world around us. It is used to study everything from the smallest viruses to the largest galaxies.

2.4.6 Mechanical properties

The ECM preserves the three-dimensional structure of the cells and mainly consists of collagen, elastin, and proteoglycans. Cellular function determines the relative amount of each component. It has been demonstrated that besides these components, other molecules, such as laminin, also play roles in maintaining structure, stiffness, and cell-to-cell adhesion. The ECM is a complex network of macromolecules that provides structural support and biochemical signals to cells. The mechanical properties of the ECM are important for a variety of cellular processes, including cell adhesion, migration, differentiation, and apoptosis. The ECM is a complex network of proteins and carbohydrates that provides support and structure to cells and tissues. The ECM is a viscoelastic material, meaning that it has both elastic and viscous properties. The elastic properties of the ECM allow it to store and release energy, while the viscous properties allow it to dissipate energy. The mechanical properties of the ECM can vary depending on the type of tissue, with stiffer ECMs found in tissues such as bone and cartilage and softer ECMs found in tissues such as fat and muscle.

The mechanical properties of the ECM are determined by a number of factors, including the composition of the ECM, the organization of the ECM components, and the interactions between the ECM components and cells. The major structural components of the ECM are collagen, elastin, and proteoglycans. Collagen is a fibrous protein that provides tensile strength to the ECM. Elastin is a rubbery protein that provides elasticity to the ECM. Proteoglycans are glycosylated proteins that attract water and give the ECM its gel-like consistency. The organization of the ECM components also plays a role in determining the mechanical properties of the ECM. For example, collagen fibers can be organized in a variety of ways, depending on the tissue type. In bone, collagen fibers are densely packed and organized in a parallel fashion. This gives bone its high tensile strength. In contrast, collagen fibers in the skin are more loosely organized, which gives the skin its flexibility. The interactions between the ECM components and cells also contribute to the mechanical properties of the ECM. Cells can attach to the ECM through specialized adhesion proteins, such as integrins. These adhesion proteins transmit forces between the cell and the ECM. Cells can also remodel the ECM by secreting proteases that degrade ECM components.

The mechanical properties of the ECM have a significant impact on cell behavior. For example, cells on stiff matrices tend to spread out and proliferate, while cells on soft matrices tend to be more rounded and less proliferative. The mechanical properties of the ECM can also influence cell differentiation. For example, stem cells cultured on stiff matrices tend to differentiate into bone cells, while stem cells cultured on soft matrices tend to differentiate into fat cells. The mechanical properties of the ECM are also important for tissue function. For example, the stiffness of the ECM in the heart helps to maintain the heart's pumping function. The stiffness of the ECM in the lungs also helps to maintain the lungs' gas exchange function.

Some of the key mechanical properties of the ECM include:

- Stiffness: The stiffness of the ECM is a measure of how much it resists deformation. Stiffer ECMs are more difficult to deform, while softer ECMs are easier to deform.
- Elasticity: The elasticity of the ECM is a measure of how well it returns to its original shape after being deformed. Elastic ECMs will return to their original shape quickly, while nonelastic ECMs will not.
- Viscosity: The viscosity of the ECM is a measure of its resistance to flow. Viscous ECMs are more difficult for cells to move through, while less viscous ECMs are easier for cells to move through.

Here are some examples of how the mechanical properties of the ECM can influence cellular behavior:

- Cell adhesion: Stiffer ECMs promote cell adhesion by providing more binding sites for integrins, which are cell surface receptors that connect cells to the ECM.
- Cell migration: Softer ECMs promote cell migration by being easier for cells to move through.
- Cell differentiation: The stiffness of the ECM can influence cell differentiation by activating different signaling pathways in cells. For example, stiffer ECMs can promote the differentiation of mesenchymal stem cells into osteoblasts (bone cells).
- Apoptosis: The stiffness of the ECM can also influence apoptosis (programmed cell death). For example, stiffer ECMs can promote apoptosis in cancer cells.

The mechanical properties of the ECM are an important factor in cellular behavior and tissue function. Researchers are developing new ways to manipulate the mechanical properties of the ECM to promote tissue regeneration and treat disease. Overall, the mechanical properties of the ECM play a vital role in regulating cell behavior and tissue function.

To assess the mechanical properties of ECM, there are two main categories of measurement:

- Micro-scale measurement: atomic force microscopy, micro stretching. For example, combining SHG with a micro-stretching technique presents a practical evaluation of the collagen residue in the decellularized ECM.
- Macro-scale measurement: measurement of the force-length curve of a tissue. This curve is result of a change in dimensions in a given material, such as collagen sheets or elastin bands, during applying a force stress.

2.4.7 Zymography

Zymography is a technique for detecting and characterizing hydrolytic enzymes, such as proteases, lipases, and glycosidases. It is based on the principle that enzymes can be separated by electrophoresis and then detected by their ability to degrade a substrate that is incorporated into the gel. The most common type of zymography is in-gel zymography. In this method, the enzyme sample is electrophoresed on a polyacrylamide gel containing the substrate. After electrophoresis, the gel is washed and incubated in a buffer that allows the enzyme to react with the substrate. The gel is then stained to reveal the areas of substrate degradation. These areas appear as clear bands against a darkly stained background. Zymography is a powerful tool for studying the activity and regulation of hydrolytic enzymes. It is also used in clinical diagnostics to detect the presence of certain enzymes in biological samples.

Zymography is a substrate gel electrophoresis to assess the amount of matrix metalloproteinase. Metalloproteinases are a group of proteolytic enzymes that play a key role in tissue remodeling. The zymography technique established the splitting up of proteins by nonreducing sodium dodecyl sulfate-polyacrylamide gel electrophoresis (SDS-PAGE). Most commonly, the entire gel is composed of gelatin or casein. During electrophoresis, SDS nonproteolyticly activates MMPs. After being separated by electrophoresis and a renaturation step, the gel gets incubated in a buffer of ionized calcium and zinc at 37°C that is optimized for measuring MMP activity toward distinct substrates.

MMPs are enzymes that break down the ECM, which is the network of proteins and carbohydrates that surrounds cells. MMPs play a role in cancer progression by allowing tumor cells to invade surrounding tissues and spread to other parts of the body. Zymography can be used to measure the activity of MMPs in cancer cells and tissues, which can help researchers understand how MMPs contribute to cancer progression and develop new drugs that target MMPs. Zymography is a relatively simple and inexpensive technique, and it is widely available in research laboratories. It is a valuable tool for studying and manipulating the activity of hydrolytic enzymes. Zymography can be used to detect a wide range of hydrolytic enzymes. Some common examples include proteases, such as matrix metalloproteinases (MMPs), serine proteases, and cysteine proteases; lipases; glycosidases; phosphatases; and nucleases.

2.5 Development of decellularized naturally derived biomaterials

Purohit et al. [60] developed the acellular dermal matrix from the skin of different species of animals using biological detergents and enzyme combinations. Protocols for making acellular dermal matrix from pig, goat, and buffalo skin were optimized. The optimization time for deepithelialization of pig skin is 2% trypsin (pH 6.0) for 48 hours, and for goat skin, 1% trypsin (pH 6.0) for 36 hours. Buffalo skin cannot be deepithelialized by trypsin, and the epidermis was removed by dermatome. After deepithelialization, the matrix was made acellular using different concentrations and time intervals of ionic and nonionic biological detergents (Fig. 2.1).

Singh et al. [43] developed acellular biomaterials of porcine origin for the reconstruction of abdominal wall defects in rabbits. The diaphragm, dermis, and urinary bladder were made acellular, and the ECM was used as a scaffold for tissue replacement. The materials were placed in distilled water and agitated for 1 hour using a magnetic bar to lyse the cells and release the intracellular contents. Tissues were then suspended in 4% sodium deoxycholate for 3 hours under continuous agitation, followed by treatment with deoxyribonuclease-1 (2000 Kunitz units)

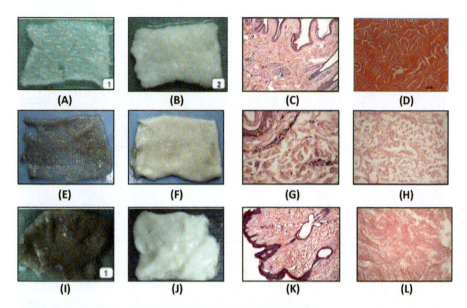

Figure 2.1 Methods of decellularization in pig, goat, and buffalo: (A) showing gross photo of epithelium removed from pig skin; (B) gross photo of acellular dermis of pig; (C) native pig skin; (D) acellular pig dermal matrix; (E) showing gross photo of epithelium removed from goat skin; (F) gross photo of acellular dermis of goat; (G) native goat skin; (H) acellular goat dermal matrix; (I) showing gross photo of epithelium removed from buffalo skin; (J) gross photo of acellular dermis of buffalo; (K) native buffalo skin; (L) acellular buffalo dermal matrix.

suspended in 1M sodium chloride solution. This process was repeated thrice to extract all the cells from the tissue. In vitro cell cytotoxicty examination in peripheral blood leukocytes of rabbits and chicken embryo fibroblasts (CEF) did not show an appreciable change in cell morphology at 24 hours postincubation in rabbit leukocytes.

Acellular buffalo small intestinal submucosa and fish swim bladder for the repair of full thickness skin wounds in rabbit model. Technique for preparation of acellular buffalo small intestinal submucosa and fish swim bladder was standardized [61,62]. Crosslinked biomaterials were successfully used with the least immunogenicity in a rabbit model for the repair of full-thickness skin wounds [63–65].

Kumar et al. [66] developed collagen-based decellularized biomaterials as 3-D scaffolds for tissue engineering of skin. The tissues used for making 3-D scaffolds included the buffalo diaphragm and pericardium, the skin of rat, and rabbit skin (Fig. 2.2A–K).

These tissues were subjected to different protocols for decellularization to minimize the immunological reaction. Mouse embryonic fibroblast cells were cultured in vitro and seeded on decellularized tissues. These decellularized tissues as well as mouse embryonic fibroblast cells-seeded tissues were subjected to histological and scanning electron microscopic examinations. The bioengineered 3-D scaffolds with

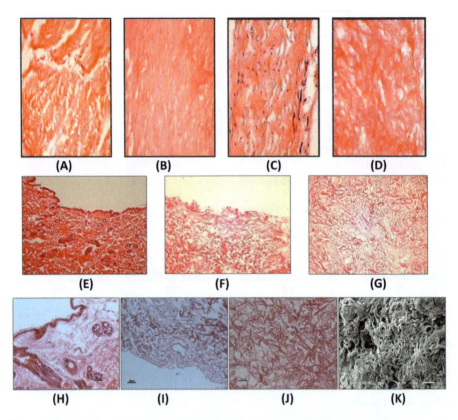

Figure 2.2 Methods of decellularization in buffalo diaphragm, buffalo pericardium, rat, and rabbit skin: (A) native diaphragm of buffalo; (B) acellular diaphragm matrix; (C) native pericardium of buffalo; (D) acellular pericardium matrix; (E) normal skin of a rat; (F) complete de-epithelialization of rat skin; (G) complete de-cellularization of rat skin using anionic detergent; (H) normal skin of a rabbit; (I) complete de-epithelialization of rabbit skin; (J) complete decellularization of rabbit skin (H&E stain); (K) complete de-cellularization (SEM photograph).

mouse embryonic fibroblast cells were developed. Evaluation of the attachment, proliferation, and migration of fibroblast cells on acellular matrices was studied to develop bioengineered collagen matrices for reconstructive surgery. These scaffolds were further used for reconstructive surgery.

Protocols for the preparation of acellular dermal matrices for wound repair were developed [67]. Decellularization of rat and rabbit skin with 1% ionic biological detergent was performed. Scanning electron and light microscopic examination revealed complete loss of cellularity and rearrangement of fibers. Singh [68] evaluated bovine omasum-derived ECM for the repair of full-thickness skin wounds in rats. The protocols for deepithelialization and decellularization of bovine omasum laminae were optimized. The treatment with a hypertonic solution for 6 hours completely removed the epithelium. The deepithelialized bovine omasum laminae were completely decellularized with 0.5% SDS for 48 hours. The treatment with 0.5% SDS for 48 hours removed 95.7% of the DNA from the ethanol-dried acellular bovine omasum matrix (Fig. 2.3A–D).

The ethanol-dried acellular bovine omasum matrix showed better healing potential among all three groups for the repair of full-thickness skin wound defects in rats. Hasan et al. [69] prepared an acellular matrix from bovine reticulum using detergents and enzymes. Treatment with 0.5% SDS for 48 hours resulted in a completely acellular matrix as evaluated by histology, SEM, and DNA quantification (Fig. 2.4A–H). Treatment with enzymes does not show any significant reduction in cellularity.

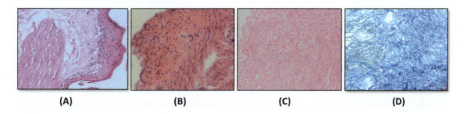

Figure 2.3 Methods of decellularization in buffalo omasum: (A) native bovine omasum; (B) delaminated bovine omasum; (C) acellular omasum matrix (H&Estaining); (D) acellular omasum matrix (Masson's trichrome staining).

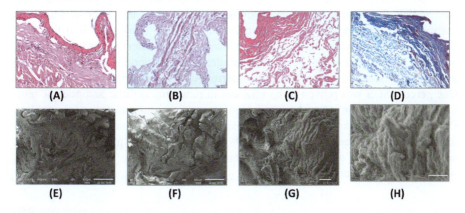

Figure 2.4 Methods of decellularization in buffalo omasum: (A) normal histological picture of native bovine reticulum; (B) delaminated bovine reticulum; (C) acellular bovine reticulum revealed loss of cellularity (H&E staining); (D) acellular bovine reticulum (Masson's trichrome staining); (E) scanning electron microscopic (SEM) observation of native bovine reticulum showing densely packed collagen fibers; (F) SEM observation of delaminated bovine reticulum showing uneven luminal surface; (G) SEM observation of acellular bovine reticulum showing loosely arranged collagen fibers; (H) SEM observation of acellular bovine reticulum showing thick collagen bundles formed of thick collagen fibrils.

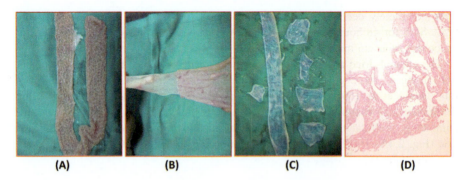

Figure 2.5 (A) Small intestine of bovine; (B) mechanical peeling of mucosa; (C) small intestine submucosa after 24 h treatment with 1% SDS; (D) complete decellularization of intestine submucosa (H&E staining, 100 ×).
From Sangetha P, Maiti SK, Kumar N, Singh K, Gopinathan A, Ninu AR, et al. Development of bioengineered porcine small intestinal submucosal scaffolds for reconstructive surgery. Trends Biomater Artif Organs. 2015; 29(2):146−50.

Sangeetha et al. [70] developed bioengineered porcine small intestinal submucosal scaffolds for reconstructive surgery. The small intestine was continuously shaken for 24 hours in 1% SDS, resulting in complete decellularization (Fig. 2.5A−D).

Porcine small intestinal submucosa has been widely used for reconstruction of various tissues and organs. It is an excellent choice for tissue engineering and clinical applications. The decellularized small intestinal submucosa was then seeded with rabbit mesenchymal stem cells (r-MSC). Sangeetha et al. [71] reported that biomaterials for corneal tissue engineering should possess characteristics such as transparency, biocompatibility, nonimmunogenicity, and integration into the host. Tissue-derived scaffolds are potentially valuable in this aspect. Acellular corneal scaffolds were prepared by chemical decellularization of porcine corneas and were subsequently seeded with rabbit bone marrow-derived mesenchymal stem cells. One percent SDS was proven to be efficient in removing cellular and nuclear remnants, which was confirmed by histopathology and scanning electron microscopy. Decellularization makes the xenogeneic tissue scaffolds nonimmunogenic. Fourteen days of culturing mesenchymal stem cells over the scaffolds provided attachment and growth of cells on the scaffolds, which was assessed by histopathology and scanning electron microscopy.

2.6 In vitro evaluation of decellularized naturally derived biomaterials

Perme et al. [72] carried out an in vitro biocompatibility evaluation of crosslinked native (cellular) and acellular bovine pericardium after croslinking with different chemical agents, namely, gluatraldehyde (GA), formadehdye (FA), polyethylene glycol (PEG), 1-ethyl-3-(3-dimethylaminopropyl)-carbodiimide (EDC), and hexamethylene diisocyanate (HMDC), taking normal saline solution (NSS) as control. In vitro evaluation revealed that a concentration of 0.5% with a duration of 48 hours at room temperature showed better crosslinking as compared to other combinations. GA treatment showed the highest resistance to enzymatic degradation at different time intervals. A significant ($P < .01$) reduction of the free amino group was observed with GA treatment at different time intervals as compared to the control. Moisture percentage after GA treatment showed a significant ($P < .01$) reductionat different time intervals when compared to control. GA treatment resulted in better crosslinking and the formation of high-molecular-weight proteins, as evaluated by SDS-PAGE.

In vitro biocompatibility testing of bladder acellular matrix graft (BAMG) was carried out after crosslinking with 0.6% GA, 1% HMDC, EDC, and 1,4-butanediol diglycidyl ether (BDDGE) for 12, 24, 48, and 72 hours at room temperature [73]. The uncrosslinked acellular bladder was used as a control. Enzymatic degradation was carried out by treating BAMG with collagenase (20 U/mL), elastase (0.1 U/mL), and trypsin (0.004 Anson U/mL). HMDC-treated biomaterials were found to be highly resistant among the crosslinked samples to enzymatic degradation. The GA and HMDC-treated bladder acellular matrix had the greatest reduction in free amino group contents in comparison to the EDC, BDDEGE, and control samples. The maximum dehydration (least moisture content) was recorded in EDC-treated BAMG, followed by HMD-, BDDGE-, and GA-treated samples at respective time intervals. In SDS-PAGE, the GA- and EDC-treated samples did not show any band pattern, whereas HMDC and BDDGE showed protein bands almost similar to those of the acellular bladder (control) but were of reduced density and intensity (Figs. 2.6 and 2.7).

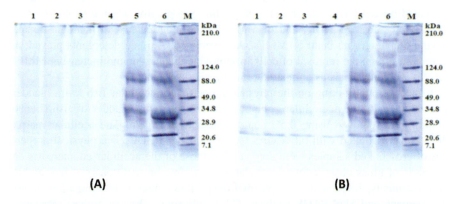

Figure 2.6 Molecular weight analysis (SDS-PAGE) after crosslinking bladder acellular matrix graft (BAMG) for 12 h with GA and HMD. (A) Crosslinked with GA: does not show any band pattern. (B) Crosslinked with HMD: three bands are visible; density is reduced with an increase in crosslinking time, Lane 1—12 h crosslinking time, Lane 2—24 h, Lane 3—48 h, Lane 4—72 h, Lane 5—acellular, Lane 6—native, Lane M—marker.

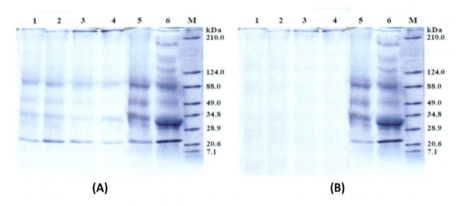

Figure 2.7 Molecular weight analysis (SDS-PAGE) after crosslinking bladder acellular matrix graft (BAMG) for 12 h with BDDEGE and EDC. (A) Crosslinking with BDDGE: four bands is visible similar to that of acellular (Lane 5) but of less intensity. (B) Crosslinking with EDC: did not show any band pattern, Lane 1—12 h crosslinking time, Lane 2—24 h, Lane 3—48 h, Lane 4—72 h, Lane 5—acellular, Lane 6—native, Lane M—marker.

Reduction in the free protein contents was minimum in HMDC crosslinked samples, while it was maximum in GA and EDC-treated samples.

Singh et al. [74] identified matrix metalloproteinases (MMPs) in vitro in the extracts obtained from a multistep detergent-enzymatic extraction process involving 2 types of anionic and nonionic detergents from bovine diaphragm and their possible involvement in the degradation of these biological materials. Anionic biological detergent-treated bovine diaphragm expressed no band in SDS-PAGE analysis but nonionic-treated bovine diaphragm expressed almost similar bands as that of untreated treated bovine diaphragm. In gelatin zymography, anionic-treated biomaterial expressed a 25 KDa band of MMPs, but in nontreated and nonionic-treated bovine diaphragm no band of MMPs was expressed. Anionic detergents played an important role in the removal of cells from bovine diaphragm and increasing MMPs activity in biomaterials as compared to non ionic detergents.

Effects of crosslinking on the physical properties of acellular fish swim bladder were studied [61]. The cellular components from the swim bladder of fish (Labeo rohita) were removed using 0.5% ionic biological detergent. The acellular matrix then was crosslinked with 0.6% GA, 1% BDDGE, and EDC. The physical properties of crosslinked samples were compared to those of the acellular counterparts on the basis of gross observations, degradation tests, free protein contents, free amino group contents, fixation index, free hydroxyproline contents determination, moisture percent, and SDS-PAGE analysis. GA fixed sample showed highest resistance to nonenzymatic, collagenase degradation, and significant decrease in free protein contents, free amino groups contents, and moisture percentage. SDS-PAGE analysis revealed that GA fixed acellular fish swim bladder expressed highest crosslinking, followed by BDDGE- and EDC-treated tissues.

In vitro evaluation of bubaline acellular small intestinal matrix was done [62]. In this study, a decellularization process was employed to remove the cellular components from small intestine of buffalo (*Bubalus bubalis*) using 0.5% ionic biological detergent. The acellular matrix then was cross linked with 0.6% GA, 1% BDDGE and EDC. The physical properties of cross linked samples were compared, in vitro, to those of the acellular counterparts on the basis of gross observations, degradation tests, free protein contents, free amino group contents, fixation index, free hydroxyproline contents determination, moisture percent, and SDS-PAGE analysis. GA-treated tissue showed highest resistance to nonenzymatic, enzymatic degradation, and significant reduction ($P < .05$) of free protein contents, free amino groups contents, and moisture percentage. SDS-PAGE analysis revealed that GA-treated acellular small intestinal matrix expressed highest crosslinking, followed by EDC- and BDDGE-treated tissues.

In vitro biocompatibility of acellular aorta of bovine origin was determined to assess its performance in vivo and further in clinical applications [75]. Native aorta of buffalo origin was decellularized using 1% ionic biological detergent for 48 hours. The acellular aortic matrix was crosslinked with glutaraldehyde (GA), 1,4-butanediol diglycidylether (BDDGE), and 1-ethyl-3-(3-dimethyl aminopropyl carbodiimide) EDC for 12, 24, 48, and 72 hours. Uncrosslinked acellular aortic matrix graft was used as control. Parameters for in vitro studies included enzymatic degradation, moisture content percentage, free amino group determination, fixation index, molecular weight analysis, and free protein content estimation. Acellular aortic matrix graft crosslinked with 0.6% GA for 48 and 72 hours and 1% EDC and BDDGE for 24 hour were found promising in vitro studies which were later used for in vivo evaluation.

In vitro evaluation of cellular and acellular caprine pericardium following crosslinking with GA, glyoxal (GO), diphenyl phosphoryl azide (DPPA), and ethylene glycol diglycidyl ether (EGDGE) for 12, 24, 48, and 72 hours was carried out [65]. On the basis of gross observations, nonenzymatic degradation, enzymatic degradation (1% cyanogen bromide, pepsin and papain), free protein determination, free amino group determination, fixation index, moisture percent, and SDS-PAGE analysis, it was found that 0.6% GA and 1% GO, DPPA, and EGDGE with a duration of 24 hours at room temperature showed better crosslinking as compared to other combinations. The GO treatment showed highest resistance to nonenzymatic and enzymatic degradation at different time interval. The GO treatment resulted in better crosslinking and formation of high molecular weight protein as evaluated by SDS-PAGE.

References

[1] Badylak S, Kokini K, Tullius B, Simmons-Byrd A, Morff R. Morphologic study of small intestinal submucosa as a body wall repair device. J Surg Res 2002;103(2):190−202.
[2] Angell D.L., Angell W.W. Heart Valve Stent. 1976; USA patent 3983581.

[3] Liotta D.S., Ferrari H.M., Pisanu A.J., Donato F.O. Low profile gulteraldehyde-fixed porcine aortic prosthetic device. 1978; USA patent 4079468.
[4] Dewanjee M.R. Mayo Foundation, Aassignee, Treatment of collagenous tissue with glutaraldehyde and aminodiphosphoate calcification inhibitor. 1985; USA patent 4553974.
[5] Van der Rest M, Garrone R. Collagen family of proteins. FASEB J 1991;5(13):2814−23.
[6] Badylak SF. The extracellular matrix as a scaffold for tissue reconstruction. Semin Cell Dev Biol 2002;13(5):377−83.
[7] Dzemeshkevich SL, Konstantinov BA, Gromova GV, Lyudinovskova RA, Kudrina LL. The mitral valve replacement by the new-type bioprostheses (features of design and long-term results). J Cardiovasc Surg 1994;35(6 Suppl 1):189−91.
[8] Klein B, Schiffer R, Hafemann B, Klosterhalfen B, Zwadlo-Klarwasser G. Inflammatory response to a porcine membrane composed of fibrous collagen and elastin as dermal substitute. J Mater Sci Mater Med 2001;12(5):419−24.
[9] Zhang MH, Chen J, Kirilak Y, Willers C, Xu J, Wood D. Porcine small intestine submucosa (SIS) is not an acellular collagenous matrix and contains porcine DNA: possible implications in human implantation. J Biomed Mater Res B Appl Biomater 2005;73(1):61−7.
[10] Allman AJ, McPherson TB, Badylak SF, Merrill LC, Kallakury B, Sheehan C, et al. Transglutaminases. Annu Rev Biochem 1980;49:517−31.
[11] O'Neill P, Booth AE. Use of porcine dermis as a dural substitute in 72 patients. J Neurosurg 1984;61(2):351−4.
[12] Patino MG, Neiders ME, Andreana S, Noble B, Cohen RE. Cellular inflammatory response to porcine collagen membranes. J Periodontal Res 2003;38(5):458−64.
[13] Konstantinovic ML, Lagae P, Zheng F, Verbeken EK, De Ridder D, Deprest JA. Comparison of host response to polypropylene and non-cross-linked porcine small intestine serosal-derived collagen implants in a rat model. Bjog 2005;112(11):1554−60.
[14] Gilbert TW, Sellaroa TL, Badylak SF. Decellularization of tissues and organs. Biomaterials 2006;27:3675−83.
[15] Schmidt CE, Baier JM. Acellular vascular tissues: natural biomaterials for tissue repair and tissue engineering. Biomaterials 2000;21(22):2215−31.
[16] Malone JM, Brendel K, Duhamil RC, Reinert RL. Detergent-extracted small diameter vascular prostheses. J Vasc Surg 1984;1:181−91.
[17] Coito AJ, Kupiec-Weglinsky JW. Extracellular matrix protein by standers or active participants in the allograft rejection cascade? Ann Transpl 1996;1:14−18.
[18] Schechter I. Prolonged retention of glutaraldehyde-treated skin allografts and xenografts: immunological and histological studies. Ann Surg 1975;182(6):699−704.
[19] Weadock K, Olson RM, Silver FH. Evaluation of collagen crosslinking techniques. Biomater Med Devices Artif Organs 1984;11(4):293−318.
[20] Yannas IV. Natural materials. In: Ratner BD, Hoffman AS, Schoen FJ, editors. Biomaterials Science: An Introduction to Materials in Medicine. Academic Press. San Diego: Academic Press; 1996. p. 84−94.
[21] Gulati AK, Cole GP. Immunogenicity and regenerative potential of acellular nerve allograft to repair peripheral nerve in rats and rabbits. Acta Neurochir Wein 1994;126:158−64.
[22] Yoganathan AP. Cardiac valve prosthesis. In: Bronzino JD, editor. The Biomedical Engineering Hand Book. Boca Raton, FL: CRC Press; 1995. p. 1847−70.

[23] Kropp BP, Cheng EY, Lin HK, Zhang Y. Reliable and reproducible bladder regeneration using unseeded distal small intestinal submucosa. J Urol 2004;172(4):1710–71 Supplement.
[24] Gulati AK. Evaluation of acellular and cellular nerve grafts in repair of rat peripheral nerve. J Neurosurg 1988;68(1):117–23.
[25] Freytes DO, Badylak SF, Webster TJ, Geddes LA, Rundell AE. Biaxial strength of multilaminated extracellular matrix scaffolds. Biomaterials 2004;25:2353–61.
[26] Woods T, Gratzer PF. Effectiveness of three extraction techniques in the development of a decellularized bone–anterior cruciate ligament–bone graft. Biomaterials 2005;26:7339–734.
[27] Hodde M, Hiles A. Virus safety of a porcine-derived medical device: evaluation of a viral inactivation method. Biotechnol Bioeng 2002;79:211–16.
[28] Hardin-Young J, Parenteau NL. Bilayered skin constructs. In: Atala A, Lanza RP, editors. Methods of Tissue Engineering. California: Academic Press; 2002. p. 1177–88.
[29] Burke JF, Yannas IVJ, Quinby WC. Successful use of a physiologically acceptable artificial skin in the treatment of extensive burn injury. Ann Surg 1981;194:413–28.
[30] Kim MS, Hong KD, Shin HW, Kim SH, Kim SH, Lee MS, et al. Preparation of porcine small intestinal submucosa sponge and their application as a wound dressing in full-thickness skin defect of rat. Int J Biol Macromolles 2005;36:54–60.
[31] Grauss RW, Hazekamp MG, Oppenhuizen F, Van Munsteren CJ. Giltenberger-de Groot AC, De Ruiter MC. Histological evaluation of decellularized porcine aortic valves: matrix changes due to different decellularization methods. Eur J Cardiothorac Surg 2005;27:566–71.
[32] Dahl SL, Koh J, Prabhakar V, Niklason LE. Decellularized native and engineered arterial scaffolds for transplantation. Cell Transpl 2003;12:659–66.
[33] Cebotari S, Tudorache I, Jaekel T, Hilfiker A, Dorfman S, Ternes W. Detergent decellularization of heart valves for tissue engineering: toxicological effects of residual detergents on human endothelial cells. Artif Organs 2010;34(3):206–10.
[34] Takami Y, Matsuda T, Yoshitake M, Hanumadass M, Walter RJ. Dispase/detergent treated dermal matrix as a dermal substitute. Burns 1996;22:182–9.
[35] Yannas IV. Natural materials. In: Ratner BD, Hoffman AS, Schoen FJ, Lemon JE, editors. Biomaterial Science. San Diego: Academic Press; 1996. p. 84–94.
[36] Allaire E, Brunewal P, Mandet C, Becquemin J, Michel J. The immunogenicity of extracellular matrix in arterial xenografts. Surgery 1997;122:73–81.
[37] Cartmell JS, Dunn MG. Effect of chemical treatment on tendon cellularity and mechanical properties. J Biomed Mater Res 2000;49:134–40.
[38] Vezzoni A, Pettazzoni M, Olivieri M, Baroni E. Use of collagen in orthopaedics and traumatology of small animals: review of literature and personal experience 55 cases. Veterinaria Cremona 2001;15:9–16.
[39] Jiang DY, Chen B, Jia CY, Tao K. An experimental study on the difference of the antigenicity of xenogeneic acellular dermal matrix. Zhonghua Shao Shang Za Zhi 2003;19:155–8.
[40] Liang HC, Chang Y, Hsu CK, Lee MH, Sung HW. Effect of cross-linking degree on an acellular biological tissue on its tissue regeneration pattern. Biomaterials 2004;25:3541–52.
[41] Wilson GJ, Courtman DW, Klement P, Lee JM, Yegar H. Acellular matrix-a biomaterial approach for coronary artery bypass and heart valve replacement. Ann Thorac Surg 1995;60:353–8.

[42] Courtman DW, Errett BF, Wilson GJ. The role of cross-linking in modification of the immune response elicited against xenogenic vascular acellular matrices. J Biomed Mater Res 2001;55:576—86.
[43] Singh J, Kumar N, Sharma AK, Maiti SK, Goswami TK, Sharma AK. Acellular biomaterials of porcine origin for the reconstruction of abdominal wall defects in rabbits. Trends Biomat Artif Organs 2008;22:30—40.
[44] Marzaro M, Conconi MT, Perin L, Giuliani S, Coppi P, Perrino GP. Autologous satellite cell seeding improves *in-vivo* biocompatibility of hoimologous acellular matix implants. Int J Mol Med 2002;10:177—82.
[45] Chen RN, Ho HO, Tsai YT, Shen MT. Process development of an acellular dermal matrix (ADM) for biomedical application. Biomaterials 2004;25:2679—86.
[46] Goncalves AC, Griffths LG, Anthony RV, Orton EC. Decellularization of bovine pericardium for tissue-engineering by targeted removal of xenoantigens. J Heart Valve Dis 2005;14(2):212—17.
[47] Mayer SR, Chiu B, Churchill TA, Zhu L, Lakey JR, Ross DB. Comparison of aortic valve allograft decellularization technique in the rat. J Biomed Mater Res A 2006;79 (2):254—62.
[48] Mete K, Fusan B, Fusan B, Mehmet-Emin B, Turan K, Selcuk Y. Use of homologous acellular dermal matrix for abdominal wall reconstruction in rats. J Investig Surg 2006;19:11—17.
[49] Medberry CJ. Central nervous system extracellular matrix as a therapeutic bioscaffold for central nervous system injury. University of Pittsburgh; 2014.
[50] Cirulli V, Beattie GM, Klier G, Ellisman M, Ricordi C, Quaranta V, et al. Expression and function of avβ3 and avβ5 integrins in the developing pancreas: roles in the adhesion and migration of putative endocrine progenitor cells. J Cell Biol 2000;150(6) 1445-146.
[51] Kezwoń A, Chromińska I, Frączyk T, Wojciechowski K. Effect of enzymatic hydrolysis on surface activity and surface rheology of type I collagen. Colloids Surf B Biointerfaces 2016;137:60—9.
[52] Conconi MT, De Coppi P, Di Liddo R, Vigolo S, Zanon GF, Parnigotto PP, et al. Tracheal matrices, obtained by a detergent-enzymatic method, support in vitro the adhesion of chondrocytes and tracheal epithelial cells. Transpl Int 2005 2005;18 (6):727—34.
[53] Fischer AH, Jacobson KA, Rose J, Zeller R. Hematoxylin and eosin staining of tissue and cell sections. CSH Protoc 2008; pdb.prot4986.
[54] Crapo PM, Gilbert TW, Badylak SF. An overview of tissue and whole organ decellularization processes. Biomaterials 2011;32(12):3233—43.
[55] Purpura V, Bondioli E, Cunningham EJ, De Luca G, Capirossi D, Nigrisoli E, et al. The development of a decellularized extracellular matrix-based bioma- terial scaffold derived from human foreskin for the purpose of foreskin reconstruction in circumcised males. J Tissue Eng 2018; 9:2041731418812613.
[56] Keikhosravi A, Bredfeldt JS, Sagar AK, Eliceiri KW. Second-harmonic generation imaging of cancer. Methods Cell Biol 2014;123:531—46.
[57] Leonard AK, Loughran EA, Klymenko Y, Liu Y, Kim O, Asem M, et al. Methods for the visualization and analysis of extracellular matrix protein structure and degradation. Methods Cell Biol 2018;143:79—95.
[58] Godwin ARF, Starborg T, Sherratt MJ, Roseman AM, Baldock C. Defining the hierarchical organisation of collagen VI microfibrils at nanometre to micrometre length scales. Acta Biomater 2017;52:21—32.

[59] Keene DR, Tufa SF. Ultrastructural analysis of the extracellular matrix. Methods Cell Biol 2018;143:1−39.
[60] Purohit S, Kumar N, Sharma AK, Sharma AK. Development of acellular dermal matrix from skin of different species of animals using biological detergents and enzymes combinations. JSM Burn Trauma1 2006;(1):1004.
[61] Kumar V, Kumar N, Singh H, Gangwar AK, Dewangan R, Kumar A, et al. Effects of crosslinking treatments on the physical properties of acellular fish swim bladder. Trends Biomater Artif Organs 2013;27(3):93−100.
[62] Kumar V, Kumar N, Singh H, Gangwar AK, Dewangan R, Kumar A, et al. In vitro evaluation of bubaline acellular small intestinal matrix. Int J Bioassays 2013;2(3):581−7.
[63] Kumar V. Acellular buffalo small intestinal submucosa and fish swim bladder for the repair of full thickness skin wound in rabbits. MVSc thesis submitted to deemed university ICAR, Indian Veterinary Research Institute, Izatnagar, Uttar Pradesh, India; 2010.
[64] Kumar V, Kumar N, Gangwar AK, Singh H. Comparison of acellular small intestinal matrix (ASIM) and 1-ethyl-3-(3-dimethylaminopropyl)carbodiimide crosslinked ASIM (ASIM-EDC) for repair of full-thickness skin wounds in rabbits. Wound Med 2014;7(1):24−33.
[65] Kumar V, Kumar N, Gangwar AK, Singh H, Singh R. Comparative histological and immunological evaluation of 1,4-butanediol diglycidyl ether crosslinked versus non-crosslinked acellular swim bladder matrix for healing of full-thickness skin wounds in rabbits. J Surg Res 2015;197(2):436−46.
[66] Kumar N, Gangwar AK, Sharma AK, Negi M, Shrivastava S, Mathew DD, et al. Extraction techniques for the decellularization of rat dermal constructs. Trends Biomater Artif Organs 2013;27(3):102−7.
[67] Mathew D.D. Development of bioengineered 3-D acellular dermal matrices seeded with mesenchymal stem like cells for dermal wounds in rats. PhD thesis submitted to deemed university ICAR, Indian Veterinary Research Institute, Izatnagar, Uttar Pradesh, India; 2014.
[68] Singh P. Evaluation of omasum derived extracellular matrix for repair of full thickness skin wound in rats. MVSc thesis submitted to deemed university ICAR, Indian Veterinary Research Institute, Izatnagar, Uttar Pradesh, India; 2015.
[69] Hasan A, Naveen Kumar, Singh K, Gopinathan A, Remya V, Mondal DB, et al. Preparation of acellular matrix from bovine reticulum using detergents and enzynmes. Trends Biomater Artific Organs 2015;29(3):231−6.
[70] Sangetha P, Maiti SK, Kumar N, Singh K, Gopinathan A, Ninu AR, et al. Development of bioengineered porcine small intestinal submucosal scaffolds for reconstructive surgery. Trends Biomater Artif Organs 2015;29(2):146−50.
[71] Sangetha P, Maiti SK, Kumar N, Singh K, Gopinathan A, Ninu AR, et al. Development of bioengineered corneal matrix for econstructive urgery of eye. Trends Biomater Artif Organs 2016;30(2):111−15.
[72] Perme H, Sharma AK, Kumar N, Singh H, Dewangan R, Maiti SK. In vitro biocompatibility evaluation of crosslinked cellular and acellular bovine pericardium. Trends Biomater Artif Organs 2009;23:65−75.
[73] Dewangan R, Sharma AK, Kumar N, Maiti SK, Singh S, Gangwar AK, et al. In vitro biocompatibility determination of bladder acellular matrix graft. Trends Biomater Artif Organs 2011;25:161−71.

[74] Singh H, Kumar N, Sharma AK, Kataria M, Munjal A. In vitro study of matrix metalloproteinases in decellularized extracellular matrix of bovine diaphragm. Indian J Anim Sci 2011;81:453−5.
[75] Devarathnam J, Sharma AK, Rai RB, Maiti SK, Shrivastava S, Sonal, et al. In vitro biocompatibility determination of acellular aortic matrix of buffalo origin. Trends Biomater Artif Organs 2014;28(3):92−8.

Crosslinking of biomaterials

Naveen Kumar[1,*], Anil Kumar Gangwar[2], Vineet Kumar[3], Dayamon David Mathew[4], Pawan Diwan Singh Raghuvanshi[5], Rahul Kumar Udehiya[6], Naresh Kumar Singh[4], Sangeeta Devi Khangembam[7], Sameer Shrivastava[8], Sonal Saxena[8] and Rukmani Dewangan[9]

[1]Division of Surgery, ICAR-Indian Veterinary Research Institute, Izatnagar, Uttar Pradesh, India, [2]Department of Veterinary Surgery & Radiology, College of Veterinary Science & Animal Husbandry, Acharya Narendra Deva University of Agriculture and Technology, Ayodhya, Uttar Pradesh, India, [3]Department of Veterinary Surgery and Radiology, College of Veterinary and Animal Sciences, Bihar Animal Sciences University, Kishanganj, Bihar, India, [4]Department of Veterinary Surgery & Radiology, Faculty of Veterinary and Animal Sciences, Banaras Hindu University, Rajiv Gandhi South Campus, Barkachha, Uttar Pradesh, India, [5]Department of Veterinary Surgery and Radiology, College of Veterinary Science and Animal Husbandry, Nanaji Deshmukh Veterinary Science University, Mhow, Indore, Madya Pradesh, India, [6]Department of Veterinary Surgery and Radiology, Faculty of Veterinary and Animal Sciences, Institute of Agricultural Sciences, Rajiv Gandhi South Campus, Banaras Hindu University, Barkachha, Uttar Pradesh, India, [7]Department of Veterinary Surgery and Radiology, Acharya Narendra Deva University of Agriculture and Technology, Ayodhya, Uttar Pradesh, India, [8]Division of Veterinary Biotechnology, ICAR-Indian Veterinary Research Institute, Izatnagar, Uttar Pradesh, India, [9]Department of Veterinary Surgery and Radiology, College of Veterinary Science and Animal Husbandry, Dau Shri Vasudev Chandrakar Kamdhenu Vishwavidyalaya, Durg, Chhattisgarh, India

3.1 Introduction

Crosslinking of biomaterials is used to improve their properties for a variety of biomedical applications. For example, crosslinking can be used to improve the mechanical strength, stability, and durability of biomaterials. Crosslinking can also be used to control the release of drugs or other molecules from biomaterials.

Here are some examples of how crosslinking is used in biomaterials:

- Crosslinked hydrogels are used in tissue engineering and regenerative medicine to create scaffolds that can support cell growth and differentiation.
- Crosslinked collagen is used in a variety of medical implants, such as heart valves and artificial skin.
- Crosslinked hyaluronic acid is used in dermal fillers and other cosmetic products.
- Crosslinked cellulose is used in wound dressings and other medical devices.

---

* Present affiliation: Veterinary Clinical Complex, Apollo College of Veterinary Medicine, Jaipur, Rajasthan, India.

- Crosslinking is a powerful tool for improving the properties of biomaterials for a variety of biomedical applications. However, it is important to carefully select the appropriate crosslinking method and crosslinking agent to ensure that the desired properties are achieved without compromising the biocompatibility or safety of the biomaterial.

Crosslinking is used in a wide variety of biomedical applications, including the following:

- Tissue engineering and regenerative medicine: Crosslinked biomaterials can be used to create scaffolds that support cell growth and differentiation.
- Drug delivery: Crosslinked biomaterials can be used to encapsulate and deliver drugs to the body in a controlled manner.
- Medical implants: Crosslinked biomaterials can be used to develop durable and biocompatible medical implants, such as artificial heart valves and stents.
- Wound healing: Crosslinked biomaterials can be used to develop dressings and other wound healing products that promote tissue regeneration.

3.2 Advantages of crosslinking biomaterials

- Improved mechanical strength: Crosslinking can make biomaterials stronger and more resistant to deformation and degradation. This is important for many biomedical applications, such as tissue engineering and drug delivery, where the biomaterial needs to withstand mechanical stresses.
- Increased stability: Crosslinking can make biomaterials more resistant to degradation and swelling.
- Controlled release of drugs and other bioactive molecules: Crosslinked biomaterials can be used to create drug delivery systems that release their cargo in a controlled manner.
- Reduced swelling and erosion: Crosslinked biomaterials are less likely to swell and erode in aqueous environments, which can improve their performance in applications such as wound dressings and drug delivery systems.
- Controlled degradation rate: Crosslinking can be used to control the degradation rate of biomaterials, which is important for applications such as tissue engineering and drug delivery. A slower degradation rate can allow the biomaterial to provide longer-term support for cell growth and tissue regeneration.
- Tailored biocompatibility: Crosslinking can be used to modify the biocompatibility of biomaterials by incorporating functional groups that promote cell adhesion and proliferation. This is important for tissue engineering applications, where the biomaterial needs to be able to support cell growth and differentiation. Crosslinking can reduce the cytotoxicity of some biomaterials.
- Improved printability: Crosslinking can be used to improve the printability of biomaterials, which is important for applications such as 3D bioprinting. Crosslinked biomaterials can be extruded through finer needles and maintain their shape after printing.
- Controlled release of drugs or growth factors: Crosslinking can be used to create biomaterials that can release drugs or growth factors in a controlled manner. This is useful for applications such as drug delivery and tissue regeneration.
- Improved cell adhesion and proliferation: Crosslinking can be used to create biomaterials that are more supportive of cell adhesion and proliferation. This is important for applications such as cell culture and tissue engineering.
- Enhanced antifouling properties: Crosslinking can be used to create biomaterials that are more resistant to protein adsorption and cell adhesion. This is useful for applications such as medical devices and implants.

3.3 Disadvantages of crosslinking biomaterials

- Cytotoxicity: Some crosslinking agents can be cytotoxic to cells, both in vitro and in vivo. This is especially a concern for applications such as tissue engineering and regenerative medicine, where cells are directly seeded onto or encapsulated within biomaterials. Some crosslinking agents, such as glutaraldehyde, can be cytotoxic. It is important to carefully select the crosslinking agent and crosslinking conditions to minimize cytotoxicity.
- Reduced biodegradability: Crosslinking can reduce the biodegradability of biomaterials, which can make them less suitable for applications where it is important for the material to be eventually resorbed by the body. Crosslinking can make biomaterials more resistant to degradation, which can be a disadvantage for some applications, such as tissue engineering.
- Difficult to control: The crosslinking process can be difficult to control, and it can be challenging to achieve consistent results. This can lead to variations in the properties of crosslinked biomaterials, which can impact their performance in vivo.
- Altered biological properties: Crosslinking can alter the biological properties of biomaterials, such as their ability to interact with cells and tissues. This can be both beneficial and detrimental, depending on the application.
- Increased cost and complexity: Crosslinking can add to the cost and complexity of manufacturing biomaterials. Some crosslinking methods can be expensive, which can limit their use in commercial applications

Here are some examples of how researchers are addressing the disadvantages of crosslinking biomaterials:

- Developing new, less cytotoxic crosslinking agents: Researchers are developing new crosslinking agents that are less cytotoxic to cells and tissues. For example, some researchers are developing crosslinkers that are derived from natural sources, such as plants and bacteria. Others are developing crosslinkers that are broken down by the body over time, reducing the risk of long-term toxicity.
- Developing more biocompatible crosslinking methods: Researchers are also developing more biocompatible crosslinking methods. For example, some researchers are developing crosslinking methods that use enzymes to catalyze the crosslinking reaction. This can reduce the need for toxic chemical crosslinkers.
- Developing more controllable crosslinking methods: Researchers are also developing more controllable crosslinking methods. For example, some researchers are developing crosslinking methods that can be controlled by light or ultrasound. This allows for greater precision and control over the crosslinking process.

By addressing these disadvantages, researchers are making crosslinking a more viable and versatile technique for developing new and improved biomaterials for a wide range of applications. Overall, crosslinking is a versatile technique that can be used to improve the properties of biomaterials for a wide variety of biomedical applications.

3.4 Types of crosslinking

Crosslinking of biomaterials is a process by which two or more biomolecules are covalently bonded together. This can be done using a variety of methods, including chemical, physical, and biological crosslinking.

3.4.1 Chemical crosslinking

Chemical crosslinking is the most common type of crosslinking and involves the use of crosslinking agents to form covalent bonds between biomolecules. Crosslinking agents can be bifunctional, meaning that they have two reactive groups that can bind to different biomolecules, or they can be multifunctional, meaning that they have multiple reactive groups that can bind to multiple biomolecules. Chemical crosslinking is a process by which two or more biomolecules are covalently bonded together using a crosslinking agent. Chemical crosslinking can be used to crosslink a wide variety of biomolecules, including proteins, carbohydrates, lipids, and nucleic acids. The type of crosslinking agent used will depend on the specific biomolecules being crosslinked and the desired properties of the crosslinked material. Chemical crosslinking is a widely used technique for modifying the properties of biomaterials. It can be used to improve the mechanical strength, stability, durability, and biocompatibility of biomaterials. Crosslinking can also be used to control the release of drugs or other molecules from biomaterials. However, it is important to carefully select the appropriate crosslinking agent and crosslinking conditions to ensure that the desired properties are achieved without compromising the biocompatibility or safety of the biomaterials.

Chemical crosslinking is a powerful tool for improving the properties of biomaterials for a variety of biomedical applications. For example, crosslinking can be used to improve the mechanical strength, stability, and durability of biomaterials. Crosslinking can also be used to control the release of drugs or other molecules from biomaterials. However, it is important to note that some chemical crosslinking agents can be cytotoxic or genotoxic. Therefore, it is important to carefully select the appropriate crosslinking agent and crosslinking method to ensure that the desired properties are achieved without compromising the biocompatibility or safety of the biomaterial. Some of the most commonly used chemical crosslinking agents include glutaraldehyde, formaldehyde, and carbodiimides.

Here are some examples of how chemical crosslinking is used in biomaterials:

- Crosslinked hydrogels are used in tissue engineering and regenerative medicine to create scaffolds that can support cell growth and differentiation. For example, collagen hydrogels can be crosslinked with glutaraldehyde to create a strong and stable scaffold for cartilage tissue engineering.
- Crosslinked collagen is used in a variety of medical implants, such as heart valves and artificial skin. For example, collagen heart valves can be crosslinked with glutaraldehyde or epoxides to improve their durability and resistance to degradation.
- Crosslinked hyaluronic acid is used in dermal fillers and other cosmetic products. For example, hyaluronic acid dermal fillers can be crosslinked with BDDE to increase their viscosity and longevity.
- Crosslinked cellulose is used in wound dressings and other medical devices. For example, cellulose nanofibers can be crosslinked with glutaraldehyde to create a strong and durable wound dressing.

There are a variety of chemical crosslinking agents available, each with its own advantages and disadvantages. Some of the most common chemical crosslinking agents include:

- Glutaraldehyde: Glutaraldehyde is a bifunctional crosslinking agent that reacts with amine and thiol groups on biomolecules. It is a relatively inexpensive and easy-to-use crosslinking agent, but it can be cytotoxic and cause other adverse effects.
- Epoxides: Epoxides are bifunctional crosslinking agents that react with amine groups on biomolecules. They are less cytotoxic than glutaraldehyde, but they can be more difficult to use.
- Isocyanates: Isocyanates are multifunctional crosslinking agents that can react with a variety of functional groups on biomolecules. They are highly versatile crosslinking agents, but they can be toxic and difficult to work with.
- Carbodiimides: Carbodiimides are bifunctional crosslinking agents that react with carboxyl and amine groups on biomolecules. They are less cytotoxic than some other crosslinking agents, but they can be unstable and difficult to use.

3.4.1.1 Advantages of chemical crosslinking

- Versatility: Chemical crosslinkers can be used to crosslink a wide variety of materials, including polymers, proteins, and carbohydrates.
- Controllability: Chemical crosslinking reactions can be carefully controlled to achieve the desired properties of the crosslinked material.
- Durability: Chemical crosslinks are typically very durable and can withstand harsh environmental conditions.
- Cost-effectiveness: Chemical crosslinking is a relatively inexpensive process.

3.4.1.2 Disadvantages of chemical crosslinking

- Toxicity: Many chemical crosslinkers are toxic and can be harmful to human health and the environment.
- Irreversibility: Chemical crosslinks are typically irreversible. This can make it difficult to modify the properties of the crosslinked material once it has been crosslinked.
- Difficult processing: Chemical crosslinking reactions can be difficult to control and can require specialized equipment.
- Limited biocompatibility: Some chemical crosslinkers are not compatible with biological materials. This can limit the use of chemical crosslinking in biomedical applications.
- Chemical crosslinking of polymers can release toxic fumes and by-products. This can be a safety hazard for workers and can pollute the environment.
- Chemical crosslinking of proteins can modify the structure and function of the proteins. This can be a disadvantage for applications where the biological activity of the proteins is important.
- Chemical crosslinking of carbohydrates can alter their nutritional value and digestibility. This can be a disadvantage for food applications.

Chemical crosslinking is a versatile and powerful tool for improving the properties of biomaterials for a wide variety of biomedical applications. However, it is important to carefully select the appropriate crosslinking agent and crosslinking method to ensure the safety and biocompatibility of the crosslinked material.

3.4.2 Physical crosslinking

Physical crosslinking is the formation of bonds between polymer chains through weak physical interactions. These interactions can include coordination bonding, hydrogen bonding, ionic interactions, and van der Waals interactions. Physical crosslinks are typically weaker and more reversible than chemical crosslinks, but they can still be used to create materials with useful properties. This can be achieved through a variety of mechanisms, such as entanglement, ionic interactions, and hydrogen bonding. Physical crosslinking is often reversible, meaning that the crosslinks can be broken under certain conditions. Some examples of physical crosslinking methods include freezing-thawing, heating, and exposure to UV light.

Physical crosslinking can be reversible or irreversible. Reversible crosslinks can be broken by changes in environmental conditions, such as temperature or pH. Irreversible crosslinks are more permanent and cannot be easily broken. Physical crosslinking can be used to create a wide range of materials with different properties. For example, physically crosslinked hydrogels can be used for drug delivery, wound healing, and tissue engineering. Physically crosslinked polymers can also be used to create strong and durable materials for use in engineering and construction. Physical crosslinking is the formation of bonds between polymer chains through weak physical interactions. These interactions can include:

- Hydrogen bonding: The formation of hydrogen bonds between polar groups on different polymer chains.
- Ionic interactions: The attraction between positively and negatively charged groups on different polymer chains.
- Van der Waals forces: Weak attractive forces that arise between all molecules due to the uneven distribution of electrons.
- Crystallinity: The formation of ordered regions in the polymer chains, which can act as crosslinks.

Some common examples of physical crosslinking include the following:

- Gelatin: Gelatin is a protein derived from collagen. It forms a gel when dissolved in water and cooled. The gelatin chains form physical crosslinks through hydrogen bonding and ionic interactions. Gelatin is a protein that can be physically crosslinked by cooling it below its gelation temperature. This causes the gelatin molecules to form hydrogen bonds with each other, creating a gel.
- Agar: Agar is a polysaccharide that can be physically crosslinked by adding calcium ions. Calcium ions form ionic bonds with the agar molecules, creating a gel.
- Natural rubber: Natural rubber is a polymer made up of isoprene units. It forms a physical crosslink through the crystallization of the isoprene units. It can be physically crosslinked by vulcanization. Vulcanization involves heating the rubber with sulfur or other crosslinking agents. This causes the sulfur atoms to form covalent bonds between the rubber molecules, creating a stronger and more durable material.
- Polyvinyl alcohol (PVA): PVA is a synthetic polymer that can be physically crosslinked through hydrogen bonding.

Here are some of the advantages and disadvantages of physical crosslinking:

3.4.2.1 Advantages of physical crosslinking

- Biocompatibility: Physical crosslinking is often more biocompatible than chemical crosslinking, as it does not require the use of toxic crosslinking agents. This makes physical crosslinking ideal for applications such as tissue engineering and drug delivery.

- Reversibility: Physical crosslinks can be reversible, which means that they can be broken and reformed without damaging the material. This can be useful for applications where the material needs to be able to change its properties in response to its environment.
- Tunability: Physical crosslinking can be tuned to control the properties of the material, such as its strength, stiffness, and degradation rate. This makes physical crosslinking a versatile technique for creating materials with a wide range of properties.
- Simplicity: Physical crosslinking is often simpler to implement than chemical crosslinking, and it does not require specialized equipment. Physical crosslinking is typically a simpler and more cost-effective process than chemical crosslinking.

3.4.2.2 Disadvantages of physical crosslinking

- Physical crosslinks are typically weaker than chemical crosslinks, so physically crosslinked materials may not be as strong or durable.
- Lower stability: Physically crosslinked materials are typically less stable than chemically crosslinked materials. This is because the physical crosslinks can be broken by changes in environmental conditions, such as temperature, pH, and solvent.
- Lower mechanical strength: Physically crosslinked materials also tend to have lower mechanical strength than chemically crosslinked materials. This is because the physical crosslinks are not as strong as the covalent bonds that are formed during chemical crosslinking.
- More difficult processing: Physically crosslinked materials can be more difficult to process than chemically crosslinked materials. This is because the processing conditions need to be carefully controlled to avoid breaking the physical crosslinks.

Physical crosslinking is used in a wide variety of materials, including:

- Hydrogels: Hydrogels are networks of polymer chains that can absorb and retain large amounts of water. Physical crosslinking is often used to create hydrogels for applications such as wound dressings, drug delivery systems, and tissue engineering scaffolds.
- Elastomers: Elastomers are materials that can undergo large deformations and then return to their original shape. Physical crosslinking is used to create elastomers for applications such as tires, gaskets, and seals.
- Biomaterials: Biomaterials are materials that are used in medical devices and implants. Physical crosslinking is used to create biomaterials with the desired mechanical properties, biocompatibility, and degradability.
- Physical crosslinking is a versatile technique for modifying the properties of materials. It can be used to create materials with a wide range of mechanical properties, biocompatibility, and degradability.

3.4.3 Biological crosslinking

Biological crosslinking involves the use of enzymes to form covalent bonds between the molecules of a biomaterial. Biological crosslinking is often used to crosslink proteins, as enzymes can be used to target specific amino acid residues on proteins. This is a relatively new type of crosslinking that is still under development, but it has the potential to produce biomaterials with improved biocompatibility and degradability. Biological crosslinking is the process of forming covalent bonds between biological molecules, such as proteins, nucleic acids, and carbohydrates. This can be done using a variety of crosslinking agents, including chemical reagents, UV light, and enzymes.

Biological crosslinking is used in a wide variety of research and clinical applications, including the following:

- Protein–protein interaction mapping: Biological crosslinking can be used to identify and map protein–protein interactions in cells and tissues. This can help researchers understand how proteins function and how they interact with each other. This information can be used to understand cellular processes and develop new drugs.
- Structural biology: Biological crosslinking can be used to study the structure of proteins and other biological molecules. This can help researchers understand how these molecules function and how they can be targeted by drugs. UV light can be used to crosslink proteins and nucleic acids. This is a particularly useful technique for studying the structure of large protein complexes and DNA–protein interactions.
- Drug discovery: Biological crosslinking can be used to develop new drugs that target protein–protein interactions. This is a promising approach to developing new treatments for a variety of diseases, including cancer, Alzheimer's disease, and HIV/AIDS. For example, some drugs are being developed that target the interactions between cancer proteins.
- Clinical diagnostics: Biological crosslinking can be used to develop new diagnostic tests for diseases. For example, crosslinking agents can be used to detect the presence of specific proteins or nucleic acids in blood or tissue samples. A crosslinking agent can be used to detect the presence of amyloid beta plaques in the brains of Alzheimer's patients.
- Tissue engineering: Crosslinking can be used to create scaffolds for tissue engineering. These scaffolds can be used to grow new tissues and organs.
- Drug delivery: Crosslinking can be used to create drug delivery systems that can release drugs in a controlled manner.
- Food science: Crosslinking can be used to improve the texture and stability of food products.

3.4.3.1 Advantages of biological crosslinking

- Biocompatibility: Biological crosslinking typically uses enzymes or other biological molecules to form crosslinks. This makes it a more biocompatible process than chemical crosslinking, which often uses toxic chemicals.
- Selectivity: Biological crosslinkers can be designed to be selective for specific types of biomolecules. This allows for more precise control over the crosslinking process.
- Mild reaction conditions: Biological crosslinking reactions can often be carried out under mild conditions, such as physiological temperature and pH. This is important for preserving the structure and function of the biomolecules being crosslinked.
- Reversibility: Some biological crosslinks are reversible. This allows for greater control over the properties of the crosslinked material.

3.4.3.2 Disadvantages of biological crosslinking

- Slower reaction rates: Biological crosslinking reactions can be slower than chemical crosslinking reactions. This can be a disadvantage for applications where rapid crosslinking is required.
- Lower stability: Biological crosslinks are often less stable than chemical crosslinks. This is because the biological crosslinkers can be degraded by enzymes or other biological molecules.

- Higher cost: Biological crosslinking can be more expensive than chemical crosslinking. This is because biological crosslinkers are often more difficult to produce and purify.

Here are some examples of biological crosslinking in nature:
- Collagen: Collagen is a protein that is found in connective tissues throughout the body. The collagen fibers are crosslinked together to form a strong and flexible structure.
- Elastin: Elastin is a protein that is found in elastic tissues, such as the skin and blood vessels. The elastin fibers are crosslinked together to allow the tissues to stretch and recoil.
- Fibrin: Fibrin is a protein that is involved in blood clotting. The fibrin molecules are crosslinked together to form a clot that stops the bleeding.

Biological crosslinking is a powerful tool that can be used to modify the properties of biomaterials and study biological processes. It has the potential to revolutionize the fields of medicine, biotechnology, and food science.

3.5 Crosslinking agents

The aim of crosslinking is to prolong the material's original structural and mechanical integrity. Another aim of crosslinking is to remove or at least neutralize the antigenic properties attributed to these materials. Methods of crosslinking concentrate mainly on creating new additional chemical bonds between the collagen molecules. These supplementary links (crosslinking) reinforce the tissue to give a tough, strong, but nonviable material that maintains the original shape of the tissue. The process of crosslinking involves the chemical agents initiating, ideally irreversible and stable intra- and intermolecular chemical bonds between collagen molecules, preferably the agent that promotes bonds between the groups of the amino acids. Bovine pericardium and diaphragm were crosslinked with glutaraldehyde (GA), formaldehyde (FA), polyethylene glycol (PEG), 1-ethyl-3-(3-dimethylaminopropyl)-carbodiimide (EDC), and hexamethylene diisocyanate (HMDC) at 0.5% and 1% concentrations. Crosslinking of biomaterials was found to be better at concentrations of 0.5% with a 48-hour duration at room temperature [1]. The efficiency and extent of crosslinking reactions depend upon the thickness of the layers of the collagenous tissue and define the magnitude of the penetration. Other parameters, like the concentration of the crosslinker, the time, and the temperature of exposure, affect the crosslinking [2]. Ideally, the treatment should also maintain much of the original character of the tissue, such as its flexible mechanical properties (biomaterial), and should not shrink the tissue. Hence, it is necessary to keep the tissue at a near-neutral pH, ensuring an aqueous media environment, and minimizing denaturation of the collagen to optimize crosslinking. Therefore, a balance must be achieved for attaining enough reliable crosslinks for the biomaterials to last the lifetime of the recipients while still permitting the biomaterial to perform in its natural state. The methods that have been developed do not and probably cannot satisfy the dual requirements. There are numerous crosslinking agents, but till now, the ideal crosslinking agent without the disadvantages of immunogenicity, cytotoxicity, and mineralization has yet to be discovered. So, there is a need to search for an ideal chemical reagent for crosslinking of collagen material.

3.5.1 Glutaraldehyde

Glutaraldehyde is a five-carbon bifunctional aldehyde; its introduction into biomedicine in the late 1960s has dominated other crosslinking agents. Glutaraldehyde's success as a fixative is due to its mixed hydrophobic and hydrophilic properties, which allow the molecule to rapidly penetrate both aqueous media and cell membranes [3].

One of the most common crosslinkers in use today is glutaraldehyde. Glutaraldehyde kills cells quickly and creates permanent crosslinks between proteins. They are bound together due to the action of the dialdehydes on the ε-amino group of the lysyl residues in the protein. If an ECM biomaterial prosthetic is fully crosslinked by a strong glutaraldehyde solution, it becomes essentially nonbiodegradable and will remain in the body until physically removed. Glutaraldehyde is also known to reduce the antigenicity of ECM biomaterials while making the prosthesis very resistant to infection [4]. Collagen crosslinking by glutaraldehyde is most rapid in an alkaline medium at pH 7–9, but an acid medium of pH 3–4 is used in the manufacture of collagen sponges since the degree of new intra- and intermolecular crosslinks can be better controlled in the acidic environment [5]. In addition to sensitivity to pH, chemical reactions and polymerization of glutaraldehyde in an aqueous solution are affected by temperature, light exposure, molar ratios of reactants, and time. Rasmussen and Albrecthsen [6] found that glutaraldehyde treatment of dermal autografts prolonged biodegradation and survival time. Acellular and GA-crosslinked diaphragms were used for the repair of experimentally created abdominal wall defects in rabbits. The acellular grafts showed faster resorption as compared to crosslinked grafts [7].

Glutaraldehyde in concentrations of 0.2%–2% exhibited no systemic toxicity when the graft was properly rinsed with saline before implantation. A higher concentration of glutaraldehyde, however, increased the stiffness of the material due to an increase in intra- and intermolecular crosslinks [8]. However, incomplete crosslinking was observed when the concentration of glutaraldehyde was less than 0.1% [9]. The fixing process in the glutaraldehyde was virtually completed within 2 hours. The fixing process produced a significant change in the mechanical behavior of tissue. The tissue became progressively stiffer as the treatment period was extended, which was especially pronounced at low levels of stress [10]. The ideal concentration of glutaraldehyde to be used in the preservation of material was 0.5%, as higher concentrations negatively affect tensile strength and elongation [11]. Unfortunately, the exact mechanism by which crosslinking occurred was complex and incompletely understood [12].

The importance of crosslinking in protecting the dermal autograft from biodegradation and prolonging its survival and permanence following implantation has been reported [13]. A collagen heterograft that was crosslinked using a low concentration of glutaraldehyde underwent the same level of neovascularization and fibroblast proliferation as a crosslinked autograft [13].

The treatment of porcine dermal collagen with glutaraldehyde increased the crosslinking between the collagen molecules and ensured an early phase of healing after implantation and encapsulation, followed by slow absorption and replacement by host tissue. The vascularity at the site of implantation has been described as an important factor in the rate of absorption and replacement [14,15]. The concentration and duration of exposure to glutaraldehyde have been reported to be important factors for the stabilization of treated tissue. In too low concentrations, complete crosslinking did not occur and it took longer to stabilize the tissue, along with some tissue degradation. Both the number and stability of collagen crosslinks increased with the increase in exposure time. The majority of the stabilization process occurred in the first 24 hours of exposure to glutaraldehyde solution. However, the process continued for a longer period to crosslink more and more of the available protein residue. It increased the resistance to proteolytic digestion [16]. Glutaraldehyde reacted with the amino group of lysyl residues in protein (collagen) and induced the formation of interchain crosslinks [17], which stabilized the tissue against chemical and enzymatic degradation depending upon the extent of crosslinking [18,19]. In crosslinking, two amine ($-NH_2$) groups in every glutaraldehyde induce the primary amine crosslink [20].

Comparative evaluation of crosslinking with glutaraldehyde (GA), formaldehyde (FA), polyethylene glycol (PEG), 1-ethyl-3-(3-dimethylaminopropyl)-carbodiimide (EDC), and hexamethylene diisocyanate (HMDC) revealed that GA was a better crosslinking agent followed by FA, PEG, EDC, and HMDC [1]. After implantation, over time, glutaraldehyde residues can leach into the host tissue and, due to their cytotoxic nature, cause the surrounding cells to die [21]. The influx of calcium ions accompanying the presence of glutaraldehyde also contributes to the calcification of the surrounding tissue [22] and can ultimately accelerate the failure of the implanted prosthesis [23].

Calcification has been described as one of the major causes of failure of bioprosthetic heart valves derived from glutaraldehyde-treated bovine pericardium or porcine aortic valves [24−26]. Primary tissue degeneration due to calcification necessitates reoperation or causes death in 20%−25% of adult recipients of porcine aortic bioprosthesis by 7−10 years after operation [27]. Bovine pericardial tissue, porcine aortic valves pretreated with glutaraldehyde, as well as type I collagen, mineralized when implanted subdermally in rats, but fresh (unfixed) implants underwent inflammatory attack and partial digestion without mineralization [28]. Glutaraldehyde-treated collagen biomaterials from intestinal submucosa of rats promoted calcification, exhibited poor host tissue incorporation, and ultimately the mechanical failure of bioprosthesis [29]. Crosslinking with glutaraldehyde has also been shown to suppress immunological recognition of the tissue [19,30], presumably by preventing the display of antigenic determinants by killing viable cells and by controlling the stability of the collagen triple helix [17].

The T lymphocytes from animals with a glutaraldehyde-tanned bovine pericardium implant responded significantly to glutaraldehyde bovine pericardium antigen.

In vitro ELISA showed that there were developments of antibodies against glutaraldehyde-treated bovine pericardium [31]. The antibody titer for glutaraldehyde-crosslinked collagen was found to be significantly lower than that for uncrosslinked collagen [32]. Residual glutaraldehyde molecules were incorporated in the collagen fibrils after crosslinking had impeded cellular growth [33]. Treatment of glutaraldehyde-fixed prostheses with L-glutamic acid followed by storage in bacteriostatic preservation had improved the biocompatibility of the implants [34]. Residual glutaraldehyde might be present on collagen fibrils or might be incorporated into dimers throughout the collagen matrix in a manner that promotes cytotoxicity [35].

Despite these drawbacks, glutaraldehyde is still cited in several publications in order to determine an upper-level number of crosslinks a collagen material may possess as well as to determine maximum enzymatic resistance and molecular weight [36–39]. Glutaraldehyde-treated bovine pericardium with homocysteic acid effectively reduced the toxicity without disturbing stability [40]. Brief immersion of human pericardial tissue in 0.65% glutaraldehyde reduced tissue stiffness and improved durability [41]. The amount of reagent, nature, and distribution are important for uniform crosslinking of biomaterials [42].

The increase in temperature had enhanced the crosslinking even at lower concentrations of glutaraldehyde. The concentrations of the free aldehyde monomer increased from 4% at room temperature to 35% at 50°C. The higher temperature had enhanced the diffusion of glutaraldehyde. However, if the time, temperature, and concentration of the reagent were optimal, the network would be uniform [43]. Degree and duration of the inflammatory response to acellular tissue were higher in glutaraldehyde-fixed acellular tissue as compared to genipin-fixed acellular tissue. However, the tissue regeneration rate for genipin-fixed acellular tissue was significantly faster than glutaraldehyde-treated tissue [21].

3.5.2 Formaldehyde

The formaldehyde treatment of natural biomaterials resulted in the formation of crosslinks in the collagen between the nitrogen atom at the end of the side chain of lysine and the nitrogen atom of a peptide linkage. The number of such crosslinks increased with time [44]. The effect of glutaraldehyde and formaldehyde on the stress/strain response of the bovine pericardium showed that the fixing process in glutaraldehyde was virtually completed within 2 hours and caused a significant change in the mechanical behavior of the tissue. The tissue became progressively stiffer as the treatment period was extended, which was especially pronounced at low levels of stress. Formaldehyde storage subsequent to fixing in glutaraldehyde was found to have no effect on the stress/strain response of the bovine pericardium. However, formaldehyde storage subsequent to fixing with glutaraldehyde was found to have no effect on the stress/strain response of the bovine pericardium [10]. Bovine pericardial tissue and tissue-derived bioprosthesis fixed in glutaraldehyde and stored in either glutaraldehyde or formaldehyde could induce cytotoxic reactions even after prolonged washing due to the slow leaching of the chemicals used for crosslinking and sterilizing [45]. Long-term aldehyde-crosslinked implants might develop foci of mineralization (calcification) [45].

Subcutaneously implanted formaldehyde-pretreated implants (type I collagen sponges) calcified more extensively than glutaraldehyde-pretreated implants in rats [28]. The formaldehyde vapor was highly effective in crosslinking the three-dimensional structure of collagen sponges. Sponges exhibited a very low susceptibility to collagenolytic degradation, a strong increase in melting temperature, and adequate tensile strength in the wet state. Fluid was more readily absorbed into formaldehyde-treated sponges than nonformaldehyde-treated sponges. It was observed that formaldehyde crosslinking required extensive aeration to meet the official limits of residual formaldehyde in the final implants [28].

3.5.3 Glyoxal

The most commonly used crosslinking agents' aldehydes, namely formaldehyde and glutaraldehyde, have arisen from an exacerbating effect on the calcification of prosthesis materials [18], cytotoxicity due to postimplantation depolymerization, and monomer release from the crosslinked materials [33]. A promising solution for biomedical purposes is glyoxal, a dialdehyde with lower toxicity when compared with other aldehyde agents [46]. Glyoxal is an organic compound with the formula OCHCHO. Glyoxal is a yellow-colored liquid that is the smallest dialdehyde (two aldehyde groups). It is commercially prepared by the gas-phase oxidation of ethylene glycol in the presence of a copper catalyst or by the liquid-phase oxidation of acetaldehyde with nitric acid. Glyoxal is a highly reactive monomer having excellent crosslinking and insolubilizing properties, as revealed by the free amine groups [47]. Glyoxal (dialdehyde) is routinely used to crosslink hydroxyethyl cellulose (HEC) particles during industrial processing to delay hydration and prevent lumping [48].

Glyoxal acts as a *crosslinking agent for collagen-based biomaterials at pH* 3.2 with a dispersion time of 5 minutes [49]. Glyoxal at different concentrations (0.25%, 0.5%, and 1%) acts as a crosslinking agent for collagen-based biomaterials when buffered with sodium hydrogen carbonate at pH 8.8 [50]. It is known that collagen crosslinked with glyoxal and other crosslinking agents at 25°C for 24 hours at pH 5.5 can act as a good *crosslinking* agent for *collagen-based biomaterials*. Agents that attract electrons more strongly than hydrogen are good crosslinking agents [51]. Vaz et al. [52] studied the in vitro degradation behavior of soy samples after crosslinking with glyoxal. The results confirmed that glyoxal crosslinking occurred via the free amine groups of the lysine (or hydroxylysine) residues, and a decrease was observed for higher amounts of the crosslinking agent. The stiffness of the materials also decreased with the increase in crosslinking degree as a result of the thermomechanical degradation that occurred during processing [52].

Yang et al. [53] used glyoxal for crosslinking chitosan fiber to improves mechanical properties. The effects of the preparation of glyoxal solution, for example, the concentration and pH, and the reaction conditions, for example, time and temperature, have been studied. Results showed that the glyoxal solution at 4%

concentration and 4.0 pH was better, while the reaction condition was better at 40°C and 60−70 minutes. Gupta and Jabrail [54] compared glutaraldehyde- and glyoxal-crosslinked chitosan microspheres for controlled delivery of centchroman. The glyoxal-crosslinked microspheres were found to be more compact and hydrophobic and showed better sustained release in comparison to chitosan microspheres and glutaraldehyde-crosslinked chitosan microspheres. Glyoxal is a highly reactive glycating agent involved in the formation of AGEs and is known to induce apoptosis, as revealed by the upregulation of caspase-3 and fractin. Osteoblasts treated with glyoxal undergo apoptosis, whereas the collagen type I (major extracellular matrix proteins) coating of titanium alloys (used for implants) has an antiapoptotic function [55]. A comparison of formaldehyde and glyoxal to evaluate the effect of chemical reticulation treatment on the mechanical properties, water vapor permeability, solubility, and color parameter of gelatin-based films revealed that films reticulated with formaldehyde and glyoxal had higher strengths, lower water vapor permeability, higher opacities, and greater color differences then the untreated films.

3.5.4 Diphenyl phosphoryl azide

It is an organic compound having the molecular formula $C_{12}H_{10}N_3O_3P$ and is widely used in the synthesis of other organic compounds. Diphenyl phosphoryl azide (DPPA) has been obtained by the reaction of phosphor chloridate with sodium azide. This is a new crosslinking agent that has been developed by a process known as the iphenyl phosphoryl azide. The technique achieved a natural crosslink between peptide chains without leaving foreign products in the crosslinked collagen [56,57]. DPPA was used to convert the carboxylic acid group into an acyl azide group in one single step. Acyl-azide-crosslinked materials showed very high values of shrinkage temperature and very good in vitro stability [58]. DPPA and the acyl-azide-crosslinked materials were found to be less toxic than their GA counterparts [2]. There was a marked reduction in calcification after DPPA treatment as compared to GA-crosslinked controls after 90 days of subcutaneous implantation in rats [58]. Glutaraldehyde treatment forms a bifunctional crosslink between collagen molecules. Due to DPPA treatment, the carboxylic functional groups of aspartic and glutamic amino acids on collagen undergo a three-step chemical reaction and, in the process, are converted to acyl azide functionalities, which in turn react with amino groups on collagen to form urea crosslinks [2]. The transformation of carboxylic acid groups into acryl azide groups is followed by one of two reactions: either a reaction with the amino acid of an enzyme or with a molecule on an adjacent collagen molecule [59].

Diphenyl phosphosphorylazide has also been described as a good chemical crosslinking agent for collagen-based scaffolds [60,61]. Reactions with DPPA take place in a nonaqueous solvent medium. The nonaqueous solvent is dimethylformamide, and the concentration of crosslinking agents varies between 0.0125% and 1.5% in volume/volume, preferably between 0.25 and 0.7%, with an incubation temperature of 0°C to 10°C for an incubation period ranging from a few hours to about one day. DPPA provides a simpler and faster crosslinking performance compared to glutaraldehyde, although it is a solvent-based reaction. In a comparison study of the cytotoxic effects of these reagents, diphenyl phosphoryl azide was found to be less cytotoxic than acyl azide, and both were better than glutaraldehyde [59]. Treatment of native bovine pericardium with 0.5% DPPA for 24 hours led to efficient crosslinking, corresponding to a 50% decrease in the free primary amino group content of the sample and raising its thermal stability from 62.8°C up to 81.3°C. DPPA-treated pericardium had a resistance to collagenase digestion similar to that of glutaraldehyde- or hydrazine-treated pericardium [59]. Petite et al. [62] obtained the best result in terms of the cytocompatibility of calf pericardium with 0.5% DPPA. The treated tissues showed a high level of crosslinking and a threefold increase in cell growth and migration over those in the nontoxic control. Collagen-based films and sponges are widely used as biomaterials. The rate of their biodegradation can be reduced by treating them with different crosslinking agents. Comparative analysis of pericardium treated with GTA and with acyl azide methods demonstrated the excellent cytocompatibility (endothelial cells) of DPPA-treated biomaterials [63].

To retard collagen membrane enzymatic degradation and increase its mechanical strength, the DPPA technique has been demonstrated to achieve natural crosslinks between peptide chains of collagen without leaving any foreign product in crosslinked molecule. In addition, DPPA-crosslinked collagen membranes are biocompatible and have handling characteristics that are suitable for guided bone regeneration applications in mandibular defects in rats, and DPPA-crosslinked collagen membranes in the treatment of human buccal soft tissue recessions resulted in predictable amounts of root coverage and clinical attachment gain [64]. Collagen-based biomaterials in the form of sponges (bovine type I collagen), both native and crosslinked by treatment with DPPA, were tested as three-dimensional scaffolds to support chondrocyte proliferation with maintenance of the phenotype in order to form neocartilage. These in vitro results indicated that both collagen matrices could support the development of tissue-engineered cartilage after chemical treatment [61]. The DPPA-treated collagen membranes were more biocompatible than the GA for crosslinking collagen biomaterials, and membranes made of collagen plus chondroitin sulfate were better than membranes made of pure collagen [65]. Gotora and Czernuszka [66] used GA and DPPA for mineralization and crosslinking techniques to improve the mechanical properties of collagen calcium phosphate composites to be used as bone analogs. DPPA-treated matrices showed favorable increase in stiffness after mineralization of thick matrices than GA-treated matrices.

3.5.5 Ethylene glycol diglycidyl ether

Epoxy compounds have been used in the past decade for the stabilization of collagen-based materials, including porcine aortic heart valves [67–72]. Due to their highly strained three-membered ring, epoxide groups were susceptible to a nucleophilic attack [73]. Predominantly, a reaction with the amine groups of (hydroxy) lysine residues occurred [74,75]. Additionally, epoxide groups could react with the secondary amine groups of histidine. Furthermore, reactions with the carboxylic acid groups of aspartic and glutamic acids existed, thereby increasing the versatility of the crosslinking [2,74]. In general, biological tissues are crosslinked in basic solutions (pH > 8.0) containing relatively high concentrations of epoxy compounds (1–5 wt%). The reaction between an epoxy group and an amino group is very slow compared to that between aldehydes such as glutaraldehyde and an amino group, but sufficient crosslinking can be achieved by adjusting the time, temperature, and concentration of hydrogen ions. A lower shrinkage temperature was obtained as compared to GA-crosslinked materials, but the in vitro stability of the crosslinked tissue was similar [72,76]. Crosslinking with an epoxy compound was carried out in a basic solution. Under these conditions, ethylene glycol diethyl ether (EGDGE) reacts with the amine groups of lysine or hydroxylysine residues present in the collagen resulting in bridges between the collagen molecules, which contain secondary amines and hydroxyl groups [77]. Vascular grafts crosslinked with epoxy compounds had higher tensile strengths and extensibilities, a lower stiffness and a better compliance as compared to GA-treated prosthesis [68,78]. In addition, these tissues retained more the original character, whereas GA-crosslinked grafts were somewhat stiffer [79]. Porcine aortic heart valves crosslinked with glycerol polyglycidyl ether were more pliable than their GA counterparts.

Subcutaneous implant studies in rats revealed that grafts crosslinked with epoxy compounds displayed lower calcification. Crosslinking of bovine pericardium with GA or a polyepoxy compound resulted in an increase in extensibility and a reduction in stress relaxation. Imamura et al. [80] did not find an influence of GA or epoxy crosslinking on the tensile strength. The effect of the crosslinking method on the mechanical behavior of the tissue showed that polyepoxy compound fixation resulted in a more flexible aortic wall as compared to GA crosslinking [79]. The group of polyglycidyl ethers has gained attention because of their good crosslinking ability and less stiffening of the tissue. Implantation of ethylene-glycol-diglycidyl-ether-crosslinked valves in 3-week-old Spraque-Dawley male rats did not result in considerably lower calcium levels than glutaraldehyde-fixed controls in both the leaflets and wall [76]. No calcification was found when polyglycerol-glycidyl-ether-crosslinked valves were implanted in the mitral position of a juvenile sheep [81,82]. Crosslinking of heart valves by several polyepoxy compounds resulted in a very low calcium content after subcutaneous implantation in 4-week-old rats. An average of 0.96 mg Ca per mg tissue was found, while 140.7 mg Ca per mg tissue was found in GA-fixed tissue [80].

Sung et al. [77] demonstrated that the tensile strength of ethylene-glycol-diglycidyl-ether-crosslinked porcine internal thoracic arteries was reduced to 40% after collagenase degradation for 24 hours at 37°C, whereas the tensile strength of its GA-fixed counterpart was reduced by 25%. Later in another study, porcine pericardium crosslinked with either ethylene glycol diglycidyl ether or (poly)glycerol polyglycidyl ether (pH 10.5) degraded at a comparable rate to GA-crosslinked controls upon exposure to collagenase or pronase, which means that epoxy compounds are suitable reagents to stabilize collagen-based tissues. Epoxy-crosslinked collagen showed a large reduction in calcification compared to GA-crosslinked materials

[67,80]. The biological tissues were crosslinked by EGDGE and GA through a fixative mechanism. After fixation, it was noted that the EGDGE-fixed tissues were more pliable than the glutaraldehyde-fixed tissues. This may be due to the fact that there were a few ether bonds (−O−) in the EGDGE, which may serve as flexible joints in the crosslinking bridge. In contrast, in glutaraldehyde, there were only carbon−carbon bonds (C−C), which were known to be relatively inflexible. The cellular compatibilities of the EGDGE-fixed vascular scaffolds were markedly increased through treating them with lysine and coating type I collagen on their luminal surface. The HUVECs seeded on the luminal surface of the EGDGE-fixed vascular scaffolds proliferated very well and reached a confluent monolayer, finally endothelializing the scaffolds [78].

3.5.6 Polyethylene glycol

Polyethylene glycol (PEG) is a family of long-chain polymers made up of ethylene glycol subunits. It is linear or branched, a neutral polyether, and soluble in water and most organic solvents. This molecule is the focus of much interest in the biotechnical and biomedical communities. PEG is produced by the interaction of a calculated amount of ethylene oxide with water, ethylene glycol, or ethylene glycol oligomers.

Various chemical techniques for grafting polyethylene glycol on bovine pericardium, their biostability, and calcification have been described. Calcification is a frequent cause of the clinical failure of bioprosthetic heart valves fabricated from glutaraldehyde-pretreated bovine pericardium [83]. Cardiovascular calcification, the formation of calcium phosphate deposits in cardiovascular tissue, is a common end-stage phenomenon affecting a wide variety of bioprosthesis [84].

PEG grafting of bovine pericardium followed by glutaraldehyde or hexamethylene diisocyanate has shown better mechanical stability compared to other grafting methods used. Surface modification of the bovine pericardium through high molecular weight PEGs via glutaraldehyde linkages provided new ways of controlling tissue biodegradation and calcification. Polyethylene glycol-grafted bovine pericardia (BP) and glutaraldehyde-treated BPs retained maximum stability in collagenase digestion compared with SDS-treated BP. Incubation of various enzymes (chymotrypsin, bromelain, esterase, trypsin, and collagenase) in these crosslinked pericardia variably reduced the tensile strength of these tissues, and chemical treatments of pericardial tissues might have altered their physical and chemical configuration and the subsequent degradation properties [83]. Porcine pericardium crosslinked with polyethylene glycol could control biodegradation and calcification and have low immunogenicity [85].

PEG-grafted tissues retained the maximum strength in trypsin buffer and calcium phosphate solutions and substantially inhibited platelet-surface attachment and their spreading. High molecular weight polyethylene glycol-grafted pericardium (a hybrid tissue) might be a suitable calcium-resistant material for developing prosthetic valves due to their stability and biocompatibility [86]. A method of reducing pericardial calcification and thrombosis via coupling polyethylene glycols (PEG) to glutaraldehyde-treated bovine pericardium via acetal linkages was developed. Grafting of the pericardium with PEG dramatically modified the surface and subsequently inhibited the deposits of calcium [86].

Treatment with aspirin or heparin has a synergistic effect on inhibiting glutaraldehyde-treated bovine pericardium calcification. The aspirin/heparin combination synergistically inhibited pericardial calcification in addition to their antithrombotic function [87]. The PEG grafting and Fe/Mg release had substantially inhibited the deposition of calcium in the bovine pericardium. The extractable alkaline phosphatase activity was reduced with PEG grafting and metal ion release to the bovine pericardium. Ferric ions might have slowed down or retarded the calcification process by inhibiting the proper formation of hydroxyapatite, while magnesium ions disrupted the growth of these crystals by replacing Ca^{2+}. In addition, it could be hypothesized that these metal ions might have inhibited the key element alkaline phosphatase, which acted as the substrate for mineralization. Hence, it was conceivable that a combination therapy via surface grafting of PEG and local delivery of low levels of ferric and magnesium ions might have prevented the bioprosthesis-associated calcification [88].

3.5.7 1-Ethyl-3-(3-dimethylaminopropyl) carbodiimide hydrochloride

Carbodiimides belong to the class of zero-length crosslinkers that modify amino acid groups to permit crosslink formation but do not remain as part of that linkage [91]. Protein crosslinks within collagen-based materials can be induced by methods other than glutaraldehyde treatment [89]. Carbodiimides such as cyanamide and 1-ethyl-3-(3-dimethylaminopropyl) carbodiimide hydrochloride (EDC) have been used with moderate success. The cytotoxic by-products found with glutaraldehyde appear to be avoided with carbodiimide, while mechanical properties that more closely match those of natural tissue can be maintained to a greater degree than are found following glutaraldehyde treatment [22,46,89,90].

The effects of EDC and glutaraldehyde on the hydrothermal, biochemical, and uniaxial mechanical properties of the bovine pericardium were studied. Both treatments increased the resistance to collagenase and cynogen bromide degradation. However, after denaturation, the EDC-treated tissue was slightly more resistant to collagenase and markedly more resistant to trypsin. EDC-treated tissue materials were more resistant, extensible, and elastic than glutaraldehyde-treated materials [91]. Small intestine collagen crosslinked with EDC provided the requisite physical properties with no immune response and without calcification and adhesion formation [92]. Differences in crosslinking of ovine carotid arteries under different pH conditions are attributed to differences in the location and types of the exogenous crosslinks formed. The location or type of crosslinks differentially affected the mechanical behavior of treated materials without affecting the increase in resistance to enzymatic degradation. Carbodiimide crosslinking of bioprosthetic materials has been shown to provide tissue stabilization equivalent to that of glutaraldehyde crosslinking but without the risk of the release of unreacted or depolymerized cytotoxic reagents after implantation [93].

Porous matrices containing collagen and hyaluronic acid were crosslinked with EDC in a range of 1–100 mM concentrations to enhance the mechanical stability of the composite matrix. In an enzymatic degradation test, EDC-treated membranes showed significant enhancement of their resistance to collagenase activity in comparison with 0.625% glutaraldehyde membranes [94]. Crosslinking characteristics, mechanical properties, and resistance against enzymatic degradation of biological tissues after fixation with genipin (a naturally occurring crosslinking agent) and carbodiimide were studied. Genipin crosslinking was comparatively slower than carbodiimide crosslinking. Tissue fixation in genipin and carbodiimide produced distinct crosslinking structures. Carbodiimide formed intrahelical and interhelical crosslinks within or between tropocollagen molecules, whereas genipin probably further introduced intermicrofibrilar crosslinks between adjacent collagen microfibrils. The stability of the fixed tissue was mainly determined by its intrahelical and interhelical crosslinks [95].

Crosslinking of collagen scaffolds by EDC in the presence of lysine was found to be an effective way to achieve a collagen scaffold with improved biostability and a more stable structure that could resist cell-mediated contraction [96]. In another study, EDC-crosslinked scaffolds-maintained strength and moderate cellularity and were more suitable for anterior cruciate ligament reconstruction than those crosslinked with ultraviolet irradiation. Crosslinking with EDC was an effective method to achieve a collagen scaffold with improved biostability and a more stable structure that could resist cell-mediated contraction [97].

3.5.8 Hexamethylene diisocyanate

Hexamethylene diisocyanate (HMDC) is a bifunctional molecule that covalently bonds with amino groups of lysine to form covalent crosslinks [98]. It was nontoxic even at higher concentrations and did not calcify the prosthesis [99]. The mechanical, thermal, and biochemical properties of HDMC-treated bovine pericardial tissues were studied. HMDC treatment produced changes in mechanical properties, denaturation temperature, and enzymatic resistance consistent with crosslinking similar to those seen in glutaraldehyde-treated tissue. However, the overall acceptability of HMDC as a crosslinking agent for biomaterial applications remained unclear. It appeared to be an interesting alternative to glutaraldehyde with many similar features [98].

The use of HMDC as a crosslinking agent for dermal sheep collagen under optimized conditions was studied. A linear relationship between the decrease in free amine group content and the increase in shrinkage temperature was observed. Crosslinking with HMDC did not influence the tensile strength of the noncrosslinked dermal sheep collagen samples but increased the elongation and decreased the high strain modulus [20]. Mechanical properties, enzymatic degradation, and cytotoxicity in pure collagen membrane after treatment with HMDC alone and in combination with PEG were studied [100]. It was found that material crosslinked with HMDC and PEG in a ratio of 5:10 had the highest tensile strength. HMDC-treated collagen materials strongly inhibited the enzymatic degradation. Weak cytotoxicity (5%) was observed during testing on cultures of human larynx carcinoma cells HEp-2. HMDC- and GA-crosslinked native diaphragm and pericardium were subcutaneously implanted in rabbits, and HMDS-crosslinked biomaterials were found better on the basis of biochemical parameters [101]. In vitro comparative evaluation of bladder acellular matrix crosslinked with glutaraldehyde, 1−4-butanediol diglyceride ether, and EDC revealed that HMDC-crosslinked matrices were highly resistant to enzymatic degradation [102].

In vitro calcification and enzymatic degradation of bovine pericardia after HMDC crosslinking and subsequent modification with PEG were studied. Incubation with alpha-chymotrypsin, bromelain, esterase, trypsin, and collagenase enzymes with crosslinked pericardia variably reduced their tensile strength. The PEG-modified tissues also indicated a substantial reduction in calcification. The PEG modification of the bovine pericardium via HMDC provided new ways of controlling tissue biodegradation and calcification. Pathologic calcification is thought to be the main cause of failure in the present generation of tissue valves fabricated from glutaraldehyde-pretreated bovine pericardium [87].

Porcine dermal collagen was permanently crosslinked with HMDC for its suitability as a dermal tissue engineering matrix [103]. The chemically crosslinked collagen had far fewer free lysine groups per collagen molecule than did the uncrosslinked matrix. The ability of the matrix to support human primary fibroblast outgrowth from explants was compared for matrices that had been presoaked in various solutions, including fibroblast media, cysteine, and phosphate buffered saline (PBS). It was found that superior cell outgrowth was obtained after soaking with fibroblast media and PBS. The collagen matrix showed the least amount of cell retention compared to the other two matrices. Acellular pericadial tissue crosslinked with HMDC was found to be better in vivo as compared to GA-crosslinked tissue [104].

Artificial extracellular matrix proteins, genetically engineered from elastin- and fibronectin-derived repeating units, were crosslinked with HMDC in dimethyl sulfoxide. It was found that the hydrogel films were transparent, uniform, and highly extensible. Their tensile moduli depended on crosslinker concentration and spanned the range characteristic of native elastin. The water content of the films was low, but the temperature-dependent swelling behavior of the crosslinked materials was reminiscent of the lower critical solution temperature property of the soluble polymers [105].

3.5.9 Physical

Physical crosslinking, comprising several steps of heating or compression, is advantageous because there are no chemical residues to cause concern for long-term in vivo stability, but the crosslinks are not always as effective as ones made with other methods. In contrast, dye-mediated photooxidation provides a more pervasive effect. Several amino acids are capable of being oxidized by light in the presence of specific photosynthesizing dyes [106,107]. No chemical residues result from photooxidation, and it has been shown that collagen matrices are stabilized by denaturation and enzymatic degradation. Crosslinking this way uses the matrix's own structure, leaving a more natural matrix after the process is completed [108]. Physiologically, photooxidized biomaterials show very little host cell infiltration but low immunogenicity and high resistance to calcification [109,110].

3.5.10 Enzymatic

Enzymatic crosslinkers provide a natural way to create crosslinks [111]. In particular, transglutaminase (TGase) is an enzyme found in many organisms that catalyzes a reaction between glutamine residues and the ε-amino groups of lysine residues [112]. TGases crosslink in a more natural way than most other methods that can be applied by researchers because it is one of the crosslinkers used by the body. One disadvantage of most TGases is that their activity is dependent on the concentration of calcium ions present. One particular TGase derived from Streptomyces mobaraensis, referred to as microbial TGase (MTGase), is calcium independent giving it an advantage over other types [113]. As a catalyst, MTGase leaves no residue and does not affect cell proliferation or attachment to the matrix. Research to improve the preparation of samples using MTGase shows promise that it is a viable way to stabilize a collagen matrix against degradation [111,114].

Ultimately, a native ECM will be completely digested by the body's enzymatic processes within several weeks. Problems can occur when all the immunogenic material is not removed from a bioprosthetic device, or cytotoxic and degradative factors negate the positive impacts of the device and lead to complications or even catastrophic failures [115]. Rigorous testing maintains the high standard of current bioprostheses, while research strives to modulate and optimize ECM biomaterials to degrade at specific rates and maintain structural integrity for specific periods while still allowing cells to grow, infiltrate, and heal in a natural way.

The influence of a wide range of crosslinking approaches (chemical, physical, biological) of *extruded* collagen fibers comprised a promising scaffold for anterior cruciate ligament and tendon reconstruction; however, the engineering of these fibers had yet to be improved to bring this material to clinical practice. Ultrastructural evaluation revealed a closely packed interfiber structure independent of the crosslinking method employed. The thermal properties were dependent on the crosslinking method employed and closely matched native tissues. The stress-strain curves were found to depend on the water content of the fibers, which was influenced by the crosslinking method. An inversely proportional relationship between both dry and wet fiber diameter and stress at break was found, which indicated that tailored-made biomaterials could be produced. Overall, the chemical stabilizations were more potent than both physical and biological approaches. Bifunctional agents such as hexamethylene diisocyanate and ethylene glycol diglycidyl ether or agents that promote matrix formation such as glutaraldehyde produce fibers with properties similar to those of native or synthetic fibers to suit a wide range of tissue engineering.

References

[1] Perme H. Invitro and in-vivo biocompatibility of crosslinked bovine pericardium and diaphragm. MVSc thesis submitted to deemed university ICAR, Indian Veterinary Research Institute, Izatnagar, Uttar Pradesh, India; 2006.
[2] Khor E, Wee A, Tan BL, Chew TY. Methods for the treatment of collagenous tissues for bio-prosthesis. Biomaterials 1997;18:95−105.
[3] Klein H, Deforest A. The inactivation of viruses by germicides. Chem Spec Manu Proc 1963;49:116−18.
[4] Schechter I. Prolonged retention of glutaraldehyde-treated skin allografts and xenografts: Immunological and histological studies. Ann Surg 1975;182(6):699−704.
[5] Chvapil M. Process for the production of collagen fibers fabrics in the form of felt-like membranes or sponges-like layer. US Patent 1974;3:823, 212.
[6] Rasmussen KE, Albrecthsen J. Glutaraldehyde: the influence of pH, temperature and buffering on the polymerization rate. Histochemistry 1974;38:19−26.
[7] Singh J, Kumar N, Sharma AK, Maiti SK, Goswami TK, Sharma AK. Acellular biomaterials of porcine origin for the reconstruction of abdominal wall defects in rabbits Trends Biomater. Artif Organs 2008;22:30−40.
[8] Woodroof EA. Use of glutaraldehde and formaldehyde to process heart valves. J Bioeng 1977;2:1−9.
[9] Rao A, Shanti C. Reduction of calcification by various treatments in cardiac valves. Biomater Appl 1999;13:238−68.
[10] Van-Noort R, Yates SP, Martin TR, Barker AT, Black MM. A study on the effects of glutaraldehyde and formaldehyde on the mechanical behaviour of bovine pericardium. Biomaterials 1982;3:21−6.
[11] Santillan-Doherty VJ, Sotres-Vega A, Olmos R, Arrelo JL, Garcia D, Vanda B, et al. Thoracoabdominal wall repair with preserved bovine pericardium. J Invest Surg 1996;9:45−55.
[12] Cheung DT, Nimni ME. Mechanism of crosslinking of protein by glutaraldehyde-I: reaction with monomeric and polymeric and polymeric collagen. Connect Tissue Res 1982;10:655−69.
[13] Oliver RF, Grant RA, Cox RW, Crooke A. Effects of aldehyde cross linkage on human dermal collagen implants in rats. Br J Exp Pathol 1980;62:544−9.
[14] Sarmah BD, Allen RTJ. Porcine dermal collagen repair of incisional hernia. Br J Surg 1984;71:524−5.
[15] Frankland AL. Use of porcine dermal collagen in the repair of perineal hernia in the dogs. A preliminary report. Vet Rec 1986;199:13−14.
[16] Smith JR, Jones S, Huil M, Robson C, Kleinert E. Bio prosthesis in hand surgery. J Surg Res 1986;41:378−87.
[17] Yannas IV. Natural materials. In: Ratner BD, Hoffman AS, Schoen FJ, Lemon JE, editors. Biomaterial Science. San Diego: Academic Press; 1996. p. 84−94.
[18] Golombo G, Schoen FJ, Smith MS, Linden J, Dixon M, Levy RJ. The role of glutaraldeyde-induced crosslinks in the calcification of bovine pericardium used in cardiac valve bioprostheses. Am J Pathol 1987;127:122−30.
[19] Nimni ME, Cheung D, Strates B, Kodoma M, Sheikh K. Chemically modified collagen a natural biomaterials for tissue replacement. J Biomed Mater Res 1987;21:741−71.

[20] Olde Damink LHH, Dijkstra PJ, Van Luyn MJ, Van Wachem PB, Nieuwenhuis P, Feijen J. Glutaraldehyde as a crosslinking agent for collagen based biomaterials. J Mat Sci Mater Med 1995;6:460−72.
[21] Chang Y, Tsai CC, Liang HC, Sung HW. In vivo evaluation of cellular and acellular bovine pericardium fixed with naturally occurring cross-linking agent. Biomaterials 2002;23:2447−57.
[22] Kim KM, Herrera GA, Battarbee HD. Role of glutaraldehyde in calcification of porcine aortic valve fibroblasts. Am J Pathol 1999;154(3):843−52.
[23] Schoen FJ, Levy RJ. Calcification of tissue heart valve substitutes: progress toward understanding and prevention. Ann Thorac Surg 2005;79(3):1072−80.
[24] Schoen FJ, Harasaki H, Kim KM, Anderson HC, Levy RJ. Biomaterial-associated calcification: Pathology, mechanisms, and strategies for prevention. J Biomed Mater Res 1988;22:11−36.
[25] Potkins B, Mcintosh C, Cannon R, Roberts W. Bio prostheses in tricuspid and mitral valve position or 95 months with heavier calcific deposit on the right-sided valve. Am J Cardiol 1988;61:949.
[26] Pelletier L, Carrier M, Leclere Y, Lepage G, Deguise P, Dyrda I. Porcine versus pericardial bioprostheses: a comparison of late results in 1,593 patients. Ann Thorac Surg 1989;47:352−61.
[27] Ionescu MJ, Smith DR, Hansan SS, Chidambaran AP, Tandom AP. Clinical durabilty of the pericardial xenograft valve: ten years experience with mitral replacement. Ann Thorac Surg 1982;34:265−77.
[28] Levy RJ, Schoen FJ, Sherman FS, Nicholas J, Hawley MA, Lund SA. Calcification of subcutaneously implanted type I collagen sponges: effects of formaldehyde and glutaraldehdye pretreatments. Am J Pathol 1986;122:71−82.
[29] Owen JJ, Lantz GC, Hiles MC, Varvleet J, Martin BR, Geddes LA. Calcification potential of small intestinal submucosa in a rat subcutaneous model. J Surg Res 1997;71:179−86.
[30] O'Brien TK, Gabbay S, Parkes AC, Knight RA, Zalesky PJ. Immunological reactivity to a new tanned bovine pericardial heart valve. Trans Am Soc Artif Intern Org 1984;30:440−4.
[31] Dahm M, Layman WD, Schwell AB, Factor SM, Freater RW. Immunogenicity of glutaraldehyde tanned bovine pericardium. J Thoraco Cardiovas Surg 1990;99:1082−90.
[32] Meade KR, Silver FH. Immunogenicity of collagenous implants. Biomaterials 1990;11:176−80.
[33] Van-Luyn MJA, Van-Wachem PB, Olde-Damink LHH, Dijkstra PJ, Feijen J, Nieuwenhuis P. Secondary cytotoxicity of cross-linked dermal sheep collagens during repeated exposure to human fibroblasts. Biomaterials 1992;13:1017−24.
[34] Grabenwoger M, Grimm M, Eybl E. Decreased tissue reaction to bioprosthetic heart valves material after L-glutamic acid treatment: a morphological study. J Biomed Mater Res 1992;26:1231−40.
[35] Goissin G, Marcantonio Jr EMJ, Marcantonio RAC, Lia RCC, Cancian DCJ, De Carvacho WM. Biocompatibility studies of anionic collagen membranes with different degree of glutaraldehyde crosslinking. Biomaterials 1997;20(1):27−34.
[36] Suh H, Lee WK, Park JC, Cho BK. Evaluation of the degree of crosslinking in UV irradiated porcine valves. Yonsei Med J 1999;40(2):159−65.
[37] Charulata V, Rajaram A. Influence of different crosslinking treatments on the physical properties of collagen membranes. Biomaterials 2003;24:759−67.

[38] Kumar V, Kumar N, Singh H, Gangwar AK, Dewangan R, Kumar A, et al. Effects of crosslinking treatments on the physical properties of acellular fish swim bladder. Trends Biomater. Artif Organs 2013;27(3):93–101.
[39] Kumar V, Kumar N, Singh H, Gangwar AK, Dewangan R, Kumar A, et al. In vitro evaluation of bubaline acellular small intestinal matrix. Intern J Bioassays 2013;2(3):581–7.
[40] Stacchino C, Bona G, Bonetti E, Rinaldi S, Della Ciana L, Grgnanai A. Detoxification process for glutaraldehyde treated bovine pericardium: biological, chemical and mechanical characterization. J Heart Valve Dis 1998;7:190–4.
[41] Vincentelli A, Zedgi R, Prat A, Lajos P, Latermoulle C, Le Bret E, et al. Mechanical modification to human pericardium after brief immersion in 0.625% glutaraldehyde. J Heart Valve Dis 1998;7:29–39.
[42] Ruijgrok JM, De Wijhn JR, Boon ME. Optimizing glutaraldehdye crosslinking of collagen: effect of time, temperature and concentration as measured by shrinkage temperature. J Mater Sci Mater Med 1994;5(2):80–7.
[43] Butler CE, Navaro FA, Orgill OP. Reduction of abdominal adhesions using composite collagen-glutaraldehyde implants for ventral hernia repair. J Biomed Mater Res 2001;58:75–80.
[44] Gustavson KH. The Chemistry of Tanning Processes. New York: Academic Press; 1956.
[45] Gendler E, Gendler S, Nimni ME. Toxic reactions evoked by glutaraldehyde-fixed pericardium and cardiac valve tissue bio prosthesis. J Biomed Mater Res 1984;18 (7):727–36.
[46] Weadock K, Olson RM, Silver FH. Evaluation of collagen crosslinking techniques. Biomater Med Devices Artif Organs 1984;11(4):293–318.
[47] Whipple EB. Structure of glyoxal in water. J Am Chem Soc 1970;90:7183–6.
[48] Brandt L. Cellulose ethers. In: Campbell FT, Pfefferkorn R, Rounsaville JF, editors. Ullmann's Encyclopedia of Industrial Chemistry, Weinheim: Wiley-VCH; 2005.
[49] Rehakova M, Bakos D, Vizarova K, Soldan M, Kova MJ. Properties of collagen and hyaluronic acid composite materials and their modification by chemical cross-linking. J Biomed Mater Res Part B: Appl Biomater 1998;30(3):369–72.
[50] Vizarova K, Bako D, Rehakova M, Petrikova M, Panakova E, Koller J. Modification of layered atelocollagen: enzymatic degradation and cytotoxicity evaluation. Biomaterials 1995;16(16):1217–22.
[51] Fathima NN, Madhan B, Rao JR, Nair BU, Ramasami T. Interaction of aldehydes with collagen: effect on thermal, enzymatic and conformational stability. Int J Bio Macro 2004;34(4):241–7.
[52] Vaz CM, De Graaf LA, Reis RL, Cunha AM. In vitro degradation behaviour of biodegradable soy plastics: effects of cross-linking with glyoxal and thermal treatment. Poly Degrd Stab 2003;81:65–74.
[53] Yang QG, Dou F, Liang B, Shen Q. Studies of cross-linking reaction on chitosan fiber with glyoxal. Carbo Poly 2005;59(2):205–10.
[54] Gupta KC, Jabrail FH. Glutaraldehyde and glyoxal cross-linked chitsan microspheres for controlled delivery of centrochroman. Carb Res 2004;341(6):744–56.
[55] Tippelt S, Ma C, Witt M, Bierbaum S, Funk RHW. Collagen type I prevents glyoxal-induced apoptosis in osteoblastic cells cultured on titanium alloy. Cell Tissues Organs 2004;177:29–36.
[56] Petite H, Menasche A, Huc A. Process for cross-linking of collagen by DPPA, the cross-linked collagen obtained thereby, and biomaterials of collagen base thus cross-linked. 1990; Patent WO 9012055 France.

[57] Petite H, Rault I, Huc A. Use of the azylazide method for cross-linking collegen rich tissue such as pericardium. J Biomed Mater Res 1990;24:179–87.
[58] Anselme K, Petite H, Herbage D. Inhibition of calcification in-vivo by acryl azide crosslinking of a collagen-GAG sponge. Matrix12 1992;264–73.
[59] Petite H, Frei V. Use of di-phenyl-phosphoryl-azide for cross-linking collagen based biomaterials. J Biomed Mater Res 1994;28:159–65.
[60] Vaissiere G, Chevallay B, Herbage D, Damour O. Comparative analysis of different collagen-based biomaterials as scaffolds for long-term culture of human fibroblasts. Med Biol Eng Comput 2000;38:205–10.
[61] Roche S, Ronziere MC, Herbage D, Freyria AM. Native and DPPA cross-linked collagen sponges seeded with fetal bovine epiphyseal chondrocytes used for cartilage tissue engineering. Biomaterials 2001;22:9–18.
[62] Petite D, Duval JL, Frei V, Abdul-Malik N, Sigot-Luizard MF, Herbage D. Cytocompatibility of calf pericardium treated by glutaraldehyde and by the acylazide methods in an organotypic culture model. Biomaterials 1995;16:1003–8.
[63] Rault V, Frei D, Herbage N, Abdul-Marak, Hue A. Evaluation of different chemical methods for cross-linking collagen gels, films and sponges. J Mat Sci: Mat Med 1996;7(4):215–22.
[64] Zahedi S, Legrand R, Brunel G, Albert A, Dewe W, Coumans B, et al. Evaluation of a diphenyl phosphorylazide cross-linked collagen membrane for guided bone regeneration in mandibular defects in rats. J Periodontal 1998;69(11):1238–46.
[65] Lorella M, Cinzia L, Mario G, Salvatore B, Ennio B, Giordano SC, et al. Biocompatibility of collagen membranes cross-linked with glutaraldehyde or diphenyl phosphorylazide: An in vitro study. J Biomed Mater Res 2003;62(2):504–9.
[66] Gotora D, Czernuszka JT. Mineralization and cross-linking techniques to improve the mechanical properties of collagen calcium phosphate (Coll-CAP) composites to be used as bone analogues. Int Symposium Res Stud Mater Sci Eng 2004;.
[67] Noishiki Y, Koyangi H, Miyata T. Bioprosthetic valve. 1988; Patent EP 0 306 256 A2.
[68] Lee JM, Thyagarajan K, Pereira CA, McIntyre J, Tu R. Cross-linking of a prototype bovine artery xenograft: comparison of the effects of glutaraldehyde and poly epoxy compounds. Artif Organs 1991;15:303–13.
[69] Lee JM, Pereira CA, Kan LWK. Effect of molecular structure of poly (glycidyl ether) reagents on cross-linking and mechanical properties of bovine pericardial xenograft materials. J Biomed Mat Res 1994;28:981–92.
[70] Tu R, She SH, Lin D, Hata C, Thyagarajan K, Noishiki Y, et al. Fixation of bioprosthetic tissues with monofunctional and multifunctional poly epoxy compounds. J Biomed Mat Res 1994;28:677–84.
[71] Tang Z, Yue Y. Cross-linking of collagen by poly glycidyl ethers. ASAIO J 1995;41:72–8.
[72] Sung HW, Hsu HL, Shih CC, Lin DS. Cross-linking characteristics of biological tissue fixed with monofunctional or multifunctional epoxy compounds. Biomaterials 1996;17(14):1405–10.
[73] Solomons TWG. Organic Chemistry. Fourth ed New York, USA: John Wiley & Sons; 1988.
[74] Tu R, Quijano RC, Lu CL, Shen S, Wang E, Hata C, et al. A preliminary study of the fixation mechanism of collagen reaction with a polyepoxy fixative. Int J Art Org 1993;16(7):537–44.

[75] Wang E, Thyagarajan K, Tu R, Hata C, Shen SH, Quijano RC. Evaluation of collagen modification and modification and surface properties of a bovine artery via polyepoxy compound fixation. Int J Art Org 1993;16(7):530–6.
[76] Myers DJ, Gross J, Nakaya G. Stent less heart valves: biocompatibility issues associated with new anti-mineralisation and fixation agents. In: Piwnica A, Westaby S, editors. Stent Less Bio Prostheses. Oxford: Isis Medical; 1994. p. 100–17.
[77] Rosenberg D. Dialdehyde starch tanned bovine heterografts: development. In: Sawyer PN, Kaplitt MJ, editors. Vascular grafts. New York: Appleton-Century Crofts; 1978. p. 261–70.
[78] Xi T, Liu F. Effect of pretreatment with epoxy compounds on the mechanical properties of bovine pericardial bioprosthetic materials. J Biomater Appl 1992;7:61–75.
[79] Zhou J, Quintero LJ, Helmus MN, Lee C, Kafesjian R. Porcine aortic wall flexibility: Fresh versus denacol fixed versus glutaraldehyde fixed. ASAIO J 1997;43:M470–5.
[80] Imamura E, Sawatani O, Koyanagi H, Noishiki Y, Miyata T. Epoxy compounds as a new cross-linking agent for porcine aortic leaflets: subcutaneous implant studies in rats. J Card Surg 1989;4:50–7.
[81] Shen SH, Sung HW, Tu R, Hata C, Lin D, Noishiki Y, et al. Characterization of poly epoxy compound fixed porcine heart valve bioprosthesis. J Appl Biomater 1994;5:159–62.
[82] Sung HW, Hsu CS, Wang SP, Hsu HL. Degradation potential of biological tissues fixed with various fixatives: An in-vitro study. J Biomed Mat Res 1997;35:147–215.
[83] Vasudev SC, Chandy T, Sharma CP. Influence of polyethylene glycol grafting on the in vitro degradation and calcification of bovine pericardium. J Biomater Appl 1997;11(4):430–52.
[84] Vasudev SC, Chandy T. Polyethylene glycol-grafted bovine pericardium: a novel hybrid tissue resistant to calcification. J Mater Sci Mater Med 1999;10(2):121–8.
[85] Aravind S, Paul W, Vasudev SC, Sharma CP. Polyethylene glycol (PEG) modified bovine pericardium as a biomaterial: a comparative study on immunogenicity. J Biomater Appl 1998;13(2):158–65.
[86] Vasudev SC, Chandy T, Sharma CP. The antithrombotic versus calcium antagonistic effects of polyethylene glycol grafted bovine pericardium. J Biomater Appl 1999;14(1):48–66.
[87] Vasudev SC, Chandy T, Sharma CP, Mohanty M, Umasankar PR. Synergistic effect of released aspirin/heparin for preventing bovine pericardial calcification. Artif Org 2000;24(2):129–36 2000.
[88] Vasudev SC, Chandy T, Umasankar MM, Sharma CP. Inhibition of bio prosthesis calcification due to synergistic effect of Fe/Mg ions to polyethylene glycol grafted bovine pericardium. J Biomater Appl 2001;16(2):93–107.
[89] Giradot JM, Giradot MN. Amide crosslinking: an alternative to glutaraldehyde fixation. J Heart Valve Dis 1996;5:518–25.
[90] Buttafoco L, Engbers-Buijtenhuijs P, Poot AA, Dijkstra PJ, Daamen WF, van Kuppevelt TH, et al. First steps towards tissue engineering of small-diameter blood vessels: preparation of flat scaffolds of collagen and elastin by means of freeze drying. J Biomed Mater Res B Appl Biomater 2006;77(2):357–68.
[91] Lee JM, Edwards HHL, Pereira CA, Samii SI. Crosslinking of tissue-derived biomaterials in 1-ethyl-3-(3-dimethylaminopropyl)-carbodiimide (EDC). J Mat Sci: Mater Med 1996;7(9):531–41.
[92] Abraham GA, Murray J, Billiar K, Sullivan SJ. Evaluation of porcine intestinal collagen layer as biomaterial. J Biomed Mater Res 2000;51:442–52.

[93] Gratzer PF, Lee JM. Control of pH alters the type of cross-linking produced by 1-ethyl-3-(3-dimethylaminopropyl)-carbodiimide (EDC) treatment of acellular matrix vascular grafts. J Biomed Mater Res 2001;58:172−9.
[94] Park SN, Park JC, Kim HO, Song MJ, Suh H. Characterization of porous collagen/hyaluronic acid scaffold modified by 1-ethyl-3-(3-dimethyl aminopropyl) carbodiimide cross-linking. Biomaterials 2002;23(4):1205−12.
[95] Sung HW, Wen-Hisang C, Chiun-Yuang M, Meng-Horng L. Crosslinking of biological tissues using genipin and/or carbodiimide. J Biomed Mater Res 2003;64A:427−38.
[96] Ma L, Gao C, Mao Z, Zhou J, Shen J. Biodegradability and cell-mediated contraction of porous collagen scaffolds: the effect of lysine as a novel crosslinking bridge. J Biomed Mater Res 2004;71(2):334−42.
[97] Caruso AB, Dunn MG. Changes in mechanical properties and cellularity during long-term culture of collagen fiber ACL reconstruction scaffolds. J Biomed Mater Res 2005;73(4):388−97.
[98] Naimark WA, Pereira CA, Tsang K, Lee JM. HMDC crosslinking of bovine pericardial tissue: A potential role of the solvent environment in the design of bioprosthetic materials. J Mater Sc: Mater Med 1995;6:235−41.
[99] Oliver RF, Baker H, Crooke A, Grant RA. Dermal collagen implants. Biomaterials 1982;1982(23):38−40.
[100] Bakos D, Koniarová D. Collagen and collagen/hyaluronan complex modifications. Chem Pap 1999;53(6):431−5.
[101] Singh H, Kumar N, Sharma AK, Kataria M. Munjal A. Biochemical changes in rabbit organs after subcutaneous implantation with bovine pericardium and diaphragm. Int J Genet Eng Biotech 2011;2(1):77−89.
[102] Dewangan R, Sharma AK, Kumar N, Maiti SK, Singh S, Gangwar AK, et al. In vitro biocompatibility determination of bladder acellular matrix graft. Trends Biomater Artif Organs 2011;25:161−71.
[103] Jarman-Smith ML, Bodamyali T, Stevens C, Howell JA, Horrocks M, Chaudhuri JB. Porcine collagen crosslinking, degradation and its capability for fibroblast adhesion and proliferation. J Mater Sci Mater Med 2004;15(8):925−32.
[104] Singh H, Kumar N, Sharma R, Dewangan R, Kumar A, Kumar V, et al. In vivo biocompatibility determination of crosslinked as well as uncross linked native and acellular pericardium of buffalo. Intern J Bioassays 2012;2(2):391−7 2012.
[105] Nowatzki PJ, Tirrell DA. Physical properties of artificial extracellular matrix protein film as prepared by isocyanate crosslinking. Biomaterials 2004;25(7−8):1261−7.
[106] Gurnani S, Arifuddin M. Effect of visible light on amino acids II-Histidine. PhotochemPhotobiol 1966;5(4):341−5.
[107] Gurnani S, Arifuddin M, Augusti KT. Effect of visible light on amino acids I-Tryptophan. Photochem Photobiol 1966;5(7):495−505.
[108] Schmidt CE, Baier JM. Acellular vascular tissues: natural biomaterials for tissue repair and tissue engineering. Biomaterials 2000;21(22):2215−31.
[109] Bianco RW, Phillips R, Mrachek J, Witson J. Feasibility evaluation of a new pericardial bioprosthesis with dye mediated photo-oxidized bovine pericardial tissue. J Heart Valve Dis 1996;5(3):317−22.
[110] Moore MA, Adams AK. Calcification resistance, biostability, and low immunogenic potential of porcine heart valves modified by dye-mediated photooxidation. J Biomed Mater Res 2001;56(1):24−30.

[111] Broderick EP, O'Halloran DM, Rochev YA, Griffin M, Collighan RJ, Pandit AS. Enzymatic stabilization of gelatin-based scaffolds. J Biomed Mater Res B Appl Biomater 2005;72(1):37–42.
[112] Folk JE. Transglutaminases. Ann Rev Biochem 1980;49:517–31.
[113] Yokoyama K, Nio N, Kikuchi Y. Properties and applications of microbial transglutaminase. Appl Microbiol Biotechnol 2004;64(4):447–54.
[114] Chen RN, Ho HO, Sheu MT. Characterization of collagen matrices crosslinked using microbial transglutaminase. Biomaterials 2005;26(20):4229–35.
[115] Halloran DMO, Collighan RJ, Griffin M, Pandit AS. Characterization of a microbial transglutaminase cross-linked type II collagen scaffold. Tissue Eng 2006;12(6):1467–74.

Rumen-derived extracellular matrix scaffolds and clinical application

Ajit Kumar Singh[1], Naveen Kumar[2,*], Pawan Diwan
Singh Raghuvanshi[1], Harendra Rathore[1], Anil Kumar Gangwar[3],
Sameer Shrivastava[4], Sonal Saxena[4], Mohar Singh[1], Dayamon
David Mathew[5] and Karam Pal Singh[1]

[1]Department of Veterinary Surgery and Radiology, College of Veterinary Science and Animal Husbandry, Nanaji Deshmukh Veterinary Science University, Mhow, Indore, Madya Pradesh, India, [2]Division of Surgery, ICAR-Indian Veterinary Research Institute, Izatnagar, Uttar Pradesh, India, [3]Department of Veterinary Surgery & Radiology, College of Veterinary Science & Animal Husbandry, Acharya Narendra Deva University of Agriculture and Technology, Ayodhya, Uttar Pradesh, India, [4]Division of Veterinary Biotechnology, ICAR-Indian Veterinary Research Institute, Izatnagar, Uttar Pradesh, India, [5]Department of Veterinary Surgery & Radiology, Faculty of Veterinary and Animal Sciences, Banaras Hindu University, Rajiv Gandhi South Campus, Barkachha, Uttar Pradesh, India

4.1 Introduction

The ruminant refers to a mammal having a stomach with four chambers. These include a forestomach, consists of a rumen, a reticulum and an omasum, and a fourth chamber known as an abomasum. Examples of ruminants include mammals belonging to the genus Copra, Bos, Cervus, and Ovis. The rumen underpins much of our agricultural industry. Without this stomach chamber, cows and other ruminants would be much less efficient at turning grass into milk, meat and wool. A cow's rumen has a capacity of up to 95 L and contains billions of bacteria and other microbes. These microbes produce the enzymes that digest cellulose into sugars and fatty acids for their hosts to use. A less desirable by-product is the potent greenhouse gas, methane; a single cow can produce up to 280 L of methane a day. Collectively, these organs occupy almost three-fourths of the abdominal cavity, filling virtually all of the left side and extending significantly into the right. The reticulum lies against the diaphragm and is joined to the rumen by a fold of tissue. The rumen is the largest of the forestomachs and is itself sacculated by muscular pillars into what are called the dorsal, ventral, caudodorsal, and caudoventral sacs. In many respects, the reticulum can be considered a

---

*Present affiliation: Veterinary Clinical Complex, Apollo College of Veterinary Medicine, Jaipur, Rajasthan, India.

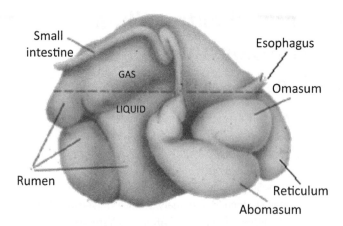

Figure 4.1 The ruminant stomachs, as seen from the right side.

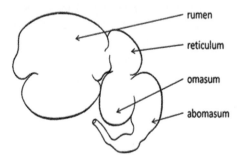

Figure 4.2 Relative size of ruminant stomach chambers.

"cranioventral sac" of the rumen; for example, ingesta flow freely between these two organs. The reticulum is connected to the spherical omasum by a short tunnel. The abomasum is the ruminant's true or glandular stomach (Figs. 4.1 and 4.2).

Histologically, abomasum is very similar to the stomach of monogastrics. The interior of the rumen, reticulum, and omasum is covered exclusively with stratified squamous epithelium similar to what is observed in the esophagus. Each of these organs has a very distinctive mucosa structure, although within each organ, some regional variation in morphology is observed. The anatomic features described above are exemplified by cattle, sheep, and goats. Certain other animals are also generally called ruminants but have slightly different forestomach anatomy. Camelids (camels, llamas, alpacas, vicunas) have a reticulum with areas of gland-like cells, and an omasum that is tubular and almost indistinct. These animals are occasionally referred to as pseudo ruminants or as having "three stomachs" rather than four. Stratified, squamous epithelium such as one found in the rumen is not usually considered an absorptive type of epithelium. Ruminal papillae are however very richly vascularized, and the abundant volatile fatty acids produced by fermentation are readily absorbed across the epithelium. Venous blood from the fore stomachs, as well as the abomasum, carries these absorbed nutrients into the portal vein, and hence, straight to the liver.

4.2 Anatomy of ruminant stomach

The ruminant stomach has four parts: (1) rumen ("paunch"), (2) reticulum ("honeycomb tripe"), (3) omasum ("book"), and (4) abomasum. Collectively the rumen, reticulum, and omasum are known as the forestomach (proventriculus). The rumen and reticulum are intimately related in structure and function and are often referred to as rumeno-reticulum. The abomasum is structurally analogous to the simple glandular stomach. The anatomical differences between the forestomach and the simple glandular stomach reflect their distinct functional roles. The primary functions of the forestomach are storage, fermentation, and absorption, while the simple glandular stomach performs secretary and digestive functions. Consequently, the forestomach has gross anatomical and histological features that are quite distinct from those of the glandular stomach. The forestomach does not contain a glandular mucosa but instead consists of a nonglandular keratinized stratified squamous epithelium, which consists of the stratum corneum, stratum granulosum, stratum spinosum, and stratum basale, and appears in many respects analogous to the structure of the skin. The epithelium is located on the luminal side of the forestomach and is separated from the underlying propria-submucosa by a basement membrane. The abluminal side of the forestomach contains a muscle layer known as the tunica muscularis.

Stomach submucosa compositions are derived from the wall of the glandular stomach, which contains the following layers: the tunica mucosa (including an epithelium layer, a lamina propria layer consisting of reticular or fine areolar tissue, and a glandular layer), the tunica submucosa layer (composed of areolar tissue and lacking glands), the tunica muscularis layer (composed of three layers of muscle), and the serosa (a layer of mesothelium outside the loose connective tissue which invests the muscle layers). The presence of the glandular layer within the stomach wall is characteristic of the glandular, gastric, or simple stomach of monogastric mammals. Only the last chamber of the complex stomach of ruminants, the abomasum, contains this glandular layer.

Two unique features of the forestomach relative to the glandular stomach are that the lamina propria of the forestomach is much denser and does not include glands or a glandular layer. In addition, the lamina muscularis mucosa, a fine muscle layer in the basal region of the tunica mucosa layer of the glandular stomach, is absent from the rumen and most of the reticulum. In the absence of the lamina muscularis, the lamina propria blends with the submucosa to form a layer that is collectively referred to as the propria-submucosa. Also unique to the forestomach is an unusually thick and dense band of extracellular matrix (ECM) within the lamina propria, which runs parallel to the epithelial surface. This band of tissue contains collagen IV and laminin, which play a critical role in cell growth, differentiation, and migration during tissue development and reconstruction. Beneath this band of tissue, the ECM has a more typical open reticular pattern. Forestomach tissue also includes surface protrusions known as papillae in the rumen, reticular crests in the reticulum, and lamellae in the omasum. The propria-submucosa extends into these protrusions. The mucosa of reticulum contains rows of folds (laminae) that form square honeycomb-like compartments called cells. On the surface of the laminae are short, conical projections called papillae. Near the margins of the folds are strands of smooth muscle fibers forming the muscularis mucosae. Contractions of the honeycomb cells, with the purse-string action of the smooth muscle strands, help the mechanical digestion of the ingesta. This is the most cranial part of the forestomach. ECMs can be derived in whole or in part from tissues, such that they retain at least one component of the tissue, such as the propria-submucosa.

4.3 Decellularization of forestomach matrix

Methods for decellularization of the bovine rumen have not been reported in the literature. In the search for biomaterials Lun et al. [1] have identified ovine forestomach matrix (OFM), a thick, large format ECM, which is biochemically diverse and biologically functional. Ovine forestomach matrix was purified using an osmotic process that was shown to reduce the cellularity of the ECM and aid tissue delamination. Ovine forestomach matrix produced using this technique was shown to retain residual basement membrane components, as evidenced by the presence of laminin and collagen IV. The collagenous microarchitecture of OFM retained many components of native ECM including fibronectin, glycosaminoglycans, elastin, and fibroblast growth factor. Ovine forestomach matrix was nontoxic to mammalian cells and supported fibroblast and keratinocyte migration, differentiation and infiltration. Ovine forestomach matrix is a culturally acceptable alternative to current collagen-based biomaterials and has immediate clinical applications in wound healing and tissue regeneration. ECM-based biomaterials have an established place as medical devices for wound healing and tissue regeneration.

Collagen IV forms the basis of the collagenous matrix that defines the basement membrane and participates in cell adhesion, especially via laminin, which in turn binds heparin sulfate and integrin receptors. It was found that the presence of residual basement membrane components would direct keratinocyte adhesion and migration along the luminal surface of OFM, eventually leading to cellular infill of the matrix [2]. Studies demonstrated that keratinocytes preferentially bound to the luminal surface, most likely via laminin present in the remnants of the basement membrane [1]. This was evidenced by collagen IV and laminin immunoreactivity on the abluminal surface of the lamina epithelialis, consistent with fracturing of the basement membrane during delamination of the lamina epithelialis and propria-submucosal layers. Basement membrane components were also evident in the remnants of vascular channels within OFM and would therefore be available during repopulation of the channels by endogenous endothelial cells.

Ovine forestomach matrix is significantly thicker (300 mm) than intestinal submucosa-derived SIS (100 mm) [3]. These structural features impart excellent biophysical characteristics to the OFM, which make it an ideal biomaterial for load-bearing implant applications (e.g., hernioplasty or breast reconstruction) where the biomaterial must function in a supportive role during tissue regeneration. This suggests that there may be additional components in the OFM extract (e.g., growth factors or GAGs) that also contribute to the observed stimulatory effect.

Further studies are underway to understand the contribution of these additional components to the observed bioactivity of OFM. The presence of important ECM cofactors, as well as the demonstrated bioactivity of OFM in culture, supports the functional properties of OFM. The functional properties of OFM, as well as its structural features, suggested that this biomaterial is ideally suited to wound healing and soft tissue regeneration. Clinical applications for native ECM in wound healing and tissue regeneration span a range of size, format, and strength requirements. The forestomach is substantially larger than the glandular stomach and has gross anatomical and histological features that are quite distinct also. The forestomach does not contain a glandular mucosa but instead consists of a nonglandular keratinized stratified squamous epithelium, which appears in many respects analogous to the structure of human skin. One feature of the forestomach that was retained in OFM was an especially dense layer of connective tissue present in the propria-submucosa. This layer has previously been described in the forestomach of several species as a "condensed fibrous layer" [4].

Forestomach is one of the largest available organs for the production of intact ECMs. In this way, it is possible to produce large format single sheets of OFM (approx. 40 × 40 cm), which would be especially suitable for application to large tissue deficits, for example, pressure ulcers and burns. Native and reconstituted animal collagens are well tolerated in humans due to the close sequence homology between mammalian collagens. As such, the biocompatibility of collagens from a variety of mammalian sources is well established in a number of clinical applications. An additional attractive feature of ovine collagens is the low disease transmission risk these present relative to human-, bovine-, or porcine-derived collagens.

4.4 Preparation of acellular matrices from bubaline rumen

Fresh rumen tissue of water buffalo (*Bubalus bubalis*) was procured from the local abattoir and immediately preserved in chilled $1 \times$ phosphate buffer saline (PBS), (pH 7.4) solution containing 0.1% amikacin, and 0.02% EDTA. Ruminal tissue was thoroughly washed with PBS to remove adhered ruminal contents. The tissue was cut into 2×2 cm^2 pieces and was treated with different concentrations of decellularizing solutions such as tri-n-butyl phosphate (TNBP), triton X-100, sodium dodecyl sulfate (SDS), tween-20, and trypsin under constant agitation (180 rpm/min) at room temperature (Fig. 4.3).

Solutions were changed at every 24 hours. Decellularization protocols were optimized based on the principle of maximum removal of cellular contents with minimum damage to basic tissue architecture. Degree of decellularization was evaluated by microscopic examination. DNA quantification of samples was done for optimized protocol. Details of the decellularization protocols are present in Table 4.1

Native and decellularized tissue samples were fixed in 10% formalin. The samples were dehydrated with graded ethanol, embedded in paraffin wax, and sections of size 5 μm were prepared. Acellularity and basic tissue architecture of the samples were evaluated after hematoxylin and eosin (H&E) staining [5,6].

Figure 4.3 Constant agitation of ruminal tissue at room temperature.

Table 4.1 Protocols for decellularization of bovine ruminal tissue matrix.

Groups	Chemical/enzyme used	Strength of solution (%)		Decellularization time (h)			
1	Tri-N-butyl phosphate	0.5%	1%	24	48	72	96
2	Triton X-100	0.5%	1%	24	48	72	96
3	Sodium dodecyl sulfate (SDS)	0.5%	1%	24	48	72	96
4	Tween-20	0.5%	1%	24	48	72	96
5	Trypsin	0.5%	1%	24	48	72	96

4.4.1 Macroscopic observations

Protocols for delamination and decellularization of bubaline rumen were optimized. Macroscopic observations showed that the keratinized mucosal layer was easily scrubbed off and serosal layer was separated with slight mechanical assistance. The isolated delaminated and decellularized bubaline rumen was kept in 70% ethanol for sterilization. The bovine rumen tissue matrix after treatment with biological detergents appeared soft, whiter, and slightly spongy in consistency then native tissue (Fig. 4.4).

4.4.2 Microscopic observations

Microscopically, native bovine rumen showed keratinized epithelium on mucosal surface. Lamina propria is the luminal portion of the propria-submucosa, which includes a dense layer of ECM and serosal layer. The delaminated propria-submucosal layer showed cellularity, tunica muscularis, and thick collagen fibers. Native bovine rumen showed dense compact arrangements of collagen fibers (Fig. 4.5).

The microscopic observations of the delaminated bovine rumen matrix subjected to ionic, nonionic, zwitterionic biological detergents and enzyme treatment at 1% concentration for 24 and 72 hours are presented in Table 4.2

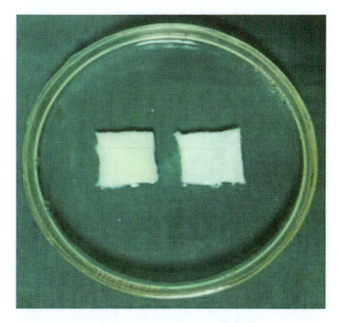

Figure 4.4 Showing decellularized bubaline ruminal tissue matrix.

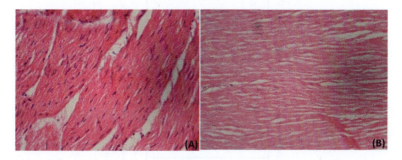

Figure 4.5 Histological picture of bovine rumen: (A) native tissue and (B) decellularized tissue matrix

4.4.2.1 Treatment with zwitter ionic detergent (tri-N-butyl phosphate)

The delaminated rumen treated with zwitter ionic detergent, that is, 0.5% tri-N-butyl phosphate (TNBP), for 24 hours under constant agitation showed 45%—50% loss of cellularity along with presence of cellular debris (Fig. 4.6A). At 48 hours time interval, no further effect on loss of cellularity was observed. Cellular contents were higher at 24 hours as compared to 48 hours (Fig. 4.6B). At 72 hours nearly 70% of decellularization was observed (Fig. 4.6C). At 96 hours more than 70% decellularization and loss of cellular debris with no damaged to collagen fibers was observed (Fig. 4.6D) [5,6].

Table 4.2 Showing microscopic observations by different protocols for the decellularization of the delaminated bovine rumen.

S. No.	Observations	1% SDS		1% Triton X-100		1% Trin BP		1% Trypsin		1% Tween-20	
		24	72	24	72	24	72	24	72	24	72
1	Cellular contents	+	−	++	+	++	+	+++	++	+++	+
2	Cellular debris	++	−	++	++	++	+	+++	++	+++	++
3	Collagen fiber compactness	+++	+	++	−	++	+	+++	++	+++	+
4	Collagen fiber morphology	Thick	Thin	Thick	Thin	Thick	Thick	Thick	Thin	Thick	Thick
5	Porosity	+	++	+	++	+	++	+	++	+	++

Cellular contents: +++, normal cellularity; ++, moderate no of cells; +, mild no of cells; −, no cellular material. Debris: +++, more; ++, moderate; +, mild; −, no debris. Collagen fibers arrangement: +++, compact; ++, mildly loose; +, moderately loose; −, heavily loose. Collagen fibers morphology: Thin and thick. Porosity: ++, highly porous; +, moderate; +, mild.

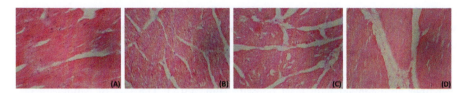

Figure 4.6 Histological picture of bovine rumen treated with 0.5%TNBP: (A) 24 h, (B) 48 h, (C) 72 h, (D) 96 h posttreatment.

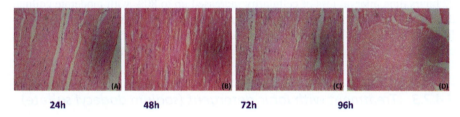

Figure 4.7 Histological picture of bovine rumen treated with 1% TNBP: (A) 24 h, (B) 48 h, (C) 72 h, (D) 96 h posttreatment.

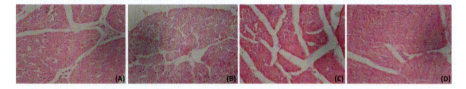

Figure 4.8 Histological picture of bovine rumen treated with 0.5% triton X-100: (A) 24 h, (B) 48 h, (C) 72 h, (D) 96 h posttreatment.

Treatment with 1%TNBP for 24 hours under constant agitation showed 55%–60% loss of cellularity along with presence of cellular debris, which was higher as compared to 0.5% at the same time interval (Fig. 4.7A). At 48 hours time interval, further loss of cellularity was observed (Fig. 4.7B). At 72 hours nearly 80% of decellularization was observed (Fig. 4.7C). At 96 hours more than 80% decellularization and loss of cellular debris along with damage to collagen fibers was observed (Fig. 4.7D) [5,6].

4.4.2.2 Treatment with nonionic detergent (triton X-100)

Treatment with 0.5%triton X-100 for 24 hours under constant agitation showed 40%–45% loss of cellularity with high cellular contents, thick and compact collagen fiber arrangement with least porosity (Fig. 4.8A). At 48 hours no further loss of cellularity was observed. High cellular content, thick, and compact collagen fiber arrangement with least porosity was observed (Fig. 4.8B). At 72 hours time interval, loss of cellularity increased up to 70% with moderate cellular debris and porosity

(Fig. 4.8C). At 96 hours time interval, no further increase in cellularity and porosity was observed. Mild cellular debris and damage in collagen fiber arrangement was observed at this stage (Fig. 4.8D) [5,6].

Treatment with 1%triton X-100 for 24 hours under constant agitation showed 50%–55% loss of cellularity with high cellular contents, thick, and compact collagen fiber arrangement with least porosity (Fig. 4.9A). The loss of cellularity was higher as compared to 0.5% concentration at same time interval. At 48 hours no further loss of cellularity was observed; high cellular content, thick, and compact collagen fiber arrangement with least porosity was observed (Fig. 4.9B). At 72 hours time interval loss of cellularity increased up to 80% with moderate cellular debris and porosity (Fig. 4.9C). At 96 hours time interval, slight further loss in cellularity and increase in porosity were observed. Mild cellular debris and damage to collagen fiber arrangement was observed at this stage (Fig. 4.9D) [5,6].

4.4.2.3 Treatment with ionic detergent (sodium dodecyl sulfate)

The delaminated rumen matrix treated with 0.5% sodium dodecyl sulfate (SDS) for 24 hours under constant agitation showed about 75% loss of cellularity along with presence of cellular debris (Fig. 4.10A), which further increased up to 80% at 48 hours time interval (Fig. 4.10B). At 72 hours time interval submucosal layer was 90% acellular. The collagen fibers were compact having mild porosity than the native tissue (Fig. 4.10C). Slight cellular debris was observed between the spaces of thick collagen fibers. At 96 hours time interval, ECM showed thin, moderately loose arranged collagen fibers with moderate porosity. Mild cellular debris was evident at this stage (Fig. 4.10D) [5,6].

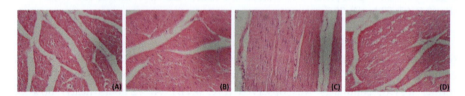

Figure 4.9 Histological picture of bovine rumen treated with 1% triton X-100: (A) 24 h, (B) 48 h, (C) 72 h, (D) 96 h posttreatment.

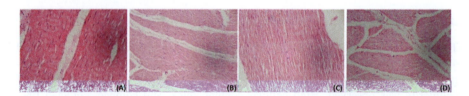

Figure 4.10 Histological picture of bovine rumen treated with 0.5%SDS: (A) 24 h, (B) 48 h, (C) 72 h, (D) 96 h posttreatment.

The treatment with 1% SDS for 24 hours under constant agitation showed 85%−90% loss of cellularity along with presence of cellular debris (Fig. 4.11A). Treatment for 48 hours showed more than 90% loss in cellularity along with slight cellular debris was noted (Fig. 4.11B). At 72 hours time interval, submucosal layer was completely acellular (Fig. 4.11C). The collagen fibers were compact and no cellular debris was observed between the spaces of thick collagen fibers. At 96 hours time interval, ECM showed thin, heavily loose arranged collagen fibers with high porosity. No cellular debris was evident at this stage (Fig. 4.11D) [5,6].

4.4.2.4 Treatment with nonionic detergent (tween-20)

Treatment with 0.5% tween-20 for 24 hours under constant agitation showed inefficient decellularization of tissue (Fig. 4.12A). At 48 hours time interval, the delaminated rumen matrix showed 40%−45% loss of cellularity with high cellular contents, thick, and compact collagen fiber arrangement with least porosity. At 48 hours no further loss of cellularity was observed (Fig. 4.12B). At 72 hours time, interval loss of cellularity was increased up to 75% with moderate cellular debris and porosity (Fig. 4.12C). At 96 hours time interval, no further increase in cellularity and porosity was observed. Mild cellular debris and damage to collagen fibers with mild compactness was observed (Fig. 4.12D) [5,6].

Treatment with 1% tween-20 for 24 hours under constant agitation showed inefficient decellularization of tissue (Fig. 4.13A). At 48 hours time interval, the delaminated rumen matrix showed 45%−50% loss of cellularity with thick and compact collagen fiber arrangement. At 48 hours no further loss of cellularity with thick and compact collagen fiber arrangement and least porosity were observed (Fig. 4.13B). At 72 hours time interval, loss of cellularity increased up to 80% with

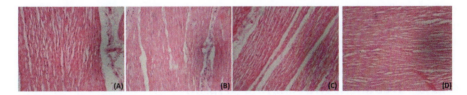

Figure 4.11 Histological picture of bovine rumen treated with 1% SDS: (A) 24 h, (B) 48 h, (C) 72 h, (D) 96 h posttreatment.

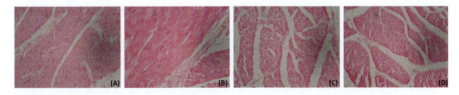

Figure 4.12 Histological picture of bovine rumen treated with 0.5% tween-20: (A) 24 h, (B) 48 h, (C) 72 h, (D) 96 h posttreatment.

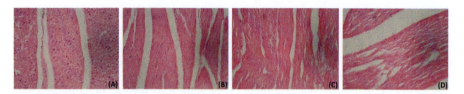

Figure 4.13 Histological picture of bovine rumen treated with 1% tween-20: (A) 24 h, (B) 48 h, (C) 72 h, (D) 96 h posttreatment.

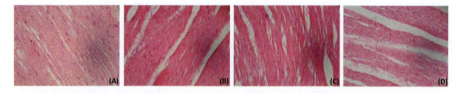

Figure 4.14 Histological picture of bovine rumen treated with 0.5% trypsin: (A) 24 h, (B) 48 h, (C) 72 h, (D) 96 h posttreatment.

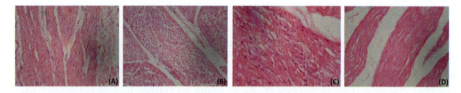

Figure 4.15 Histological picture of bovine rumen treated with 1% trypsin: (A) 24 h, (B) 48 h, (C) 72 h, (D) 96 h posttreatment.

moderate cellular debris and porosity (Fig. 4.13C). At 96 hours time interval, no further increase in cellularity and porosity was observed. Mild cellular debris and damage in collagen fiber arrangement was observed at this stage (Fig. 4.13D) [5,6].

4.4.2.5 Treatment with enzyme (trypsin)

Treatment with 0.5% trypsin was inefficient in decellularization at different time intervals. No loss in cellular contents and collagen fiber arrangement was observed at 24, 48, and 72 hours time intervals (Fig. 4.14A−C). At 96 hours slight loss of cellularity, high cellular contents, thick, and compactly arranged collagen fibers with least porosity were seen (Fig. 4.14d) [5,6].

Treatment with 1% trypsin was also found inefficient in decellularization during different time internals (Fig. 4.15A−C). However, some loss of cellularity was observed at 96 hours time interval (Fig. 4.15D) [5,6].

Optimization of protocols for preparation of acellular rumen ECM from bovine rumen showed that the best scaffold was prepared by treatment with 1% SDS for 96 hours.

The rumen matrices treated with TNBP and triton X-100 was found more or less similar. Tween-20-treated specimens were ranked third. The trypsin-treated matrix showed high cellular contents, thick, and compactly arranged collagen fibers, which appeared digested even at 48 hours of treatment and ranked fourth. Different biological detergents treatment for 96 hours causes more than 80% of decellularization [5,6].

4.4.3 DNA quantification

The native and decellularized tissues DNA was extracted using DNA isolation kit for mammalian tissues (Thermo Scientific, USA) and quantified as per method reported by Pellegata et al. [7]. In brief 40 mg wet weight samples of both native and decellularized tissues were subjected to digestion via cell lysis buffer and proteinase K. The protein fraction was removed by a precipitating the protein, followed by centrifugation. To the supernatant, isopropanol and ethanol were added and centrifuged. The pellet of isolated DNA was rehydrated and the amount of total DNA was detected using Nanodrop technique. The quantification was a direct measure to confirm the effectiveness of the processes, since the DNA is present in the active nucleus of cell. In the native rumen matrix, abundant cell components and nucleic acids were present. However, after the decellularization, cells and nucleic acids were hardly observed in ECM. The DNA concentration in native bovine rumen matrix was 48.09 ng/μL. After decellularization by different protocols, the DNA concentration decreased and ranged from 12.27 ± 0.07 ng/μL to 38.17 ± 0.05 ng/μL in different detergents treatment groups. In enzyme-treated group, the decellularization was inefficient and the DNA concentration ranged from 44.08 ± 0.05 to 46.07 ± 0.04 at different time intervals. There was significant difference ($P < .01$) between all treatment groups at different time intervals. Treatment with 1% SDS showed lowest values (12.27 ± 0.07 ng/μL) of DNA contents [5,6].

4.4.4 Cytocompatibility analysis

ECM scaffolds were tested for its cytocompatibility with xenogenic cellular environment by cytotoxicity analysis through MTT colorimetric assay as per technique described by Xu et al. [8]. Toxicity to 5% SDS was set as reference standard. ECM extracts of all the samples were prepared by incubating decellularized samples with standard culture media (DMEM supplied with 10% FBS) at 37°C for 3 days. Rat bone marrow-derived mesenchymal stem cells (rMSCs) were seeded in 96 well plates at a cell density of 1×10^3 /well in 200 μL of standard culture media and kept for incubation at 37°C and 5% CO_2 level. After 24 hours of incubation the existing media was replaced with 200 μL of ECM extract and incubated for another 48 hours. After this incubation, the ECM extracts were replaced with DPBS containing 5% MTT and incubated for 4 more hours. Formazan product was quantified using a microplate reader at 570 nm after dissolving the formazan in DMSO. The quantity of formazan product formed should be proportional to viable cells. The rMSCs were cultured in standard DMEM and exposed to SDS (cytotoxic) were used as positive and negative control, respectively. All the cellular work was done under biosafety cabinet type II [5,6].

The goal of decellularization was to efficiently remove all cellular and nuclear materials while minimizing any adverse effect on the composition, biological activity, and mechanical integrity of the native ECM [9]. Decellularization can be brought by physical, chemical, and enzymatic methods, which leaves a material composed of ECM components. In the present study, physical and chemical methods were used to prepare acellular matrices. These acellular matrices retained their natural mechanical properties and promote remodeling by neovascularization and recellularization by the host [10]. The cellular antigens are predominantly responsible for the immunological reaction associated with allograft [11]. The remnants of cell components in xenograft may contribute to calcification and immunogenic reaction in particular; lipids and DNA fragments play a role in calcification. The acellular tissue matrices possess the appropriate mechanical properties [12,13] and induced appropriate interaction with the host cells that resulted in the regeneration of functional tissues [14]. The acellular tissue matrices are reported to be biocompatible, slowly degraded upon implantation, and are replaced and remodeled by the ECM proteins synthesized and secreted by in growing host cells, which reduces the inflammatory response. The acellular matrices support the regeneration of the tissues with no evidence of immunogenic reaction [15]. Five different protocols were used to obtain the acellular bubaline rumen scaffold from the delaminated bubaline rumen. The delaminated bubaline rumen was subjected to ionic (sodium dodecyl sulfate, 0.5% and 1%), nonionic (triton X-100 and tween-20, 0.5%, and 1%), and zwitterionic biological detergents (tri-N-butyl phosphate, 5% and 1%) and enzyme (trypsin, 0.5% and 1%). The time of reaction (24, 48, 72, and 96 hours) was optimized to obtain the acellular bubaline rumen matrix. The delaminated bubaline rumen subjected to the ionic detergent (1% SDS). Treatment for 48 hours showed more than 90% cellularity. At 72 hours time interval, submucosal layer was completely acellular. At 96 hours time interval, ECM showed thin, loose arranged collagen fibers with very high porosity. No cellular debris was evident. The submucosal layer was completely acellular. The collagen fibers were loosely arranged as compared to the native tissue. Compared to other detergents, SDS yields complete removal of nuclear remnants and cytoplasmic proteins, such as vimentin [16]. Ionic detergents are effective for solubilizing both cytoplasmic and nuclear cellular membranes, but tend to denature proteins by disrupting protein-protein interactions [17]. In general, ionic detergents are used extensively in decellularization protocols due to their mild effects on tissue structure. These surfactants disrupt lipid-protein and lipid-lipid interactions but generally leave protein-protein reactions intact with the result of maintaining their functional conformations [9]. The cell extraction was effectively achieved without significant disturbances in ECM morphology and strength. Ionic detergents usually separate subunits of complex proteins.

4.5 Development of 3-D bioengineered scaffolds from bubaline rumen

The concept of tissue engineering allows in vitro expansion of isolated cells using culture techniques and their transplantations. The cells grown on the scaffold mechanically support the cells and regulate the functions of the cells in a manner analogous to ECM of mammalian tissue. The scaffolds allow cell invasion and proliferation and secrete their own ECM, thereby leading to complete and natural tissue replacement.

For the development of 3-D bioengineered scaffolds, rat bone marrow-derived mesenchymal stem cells (rMSCs) were seeded on the acellular bubaline rumen matrix to get 3-D bioengineered scaffolds. Rat bone marrow mesenchymal stem cells were isolated as per standard protocols. After euthanizing the animal, femur and tibia bones were collected under strict asepsis and brought to laminar hood. The metaphyseal regions of these long bones were cut and flushed out bone marrow was mixed with complete DMEM. The nucleated cell fraction of the bone marrow was enriched for stem cells. Cells were separated by density gradient centrifugation and then cultured in complete medium. The cells were incubated at 37°C in a humidified 5% CO_2 environment. The adherent cells were further isolated and recultured in laminin coated plates with complete DMEM.

After initial seeding, the flasks were observed at every 3 days. At first, the cells were round in shape. After 72 hours, cells got attached and nonadherent cells were removed by decanting the media. Supplementation of media was done at every 2 days interval. Gradual change in the morphology of cells could be observed (Fig. 4.16) [5].

By 18 to 21 days, P_0 flask became confluent and was passaged with 0.25% trypsin EDTA solution. Addition of trypsin detached the monolayer and cell became rounded again. Centrifuged cell pellet was seeded into T-75 flask, mixed with complete media, and maintained in a mummified 5% CO_2 incubator. The third passed bone marrow-derived mesenchymal stem cells (rMSCs) were used for seeding on decellularized rumen matrix. After 80%−90% confluency, they were passaged into new culture flasks. The decellularized matrices were washed 4−5 times with antibiotic containing RPMI and then placed in wells of the culture plate. The cells were trypsinized to detach the monolayer of cells from the flask. The growth medium, that is, RPMI containing 10% fetal bovine serum, was added to stop the activity of trypsin and mixed properly to get single cell suspension. The cells were statically seeded on the matrices at the rate of 10^6 cells/cm^2 to prepare 3-D bioengineered scaffolds (Fig. 4.17). It was maintained at 37°C in a humidified atmosphere of 5% CO_2 and 95% air in a CO_2 incubator [5].

4.6 Testing the efficacy of 3-D bioengineered scaffolds in a diabetic rat model

Diabetes mellitus is a major metabolic disorder, characterized by hyperglycemia, free radical production, abnormal insulin secretion or insulin receptor or postreceptor

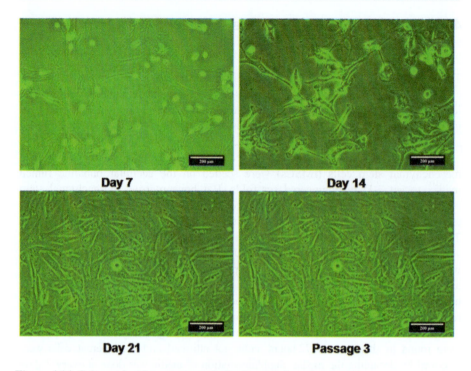

Figure 4.16 Culture of rat bone marrow-derived mesenchymal stem cells.

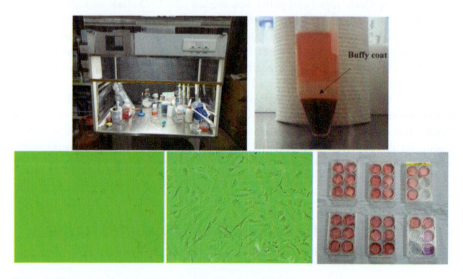

Figure 4.17 Development of 3-D bioengineered scaffolds from bubaline rumen using rat bone marrow-derived mesenchymal stem cells (rMSCs).

events affecting metabolism involving carbohydrates, proteins, and fats causing serious global health problems. Wound healing is impaired in diabetic patients with infection or hyperglycemia. There is impairment in host immunity, angiogenesis, neovascularization, failure in matrix metalloproteineases, keratinocyte, and fibroblast functions, which affects wound healing mechanism. In such cases, alternative lifesaving approaches are required. This may include the use of bioengineered skin substitutes. ECM provides a three-dimensional scaffold into which seeded stem cells could incorporate and help to build the foundation for the integration of local tissue. The scaffold allows cell invasion and proliferation and secrete their own ECM, thereby leading to complete and natural tissue replacement. The ECM is a complex mixture of structural and functional proteins, glycoproteins, and proteoglycans arranged in a unique tissue-specific 3-D ultrastructure. ECM helps in angiogenesis and vasculogenesis, cell migration, cell proliferation, orientation, and immune responsiveness in wound healing. ECM induces a host cellular response that supports constructive remodeling rather than default scar tissue formation. So the efficacy of developed 3-D bioengineered scaffold for the repair of full thickness skin wounds in diabetic rats was tested [5].

Diabetes in rats was induced by intraperitonial (i/p) administration of streptozotocin (60 mg/kg body wt.) in freshly prepared citrate buffer with pH 4.5. Blood sugar level was recorded using Acu Check glucometer before and after administration of streptozotocin (normal blood sugar level recorded 80–100 mg/dL). Rats with blood sugar level more than 280 were considered as diabetic.

Total 48 Wistar rats were equally divided into four groups of 12 animals in each group. After surgical anesthesia, a full thickness 20×20 mm^2 skin defect was created on dorsal thoracic area in all the rats. The animals of group I were treated as control where wound was left open. The wounds of group II animals were dressed with standard dressing material. The defects of group III was repaired with acellular bubaline rumen matrix. The animals of group IV were treated with 3-D bioengineered scaffolds. All the wounds were bandaged with paraffin gauze and studied for 42 days or until the completion of healing (Fig. 4.18). The efficacy of 3-D bioengineered scaffold was evaluated on the basis of clinical, gross, hematological microscopic, hematological observations [5].

The animals of all the groups remained dull on first day and assumed a hunched back posture while resting in their cages. Surgical wounds appeared apparently healthy in all the animals of different groups throughout the observation period. A significant decrease ($P < .05$) in wound area was recorded in all the groups up to day 28, except in bioengineered group where complete healing was observed on day 28. Complete healing was recorded on day 42 in other groups. A significant increase ($P < .05$) in percent contraction was observed in all the groups during the observation period. A significant ($P < .05$) degree of exudation was observed in animals of groups I (control) up to day 6 postoperatively. No exudation was observed in remaining groups after day 5 onward. Postoperative pain persisted up to day 3 in all the groups. However, significantly higher ($P < .05$) postoperative pain scores persisted up to day 6 in control group. No pain was observed in all other groups on day 5 and onward [5].

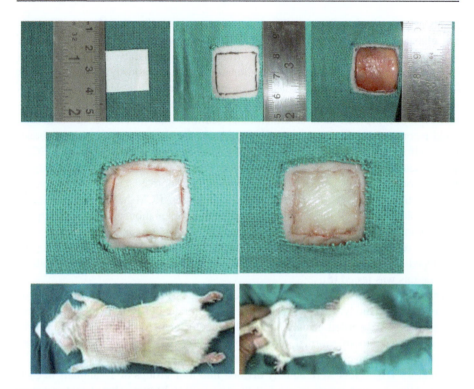

Figure 4.18 Evaluation of 3-D bioengineered rumen matrix in rat model.

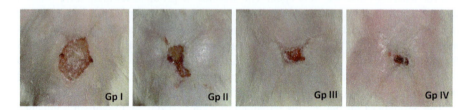

Figure 4.19 Computerized planimetry of wounds at day 28 in different groups.

Computerized planimetry showed that on day 14, the acellular rumen matrix was firmly adhered with the underlying pinkish granulation tissue and on day 42, the wound healed up completely with no scar. However, in 3-D bioengineered rumen matrix group, the wound healed up completely on day 28. In groups I and II, on day 42 the wound healed completely by severe contraction leaving a large scar (Fig. 4.19).

Gross observations on day 14 postoperatively showed better wound healing and incorporation blood vessels into graft in groups III and IV. Whereas, groups I and II showed less healing and irregular blood supply (Fig. 4.20).

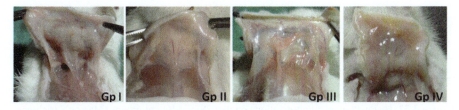

Figure 4.20 Growth of blood vessels on day 14 into the graft in different groups.

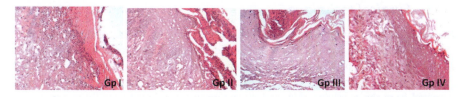

Figure 4.21 Histopathological observations on day 28 in different groups (H&E stain, × 150).

A significant increase ($P < .05$) in neutrophils was observed up to day 14 in all the groups. At day 21, the neutrophil count decreased in all groups; however, on day 28, slightly elevated values ($P > .05$) were recorded in groups I and II. On day 42 neutrophil counts were close to base line values in all the groups. Contrary to increased neutrophil count, a significant ($P < .05$) decrease in lymphocyte count was observed up to 14 days whereas in groups I and II, significant ($P < .05$) lymphocytopaenia was observed up to 21 days. Nonsignificant ($P > .05$) change was observed in eosinophil and monocyte count. The values of DLC were found within normal limits on day 42 postimplantation in all the groups of animals [5].

The biopsy specimens from the implantation site were collected on days 3, 7, 14, 21, 28, and 42, postoperatively. The histopathological scores were minimum in group IV followed by group III on days 7, 14, 21, and 42. In bioengineered rumen matrix group (group IV), on day 28, collagen fibers were denser, thicker, and better arranged. The collagen of the graft was under the process of resorption with organization of granulation tissue. On day 28, epithelization was complete and fibroplasia resembled normal skin with mild neovascularization (Figs. 4.21 and 4.22) [5].

4.7 Evaluation of bubaline rumen matrix in clinical cases

The prepared acellular rumen matrices were successfully tested in 26 buffalo calves, aged between 3 and 27 months and weighing 50–115 kg, suffering from umbilical hernia with an average diameter (6–12.5 cm) of hernial ring. The calves were sedated using xylazine HCl @ 0.1 mg/kg body wt. and circular infiltration at the site using 2% lidocaine HCl. The animals were controlled in lateral or dorsal recumbency as per the convenience [18].

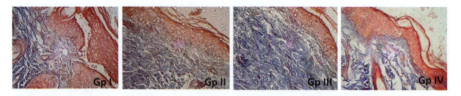

Figure 4.22 Histopathological observations on day 28 in different groups (Masson's trichome stain, × 150).

4.7.1 Surgical technique

The acellular matrix was thawed and prepared according to the size of the hernial ring. The sites of operation were prepared for aseptic surgery. An elliptical skin incision was given to expose the hernial sac, which spanned the length of the hernia and extended 2 cm beyond the cranial and caudal margins of the hernial ring. The hernial sac was dissected from the overlying skin, and dissection was continued laterally to expose the hernial ring and the external sheath of the rectus abdominis muscle. Underlay technique was used for the implantation of buffalo rumen acellular matrix. An appropriately sized acellular matrix with a preplaced horizontal mattress suture of braided silk with long ends, attached to its cranial, caudal, and mid-lateral edges, was introduced into the abdomen through the hernial ring. Each of the sutures was tied, with the knots resting on the external sheath of the rectus abdominis muscle, thus securing the acellular rumen matrix to the internal sheath of the rectus abdominis muscle. The subcutaneous tissues was closed using number 2–0 polyglactin 910 placed in a simple continuous suture pattern. The skin incision was then closed with silk sutures in a horizontal mattress suture pattern (Fig. 4.23A–F). Postoperatively, the animals were administered with antibiotic and analgesics. The skin wounds were dressed routinely and skin sutures were removed on postoperative day 12. The operated animals were followed either by regular visits or calling the owner during 1 year postsurgery [18].

4.7.2 Gross and clinical observations

Umbilical hernia was diagnosed in 26 buffalo calves. The hernial swelling was reducible in congenital cases, while in acquired ones there were varying degrees of irreducibility and adhesions. Wound dehiscence and hernia recurrence were not recorded in any of the operated cases. The fibrosed edges of the hernial rings were refreshed before implantation of the prosthetic materials in order to facilitate a better mesh incorporation with the host tissue [19]. At the end of the first postimplantation week, a thick and slightly hard mass resembling the size of implanted matrix was palpated at the site of the hernioplasty. The thick and hard mass of tissue at the site of implantation was palpable until the end of the second postimplantation week. These masses subsided by the end of the third postimplantation week. After the fifth postimplantation week, the masses become thinner and fibrous and were difficult to feel on palpation. The umbilical region looked normal on inspection [18]. All the animals recovered uneventfully. Clinical signs like wound dehiscence,

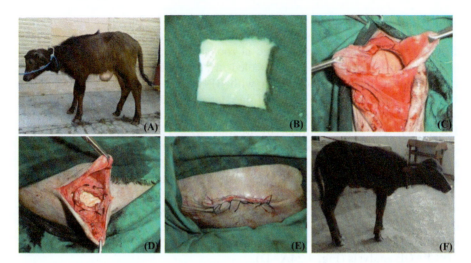

Figure 4.23 (A) Buffalo calf having umbilical hernia, (B) decellurized buffalo rumen matrix, (C) hernial ring observed intraoperatively, (D) repair of umbilical hernia with buffalo acellular rumen matrix by placing it as inlay graft, (E) repaired hernial defect and skin sutures, (F) animal after correction of hernia on the day of suture removal.

infection, or hernial recurrence were not recorded in any of the cases. All the animals were observed up to 12 months after hernioplasty. After decellularization, rumen matrices were used for umbilical hernioplasty in buffalo calves and the results demonstrated the uncomplicated healing of the repaired area, without hernial recurrence or rejection. Similar findings were reported in other studies in which acellular dermal matrix had been successfully used for the reconstruction of hernias in goats [20] and cellular aortic matrix was used for the reconstruction of abdominal hernias in calves [21]. Decellularized collagenous materials of animal origin (extracellular matrices) are preferred in hernial repair because of their inherent low antigenicity and ability to integrate with surrounding tissue. Furthermore, these matrices are biocompatible, slowly degraded upon implantation and are replaced and remodeled by the ECM proteins, synthesized and secreted by in growing host cells [22]. In this study, no postoperative complications were observed during the observation period after the hernioplasty. Similar results were reported in bubaline studies [23]. Mechanical properties that support cell in growth and induce interaction with the host cells that results in functional tissues tissue regeneration [24].

4.8 Preparation of acellular matrices from caprine rumen

Rumen of caprine (goat) was collected from the institutional abattoir and was cleaned thoroughly by washing in sterile physiological normal saline. Then rumen was kept in $1 \times$ phosphate buffered saline containing antibiotic (amikacin 1 mg/mL and 0.02% EDTA) at $-20°C$ for 1—2 hours. Preservation at $-20°C$ formed ice crystals between rumen layers facilitating in separation of different layers. Rumen was thawed at room temperature for 3—4 hours under running water. Serosa was easily peeled off, and mucosa and muscular layers were removed by gentle scrapping with BP handle. The rumen was cut into 4×4 cm^2 size pieces and placed in sterile $1 \times$ PBS containing antibiotic (penicillin, streptomycin and amphotericin) for 2 hours at room temperature. The rumen was made acellular by treating with 1% sodium dodecyl sulfate (SDS) and 1% sodium deoxylate (SDC) in wrist action shaker for 96 hours. To assess the decellularization, samples were collected at 0, 24, 48, 72, and 96 hours time intervals. After completion of procedure, the tissues were thoroughly washed with $1 \times$ PBS and then placed in 70% ethanol for 12 hours. It was again washed in $1 \times$ PBS until the froth disappeared and stored at $-20°C$ in $1 \times$ PBS containing antibiotics. The tissue samples were subjected to histological examination, DNA quantification, calorimetric protein estimation, and SDS-PAGE analysis (Fig. 4.24A—D).

4.8.1 Microscopic observations

The samples were cut into small pieces and fixed in 10% (v/v) neutral buffered formalin (NBF) for 48—72 hours. The tissues were processed in routine manner and 4 μm thick sections were cut and stained with H&E as per the standard protocol. The stained sections of were examined for degree of decellularization and arrangement of collagen fibers. Decellularized samples showed absence of lamina propria, tunica muscularis and decrease in cellular components. H&E-stained slides reflected continuous and faster decellularization in 1% SDS and this rate of decellularization was slow in case of 1% SDC. Samples at 48 hours of different tissues were satisfactory with 90% decrease in cellularity (Figs. 4.25A—E and 4.26A—E). Masson's trichome staining of different samples selected on the basis of H&E staining revealed quite decrease in collagen density at 48 hours when compared with collagen of native tissue (Fig. 4.27A and B). The collagen fibers of the decellularized samples showed decrease in the density. So, on comparison of different time interval 48 hours samples of the rumen tissues were selected and were further evaluated as wound healing construct.

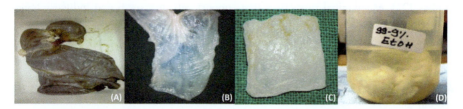

Figure 4.24 (A) Raw caprine forestomach, (B) caprine ruminal submucosa, (C) decellularized caprine rumen matrix, (D) ethanol preserved decellularized caprine rumen matrix.

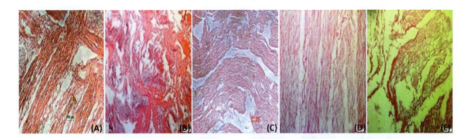

Figure 4.25 Histological picture of caprine rumen treated with 1% SDS at different time intervals: (A) native rumen, (B) 24 h, (C) 48 h, (D) 72 h, (E) 96 h posttreatment.

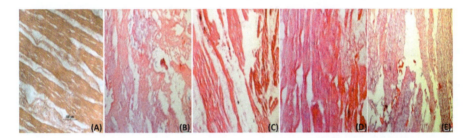

Figure 4.26 Histological picture of caprine rumen treated with 1% SDC at different time intervals: (A) native rumen, (B) 24 h, (C) 48 h, (D) 72 h, (E) 96h posttreatment.

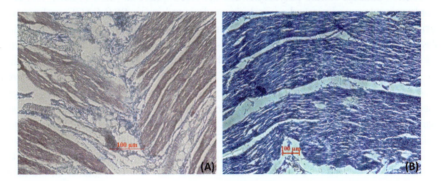

Figure 4.27 (A) Native caprine rumen and (B) decellularized caprine rumen (Massion Trichome stain, ×100).

4.8.2 DNA quantification

DNA was extracted from both from native and decellularized tissue using DNeasy Blood and Tissue Kit catalog number 69504 (Qiagen, Germany) and quantified as per the method reported by Pellegata et al. [7]. The results of DNA quantification are represented in Fig. 4.28. The values of DNA quantification showed significant difference when compared with DNA content of native ruminal tissues at every

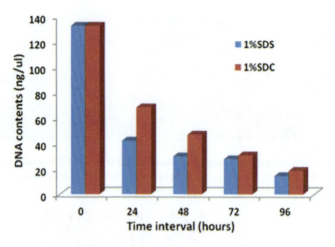

Figure 4.28 Showing DNA quantification (ng/μL).

interval during the process of decellularization. When compared biological detergents, the SDS showed significant decrease in DNA contents at each time intervals as compared to SDC.

4.8.3 SDS-PAGE analysis

For sodium dodecyl sulfate polyacrylamide gel electrophoresis (SDS-PAGE), 25 μL of sample was taken and marked properly in predesigned order. SDS-PAGE analysis was carried out in a vertical mini gel electrophoresis apparatus (Bio-Rad Mini Protean). Glass plates were cleaned and set in a gel-molding tray of electrophoresis apparatus and 5 mL of distilled water was poured in between plates to check the leakage. After few minutes, water was poured off and 4.5 mL of separating gel solution was poured between the two plates. Saturated butanol or distilled water was poured over the gel. After polymerization, supernatant was removed by tilting the plates. Stacking gel was poured over the separating gel and later on suitable comb was inserted. The polymerized gel was mounted into the electrophoresis chamber and buffer reservoir filled with 1 × Tris glycine buffer. Before loading, 7 μL loading dye was added to each of the samples. They were then heated in boiling water bath for 10 minutes. Broad-range SDS-PAGE prestain marker ranging from 6.5 to 200 KDa was loaded in the first well followed by the prepared samples of CFMs. Electrophoresis was carried out with an initial voltage of 60 V and current of 300 mA for about 30−45 minutes, till the samples crossed stacking gel, followed by 100 V and 300 mA for the rest of the gel run. After completion, gel was removed from the plates and stained with Coomassie brilliant blue for 40 minutes and then destained with destaining solution. Prepared gel was picturized by Bio-Rad Gel Documentation System.

The SDS-PAGE analysis in 10% polyacrylamide gel showed the expression of the proteins of native and decellularized rumen with different detergents at 24, 48, 72, and 96 hours time intervals (Fig. 4.29). When treated with 1% sodium dodecyl sulfate, the

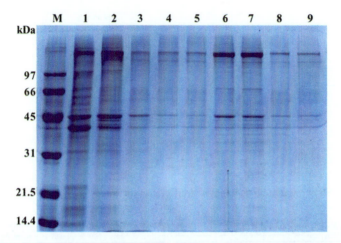

Figure 4.29 SDS-PAGE of caprine rumen: Lane M—broad range SDS-PAGE prestain marker (BioRad), Lane 1—native caprine rumen, Lane 2—1% SDS treatment for 24 h, Lane 3—1% SDS treatment for 48 h, Lane 4—SDS treatment for 72 h, Lane 5—1% SDS treatment for 96 h, Lane 6—1% SDC treatment for 24 h, Lane 7—1% SDC treatment for 48 h, Lane 8—1% SDC treatment for 72 h, Lane 9—1% SDC treatment for 96 h.

decellularized rumen showed three protein bands (thick), one of 116 KDa, another of 45 KDa and 40 KDa at 24 hours and one protein band (thin) of 45 KDa at 48 hours. At 48 hours, majority of the proteins removed. Rumen with 1% sodium deoxylate showed only two (thick) bands at 116 and 45 KDa at 24 and 48 hours.

4.8.4 Calorimetric protein estimation

Tissue homogenization buffered was prepared and 5 mg of native and decellularized rumen samples were homogenized in this buffer by electric homogenizer. All samples were vortexed and centrifuged at 10,000 rpm for 5 minutes and supernatant was collected in separate microcentrifuge tubes. For direct calorimetric protein estimation, protein for each sample was estimated with Eppendorf Bio Spectrometer at 280 nm.

Protein contents of native versus decellularized samples were not differed significantly ($P > .05$) for native rumen and acellular rumen matrix at 24 hours. For time intervals from 48 to 96 hours, samples treated with two different detergents differs significantly ($P < .05$). Treatment with 1%SDS showed significantly lower values as compared to 1% SDC at respective time intervals (Fig. 4.30).

4.8.5 Scanning electron microscopic observations

Scanning electron microscopic examination of caprine rumen revealed fine distribution of collagen matrix at lower magnification (Fig. 4.31A) and when examined in the higher magnification, the collagen matrix fibers showed even distribution and intrafibrilar spaces (Fig. 4.31B) [25].

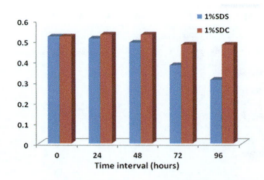

Figure 4.30 Showing protein estimation at different time intervals.

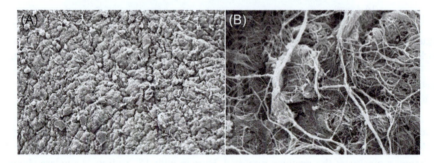

Figure 4.31 Surface characterization of goat rumen: (A) low magnification and (B) higher magnification.

4.9 Evaluation of caprine rumen matrix in clinical cases

The decellularized goat rumen matrices were applied nine clinical cases of abdominal wall defects in different species of animals. The cases having a hernial ring of more than 4 cm in diameter were included in the study. A thorough clinical examination was performed to determine the type of a hernia, ring size, adhesions, etc. Standard anesthetic protocols were used for anesthetizing the animals. Preoperative observations included signalment, etiology, and prior treatment given (if any). The animals were positioned in dorsal posture or as per the location of abdominal wall defects. The hernia sac was exposed and the contents were reduced and the prepared acellular matrix from goat rumen (Fig. 4.32) was used for closure of the hernial ring [26].

An appropriately sized acellular matrix was anchored with abdominal muscles with the underlay technique using the appropriate number of polyamide suture. The subcutaneous tissue and skin was closed using the appropriate number of Polyglactin 910. Postoperative analgesia and course of antibiotic was provided for 5 days. After removing skin sutures on completion of healing, all the animals were observed for 3 months for long-term evaluation of results (Fig. 4.33) [26].

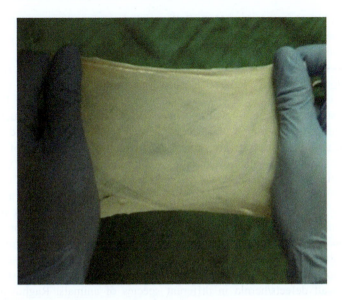

Figure 4.32 Prepared acellular goat rumen matrix.

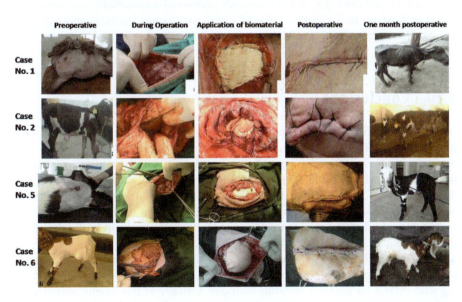

Figure 4.33 Showing animals at different time intervals repaired with acellular goat rumen matrices.

Reconstructive surgery is the only method to restore the integrity of the abdominal wall and prevent incarceration and strangulation of herniated contents [27]. Hernial ring size exceeding 3 cm in diameter requires use of prosthetic material for hernioplasty [28]. Decellularized tissues have been used in preclinical animal trials and human clinical applications [10]. Collagen derived from bovine or porcine was used in tissue engineering products. However, we used caprine collagen as domestic goat available in the subcontinent as a potential source of collagen.

In the present study, goat rumen was used as a biological material for reconstruction of abdominal wall defects because of ease of availability, ease of processing, and having sufficient tensile strength. Goat rumen decellularized using 1% SD for 48 hours resulted into retention of the three-dimensional collagen structure [26]. Acellular buffalo rumen matrix was decellularized using 1% SD and used for reconstruction of umbilical hernioplasty in buffalo calves [18]. The results of this study also support the notion that SD is a viable option for decellularization of the goat rumen. Umbilical hernia in cattle and buffalo constitutes more than half of all cases followed by ventral and lateral abdominal hernias. Normal abdominal wall integrity without any postoperative complications was observed for 3 months postoperatively. Normal healing without hernia recurrence and postoperative complications or rejection of acellular matrix was also reported using buffalo rumen [18]. Animals showed normal behavior and growth rate during follow-up period. Different biological-based biomaterials have been used for the reconstruction of abdominal wall defects/hernia in different species of animals. Raghuvansi et al. [29] reported the clinical application of goat rumen matrix in hernioplasty of cow calf, buffalo calf, and pig. All the animals recovered uneventfully and remained sound at least up to 3 months. Decellularized goat rumen matrices showed good results without complications [26].

4.10 Conclusions

In this chapter, five different protocols were used to obtain the acellular bubaline rumen matrix. The bubaline rumen was subjected to ionic (SDS, 0.5% and 1%), nonionic (triton X-100 and tween-20, 0.5% and 1%), and zwitterionic biological detergents (Trin-BP 5% and 1%) and enzyme (trypsin, 0.5% and 1%) as biological detergents for decellularization. The time of reaction (12, 24, 48, 72, and 96 hours) was adjusted to obtain optimized acellular bubaline rumen matrix. Desired results were achieved after 96 hours of treatment with 1% SDS detergent in decellularization of bubaline rumen. The concentration of DNA in native rumen matrix was 48.09 ± 0.05 ng/µL. SDS (1%)-treated matrices showed reduction in DNA contents to 12.27 ± 0.07 ng/µL. The 3-D bioengineered scaffolds were developed after seeding with rMSCs on acellular bubaline matrices. Efficacy of this bioengineered scaffold was evaluated in a diabetic rat model. Bioengineered scaffold-treated rats showed early wound healing, mimicked the properties of skin, and considered as novel biomaterial for the repair of full thickness diabetic dermal wounds. Protocols were also optimized for preparation of acellular caprine rumen matrix. Prepared acellular bubaline and caprine rumen matrices were tested in clinical cases of abdominal wall defects in different species of animals and found to be a good biocompatible biomaterial for hernia repair.

References

[1] Lun S, Irvine SM, Johnson KD, Fisher NJ, Floden EW, Negron L, et al. A functional extracellular matrix biomaterial derived from ovine forestomach. Biomaterials 2010;31:4517−29.
[2] Brown B, Lindberg K, Reing J, Stolz DB, Badylak SF. The basement membrane component of biologic scaffolds derived from extracellular matrix. Tissue Eng 2006;12:519−26.
[3] McPherson TB, Badylak SF. Characterization of fibronectin derived from porcine small intestinal submucosa. Tissue Eng 1998;4:75−83.
[4] Banks WJ. Applied Veterinary Histology.. 3rd ed St. Louis: Mosby-Year Book; 1993.
[5] Singh A.K. Development of 3-D bioengineered scaffolds from bubaline rumen and gall bladder for the regeneration of full thickness skin wounds in diabetic rats [PhD thesis]. Submitted to Deemed University Indian Veterinary Research Institute, Izatnagar, Uttar Pradesh India; 2017.
[6] Singh AK, Gangwar NK, Shrivastava AK, Saxena S, Maiti S, Raghuvansi SK, et al. Optimization of protocols for decellularization of buffalo rumen matrix. Trends Biomater Artific Organs 2021;35(1):57−64.
[7] Pellegata AF, Asnaghi M, Stefani I, Maestroni A, Maestroni S, Dominioni T, et al. Detergent-enzymatic decellularization of swine blood vessels: insight on mechanical properties for vascular tissue engineering. BioMed Res Int 2013;3:1−8.
[8] Xu L, Zhong NJ, Xie YY, Huang HL, Jiang GB, Liu YJ. Synthesis, characterization, in vitro cytotoxicity, and apoptosis-inducing properties of ruthenium (II) complexes. PLoS one 2014;9(5):e96082.
[9] Gilbert TW, Sellaroa TL, Badylak SF. Decellularization of tissues and organs. Biomaterials 2006;27:3675−83.
[10] Schmidt CE, Baier JM. Acellular vascular tissue: natural biomaterials for tissue repair and tissue engineering. Biomaterials 2000;21:2215−31.
[11] Weber M. The Sociology of Religion. Beacon Press; 1993.
[12] Courtman DW, Pereira CA, Kashef V, McComb D, Lee JM, Wilson GJ. Development of a pericardial acellular matrix biomaterial: Biochemical and mechanical effects of cell extraction. J Biomed Mater Res 1994;28:655−66.
[13] Sacks MS, Gloeckner DC. Quantification of the fiber architecture and biaxial mechanical behavior of porcine intestinal submucosa. J Biomed Mater Res 1999;46:1−10.
[14] Voytik-Harbin SL, Brightman AO, Kraine MR, Waisner B, Badylak SF. Identification of extractable growth factors from small intestinal submucosa. J Cell Biochem 1997;67:478−91.
[15] Yoo JU, Barthel TS, Nishimura K, Solchaga L, Caplan AI, Goldberg VM, et al. The chondrogenic potential of human bone-marrow-derived mesenchymal progenitor cells. J Bone Jt Surg Am 1998;80:1745−57.
[16] Woods T, Gratzer PF. Effectiveness of three extraction techniques in the development of a decellularized bone−anterior cruciate ligament−bone graft. Biomaterials 2005;26:7339−49.
[17] Seddon AM, Curnow P, Booth PJ. Membrane proteins, lipids and detergents: not just a soap opera. Biochem Biophy Acta 2004;1666:105−17.
[18] Singh AK, Kumar N, Shrivastava S, Singh KP, Dey S, Raghuvanshi PDS, et al. Evaluation of the efficacy of bubaline rumen derived extracellular matrix in umbilical hernioplasty in buffalo calves. Int J Fauna Biol. Studies 2018;5(1):29−33.
[19] Attinger CE, Bulan E, Blume PA. Surgical debridement. The key to successful wound healing and reconstruction. Clin Pediatric Med Surg 2000;17:599−630.

[20] Gangwar AK, Sharma AK, Maiti SK, Kumar N. Xenogenic acellular dermal graft for the repair of ventral hernia in a non-descript goat. Indian Vet Med J 2004;28:95−6.
[21] Kumar V, Kumar N, Gangwar AK, Saxena AC. Using acellular aortic matrix to repair umbilical hernias of calves. Australian Vet J 2013;91:251−3.
[22] Pariente JL, Kim BS, Atala A. In vitro biocompatibility assessment of naturally derived and synthetic biomaterials using normal human urothelial cells. J Biomed Mater Res 2001;55:33−9.
[23] Kumar N, Mathew DD, Gangwar AK, Remya V, Muthalavi MA, Maiti SK, et al. Reconstruction of large ventro-lateral hernia in a buffalo with acellular dermal matrix: a method for treating large hernias in animals. Vet Arh 2014;84:691−9.
[24] Badylak SF. Regenerative medicine and developmental biology: the role of the extracellular matrix. Anat Rec B N Anat 2005;287:36−41.
[25] Raghuvanshi PDS, Kumar N, Maiti SK, Mohan D, Gautam D, Shrivastava S, et al. Preparation and characterization of caprine forestomach matrix (CFM) for biomedical application. Trends Biomater Artific Organs 2019;33(1):9−14.
[26] Rathore HS, Kumar N, Singh K, Maiti SK, Shrivastava S, Shivaraju S, et al. Clinical application of acellular matrix derived from the bubaline diaphragm and caprine rumen for the repair of abdominal wall defects in animals. Aceh J Ani Sci 2019;4((2):50−60. Available from: https://doi.org/10.13170/ajas4.2.13071.
[27] Ober C, Muste A, Oana L, Mates N, Beteg F, Veres S. Using of prosthetic biomaterials in large animals: modern concepts about abdominal wall defects approach. J Cent Eur Agri 2008;9:575−80.
[28] Vilar JM, Doreste F, Spinella G, Valentin S. Double layer mesh hernioplasty for repair of incisional hernias in 15 horses. J Equine Vet Sci 2009;29:131−76.
[29] Raghuvanshi PDS, Mohan D, Gautam D, Shivaraju S, Maiti SK, Kumar N. Clinical application of animal based extracellular matrix in hernioplasty. MOJ Imunology 2018;6(4):115−18.

Reticulum-derived extracellular matrix scaffolds

Naveen Kumar[1,]*, Pawan Diwan Singh Raghuvanshi[2], Mohar Singh[1], Anwarul Hasan[1], Aswathy Gopinathan[1], Kiranjeet Singh[1], Ashok Kumar Sharma[1], Remya Vellachi[3], Sameer Shrivastava[4], Sonal Saxena[4], Swapan Kumar Maiti[1] and Karam Pal Singh[5]

[1]Division of Surgery, ICAR-Indian Veterinary Research Institute, Izatnagar, Uttar Pradesh, India, [2]Department of Veterinary Surgery and Radiology, College of Veterinary Science and Animal Husbandry, Nanaji Deshmukh Veterinary Science University, Mhow, Indore, Madya Pradesh, India, [3]Department of Veterinary Surgery and Radiology, College of Veterinary and Animal Sciences, Wayanad, Kerala, India, [4]Division of Veterinary Biotechnology, ICAR-Indian Veterinary Research Institute, Izatnagar, Uttar Pradesh, India, [5]CADRAD, ICAR-Indian Veterinary Research Institute, Izatnagar, Uttar Pradesh, India

5.1 Introduction

The term "Forestomach Matrix" (abbreviated FM) refers to an ECM scaffold containing the propria-submucosa of the forestomach of a ruminant. The term "propria-submucosa" refers to the tissue structure formed by the blending of the lamina propria and submucosa in the ruminant forestomach. Lamina propria is the luminal portion of the propria-submucosa, which includes a dense layer of ECM. ECM scaffolds can be derived from the rumen, the reticulum, or the omasum of the forestomach. Such ECM scaffolds contain the lamina propria and submucosa (propria-submucosa) layers of the forestomach wall. FM scaffolds are derived from the rumen or from individual laminae within the omasum, in addition to propria-submucosa; FM scaffolds may optionally include intact or partial layers of decellularized epithelium, basement membrane, or tunica muscularis. As a result of the unique structure and function of the forestomach, ECM tissue scaffolds derived from the forestomach have different biochemical, structural, and physical properties relative to previously described scaffolds isolated from glandular stomach, intestine, and bladder. In particular, FM includes a dense band of ECM within the lamina propria. In addition, FM optionally includes an intact or fractured basement membrane. In contrast, a scaffold derived from the glandular stomach submucosa or small intestinal submucosa will include little if any of the lamina propria, because the lamina propria is located mainly between the glands of the mucosa and is consequently removed as the mucosa is delaminated. Importantly, histology shows that the lamina propria is unusually dense, whereas the abluminal side of the FM scaffold is structured as an open reticular matrix. These differences serve an important role in epithelial regeneration, as the dense side acts as a barrier to cell migration, while the less dense side does not present a barrier and

---

* Present affiliation: Veterinary Clinical Complex, Apollo College of Veterinary Medicine, Jaipur, Rajasthan, India.

Natural Biomaterials for Tissue Engineering. DOI: https://doi.org/10.1016/B978-0-443-26470-2.00005-3
© 2025 Elsevier Inc. All rights are reserved, including those for text and data mining, AI training, and similar technologies.

therefore allows cell invasion. This structure makes the FM well suited for encouraging epithelial regeneration on the dense luminal side of the matrix, and fibroblast invasion on the less dense abluminal side of the matrix, when used as a medical device for tissue regeneration. In contrast, submucosal tissue grafts derived from the glandular stomach and the urinary bladder has a uniform density. The dense layer of ECM from the lamina propria contributes to the increased thickness and strength of FM scaffolds compared to those derived from other organs. The large surface area of the forestomach and the increased thickness and strength of scaffolds derived from the forestomach allows the isolation of larger ECM scaffolds from the forestomach than is possible from other organs. For example, ECM scaffolds of the forestomach can have a width as large as 10 cm or more. Unlike scaffolds obtained from the glandular stomach, FM scaffolds derived from the forestomach can include collagen IV and laminin from the basement membrane on the luminal surface. Surprisingly, these proteins are also present within the dense band of the lamina propria, providing important substrates for epithelial cell adhesion and growth. Glandular stomach scaffolds do not typically include the epithelium or basement membrane, or portions thereof, because these layers are fragile and do not withstand physical delamination. A glandular submucosal scaffold may include remnants of the lamina muscularis mucosa on the luminal side and tunica muscularis on the abluminal side. FM scaffolds have a contoured luminal surface, analogous to the rete ridges of the dermis. In contrast, scaffolds delaminated from small intestine, urinary bladder, and glandular stomach submucosa have a relatively smooth luminal surface. The contoured luminal surface of the FM provides a complex topology, which favors epithelial regeneration. This topology is not present in ECM scaffolds derived from small intestinal submucosa, glandular stomach submucosa, or urinary bladder submucosa.

FM scaffolds contain important regulators of wound repair, which includes growth factors FGF2, TGFbI, TGFb2, and VEGF, and the glycosaminoglycans hyaluronic acid and heparan sulfate. FGF2 plays an important role in wound healing by signaling cell migration and differentiation required for the formation of new tissue and vasculature. Heparan sulfate is an important cofactor that modulates bioactivity of FGF2 by acting on FGF2 receptors. Heparan sulfate is required for FGF2 activity and increases the stability of FGF2. Importantly, FGF2 and heparan sulfate are not present on stomach submucosa. FM additionally contains fibrillar proteins including collagen I, collagen III, and elastin, as well as adhesive proteins including fibronectin, collagen IV, and laminin. These proteins, in particular collagen and elastin, contribute to the high tensile strength and resilience of FM scaffolds.

Forestomach ECM was produced by using a process, termed STOF "sealed transmural osmotic flow," that employed an osmotic gradient and aqueous detergent solutions to facilitate removal of the epithelial and muscle layers and decellularization of forestomach tissue [1]. The resultant OFM was chemically disinfected according to established procedures [2], lyophilized and terminally sterilized using ethylene oxide. "Forestomach Matrix" (FM) scaffolds derived from a ruminant belonging to the genus Copra, Bos, Cervus, or Ovis tissue comprise a keratinized stratified squamous epithelium successfully decellularized by STOF technique [3].

5.2 Preparation of acellular bovine reticulum extracellular matrix

Reticulum of buffalo was procured from the local abattoir. Immediately, after collection, the reticulum was be kept in cold physiological saline solution containing 0.02% EDTA and antibiotic (amikacin @ 1 mg/mL). The tissue was thoroughly rinsed with normal saline before the start of protocol. The maximum time period between retrieval and the initiation of protocol was less than 4 hours. Tissue was cut from sides to obtain a flat sheet. The inner keratinized layer of reticulum was scrubbed off with a blunt edge and outer serosal layer was delaminated by mechanical assistance. The tissue was cut into 4 × 4 cm^2 size pieces and treated with 70% ethanol for 4 hours and later on with distilled water for 24 hours and was made acellular using following detergents and enzymes (Fig. 5.1A–I).

Sodium deoxycholate (ionic detergent) [4], Triton X-100 (nonionic detergents) [5], tri (n-butyl) phosphate (zwitterionic detergent) [6], and trypsin (enzyme) [7]

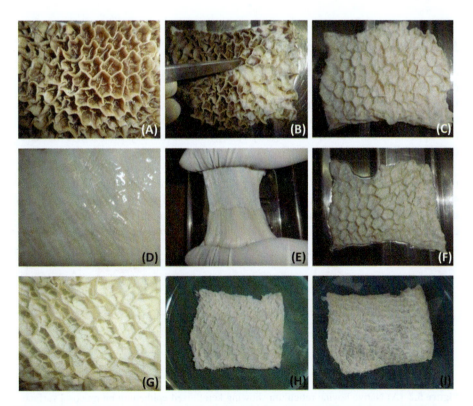

Figure 5.1 (A) Native reticulum (luminal surface), (B) removal of keratinized epithelium, (C) deepithelized reticulum, (D) serosal surface (abluminal), (E) mechanical removal of serosal layer, (F) deepithelized and delaminated reticulum, (G) decellularized reticulum, (H) ethanol-dehydrated reticulum, (I) vacuum-dried reticulum.

were used in 0.5% concentration for decellularization. Treatment with detergent was done up to 72 hours, and the solution was changed at every 12-hour time intervals and samples were collected at 12-, 24-, 48-, and 72-hour time intervals for histopathological, scanning electron microscopic evaluation and DNA quantification.

The tissue samples collected at different time intervals were fixed in 10% formal saline solution, dehydrated in ethanol, cleared in xylene, and embedded in paraffin to get 5 micron thin sections. The sections were stained with hematoxylin and eosin (H&E) staining. Masson's trichrome staining was done for assessing collagen fiber arrangement. On the basis of histological examination, the best protocol was optimized and further used for making acellular reticular-derived ECM. The samples for SEM examination were fixed in 2.5% glutaraldehyde. They were dehydrated using ascending series of ethanol starting from 35% for 30 minutes each and lastly chemical drying was done by hexamethyldisilazane. The tissues were kept for 24 hours for air-drying. The DNA contents analysis before the start of protocols and after completion of protocols was done as per method described by Gilbert et al. [8].

5.2.1 Gross and microscopic observations

After thorough cleaning of the bovine reticulum (Fig. 5.1B), keratinized epithelium of reticulum was easily scrubbed off by blunt surface of BP handle. The serosal layer was separated mechanically with forceps (Fig. 5.1E). The bovine reticulum tissue treated with different biological detergents and enzymes at 0.5% concentration appeared soft, whiter, and slightly spongy in consistency than native tissue at different time intervals (Fig. 5.1G).

Microscopically, native bovine reticulum showed keratinized epithelium on mucosal surface (Fig. 5.2A). Lamina propria is the luminal portion of the propria-submucosa, which includes a dense layer of ECM and serosal layer. The delaminated propria-submucosa, layer showed cellularity, tunica muscularis, and thick collagen fibers (Fig. 5.2B). Native bovine reticulum showed dense compact

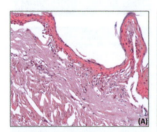
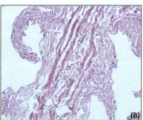
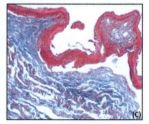

Figure 5.2 (A) Native bovine reticulum showing keratinized epithelium on mucosal surface (H&E stain, ×100); (B) delaminated propria-submucosa, showing cellularity, tunica muscularis and thick collagen fibers (H&E stain, ×100); (C) native bovine reticulum showing clearly dense compact arrangement of collagen fibers (Masson's trichrome stain, 100×).

arrangement of collagen fibers clearly visible in Masson's trichrome staining (Fig. 5.2C) [9].

The delaminated reticulum treated with 0.5% sodium deoxycholate for 12 hours under constant agitation showed 85%–90% loss of cellularity along with presence of cellular debris. Treatment for 24 hours showed slight increase in loss of cellularity, that is, more than 90%, and cellular debris was noted. At 48-hour time interval, submucosal layer was completely acellular. The collagen fibers were compact with moderate porosity than the native tissue. No cellular debris was observed between the spaces of thick collagen fibers. At 72-hour time interval, rECM showed thin, heavily loose arranged collagen fibers with very high porosity. No cellular debris was evident (Fig. 5.3A–D) [9].

Treatment with 0.5% Triton X-100 for 12 hours under constant agitation showed 50%–60% loss of cellularity with high cellular content, thick and compact collagen fiber arrangement with least porosity. At 24 hours, no further loss of cellularity was observed. High cellular contents, thick and compact collagen fiber arrangement with least porosity was observed. At 48-hour time interval, loss of cellularity increased up to 80% with moderate cellular debris and porosity. At 72-hour time interval, no further increase in cellularity and porosity was observed. Mild cellular debris and damage in collagen fiber arrangement was observed at this stage (Fig. 5.4A–D) [9].

Treatment with 0.5% tri (n-butyl) phosphate for 12 and 24 hours showed no effect on loss of cellularity. High cellular contents were noted at 12 hours as compared to 24 hours. At 48 hours, nearly 80% of decellularization was observed.

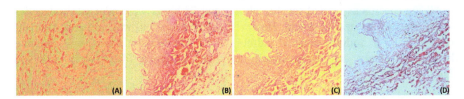

Figure 5.3 Delaminated reticulum treated with 0.5% sodium deoxycholate for different time intervals under constant agitation: (A) 12 h, (B) 24 h, (C) 48 h, (D) 72 h (H&E stain, 100×).

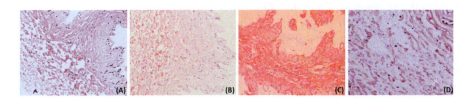

Figure 5.4 Delaminated reticulum treated with 0.5% Triton X-100 for different time intervals under constant agitation: (A) 12 h, (B) 24 h, (C) 48 h, and (D) 72 h (H&E stain, 100×).

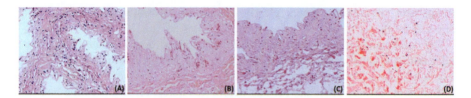

Figure 5.5 (A–D) Delaminated reticulum treated with 0.5% tri (n-butyl) phosphate for different time.

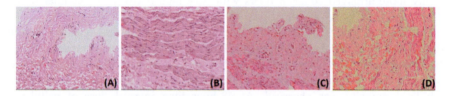

Figure 5.6 Delaminated reticulum treated with 0.5% trypsin enzyme for different time intervals under constant agitation: (A) 12 h, (B) 24 h, (C) 48 h, and (D) 72 h (H&E stain, 100 ×).

At 72 hours, more than 80% decellularization and loss of cellular debris was seen (Fig. 5.5A–D) [9].

Treatment with trypsin enzyme was inefficient in decellularization during various time intervals. No loss in cellular content with any effect on collagen fiber arrangement was observed at different time intervals (Fig. 5.6A–D) [9].

All the specimens treated for 12, 24, 48, and 72 hours were adequately decellularized, except enzyme treatment. However, the best scaffold was prepared by treatment with 0.5% sodium dodecyl sulfate (SDS). On Masson's trichrome staining thick, transversely and longitudinally arranged collagen fibers were observed (Fig. 5.7A–D) [9].

5.2.2 Scanning electron microscopic observations

Scanning electron microscopic observations revealed loss of cellularity. The native bovine reticulum matrix consisted of dense and closely packed collagen fibers (Fig. 5.8A–D). Delaminated reticulum showed uneven luminal surface and aggregated collagen bundles (Fig. 5.9A–D).

Collagen fibers were loosely arranged in decellularized bovine rECM as compared to native bovine reticulum tissue (Fig. 5.10A–D). The collagen bundles were formed of many thick collagen fibrils. The ethanol-dehydrated and vacuum-dried bovine reticulum showed denser and closely packed collagen fibers (Fig. 5.11A–D). The collagen bundles were seen to be formed of many fine threads like collagen fibrils. A meshwork of collagen bundles forms the bulk of the tissue, yet smaller fibers not associated with larger collagen bundles form a fine mesh-like fibrous component of the tissue. The fine filamentous noncollagenous content of the rECM was also clear,

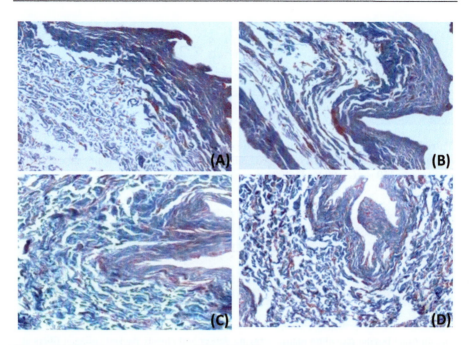

Figure 5.7 Delaminated reticulum at 72 h with different treatments: (A) sodium deoxycholate, (B) Triton X-100, (C) tri (n-butyl) phosphate, and (D) trypsin (Masson's trichrome staining, 100 ×).

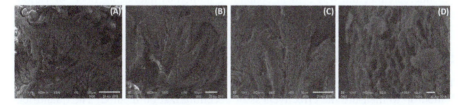

Figure 5.8 (A–D) Scanning electron microscopic observations of native bovine reticulum matrix showing dense and closely packed collagen at different magnifications (A, 50 ×; B, 150 ×; C, 500 ×; D, 1000 ×).

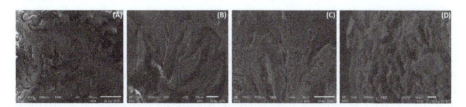

Figure 5.9 (A–D) Scanning electron microscopic observations of delaminated bovine reticulum matrix showing uneven luminal surface and aggregated collagen bundles at different magnifications (A, 50 ×; B, 150 ×; C, 500 ×; D, 1000 ×).

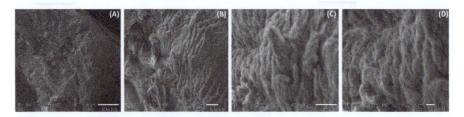

Figure 5.10 (A–D) Scanning electron microscopic observations of decellularized bovine reticulum matrix showing loosely arranged collagen fibers at different magnifications (A, 50×; B, 150×; C, 500×; D, 1000×).

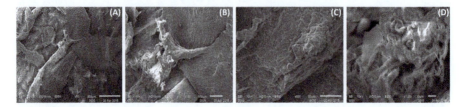

Figure 5.11 (A–D) Scanning electron microscopic observations of ethanol-dehydrated and vacuum-dried bovine reticulum matrix showing denser and closely packed collagen fibers at different magnifications (A, 50×; B, 150×; C, 500×; D, 1000×).

forming a fine weave between the collagen bundles. Within the matrix detailed collagen microarchitecture was evident containing numerous interlacing fibers of various thickness, pores, and open channels [10,11].

5.2.3 DNA contents analysis

The DNA from the samples was isolated and concentration was measured using Nanodrop to measure the effectiveness of decellularization method. The quantification was a direct measure to confirm the effectiveness of the processes, since the DNA is present in active nucleus of cell. In the native reticulum, abundant cell components and nucleic acids were present. However, after the decellularization, cells and nucleic acids were hardly observed in ECM. The DNA contents ($ng/\mu L$) before and after decellularization with sodium deoxycholate, Triton X-100, tri (n-butyl) phosphate, and trypsin are presented in Table 5.1 and Fig. 5.12.

The DNA concentration in native bovine reticulum was 47.08 $ng/\mu L$. After decellularization by different protocols, the DNA concentration decreased and ranged from 11.23 ± 0.06 $ng/\mu L$ to 37.13 ± 0.04 $ng/\mu L$. In enzyme-treated group, the extent of decellularization was inefficient and value ranged from 43.07 ± 0.03 to 45.08 ± 0.02 at different time intervals. Significant difference ($P < .01$) was recorded between all treatment's groups at different time intervals. SDS (0.5%) at 48-hour treatment showed lowest values (11.23 ± 0.06 $ng/\mu L$) of DNA content than all other groups and was further selected for evaluation in rat models.

Table 5.1 Mean ± SE values of DNA content (ng/μL) after decellularization.

Native	Treatment groups							
	Time intervals							
	24 h				48 h			
	0.5% SDS	0.5% Triton	0.5% Trin-BP	0.5% Trypsin	0.5% SDS	0.5% Triton	0.5% Trin-BP	0.5% Trypsin
47.08 ± 0.02	11.30 ± 0.01**	14.04 ± 0.05**	37.13 ± 0.04**	43.07 ± 0.03**	11.23 ± 0.06**	13.13 ± 0.04**	13.96 ± 0.07**	45.08 ± 0.02**

**Differ significantly ($P < .01$). *SDS*, Sodium dodecyl sulfate.

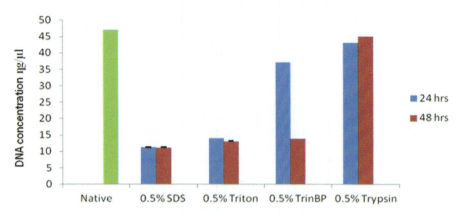

Figure 5.12 Mean ± SE values of DNA concentration (ηg/μL) after decellularization.

5.3 Preparation of acellular caprine reticulum extracellular matrix

Reticulum of caprine was collected from the institutional abattoir and was cleaned thoroughly by washing in sterile physiological normal saline. The reticulum was preserved in 1× phosphate buffered saline containing antibiotic (amikacin 1 mg/mL and 0.02% EDTA) at $-20°C$ for 1–2 hours. Preservation at $-20°C$ formed ice crystals between rumen layers facilitating in separation of different layers. Reticulum was removed from $-20°C$ and thawed at room temperature for 3–4 hours under running water. Serosa was easily peeled off and mucosa and muscular layers were removed by gentle scrapping with BP handle. The reticulum was cut into $4 \times 4\ cm^2$ size pieces and placed in sterile 1× PBS containing antibiotic (penicillin, streptomycin, and amphotericin) for 2 hours at room temperature. The reticulum was made acellular by treating with 1% sodium dodecyl sulfate (SDS) and 1% sodium deoxylate (SDC) in wrist action shaker for 96 hours. To assess the decellularization, samples were collected at 0-, 24-, 48-, 72-, and 96-hour time intervals. After completion of procedure, the tissues were thoroughly washed with 1× PBS and then placed in 70% ethanol for 12 hours. It was again washed in 1× PBS until the froth disappeared and stored at $-20°C$ in 1× PBS containing antibiotics. The tissue samples were subjected to histological examination, DNA quantification, calorimetric protein estimation, and SDS-PAGE analysis. For histological examinations, the samples were kept in 10% (v/v) neutral buffered formalin (NBF) solution, and for other parameters evaluation, tissue samples were preserved in PBS at $-80°C$.

5.3.1 Microscopic observations

The samples were cut into small pieces and fixed in 10% (v/v) neutral buffered formalin (NBF) for 48–72 hours. The tissues were processed in routine manner and

4 μm thick sections were cut and stained with H&E as per the standard protocol. The stained sections of were examined for degree of decellularization and arrangement of collagen fibers.

Histological evaluation of caprine reticulum showed that the cellular component embedded between collagen matrices decreased at different rate when compared both biological detergents. H&E-stained slides reflected continuous and faster decellularization in 1% SDS and this rate of decellularization was slow in case of 1% SDC (Figs. 5.13A–E and 5.14A–E). Samples at 48 hours of different tissues were satisfactory with 90% decrease in cellularity. Masson trichrome staining of different samples selected on the basis of H&E staining revealed quite decrease in collagen density at 48 hours when compared with collagen of native tissue.

5.3.2 DNA contents analysis

DNA was extracted from both from native and decellularized tissue using DNeasy Blood and Tissue Kit catalog number 69504 (Qiagen, Germany) and quantified as per the method reported by Pellegata et al. [12]. The DNA contents at different time intervals treated with 1% SDS and 1% SD are presented in Fig. 5.15. Results showed that there was significant ($P < .01$) difference in DNA contents as compared to native at different time intervals.

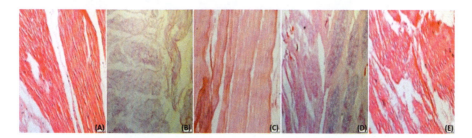

Figure 5.13 Histological picture of caprine reticulum treated with 1% SDS at different time intervals: (A) native reticulum, (B) 24 h, (C) 48 h, (D) 72 h, (E) 96 h posttreatment. *SDS*, Sodium dodecyl sulfate.

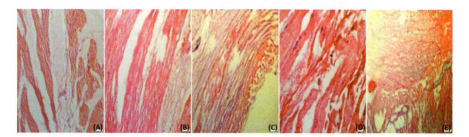

Figure 5.14 Histological picture of caprine reticulum treated with 1% SDC at different time intervals: (A) native reticulum, (B) 24 h, (C) 48 h, (D) 72 h, (E) 96 h posttreatment.

5.3.3 Calorimetric protein estimation

Tissue homogenization buffered was prepared and 5 mg of native and decellularized reticulum samples were homogenized in this buffer by electric homogenizer. All samples were vortexed and centrifuged at 10,000 rpm for 5 minutes and supernatant was collected in separate micro centrifuge tubes. For direct calorimetric protein estimation, protein for each sample was estimated with Eppendorf Bio Spectrometer at 280 nm. Results of protein estimation represented in Fig. 5.16. At 24, 48, 72, and 96 hours, the samples differ significantly ($P < .05$) with the native samples.

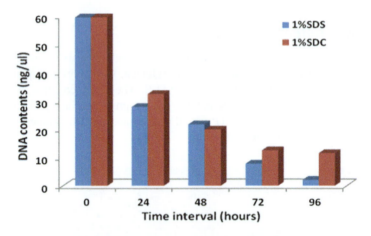

Figure 5.15 Mean ± SD of DNA quantification (ng/μL) at different time intervals.

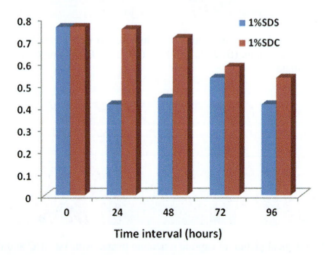

Figure 5.16 Protein contents at different time intervals.

5.3.4 Sodium dodecyl sulfate polyacrylamide gel electrophoresis analysis

For SDS-PAGE, 25 μL of sample was taken and marked properly in predesigned order. SDS-PAGE analysis was carried out in a vertical mini gel electrophoresis apparatus (Bio-Rad Mini Protean) as per standard procedure. SDS-PAGE analysis in 10% polyacrylamide gel showed the expression of the proteins of native and decellularized reticulum with different detergents at 24-, 48-, 72-, and 96-hour time intervals. The results are presented in Fig. 5.17. Reticular tissue showed only 45 KDa band in native, which was disappeared in decellularized tissue.

5.4 Evaluation of bovine reticulum extracellular matrix in rat model

The evaluation of bovine reticulum acellular matrix was done 36 clinically healthy Sprague Dawley rats of either sex, weighing from 250 to 300 g, and between 3 and 4 months of age. Animals were anesthetized using xylazine (5 mg/kg body weight) and ketamine (50 mg/kg body weight) combination. Using a sterile plastic template, the vertices of the experimental wounds of 20 × 20 mm² dimensions were outlined on the dorso thoracic region of the rats. A full thickness skin defect including the *panniculus carnosus* was created with a #11 BP blade on each animal. Hemorrhage, if any, was controlled by applying pressure with sterile cotton gauze. The animals

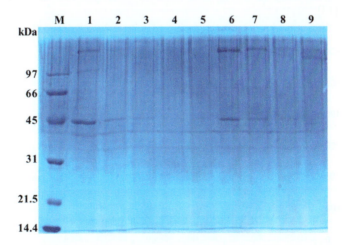

Figure 5.17 SDS-PAGE of caprine reticulum. Lane M—broad-range SDS-PAGE prestain marker (Bio-Rad), Lane 1—native caprine reticulum, Lane 2—1% SDS treatment for 24 h, Lane 3—1% SDS treatment for 48 h, Lane 4—% SDS treatment for 72 h, Lane 5—1% SDS treatment for 96 h, Lane 6—1% SDC treatment for 24 h, Lane 7—1% SDC treatment for 48 h, Lane 8—1% SDC treatment for 72 h, Lane 9—1% SDC treatment for 96 h. *SDC*, Sodium deoxylate; *SDS*, sodium dodecyl sulfate.

were randomly divided into three equal groups. The defect in group I (C) was kept as open wound and was taken as control (C). In group II, the defect was repaired with commercially available collagen sheet (b-CS). In group III, the defect was repaired with acellular b-REM. The scaffolds were secured to the edge of the skin wounds with eight simple interrupted 4/0 nylon sutures. The wounds were covered with paraffin surgical dressings and parameters were recorded at different time intervals.

5.4.1 Wound area and percent contraction

Wound area of all groups was measured by tracing its contour using transparent sheets with graph paper on postoperative days 0, 3, 7, 14, 21, and 28. The wound area (mm^2) and percent contraction were measured. A gradual decrease in wound area was observed in all the groups during the entire observation period except in group I (control) where it showed slight increase in wound area on day 3 postoperatively. The wound area decreased significantly ($P < .01$) on days 14, 21, and 28 in different groups. On day 28, complete healing was observed in b-REM group (Fig. 5.18) [13].

A gradual increase in percent wound contraction was observed in all the groups during the observation period. Although the original wound area created was 400 mm^2 but the wounds were expanded to various extent in different groups. Highly significant difference ($P < .01$) in wound contraction was observed between control and b-REM group on day 7. As the healing progressed, the wound area decreased significantly ($P < .05$) at different time intervals in all the groups (Fig. 5.19) [13].

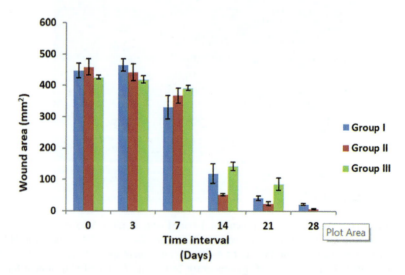

Figure 5.18 Mean ± SE of the total wound area (mm^2) of the skin wounds at different time intervals.

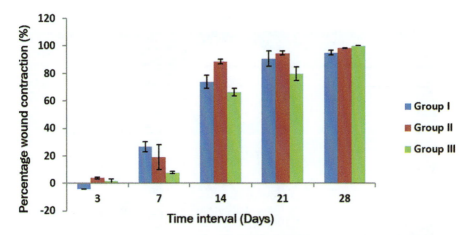

Figure 5.19 Percent contraction (mean ± SE) of wound area at different time intervals in different groups.

In control group, no matrix was used and the wound healed as denuded wound. Maximum percent wound contraction was recorded in this group and healing completed by day 28, leaving a large scar indicating the existence of severe contraction. In group b-CS, commercially available collagen sheet was applied, as it is very fragile in nature so suturing was not possible, but wound area was not expanded. The acellular matrix implanted group took more days for complete healing, but it was with minimum contraction. In b-REM group, the reticular matrix was sutured over the wound area to prevent drying and desiccation of the inner matrix. The rate of healing with minimum contraction was higher in b-REM-grafted group than group treated with collagen sheet as the matrix sutured with wound edges in these groups whereas collagen sheet was simply applied over the wound area [13]. The wound contraction is the centripetal displacement of the wound edges that facilitates its closure after trauma. This process is carried out by myofibroblasts that contain a-actin from smooth muscle and is mediated by contractile forces produced by granulation tissue from wound [14]. The wound healing rate is defined as the gross epithelialization of the wound bed. The wound contraction was assessed by percent retention of the original wound area as reported by Schallaberger et al. [15]. The wound contraction has been used to monitor wound healing. The wound area decreased gradually as the healing progressed.

5.4.2 Gross observations

Color photographs taken on 0, 3, 7, 14, 21, and 28 days postimplantation were used for analysis of shape, irregularity, and color of the lesions (Fig. 5.20). During the entire period of the study, none of the wound showed any visible suppurative inflammation. Furthermore, none of the animals became sick or died. In control group on day 3, the wounds were covered with soft and fragile pinkish mass above

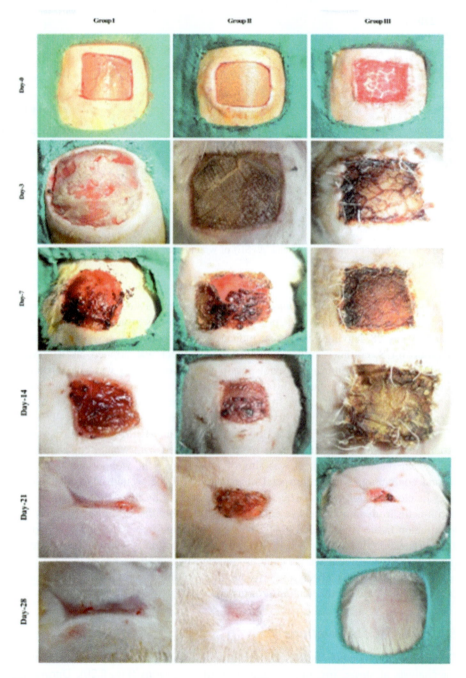

Figure 5.20 Gross observations of wound area at different time intervals in different groups.

the wound surface area, and later on, the surface became more desiccated and necrosed. By day 14, the thick crust starts detaching leaving a raw granular pink tissue, and by day 28, wound healed completely by a large scar. On day 3, in collagen sheet implanted group, the wound was covered with soft and fragile pinkish mass with necrosed margin and moderate exudates. By day 7, the top layer of collagen sheet got dried and major portion was sloughed off. By day 14, the upper layer became more desiccated, and by day 28, the wound was completely healed leaving a large scar. In b-REM group, on day 3 the top layer was dried and dark brown in color. On day 14, it was in a stage of detachment from the underlying tissue and newly formed granulation tissue covered the entire surface of the wound. By day 28, the wound healed up completely with no scar and the implanted area appeared as normal skin [13].

Graft-assisted healing is an important strategy for treating full thickness skin wounds. The forestomach-derived scaffold was rich in natural biomolecules like laminin, elastin, collagen IV, heparin sulfate, and glycosaminoglycans (GAGs), and when used, it promoted healing with excess cell proliferation at early phases and acceptable collagen deposition in the later remodeling phases [11]. After application of rECM matrix, the color of the implanted samples changed from white to dark brown and finally to dark revealed that on subsequent time intervals, decrease in vascularization and continuous loss of moisture contents of the matrix lead to change its color in to black. Kaarthick [16] also reported change in matrix color from white to brown later on black color after the repair of full thickness skin defects in rats. Upper matrix layer detached from the wound in form of scar and the underlying layer of rECM matrix were completely absorbed and newly formed granulation tissue within the matrix covered the whole surface of the wound in b-REM group.

5.4.3 Hematological observations

Blood smears for total leukocyte count (TLC) and differential leukocyte count (DLC) were prepared on 0, 3, 7, 14, 21, and 28 postoperative days and counts were expressed in percent. A significant ($P < .05$) increase in neutrophils percentage after operation was observed up to day 3 postoperatively in all the groups. Thereafter, they started decreasing and reached within normal limits on day 21. Contrary to increased neutrophil count, a significant ($P < .05$) decrease in lymphocyte count was observed in animals of all the groups up to day 7 postoperatively. The values fluctuated within the normal physiological range (65%–85%) at different time intervals. Total leukocyte count was within normal physiological range (6%–18%) in all groups at different time intervals [13]. The neutrophils have been recognized as the first cellular defense of the body and neutrophilia along with leukocytosis is generally associated with inflammation [17,18]. Besides the increase in neutrophil count, there was also decrease in lymphocyte count, which returned near to base line value by day 28 in all the animals of different groups. Moreover, DLC is proportional count of different WBCs where neutrophil and lymphocyte are major constituents. The DLC value revealed a significant increase in neutrophil percentage in early postoperative days in all the groups with gradual return nearer to normalcy, probably due to inflammatory response consequent to surgical trauma and foreign body reaction. Neutrophilia for short duration suggested the effect of surgical trauma rather than the implant provoked response. Similar observations were also reported by Gangwar et al. [19] following application of acellular dermal matrix for full thickness skin graft in rats. The leukocyte counts have been used as an indicator to monitor progress of healing; the purpose of this study was to examine effects of graft on blood leukocyte numbers before an injury and during healing [20].

5.4.4 Immunological observations

For the ELISA procedure, serum was collected on days 0 and 28 from each rat to assess the extent of antibodies titer generated toward the implant. The cell-mediated immune response was assessed by 3-(4,5-dimethylthiazol-2-yl)-2,5-diphenyltetrazolium bromide (MTT) colorimetric assay. The stimulation of rat lymphocytes with phytohemagglutinin (PHA) was considered as positive control, whereas unstimulated culture cells were taken as negative control. Blood collected on postimplantation days 0, 21, and 28 was used for lymphocyte culture as per standard protocols. The plates were immediately read at 570 with 620 nm as reference wavelength. The stimulation index (SI) was calculated using formula: SI = OD of stimulated cultures/OD of unstimulated cultures.

Commercially available collagen sheet has highest protein contents (63.3 mg/mL) followed by native bovine reticulum (34.12 mg/mL), and decellularized bovine reticulum showed lowest protein contents (9.71 mg/mL). The levels of antibodies present in serum prior to implantation were taken as basal values. The scaffold-specific antibodies were expressed as mean $\pm$ SE absorbance at 492 nm wavelength (OD 492). In control group, no rise in antibody titer was observed as the wound healed denuded without any matrix. In the treatment groups II and III, there were relative rise in antibody titer from day 0 to day 28 postimplantation. The B-cell response was significantly higher ($P < .05$) in the groups II and III on day 28 as evidenced by higher absorbance values when compared with day 0 values [13].

The nature and degree of immunological response to a foreign material is a crucial variable affecting the outcome and success or failure of implanted biomaterial. The immunological response to biomaterial depends on the nature of processing and presence of potentially foreign antigen to various lymphocyte populations. The biomaterial could stimulate an antibody response, a cell-mediated sensitization, or minimal to no response. The immune response to xenogenic transplantation included both natural and induced humoral components [21]. ELISA was performed to check the extent of antibody generated toward the graft components. In this study, the animals of biomaterials groups (group II and III) showed slightly higher immune response on days 21 and 28 as compared to day 0. The least immune response was in control group where no biomaterial was used and wound healed denuded. The b-REM group elicits a more or less similar immune response as animals of group b-CS.

The cell-mediated immune response toward cutaneously implanted scaffolds in all the experimental animals was assessed by lymphocyte proliferation assay/MTT colorimetric assay. Lymphocyte proliferation assay (LPA) measures the ability of lymphocytes placed in short-term tissue culture to undergo a clonal proliferation when stimulated in vitro by a foreign molecule (antigen/mitogen). Th1 lymphocytes produce cytokines such as IL-2, IFN-ϒ, and TNF-ß leading to macrophage activation, stimulation of complement fixing Ab isotypes (IgG2a and IgG2b in mice), and differentiation of $CD8^+$ cells to a cytotoxic T cells [22]. Activation of this pathway is associated with both allogeneic and xenogeneic transplant rejection. Th2 lymphocytes produce IL-4, IL- 5, IL-6, and IL-10, cytokines that do not activate macrophages and that lead to production of noncomplement fixing Ab isotypes (IgG1 in mice). Activation of the Th2 pathway is associated with transplant acceptance. $CD4^+$ lymphocytes proliferate in response to antigenic peptides in association with class II major histocompatibility complex (MHC) molecules on antigen-presenting cells (APCs). In the present study, high stimulation index was observed against bovine reticulum antigen in comparison to collagen antigen on day 45.

5.4.5 Histopathological observations

The biopsy specimens from the implantation site were collected on 7, 14, 21, and 28 days postimplantation. The samples were processed by routine histological examination. The host inflammatory response, neovascular tissue formation (fibroblasts, fine capillaries), deposition of neocollagen, and penetration of host inflammatory responses in the implanted matrix were evaluated as per Schallberger et al. [15]. Special staining for collagen fibers was done by using Masson's trichrome stain to observe the deposition of collagen fibers, and scoring was done as per the method of Borena et al. [23]. The group having less histopathological score was considered the best. The results of histological observations are presented in Figs. 5.21 and 5.22. The group having least number of scores was considered better [13].

In control group on day 7, the wound was covered with necrotic debris and admixed with high degree inflammatory cell infiltration, edema, and congestion. The underlying stoma contained moderately proliferated fibroblasts and high neovascularization along with spilled neutrophils. The collagen fibers were less dense, thin, and worst arranged. Angiogenesis increased and there was initiation of epithelialization at the margin of wound area covering the granulation tissue. The histopathological score was 31 at this stage. On day 14, moderately proliferation of fibroblasts and high neovascularization along with moderate degree of inflammatory cell infiltration was observed. Granulation tissue was wide and initiation of epithelialization with moderate thickness. Collagen fibers were less dense, thin with worst arrangement. The score recorded was 29. On day 21, surface epithelium was incomplete with moderately thick than normal. Granulation tissue was moderately wide. Inflammatory cell infiltration was mild with moderate neovascularization and fibroblastic proliferation. The collagen fibers were dense and thick but with worse arrangement and the histopathological score was 23 at this stage. On day 28, surface epithelium resembled to normal skin, but slightly thicker. Mild inflammation

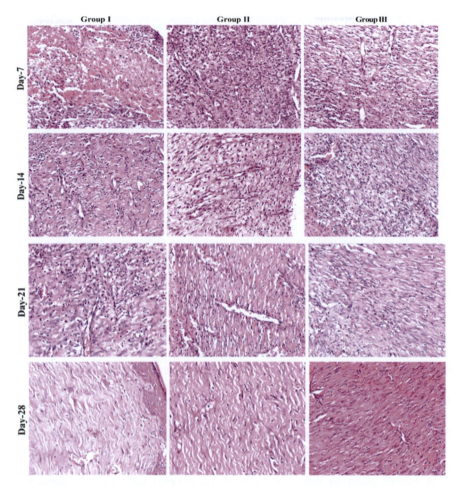

Figure 5.21 Histopathological observations of wound healing of different groups at various time intervals (H&E staining).

with narrow granulation tissue width and fibroblast proliferation resembling normal skin was observed. Collagen fiber were dense, thick with worse arrangement and the histopathological score was 16 at this stage [13].

In b-CS group, high inflammatory changes were observed on day 7 postoperatively. There was necrosis and sloughing of the superficial layer and initiation of epithelialization at the edges of wound. The surface epithelium was thicker than normal skin. Granulation tissue width was moderately wide. Collagen fiber was less dense, thin with worst arrangement. The histopathological score was 32. On day 14, the deeper layer of dermis revealed deposition of new collagen fibers. Epithelialization was initiated with moderate thickness and inflammation reduced to mild degree. Collagen fibers were dense, thin with worse arrangement. The score was 25 at this stage. By day 21, epithelialization started but was incomplete with

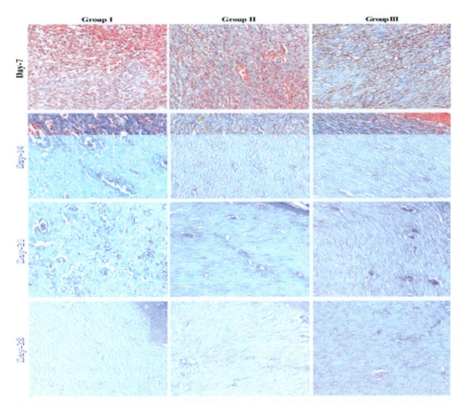

Figure 5.22 Histopathological observations of wound healing of different groups at various time intervals (H&E staining).

moderate degree of thickness. The inflammatory changes and neovascularization reduced to mild degree. Collagen tissue was dense, thick, and better arranged. The histopathological score further reduced to 20 at this stage. On day 28, superficial epithelium was complete, with thickness resembling normal skin. Inflammatory cells were absent and neovascularization reduced to mild form. Granulation tissue was slightly wide. Collagen fibers were denser, thicker, and better arranged. No hair follicle and skin gland were observed. The histopathological score was 15 at this stage [13].

In b-REM group on day 7, moderate inflammatory changes were seen. The moderate fibroblast proliferation and neovascularization was observed in matrix tissue. The necrosis and sloughing of the superficial matrix were observed. Initialization of surface epithelium with thickness more than normal was also observed. The superficial acellular matrix was still present at the wound site and epithelialization was started at the edges of wound area. The total histopathological score was 25. The neoepidermis was thicker and moderately wide granulation tissue was very prominent in the dermis. The newly formed keratinocyte was seen in both matrix-induced healing reactions, and the collagen fibers were further organized into thick bundles.

The collagen fibers were dense and thin with worse arrangement. The histopathological score was 27. On day 14, the reepithelialization was incomplete and marginal epithelialization was slightly thick than normal skin. Neovascularization, fibroblast proliferation, and inflammatory cells infiltration were moderate. Dense and thick collagen formation with better arrangement was observed and the histopathological score was 22. On day 21, the collagen formation was appreciable. The hair follicles and skin glands were observed. Epithelium was similar to normal skin. The histopathological score was 17. By day 28, complete epidermis was formed. The dermis was well developed with newly formed blood vessels and well-organized collagen fibers. The process of remodeling was evident on these samples with reorganization and disappearance of collagen. Densely packed thick collagen bundles were replaced by loosely placed thin strands and the histopathological score was 10 [13].

Advantages of using biological materials are their capacity to resist infection and induce a milder inflammatory response, angiogenesis, and host cell migration [24]. Erkin et al. [25] observed the revascularization potential of xenogenic acellular dermal matrix (alloderm) in rats. Full thickness skin wounds are characterized by a complete destruction of the epithelial regenerative elements that reside in the dermis. Full thickness skin wound healing occurs by granulation tissue formation, contraction, and epithelialization [26]. The epithelialization occurs by migration of undamaged epidermal cells from the wound margins across the granulation bed [27]. The exogenous collagen supplementation enabled faster migration of cells that are involved in cutaneous wound healing. Since the exogenous collagen is molecular in nature [28] and supplies endogenous collagen in vivo, it readily integrates with the wound tissue and facilitates the attachment, migration, and proliferation of cells on the wound site [29]. The b-REM acellular matrix showed necrosis of the superficial matrix and underlying matrix showed moderate fibroblastic proliferation with inflammatory cells. The proliferation of fibroblasts was more in the b-REM group as compared to control and b-CS group. Increase in the extent of collagen deposition by 7 days (about 20%) and 14 days (about 40%) was reported by Ref. [30]. Control group exhibited severe inflammatory cells and fibroblast proliferation in the wound area on day 7. Moderate degree of neovascularization was detected in all the three groups.

New collagen formation was not detected in any of the groups. But the graft implanted sample showed the existence of mature collagen of b-REM origin. Although inflammation is necessary for healing by fighting infection and inducing the proliferation phase, healing proceeds only after inflammation is controlled [31]. During the initial of the wound healing, fibroblasts from the surrounding normal area are known to migrate and proliferate into the wound site and, within 3–4 days, get converted to myofibroblasts [32]. The main function of myofibroblasts in a healing wound is contraction by synthesizing of ECM proteins, notably, collagen types I–VI and XVIII, laminin, thrombospondin, glycoproteins, and proteoglycans for the dermal repair. The myofibroblasts usually go on increasing from the inflammatory stage (3–4 days). During the onset of healing, the proliferating fibroblast starts to synthesize collagen, and the total collagen content increases preferentially.

On day 7, minimum score observed in b-REM group showed better wound healing in comparison to control and b-CS group. On day 14, least scores observed in b-REM group (22) may be due to early acceptance of matrix. The sloughing of the upper layer of matrix was observed in b-CS and b-REM group, except control group where wound remained open. It may be either due to the desiccation of the graft in high environmental temperature or due to an impaired formation of new blood vessels [33]. The myofibroblasts proliferate till the end of proliferation phase (14−15 days).

Meanwhile, the underlying granulation tissue increased in mass that pushed up the graft upward. The animals of b-CS and b-REM groups showed well-formed collagen and neovascularization with superficial epithelialization. The epithelialization and neovascularization were faster in b-REM group as compared to other groups. On day 21, the score was minimum in b-REM group (17), as compared to b-CS group (20) and control group (23). The epithelialization was more similar to the normal skin in the wounds of these groups. In the remodeling phase (around 21−30 days), these myofibroblasts undergo apoptosis [34]. At 21 days of postimplantation, the acellular matrix groups showed higher collagen synthesis as compared to the control group as appreciated by the Masson's trichrome staining. Purohit [35] on day 21 found that the acellular dermal matrix throughout the width and length was replaced by mature collagenous connective tissue in experimentally created wounds in rabbits. On day 28, postimplantation complete healing was observed in all the three groups. In control group abundant scar tissue was observed.In b-CS group contraction and scar formation was also observed. In b-REM group, the collagen fiber arrangement was almost similar to normal skin. Marked fibroblastic response associated with an abundant new collagenous fibrous tissue was observed in this group. The deposition of new collagen was oriented parallel to the skin surface. In acellular groups hair follicle and skin glands could be detected as in case of normal skin indicating the culmination of repair process. Here the healing was with minimum contraction and scar formation.

5.5 Conclusion

In this study, acellular matrix were prepared from bovine and caprine reticulum using biological detergents. Treatment of bovine reticulum with 0.5% SDS for 48 hours resulted in completely acellular matrix as evaluated by histology, scanning electron microscopy, and DNA quantification. The goat reticulum was treated with 1% SDS and 1% SDC and was subjected to histological evaluation, DNA quantification, calorimetric protein estimation, and SDS-PAGE analysis. H&E-stained slides reflected continuous and faster decellularization in 1% SDS and this rate of decellularization was comparatively slow in 1% SD treatment. The optimized results were obtained at 48 hours of treatment using 1% SDS. In the study, we have demonstrated that b-REM has healing potential and have comparable healing response to that of commercially available collagen sheet to reconstruct full thickness skin wounds in rats. Histological observations showed that b-REM augmented wound healing activity by increasing cellular proliferation, neovascularization, synthesis of collagen, epithelialization, and early histological maturation in excisional wounds.

References

[1] Ward BR, Johnson KD, May BCH. Tissue scaffolds derived from forestomach extracellular matrix. US Patent No 12/512, 2009;835.
[2] Badylak SF, Freytes DO, Gilbert TW. Extracellular matrix as a biological scaffold material: structure and function. Acta Biomater 2009;5:1−13.
[3] Ward BR, Johnson KD, Charles B, May H. Tissue scaffolds derived from forestomach extracellular matrix. Patent No US 8, 2014;758,781 B2, Date of patent June 24, 2014.
[4] Rieder E, Kasimir MT, Silberhumer G, Seebacher G, Wolner E, Simon P. Decellularization protocols of porcine heart valves differ importantly in efficiency of cell removal and susceptibility of the matrix to recellularization with human vascular cells. J Thorac Cardiovasc Surg 2004;127:399−405.
[5] Lin P, Chan WC, Badylak SF, Bhatia SN. Assessing porcine liver-derived biomatrix for hepatic tissue engineering. Tissue Eng 2004;10:1046−53.
[6] Woods T, Gratzer PF. Effectiveness of three extraction techniques in the development of a decellularized bone-anterior cruciate ligament-bone graft. Biomaterials 2005;26:7339−49.
[7] Gamba PG, Conconi MT, Lo Piccolo R, Zara G, Spinazzi R, Parnigotto PP. Experimental abdominal wall defect repaired with acellular matrix. Pediatr Surg Pediatr Surg Int 2002;18:327−31.
[8] Gilbert TW, Freund J, Badylak SF. Quantification of DNA in biologic scaffold materials. J Surg Res 2009;152:135−9.
[9] Hasan A, Kumar N, Singh K, Gopinathan A, Remya V, Mondal DB, et al. Preparation of acellular matrix from bovine reticulum using detergents and enzynmes. Trends Biomat Artific Organs 2015;29(3):231−6.
[10] Gilbert TW, Sellaro TL, Badylak SF. Decellularization of tissues and organs. Biomaterials 2006;27:3675−83.
[11] Lun S, Irvine SM, Johnson KD, Fisher NJ, Floden EW, Negron L. A functional extracellular matrix biomaterial derived from ovine forestomach. Biomaterials 2010;31: 4517−29.
[12] Pellegata AF, Asnaghi M, Stefani I, Maestroni A, Maestroni S, Dominioni T, et al. Detergent-enzymatic decellularization of swine blood vessels: insight on mechanical properties for vascular tissue engineering. Bio Med Res Int 2013;1−8.
[13] Hasan A, Kumar N, Gopinathan A, Singh K, Sharma AK, Remya V et al.. Bovine reticulum derived extracellular matrix (b-REM) for reconstruction of full thickness skin wounds in ratsAvailable from: http://doi.org/10.1016/j.wndm.2016.02.003.
[14] Neagos D, Mitran V, Chiracu G, Ciubar R, Iancu C, Stan C, et al. Skin wound healing in a free-floating fibroblast populated collagen latice model. Romanian J Biophys 2006;16(3):157−68.
[15] Schallberger SP, Stanley BJ, Hauptaman JG, Steficek BA. Effect of porcine small intestine submucosa on acute full-thickness wounds in dogs. Vet Surg 2008;37:515−24.
[16] Kaarthick D.T. Repair of cutaneous wounds using acellular diaphragm and pericardium of buffalo origin seeded with in-vitro cultured mouse embryonic fibroblast cells in rat

model. MVSc thesis submitted to Deemed University, Indian Veterinary Research Institute, Izatnagar, Bareilly, Uttar Pradesh, India; 2010.
[17] Coles EH. Leukocyte. 2nd edn Veterinary clinical pathology, 40. Philadelphia: WBSaunders Co; 1974. p. 98.
[18] Schalm OW, Jain NC, Carrol EJ. Veterinary haematology. 3rd edn Philadelphia: Lea and Febiger; 1975. p. 124.
[19] Gangwar AK, Kumar N, Sharma AK, Sangeeta Devi Kh, Negi M, Shrivastava S, et al. Bioengineered acellular dermal matrix for the repair of full thickness skin wounds in rats. Trends Biomat Artific Organs 2013;27(2):67−80.
[20] Haffor AS. Effect of myrrh (*Commiphora molmol*) on leukocyte levels before and during healing from gastric ulcer or skin injury. J Immunotoxicol 2010;7(1):68−75.
[21] Ruszczak Z. Effect of collagen matrices on dermal wound healing. Adv Drug Deliv Rev 2003;55:1595−611.
[22] Abbas AK, Murphy KM, Sher A. Functional diversity of helper T lymphocytes. Nature 1996;383:787−93.
[23] Borena BM, Pawde AM, Amarpal, Aithal HP, Kinjavdekar P, Singh R, et al. Autologous bone marrow-derived cells for healing excisional dermal wounds of rabbits Vet Rec 2009;165(19):563−82009. Available from: https://doi.org/10.1136/vr.165.19.563.
[24] Bellows CF, Albo D, Berger DH, Awad SS. Abdominal wall repair using human acellular dermis. Am J Surg 2007;194:192−8.
[25] Erkin UR, Kerem M, Tug M, Orbay H, Sensöz O. Prefabrication of a conjoint flap containing xenogenic tissue: a preliminary report on an experimental model. J Craniofac Surg 2007;18:1451−6.
[26] Fossum TW, Hedlund CS, Johnson AL, Schulz KS, Seim HB, Willard MD, et al. Surgery of integumentary system. Manual of Small Animal Surgery. Mosby; 2007. p. 159−75.
[27] Swaim SF, Henderson RA. Wound management. Small animal wound management, Vol., 9. Lea and Febiger; 1990. p. 33.
[28] Nithya M, Suguna L, Rose C. The effect of nerve growth factor on the early responses during the process of wound healing. Biochemica et Biophysica Acta 2003;1620:25−31.
[29] Judith R, Nithya M, Rose C, Mandal AB. Application of a PDGF containing novel gel for cutaneous wound healing. Life Sci 2010;87:1−8.
[30] Revi D, Vadavanath PV, Jaseer M, Akhila R, Anilkumar TV. Porcine cholecyst−derived scaffold promotes full-thickness wound healing in rabbit. J Biomater Tissue Eng 2013;3:261−72.
[31] Midwood KS, Williams LV, Schwarzbauer JE. Tissue repair and the dynamics of the extracellular matrix. Int J Biochem Cell Biol 2004;36:1031−7.
[32] Hinz B, White ES. The myofibroblast matrix: implications for tissue repair and fibrosis. J Pathol 2013;229:298−309.
[33] Boyce ST. Cultured skin substitutes: a review. Tissue Eng 1996;2:255−66.
[34] Li B, Wang JHC. Fibroblasts and myofibroblasts in wound healing: force generation and measurement. J Tissue Viability 2011;20:108−20.
[35] Purohit S. Biocompatibility testing of acellular dermal grafts in a rabbit model: an in-vitro and in-vivo study. PhD thesis submitted to Deemed University, Indian Veterinary Research Institute, Izatnagar, Bareilly, Uttar Pradesh, India; 2008.

Omasum-derived extracellular matrix scaffolds

Ashok Kumar Sharma[1], Naveen Kumar[1,]*, Priya Singh[2], Pawan Diwan Singh Raghuvanshi[3], Mohar Singh[1], Sangeetha P.[1], Sameer Shrivastava[4], Sonal Saxena[4], Swapan Kumar Maiti[1] and Karam Pal Singh[5]

[1]Division of Surgery, ICAR-Indian Veterinary Research Institute, Izatnagar, Uttar Pradesh, India, [2]Department of Veterinary Surgery and Radiology, College of Veterinary Science and Animal Husbandry, Nanaji Deshmukh Veterinary Science University, Jabalpur, Madhya Pradesh, India, [3]Department of Veterinary Surgery and Radiology, College of Veterinary Science and Animal Husbandry, Nanaji Deshmukh Veterinary Science University, Mhow, Indore, Madya Pradesh, India, [4]Division of Veterinary Biotechnology, ICAR-Indian Veterinary Research Institute, Izatnagar, Uttar Pradesh, India, [5]CADRAD, ICAR-Indian Veterinary Research Institute, Izatnagar, Uttar Pradesh, India

6.1 Introduction

Wound healing is a fundamental response to tissue injury that ultimately results in restoration of tissue integrity. This response is achieved mainly by the synthesis of the connective tissue matrix [1]. Most wounds can heal naturally, but full-thickness skin wounds greater than 1 cm in diameter in human being needs a skin graft to prevent scar formation, morbidity, and cosmetic deformities [2]. The tissue engineering approach for skin substitutes has relied upon the creation of three-dimensional scaffolds as extracellular matrix (ECM) analog to guide cell adhesion, growth, and differentiation to form skin—functional and structural tissue [3]. The three-dimensional scaffolds not only cover wound and provide a physical barrier against external infection as wound dressing but also provide support both for dermal fibroblasts and the overlying keratinocytes for skin tissue engineering. A successful tissue scaffold should exhibit appropriate physical and mechanical characteristics and provide an appropriate surface chemistry and nano- and microstructures to facilitate cellular attachment, proliferation, and differentiation. The use of acellular scaffolds of xenogeneic origin, ECM resulting after removal of cells, as skin substitute is an acceptable modality for treating dermal wounds. Biological scaffold derived from decellularized tissues is in use as surgical implants and scaffolds for regenerative medicine because ECM secreted from resident cells of each tissue and organ can provide favorable microenvironment that affects cell migration, proliferation, and differentiation [4,5].

---

*Present affiliation: Veterinary Clinical Complex, Apollo College of Veterinary Medicine, Jaipur, Rajasthan, India.

Natural Biomaterials for Tissue Engineering. DOI: https://doi.org/10.1016/B978-0-443-26470-2.00006-5
© 2025 Elsevier Inc. All rights are reserved, including those for text and data mining, AI training, and similar technologies.

The forestomach matrix (FM) provides a number of advantages over other scaffolds and is useful in a variety of clinical and therapeutic applications, including wound repair and tissue regeneration [6]. Bovine omasum matrix histology shows that the lamina propria is unusually dense, whereas the abluminal side of the scaffold is structured as an open omasum matrix. The dense layer of ECM from the lamina propria contributed to the increased thickness and strength of FM scaffolds compared to those derived from other organs. This structure makes the FM well suited for encouraging epithelial regeneration on the dense luminal side of the matrix and fibroblast invasion on the less dense abluminal side of the matrix, when used as a medical device for tissue regeneration. These differences serve an important role in epithelial regeneration, as the dense side of abomasums matrix acts as a barrier to cell migration, while the less dense side (abluminal) does not present a barrier and therefore allows cell invasion. The FM matrix scaffolds can be used to promote, stimulate, or increase proliferation of cells near by the scaffold attachment site as well as increase vascularization of a tissue or organ [7]. The relative size and thickness of ECM in tissue offers a solution to generate relatively large format ECM-based biomaterial with good performance characteristics [8].

6.2 Preparation of acellular matrix from buffalo omasum

Omasum of buffalo was procured from the local abattoir. Immediately, after collection the omasum was be kept in cold physiological saline solution containing 0.02% EDTA and antibiotic (Amikacin @ 1 mg/mL). The tissue was thoroughly rinsed with normal saline before the start of protocol. The maximum time period between retrieval and the initiation of protocol was less than 4 hours. After thorough cleaning of the bovine omasum, deepithelialization was done in hypertonic solution 2 M NaCl for 6 hours. The keratinized epithelium of omasum was easily scrubbed off by blunt surface of BP handle. The serosal layer was separated mechanically with forceps. The omasum after deepithelialization and delaminationwas cut into 4×4 cm^2 size pieces and placed in biological detergent and enzymes to carry out decellularization protocols. They were subjected to microscopic examination at 12, 24, 48, and 72 hours intervals. Ionic detergent (Sodium dodecyl sulfate) [9], nonionic detergents (Triton X-100) [10], Zwitterionic detergent (Tri (n-butyl) phosphate) [11], and enzyme (Trypsin) [12] were used in 0.5% concentration for decellularization. Treatment with chemical was done up to 72 hours and tissues were subjected to continuous agitation at 37°C on orbital shaker at the speed of 200 rpm. The solutions were changed at every 12 hours intervals. Finally, the tissues were thoroughly rinsed thrice (2 hours each) with sterile phosphate buffer saline (PBS) on orbital shaker and samples were collected at 12-, 24-, 48-, and 72-hour time intervals for evaluation of parameters. The prepared acellular omasum matrix was stored in PBS solution containing 1 mg/mL amikacin at 4°C until use (Fig. 6.1A−I) [13].

6.2.1 Gross observations

The tissue samples collected at different time intervals during standardization of decellularization protocols were subjected to gross observations. Gross examination at 6-hour time interval revealed that the keratinized epithelium was easily scrubbed off by blunt edge of BP handle while the serosal layer was delaminated by applying mechanical assistance and results in complete deepithelialization and delamination (Fig. 6.1B and D).

6.2.2 Microscopic observations

The samples were fixed in 10% formal saline solution, dehydrated in ethanol, cleared in xylene, and embedded in paraffin to get 5 micron thin sections. The sections were stained with hematoxylin and eosin (H&E) staining and Masson's trichrome staining. Microscopic examination revealed complete removal of keratinized epithelium and serosal layer. Native buffalo omasum showed keratinized epithelium on mucosal surface. Lamina propria is the luminal portion of the propria-submucosa, which includes a dense layer of ECM and serosal layer. Delaminated omasum treated with 0.5% sodium dodecyl sulfate for 12 hours under constant agitation showed 85% to 90% loss of cellularity along with presence of cellular debris are presented in (Fig. 6.2A). Treatment for 24 hours showed slight increase in loss of cellularity; that is, more than 90% and cellular debris is presented in (Fig. 6.2B). At 48-hour time interval,

Figure 6.1 Preparation of acellular matrix from buffalo omasum: (A) buffalo omasum laminae, (B) deepithelisation, (C) deepitheilised laminae, (D) removal of serosal layer, (E) serosal layer, (F) 70% ethanol treatment, (G) decellularized omasum laminae, (H) ethanol dehydrated, (I) vacuum dried.

submucosal layer was completely acellular. The collagen fibers were compact with moderate porosity than the native tissue. No debris was observed between the spaces of thick collagen fibers are presented in (Fig. 6.2C). At 72-hour time interval, ECM showed thin, loosely arranged collagen fibers with very high porosity. No cellular debris was evident are presented in (Fig. 6.2D) [14].

The treatment with 0.5% Triton X-100 for 12 hours under constant agitation showed 30% to 40% loss of cellularity with high cellular content, thick, and compact cellular fiber arrangement with least porosity (Fig. 6.3 ×). At 24 hours, no further loss of cellularity was observed. High cellular content, thick, and compact collagen fiber arrangement with least porosity was observed (Fig. 6.3C). At 48-hour time interval, loss of cellularity increased up to 60% with moderate cellular debris and porosity (Fig. 6.3C). At 72-hour time interval, there was no further decrease in cellularity and porosity was observed. Mild cellular debris and damage in collagen fiber arrangement were observed at this stage (Fig. 6.3C) [14].

The treatment with 0.5% Tri (n-butyl) phosphate for 12 and 24 hours showed no effect on loss of cellularity (Fig. 6.4A and B). The high cellular contents were noted

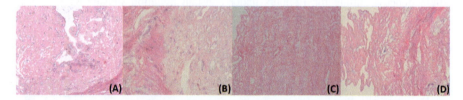

Figure 6.2 Delaminated omasal treated with 0.5% sodium dodecyl sulfate for different time intervals under constant agitation: (A) 12 h, (B) 24 h, (C) 48 h, (D) 72 h (H&E stain, × 100).

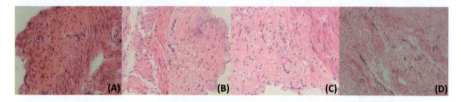

Figure 6.3 Delaminated omasal treated with 0.5% Triton X-100 for different time intervals under constant agitation: (A) 12 h, (B) 24 h, (C) 48 h, (D) 72 h (H&E stain, × 100).

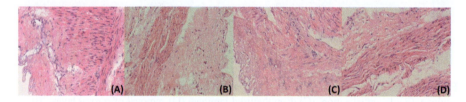

Figure 6.4 Delaminated omasal treated with 0.5% Tri (n-butyl) phosphate for different time intervals under constant agitation: (A) 12 h, (B) 24 h, (C) 48 h, (D) 72 h (H&E stain, × 100).

at these time intervals. At 48 and 72 hours, nearly 80% decellularization with loss of cellular debris and damage in collagen fiber arrangement was observed (Fig. 6.4C and D) [14].

Treatment enzyme with (trypsin) was inefficient in decellularization during various time intervals. No loss in cellular content with any effect on collagen fiber arrangement was observed at different time intervals (Fig. 6.5A—D) [14].

Microscopically, native buffalo omasum showed keratinized epithelium on mucosal surface (Fig. 6.6A). Lamina propriais, the luminal portion of the propria-submucosa, includes a dense layer of ECM and serosal layer. The delaminated propria-submucosa layer showed cellularity, tunica muscularis, and thick collagen fibers. The native bovine omasum laminae showed dense compact arrangement of collagen fibers clearly visible in Masson's trichrome staining (Fig. 6.6B). There was no cellular debris between the spaces of thick collagen fibers in H&E staining and in Masson's trichrome staining (Fig. 6.7A and B) [14].

Four different protocols were used to obtain the acellular ECM from the delaminated buffalo omasum. The delaminated bovine omasum was subjected to ionic, nonionic, zwitterionic detergents, and enzyme treatment. The time of reaction (12, 24, 48, and 72 hours) was optimized to obtain the acellular ECM. The delaminated buffalo omasum was subjected to 0.5% sodium dodecyl sulfate treatment for

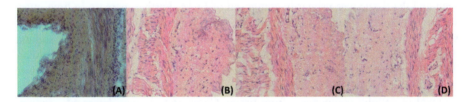

Figure 6.5 Delaminated omasal treated with 0.5% Trypsin for different time intervals under constant agitation: (A) 12 h, (B) 24 h, (C) 48 h, (D) 72 h (H&E stain, x-100).

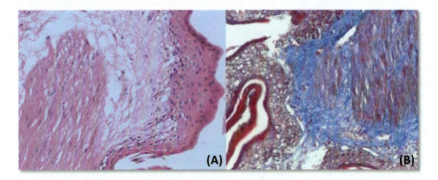

Figure 6.6 (A) Native buffalo omasum after delamination showing cellularity, loose muscular layer, and collagen fibers (H&E stain, × 200). (B) Showing dense compact arrangement of collagen fibers (Masson's trichrome staining, × 200).

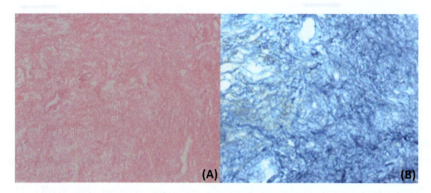

Figure 6.7 (A) Microscopic photograph showing no cellular debris between the spaces of thick collagen fibers (H&E stain, ×200). (B) Showing no cellular debris between the spaces of thick collagen fibers (Masson's trichrome staining, ×200).

24 hours became 90% acellular with mildly thick collagen fibers. The cellular debris was seen in between the void spaces of collagen fibers in the samples. Therefore, time intervals were increased to remove this cellular debris. At 48 hours, complete acellularity with no cellular debris was observed. No nuclear bodies were seen and the tissue was primarily composed of ECM. Further increase in time interval to 72 hours resulted in distributed and damaged collagen fibers. In the present study, desired results were achieved after 48 hours of treatment with 0.5% sodium dodecyl sulfate detergent. The propria-submucosa layer was completely acellular. The collagen fibers were thick and arranged in longitudinal and transverse manner as compared to the native tissue. Sodium dodecyl sulfate is very effective for removal of cellular components from tissue. Compared to other detergents, sodium dodecyl sulfate yields complete removal of nuclear remnants and cytoplasmic proteins, such as vimentin [11]. Ionic detergents are effective for solubilizing both cytoplasmic and nuclear cellular membranes but tend to denature proteins by disrupting protein–protein interactions [15]. These surfactants disrupt lipid–protein and lipid–lipid interactions but generally leave protein–protein reactions intact with the result of maintaining their functional conformations [16]. The cell extraction was effectively achieved without significant disturbances in ECM morphology and strength. The extraction protocol for decellularization of delaminated buffalo reticulum by 0.5% sodium dodecyl sulfate treatment for 48 hours has been documented by Hasan et al. [17]. Nonionic biological detergents (Triton X-100) do not usually denature proteins and interaction between peptides remains intact; therefore, they are considered most suitable to investigate the subunits structure of membranes proteins. Sodium dodecyl sulfate differs very much from nonionic detergent Triton X-100. It is relatively nondenaturing. It inhibits protein–protein and protein–lipid binding [18]. Zwitterionic detergents exhibits properties of nonionic and ionic detergents are efficient cell removal with ECM disruption similar to that of Triton X-100. Treatment with 0.5% Tri (n-butyl) phosphate for 48 and 72 hours showed

damage in collagen fiber arrangement and up to 80% loss of cellularity was observed. The enzymatic decellularization using trypsin cleaved peptide bonds on the C side of Arg and Lys but not completely decellularized the tissue [19].

6.2.3 DNA quantification

To measure the effectiveness of decellularization method, the DNA from cells of the samples was isolated and the DNA concentration was measured using Nanodrop. This quantification was an indirect measure to confirm the effectiveness of the processes, since the DNA is present in active nucleus of cell. The DNA contents analysis before the start of decellularization protocols and after completion of protocols was done as per the method described by Gilbert et al. [16]. The DNA from the samples was isolated and concentration was measured using Nanodrop to measure the effectiveness of decellularization method. In the native omasum, abundant cell components and nucleic acids were present. However, after the decellularization, cells and nucleic acids were hardly observed in ECM. The DNA contents (ng/μL) before and after decellularization with different protocols is presented in Fig. 6.8.

The concentration of native buffalo omasum DNA ($P < .01$) was 82.40 ± 1.41 ng/μL. Treatment with 0.5% SDS showed lowest values 4.30 ± 0.14 ng/mL of DNA contents. Significant decrease ($P < .01$) in DNA contents showed the effectiveness of treatment for decellularization. Significant decrease ($P < .05$) in DNA contents was also observed by Shakya et al. [20], while preparing acellular cholecyst matrices from bovine and porcine origin. Treatment with SDS showed effective removal of cells. Quantification of residual DNA in animal derived biological scaffold materials is one of technical specifications for evaluating decellularization process and immunotoxicity risk [21].

6.2.4 Sodium dodecyl sulfate polyacrylamide gel electrophoresis

The expression of protein bands was determined by the sodium dodecyl sulfate polyacrylamide gel electrophoresis (SDS-PAGE). The SDS-PAGE analysis was carried out in a vertical mini gel electrophoresis apparatus. The glass plates were cleaned and set in a gel molding tray of electrophoresis apparatus and 10% separating gel solution

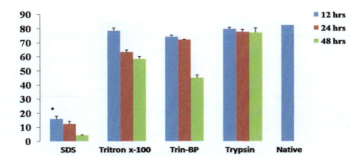

Figure 6.8 DNA content (ng/μL) after decellularization.

was poured between the two plates and 1 mL saturated butanol was poured over the gel. After polymerization, butanol was removed by tilting the plates and 5 mL of 3% stacking gel was poured over the separating gel and later on suitable comb was inserted. The polymerization gel was mounted into the electrophoresis chamber and buffer reservoir was filled with 1× Tris glycine buffer. Before loading, 30 μL sample buffer was added to each sample. The samples were boiled for 10 minutes and kept on ice. The samples were briefly centrifuged before loading. A standard protein marker was included along with the samples. Electrophoresis was carried out at a constant current of 150 V and 150 mA, until the tracking dye reached the bottom of the gel. The gel was removed from the plates and stained with Coomassie brilliant blue for 1 hour and then destained with destaining solution [22].

The SDS-PAGE analysis in 10% polyacrylamide gel showed the expression of the proteins of the tissues namely native bovine omasum, decellularized bovine omasum with 0.5% SDS at 48 hours interval and its comparison with different biological detergents at same percentage and at 24 hours interval. The decellularized buffalo omasal laminae showed two protein bands (thick), one of 31 KDa and another of 116 KDa at 24 hours, and one protein band (thin) of 31 KDa at 48 hours is presented in Fig. 6.9A and B. At 48 hours, majority of the proteins removed and leave only the collagen [13].

SDS-PAGE analysis for native buffalo omasum laminae revealed presence of different proteins and after decellularization of it, the majority of the proteins were removed and only two protein bands (thick), one of 31 kDa for 24 hours and another of 116 kDa, and one protein band (thin) of 31 kDa at 4 hours were present. This reveals that the decellularization procedure removed most of the proteins and leaves only the collagen, which was the main component of the ECM. SDS-PAGE analysis of native rat and rabbit skin for tissue proteins revealed two protein bands (one thin and one thick) above 66 kDa [23].

6.3 Preparation of acellular matrix from goat omasum

Omasum of goat was collected from the institutional abattoir and was cleaned thoroughly by washing in sterile physiological normal saline. The omasum was cut into 2×2 cm^2 size pieces and placed in sterile 1× PBS containing antibiotic (penicillin, streptomycin, and amphotericin) for 2 hours at room temperature. The

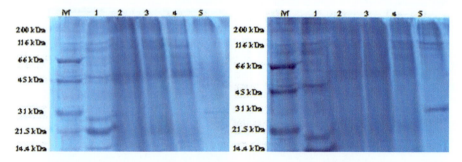

Figure 6.9 SDS-PAGE of buffalo omasum laminae at (A) 24 h and (B) 48 h posttreatment.

omasum was made acellular by treating with 1% sodium dodecyl sulfate (SDS) and 1% sodium deoxylate (SDC) in wrist action shaker for 96 hours. To assess the decellularization, samples were collected at 0-, 24-, 48-, 72-, and 96-hour time intervals. After completion of procedure the tissues were thoroughly washed with $1 \times$ PBS and stored at $-20°C$ in $1 \times$ PBS containing antibiotics. The tissue samples were subjected to histological examination, DNA quantification, calorimetric protein estimation, and SDS-PAGE analysis. For histological examinations, the samples were kept in 10% (v/v) neutral buffered formalin (NBF) solution and for other parameters evaluation, tissue samples were preserved in PBS at $-80°C$ [24].

6.3.1 Microscopic observations

The samples were cut into small pieces and fixed in 10% (v/v) neutral buffered formalin (NBF) for 48–72 hours. The tissues were processed in routine manner and 4 μm thick sections were cut and stained with H&E as per the standard protocol. The stained sections of were examined for degree of decellularization and arrangement of collagen fibers. Histological evaluation of omasum showed that the cellular component embedded between collagen matrices decreased at different rate when compared both biological detergents. H&E-stained slides reflected continuous and faster decellularization in 1% SDS and this rate of decellularization was slow in case of 1% SDC. Samples at 48 hours of different tissues were satisfactory with 90% decrease in cellularity. Masson trichome staining of different samples selected on the basis of H&E staining revealed quite decrease in collagen density at 48 hours when compared with collagen of native tissue (Figs. 6.10 and 6.11) [24].

6.3.2 DNA quantification

DNA was extracted from both from native and decellularized tissue using DNeasy Blood and Tissue Kit catalog number 69504 (Qiagen, Germany) and quantified as per the method reported by Pellegata et al. [25]. The results of DNA quantification are presented in Table 6.1. Omasal tissue showed fast deterioration in cellular components treated with biological detergents and histopathological examination was also supported by quick loss in cellular and collagen microstructures. The decrease in DNA contents was significant [24].

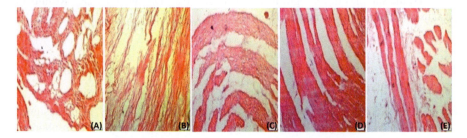

Figure 6.10 Histological picture of goat omasum treated with 1% SDS at different time intervals: (A) native omasum, (B) 24 h, (C) 48 h, (D) 72 h, (E) 96 h posttreatment.

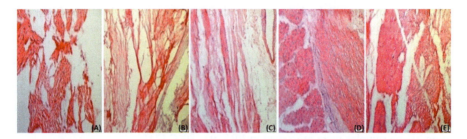

Figure 6.11 Histological picture of goat omasum treated with 1% SDC at different time intervals: (A) native omasum, (B) 24 h, (C) 48 h, (D) 72 h, (E) 96 h posttreatment.

Table 6.1 Mean ± SD of of DNA quantification (ng/μL).

Hours	Caprine omasum	
	(1% SDS)	(1% SDC)
Native	128.70 ± 0.30^b	128.70 ± 0.30^b
24	21.93 ± 1.43a**	43.23 ± 0.75d**
48	16.93 ± 0.11a**	31.43 ± 0.51e**
72	17.63 ± 2.65c**	25.86 ± 0.23d**
96	16.93 ± 4.90c**	27.26 ± 3.75d**

Different superscripts indicated mean value differ significantly at $P < .05$ between groups. **Mean value differ extremely significantly $P < .01$ within the group.

6.3.3 Calorimetric protein estimation

Tissue homogenization buffered was prepared and 5 mg of native and decellularized omasal samples were homogenized in this buffer by electric homogenizer. All samples were vortexed and centrifuged at 10,000 rpm for 5 minutes and supernatant was collected in separate microcentrifuge tubes. For direct calorimetric protein estimation, protein for each sample was estimated with Eppendorf Bio Spectrometer at 280 nm. The results of protein estimation are presented in Table 6.2. Protein content of native versus decellularized samples was not differed significantly. At 24 and 48 hours, the samples differ significantly between 1% SDS and 1% SDC. Results of protein estimation represented in Table 6.2 [24].

6.3.4 Sodium dodecyl sulfate polyacrylamide gel electrophoresis analysis

For SDS-PAGE, 25 μL of sample was taken and marked properly in predesigned order. SDS-PAGE analysis was carried out in a vertical mini gel electrophoresis apparatus (Bio-Rad Mini Protean) as per standard procedure. The results of SDS-

Table 6.2 Mean ± SD of protein estimation.

Hours	Caprine omasum	
	(1% SDS)	(1% SDC)
Native	1.01 ± 0.01[d]	1.02 ± 0.00[d]
24	0.48 ± 0.05[b]	0.40 ± 0.02[a]
48	0.45 ± 0.03[b]	0.39 ± 0.07[a]
72	0.49 ± 0.06[a]	0.34 ± 0.01[a]
96	0.32 ± 0.00[a]	0.31 ± 0.00[ab]

Different superscripts indicated mean value differ significantly at $P < .05$ between groups.

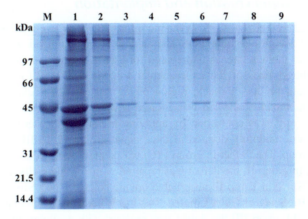

Figure 6.12 SDS-PAGE of goat omasum. Lane M—broad range SDS-PAGE prestain marker (BioRad), Lane 1—native goat omasum, Lane 2—1% SDS treatment for 24 h, Lane 3—1% SDS treatment for 48 h, Lane 4—% SDS treatment for 72 h, Lane 5—1% SDS treatment for 96 h, Lane 6—1% SDC treatment for 24 h, Lane 7—1% SDC treatment for 48 h, Lane 8—1% SDC treatment for 72 h, Lane 9—1% SDC treatment for 96 h.

PAGE analysis are presented in Fig. 6.12. Omasal tissues showed 116, 45, and 40 KDa bands in native tissue, which was apparent at 24 hours with 1% SDS not in 1% SDC. At 48 hours omasal tissue showed thin band of above protein [24].

6.4 Experimental evaluation of acellular buffalo omasum in wound healing in rat model

Developed acellular buffalo omasum matrices were experimentally evaluated for its healing potential in rat model and compared with commercially available collagen sheet.

6.4.1 Animals and ethics statement

Protocols for this study were approved by the Institute Animal Ethics Committee of the Indian Veterinary Research Institute, Izatnagar, Uttar Pradesh, India according to guidelines for the care and use of laboratory animals. Thirty-six clinically healthy Sprague Dawley rats of either sex, weighing from 250 to 300 g, and of 3 to 4 months of age were used for evaluating healing potential of acellular buffalo omasal matrix. Animals were housed individually in cages, provided with commercial diet and water ad libitum in a temperature- and humidity-controlled environment. The animals were acclimatized to approaching and handling for a period of 10 days before the start of study [26].

6.4.2 Skin wound creation and implantation

Animals were anesthetized using xylazine (5 mg/kg body weight) and ketamine (50 mg/kg body weight) combination. The animals were restrained in sternal recumbency and dorsal thoracic area was prepared for aseptic surgery. Using a sterile plastic template, the vertices of the experimental wounds of 20×20 mm^2 dimensions were outlined on the dorsothoracic region of the rats. A full thickness skin defect including the *panniculus carnosus* was excised with a #11 BP blade on each animal. Hemorrhage, if any, was controlled by applying pressure with sterile cotton gauze. The animals were randomly divided in three equal groups (I, II, III) having 12 animals in each group. The defect in group I (C) was kept as open wound and was taken as control (C). In group II, the defect was repaired with commercially available collagen sheet (b-CS) (Skin Temp II, absorbable bovine collagen, polyethylene nonwoven mesh, Human BioSciences India Limited, 142/143 P Vasna Chacharvadi, Ahmedabad 382213, India). In group III, the defect was repaired with acellular omasal matrix of buffalo origin (b-AOM). The scaffolds were secured to the edge of the skin wounds with eight simple interrupted 4/0 nylon sutures. The repaired wounds were covered with parafin surgical dressings. After recovery from anesthesia, rats were housed individually in properly disinfected cages. Further, antimicrobial treatment (Enrofloxacin 10 mg/kg intramuscularly once a day) was continued for 5 days and meloxicam (0.3 mg/kg intramuscularly once a day) for 3 days [26]. The healing potential was evaluated on the basis of the following parameters:

6.4.3 Wound area and wound contraction

Wound area of all groups was measured by tracing its contour using transparent sheets with graph paper on postoperative days 0, 3, 7, 14, 21, and 28. The area (mm^2) within the boundaries of each tracing was determined and percent contraction was measured by a rate of wound reduction, as well as percent reduction in wound area.

Mean ± SD of the total wound area (mm^2) of the skin wounds at different time intervals is presented in Fig. 6.13. A gradual decrease in wound area (mm^2) was observed in all the groups during the entire observation period except in group I where a significant increase in wound area was recorded on day 3 postoperatively.

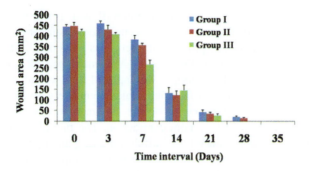

Figure 6.13 Mean + SD of the wound area (mm^2) at different time intervals in various groups.

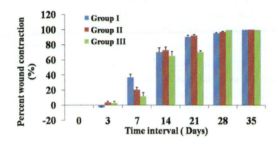

Figure. 6.14 Mean + SD of percent contraction of wound (%) at different time intervals in various groups.

The decrease in the wound area was significant ($P < .01$) at all the time intervals in all the groups. However, the rate of decrease in wound area was faster in group II as compared to the other groups on day 14. On day 21, group III showed significantly reduced ($P < .05$) wound area as compared to other two groups. On day 28, complete healing was observed in group III whereas groups I and II were yet to be healed completely. Complete healing was observed on day 45 in groups I and II [26].

A gradual increase in percent wound contraction was observed in all the groups during the observation period. Percent wound area contraction (mean ± SD) at different time intervals in different groups are presented in Fig. 6.14. Although the original wound area created was 2×2 cm^2 (400 mm^2) but the wounds were expanded to various extent in different groups. In group I, no matrix was used and the wound healed as open wound. In group II, commercially available collagen sheet was applied, as it is very fragile in nature and suturing was not possible, but wound area was not expanded to that extent as in group I. In group III, acellular omasal matrix was sutured over the wound area. As the healing progressed, the wound area decreased significantly ($P < .05$) at different time intervals in all the groups [26].

Highly significant ($P < .05$) difference in wound contraction was observed between groups I and III on day 7. Percent wound contraction was highly significant ($P < .01$) from day 14 onward between the groups. There was no significant difference ($P > .05$) in percent contraction between all the groups in later stages (on day 14 onward).

On day 28, complete healing was recorded in group III animals where acellular buffalo omasum laminae was used. The rate of wound contraction was faster in group III.

As the healing of the wound progressed, there was reduction in the wound area at different time intervals in all the groups. Maximum percent contraction was recorded in group I as no biomaterials was applied and wound healed as open wound. The wound in animals of this group healed completely by 35 days leaving a large scar indicating the existence of severe contraction. The acellular matrix implanted group healed with minimum contraction. The rate of healing with minimum contraction was more in acellular buffalo omasal grafted group than collagen sheet. The commercially available collagen sheet was simply applied over the dorsum of the wound created in the animals as the suturing with wound edges was not possible due to brittle nature of sheet [26].

After trauma, a centripetal displacement of the wound edges occurs that facilitates its closure and termed as wound contraction. This process is carried out by myofibroblasts that contain ß-actin from smooth muscle and is mediated by contractile forces produced by granulation tissue from wound [27]. The wound healing rate defined as the gross epithelialization of the wound bed. The wound contraction was assessed by percent retention of the original wound area as reported by Schallaberger et al. [28]. The wound contraction has been used to monitor wound healing.

6.4.4 Gross observations/planimetry

Color photographs were taken on postimplantation days 7, 14, 21, and 28 with digital camera at a fixed distance. Shape, irregularity, and color of the lesion were determined. The color digital photographs of wounds repaired with different biomaterials at different postimplantation days are presented in Fig. 6.15. In control group, on day 3, wounds were covered with soft and fragile pinkish mass above the wound surface area with serous exudates oozing out. On day 7, the surface became more desiccated and necrosed with some amount of exudates. By day 14, the wound size markedly decreased and a thick crust developed, which starts detaching and leaving a raw granular pink tissue. By day 28, a large scar was observed and wound healed completely. In collagen sheet group on day 3, the top layer of collagen sheet was covered with soft and fragile pinkish mass along with necrosed margin and moderate amount of exudate. By day 7, the top layer of collagen sheet was dry, brownish, and major portion of sheet was sloughed off. By day 14, the upper layer became more desiccated and shriveled. On day 28, the wound healed partially leaving a large scar. In omasal matrix group on day 3, the top layer of acellular matrix appeared chocolate brown in color, which converted into dark brown on day 7. On day 14, the top layer was dry and was seen in stage of detachment from the underlying tissue as only the ends sutured holding it in place. After day 14, the dried up top layer was sloughed off completely exposing the healing tissue. By day 21, wound margins were healed leaving a small central area. By day 28, the wound healed up completely with no scar and the implanted area appeared similar to the normal skin [26].

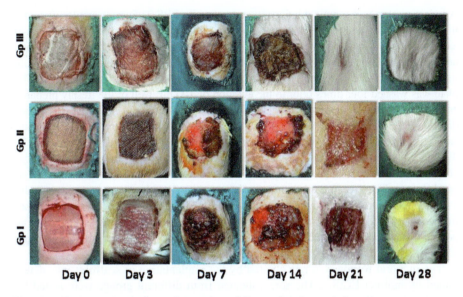

Figure 6.15 Gross observations of wounds at different time intervals in various groups.

The graft-assisted healing is an important strategy for treating full thickness skin wounds. The omasal scaffold was rich in natural biomolecules like laminin, elastin, collagen IV, heparin sulfate, and glycosaminoglycans (GAGS) and was used as a xenograft, with excess cell proliferation at early phases and acceptable collagen deposition in the later remodeling phases [6]. After application acellular buffalo omasum matrix, the color of the implanted samples changed from white to dark brown and finally to dark. The decrease in vascularization and continuous loss of moisture contents of the graft lead to change in its color and desiccation in due course of time [20,29,30]. Upper matrix layer detached from the wound in form of scar and the underlying layer of omasal matrix was completely absorbed, and newly formed granulation tissue within the matrix covered the whole surface of the wound.

6.4.5 Immunological observations

6.4.5.1 Humoral response

For conducting ELISA procedure, serum was collected on days 0, 21, and 28 from each rat. Harvested sera were assessed for the extent of antibodies titer generated toward the implant. Briefly, 96-well flat bottom polystyrene plate was coated with 0.25 mg protein (derived from implant material) in 100 μL/well of 0.05 M sodium carbonate/sodium bicarbonate coating buffer (pH 9.6) per well. The plate was covered with aluminum foil and incubated at 4°C overnight and washed four times using washing buffer (PBS with 0.05% Tween-20; PBS-T; pH 7.4). The wells were blocked with 5% skimmed milk powder (200 μL/well) for 3 hours and washed four times.

All plate washes and sample dilutions were made with PBS-T. Serum samples were run in triplicate. One hundred microliter (100 μL) of diluted (1:100) test serum was added to each of antigen-coated well and incubated for 2 hours at 37°C. The plate was washed and reincubated for 2 hours with 100 μL/well of the secondary antibody (goat antirat immunoglobulin G conjugated with horseradish peroxidase) in a 1:30,000 dilution. The plate was washed four times prior to the addition of 100 mL/well of the freshly prepared enzyme substrate solution (citric acid 0.1%, Na_3PO_4 0.1 M, 0.4 mg/mL o-phenylene diamine, 0.03% hydrogen peroxide) and was incubated for 15 minutes at 37°C and stopped with 3 M solution of sulfuric acid. Absorbance (OD) values were recorded at 492 nm wavelength using an automated ELISA plate reader (ECIL, Hyderabad, India) [13]. The antibodies titer present in serum samples harvested on day zero was taken as basal values. The free protein contents of the native buffalo omasum, commercially available collagen sheet, and decellularized buffalo omasum were estimated as per method of Lowry et al. (1951) using bovine serum albumin (BSA) as a standard. The protein contents of native buffalo omasum, decellularized buffalo omasum, and collagen sheet are 21.13, 2.38, and 38.23 mg/mL, respectively [26].

The humoral response in rats elicited by the implantation of scaffolds was determined by indirect ELISA. The sera collected from different groups were tested by ELISA using collagen sheet (38.23 mg/mL) and native buffalo omasum (21.13 mg/mL) as coating antigen. The sera collected on days 0, 21, and 28 postimplantation from different implanted groups were evaluated for antibody titer generated. The sera collected from animals implanted with different matrices were tested in ELISA to assess B-cell immune response. The scaffolds specific antibodies were expressed as mean ± SD absorbance at 492 nm wavelength (OD 492). ELISA reaction is presented in Fig. 6.16.

The hyperimmune serum rose against the collagen sheet and native buffalo omasum was used as standard positive control. A fixed dilution of the antibody of 1:40 (collagen sheet) and 1: 10 (native buffalo omasum) was used throughout the experiment. The absorbance at this dilution was found optimum. Below this dilution, the absorbance was too low and above this dilution there was no appreciable change in absorbance. The levels of antibodies present in the serum prior to implantation

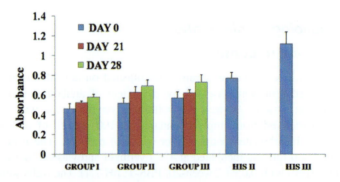

Figure 6.16 Mean ± SD of absorbance values at 492 ηm wavelength (OD 492) of ELISA reaction at days 0, 21, and 28 in different groups.

(0 day) were recorded as basal values. In control group, no rise in antibody titer was observed as the wound healed denuded without any matrix. In the treatment groups (groups II and III), there were relative rise in antibody titer from day 0 to day 28 postimplantation. The antibody titer was significantly higher ($P < .05$) in the groups II and III on day 28 as evidenced by higher absorbance values when compared with day 0 values [26].

6.4.5.2 Cell-mediated immune response

The cell-mediated immune response was assessed by 3-(4,5-dimethylthiazol-2-yl)-2,5- diphenyltetrazolium bromide (MTT) colorimetric assay. The stimulation of rat lymphocytes with phytohemagglutinin (PHA) was considered as positive control, whereas unstimulated culture cells were taken as negative control. Blood collected from rats with implants (b-CS and b-AOM) on postimplantation days 0 and 28 was used for lymphocyte culture as per standard method. Briefly, 1 mL of blood was aseptically collected in a heparinized tube and mixed with equal volume of $1 \times$ PBS. It was carefully layered over 2 mL of lymphocyte separation medium (Histopaque 1077) and centrifuged at 2000 rpm for 30 minutes. The buffy coat was harvested in a fresh tube and two washings were done with $1 \times$ PBS at 1000 rpm for 5 minutes. Supernatant was discarded and pellet was resuspended in RPMI1640 growth medium supplemented with 10% fetal bovine serum. The cells were adjusted to a concentration of 2×10^6 viable cells/mL in RPMI1640 growth medium and seeded in 96-well tissue culture plate (100 μL/well). The cells were incubated at 37°C in 5% CO_2 environment. Cells from each rat were stimulated with antigens (b-CS and b-AOM) (10–20 μg/mL) and PHA (10 μg/mL). A T-cell mitogen in triplicates and three wells were left unstimulated for each sample. After 45 hours, 40 μL of MTT solution (5 mg/mL) was added to all the wells and incubated further for 4 hours. The plates were then centrifuged for 15 minutes in plate centrifuge at 2500 rpm. The supernatant was discarded, plates dried, and 150 mL dimethyl sulfoxide was added to each well and mixed thoroughly by repeated pipetting to dissolve the formazan crystals. The plates were immediately read at 570 nm with 620 nm as reference wavelength. The stimulation index (SI) was calculated using formula:

Stimulation index (SI) = OD of stimulated cultures/ OD of unstimulated cultures

The cell-mediated immune response toward implanted scaffolds was assessed by MTT colorimetric assay. The stimulation of rat lymphocytes with concanavalin A (Con A) was considered as positive control, whereas unstimulated cultured cells were taken as negative control. Mean ± SD stimulation index (SI) values of different groups on 28 days postimplantation are presented in Fig. 6.17.

The animals of groups II and III showed significant ($P < .05$) amount of stimulation against the collagen sheet antigen and native buffalo omasum antigen. Significantly higher SI values ($P < .05$) with Con A were observed in all the groups. The T-cell response was lowest in the group I (control) and higher in groups II and III, as evidenced by lower SI values in comparison to other groups. Significant

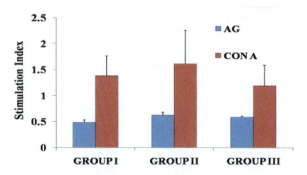

Figure 6.17 Mean ± SD of stimulation Index (SI) on stimulation with collagen sheet, native buffalo omasum antigen and Con A in various groups.

difference ($P < .05$) in SI value between groups II and III were observed as compared to groups I (control group). The stimulation was lowest with acellular buffalo omasal matrix (group III) followed by commercially available collagen (group II) [26].

The nature and degree of immunological response to a foreign material is a crucial variable affecting the outcome and success or failure of implanted biomaterial. The immunological response to biomaterial depends on nature of processing and presence of potentially foreign antigen to various lymphocyte populations. The biomaterial could stimulate an antibody response, a cell-mediated antibody sensitization, or minimal to no response. The immune response to xenogenic transplantation included both natural and induced humoral components [31]. The biological scaffolds are mainly composed of mammalian ECM and can be used for the reconstruction of various tissues and organs. These scaffolds are typically allogic or xenogenic in origin and derived from tissues such as gallbladder, dermis, pericardium, diaphragm, small intestine submucosa, etc. The cells in the ECM have class I and II histocompatibility antigens capable of eliciting rejection reactions [31]. Therefore, if these substances are eliminated from ECM, rejection reactions can be prevented. However, complete elimination of allo-antigens is considerably difficult to perform and verify [32]. Conventionally, two strategies have been used to reduce graft rejection. One of these strategies is to reduce the immune reaction of the hosts and the other is to reduce the antigenicity of allografts or xenografts [33]. In the present study, the antigenicity of the graft was reduced by making the tissue acellular. The acellular grafts were less immunogenic having better tolerance by allogenic hosts and equally effective as isograft [34]. In addition, removal of cellular components may reduce the antigenicity and potential for protracted inflammation. The tissue was processed by decellularization and/or crosslinking to remove or mask antigenic epitopes and DNA [16]. The processing method for the preparation of biological scaffolds plays an important role in determining the host response.

An immune response against nonself and self-antigens is initiated by presentation of the antigen in a suitable form of T cells. The antigen can only be presented to T cells in the context of molecules of MHC [35]. Hence, practically each nucleated cell of the body is able to present antigen, first by virtue of a constructive MHC class I expression and second by a de novo expression of MHC class II molecules on the surface of the cell. It is well known that when an antigen enters the body, two different types of immune response may occur. The first type, known as the "humoral immune response," involves the synthesis and release of free antibody into the blood and other body fluids. The second type of immune response, known as the "cell-mediated immune response," involves the production of "sensitized" lymphocytes, which are the effectors of this type of immunity.

The ELISA was performed to check the extent of antibody generated toward the graft components. The absorbance values were taken as measure to compare the magnitude of immune response. The antibody titer response in treated groups was slightly higher on days 21 and 28 as compared to day 0. The least immune response was in animals of control group, where no graft was used and wound healed denuded or open wound. Whatever, the immune response present in control group might be due to bacterial and viral contamination in the wound as it was open. Immune response in group II animals might be due to some immunogenic nature of decellularized ECM, and acellular omasal matrix group (group III) elicits more or less similar immune response as animals of group II. Zheng et al. [36] observed the inflammatory reaction caused by remnant porcine DNA with biologic scaffold materials after decellularization following the implantation of porcine derived scaffolds for orthopedic implantations. The host response to biologic scaffold materials composed of ECM involves both the innate and acquired immune system, and the response is affected by source of raw material/tissue from which the ECM is harvested and the processing steps.

Lymphocyte proliferation assay (LPA) measures the ability of lymphocytes placed in short-term tissue culture to undergo a clonal proliferation when stimulated in vitro by a foreign molecule (antigen/mitogen). The lymphocytes consist of various subpopulations with distinctive functions, which play important roles in immune response [37]. The activation and proliferation of these subpopulations can be achieved by treating them with antigens. Th1 lymphocytes produce cytokines such as IL-2, IFN-γ, and TNF-β leading to macrophase activation, stimulation of complement fixing Ab isotypes, and differentiation of $CD8^+$ cells to a cytotoxic T cells [38]. Activation of this pathway is associated with both allogenic and xenogenic transplant rejection [18]. Th2 lymphocytes produce IL-4, IL-5, IL-6, and IL-10, cytokines that do not activate macrophases and that lead to production of non-complement fixing Ab isotypes. Activation of Th2 pathway is associated with transplant acceptance [39]. The graft rejection is usually mediated by activity of T cells, especially cytotoxic T cells.

Lowest stimulation index was observed with animals treated as open wound (Group I) followed by acellular buffalo omasal matrix (Group III) and then by collagen sheet-treated group (Group II). The acellular buffalo omasal matrix showed minimum antigenicity. The ability of this antigen to stimulate the lymphocytes in vitro may be attributed to the fact that on treatment with biological detergent, the bonds between protein molecules are broken and result into a change from quaternary and tertiary structure to primary and secondary structures. Therefore, the acellular antigen had ability to trigger CMI response in host because of presence of shorts peptide fragments, which can be presented to the immune system by MHC class II pathway and stimulate the $CD4^+$ lymphocytes.

6.4.6 Histological observations

The biopsy specimen from the implantation site was collected on 7, 14, 21, and 28 days postimplantation days for histopathological evaluation. Biopsy samples were fixed in 10% formalin saline. The samples were then processed for paraffin embedding technique to get 5 micron thick paraffin sections. The sections were stained by H&E as per standard protocol. The host inflammatory response, neovascular tissue formation (fibroblasts, fine capillaries), deposition of neocollagen, and penetration of host inflammatory responses in the implanted matrix were evaluated as per Schallberger et al. [28]. Special staining for collagen fibers was done by using Masson's Trichrome stain [40] to observe the deposition of collagen fibers and scoring was done as per the method of Borena et al. [41]. The group having less histopathological scores was considered the best. The histopathological observations are presented in Figs. 6.18 and 6.19.

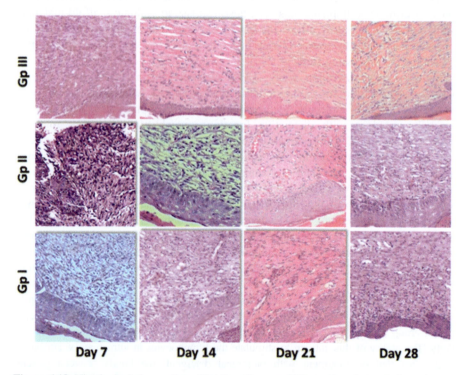

Figure 6.18 Histological observation of healing tissue at different time intervals in various groups (H&E stain, × 200).

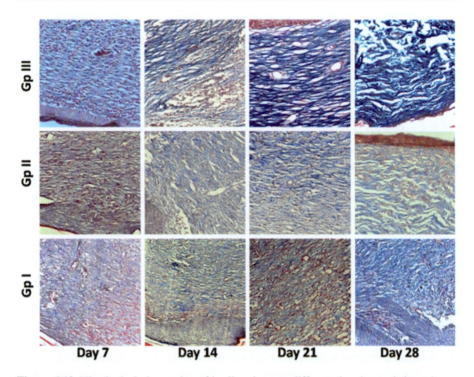

Figure 6.19 Histological observation of healing tissue at different time intervals in various groups (Masson's trichrome stain, ×200).

6.4.6.1 Group I (open wound)

On day 7, the wound was covered with necrotic debris and admixed with high degree inflammatory cell infiltration, edema, and congestion. The underlying stoma contained severely proliferated fibroblasts and severe neovascularization along with spilled neutrophils. The angiogenesis increased, and there was initiation of epithelialization at the margin of wound area covering the granulation tissue, and surface epithelium was thicker than the normal skin. The granulation tissue (fibroblast and neovascularization) was very wide and showing presence of severe neutrophils and mononuclear cells (lymphocytes and macrophages). The collagen fibers were less dense, thin, and worst arranged. The mean histopathological score was 33 at this stage. On day 14, moderately proliferated fibroblasts and severe neovascularization along with moderate degree of inflammatory cell infiltration were observed. The granulation tissue was wide, and there was initiation of epithelialization, which was thicker than normal skin. The collagen fibers were less dense, thin with worst arrangement. The score recorded was 30. On day 21, inflammatory cells infiltration was mild with moderate neovascularization and moderate proliferation of fibroblasts. There was incomplete surface epithelization, which was thicker than normal skin. The granulation tissue was moderately wide. The collagen fibers were dense and thick with worse arrangement and the histopathological score was 25 at this stage. On day 28, there was complete formation of surface epithelium, but slightly thicker than normal. Mild inflammation with neovascularization and mild fibroblasts proliferation was observed. The Granulation tissue was moderately wide. The collagen fibers were dense, thick but worse arrangement was observed. The surface epithelium was complete but more contraction was seen in epithelium. The histopathological score was 19 at this stage [26].

6.4.6.2 Group II (wound with commercially available collagen sheet)

On day 7, severe fibroblasts proliferation along with severe neovascularization was observed. Inflammatory cells infiltration was severe. There was necrosis and sloughing of the superficial layer, and initiation of the formation of surface epithelium was observed at the edges of wound. The surface epithelium was thicker than normal skin. The granulation tissue width was moderately wide. The collagen fibers were less dense, thin with worst arrangement. The histopathological score was 32. On day 14, the deeper layer of dermis revealed deposition of new collagen fibers in which fibroblasts were dispersed. The granulation tissue was moderately wide, and there was initiation of epithelialization with moderate thickness. The inflammatory cells infiltration was reduced to mild degree with moderate neovascularization and moderate proliferation of fibroblasts. The collagen fibers were dense, thin with worst arrangement. The histopathological score was 26 at this stage. By day 21, epithelialization was incomplete with slightly thicker skin. The inflammatory changes and neovascularization reduced to mild degree with mild fibroblasts proliferation. The granulation tissue was moderately wide. The collagen tissue was dense, thick, and better arranged. The histopathological score was further reduced to 19 at this stage. On day 28, superficial epithelialization was complete, with thickness resembling normal skin. Inflammatory cells were absent and neovascularization reduced to mild form with slightly wide granulation tissue. The collagen fibers were dense, thick, and better arranged. No hair follicles and skin glands were observed. The epithelialization was complete with contraction. The total histopathological score was 15 at this stage [26].

6.4.6.3 Group III (wound with acellular buffalo omasum laminae)

On day 7, the surface epithelialization was initiated and surface epithelium was thicker than normal skin. Moderate inflammatory cells infiltration along with moderately wide granulation tissue was very prominent in the dermis. Moderate fibroblasts proliferation and neovascularization was observed. The necrosis and sloughing of the superficial matrix acellular matrix were observed. The superficial acellular matrix was still present at the wound site and epithelialization was started at the edges of wound area. The newly formed keratinocyte were seen in both matrix induced healing reactions, and the collagen fibers were further organized into thick bundles. The collagen fibers were dense and thin along with worse arrangement. The histopathological score was 27. On day 14, the surface neoepithelialization was incomplete and slightly thicker than normal skin. Mild neovascularization along with slightly wide granulation tissue width was observed at this stage.

The fibroblasts proliferation and inflammatory cells infiltration were moderate. The dense and thick collagen formation with better arrangement of collagen fibers was observed. The total histopathological score was 20 at this stage. On day 21, the collagen formation was appreciable. The hair follicles and skin glands were observed. The surface epithelization was complete with thickness resembling to normal skin. The histopathological score was 14. By day 28, there was complete formation of surface epithelium with no contraction. The dermis was well developed with newly formed blood vessels and well-organized collagen fibers. The process of remodeling was evident with reorganization and disappearance of collagen. The densely packed thick collagen bundles were replaced by loosely placed thin strands and the histopathological score was 9 [26].

The full thickness skin wounds are characterized by a complete destruction of epithelial regenerative elements that reside in the dermis. In these wounds, only epithelialization from the edges of the wound is possible. The critical factor for epithelialization is the size of the wound. The full thickness skin wounds in human being having more than 1 cm diameter need skin grafting to prevent scar formation [42]. The acellular graft should not stimulate immune response, but it covers and protects the wound bed, enhances the healing process, lessens the pain of the patient and results in little or no scar formation. In full thickness skin grafting, many grafts became necrotic due to an impaired formation of new blood vessels.

The full thickness skin wounds healing occur by granulation tissue formation, contraction, and epithelialization [43]. The epithelialization occurs by migration of undamaged epithelial cells from the wound margins across the granulation bed [44]. The exogenous collagen supplementation enabled faster migration of cells that are involved in cutaneous wound healing. Since the exogenous collagen is molecular in nature [45] and supplies endogenous collagen in vivo, it readily integrates with the wound tissue and facilitates the attachment, migration and proliferation of cells on the wound site [46]. On day 7, minimum histological score (score = 27) was noted in group III, and this group showed better wound healing in comparison to groups I and II. The buffalo omasal ECM applied in group III showed necrosis of the superficial matrix and underlying matrix showed moderate proliferation of fibroblasts. The moderate to severe degree of inflammation, fibroplasia, and neovascularization along with spilled neutrophils was observed in all the groups, but it was minimum in group III. The early reduction of inflammation as in case of group III might facilitate the progress to the next phase of wound healing. Although inflammation is necessary for healing by fighting infection and inducing the proliferation phase, healing proceeds only after inflammation is controlled [47]. The acellular implants were invaded by the fibroblasts and underwent neovascularization without any evidence of rejection and the initialization of surface epithelium formation. The epithelialization process had started by 2 weeks in full thickness skin wounds treated with small intestinal submucosal sponges in rat model [48]. New collagen formation was not detected in any of the groups, but the graft implanted sample showed the existence of mature collagen of omasum ECM origin. Perme et al. [49] observed more proliferation of fibroblasts in the acellular matrices and the bioengineered acellular matrices. During the initial stages of wound healing, fibroblasts from the surrounding wound area are known to migrate and proliferate into the wound site and within 3−4 days and get converted to myofibroblasts [50]. The main function of myofibroblasts in a healing wound is contraction by synthesis of ECM proteins, notably, collagen types I−VI and XVIII, laminin, thrombospondin, glycoproteins, and proteoglycans for the dermal repair. The myofibroblasts usually go on increasing from the inflammatory stage (3−4 days). During the onset of healing, the proliferating fibroblast starts to synthesize collagen and the total collagen content increases preferentially.

On day 14, least score observed in group III (20) followed by group II may be due to early acceptance of matrix. The sloughing of upper layer of matrix was observed in all the groups except control group where wound remained open. It may be either due to the desiccation of the graft in high environmental temperature or due to an impaired formation of new blood vessels [51]. The myofibroblasts proliferate till the end of proliferation phase (14−15 days). Meanwhile, the underlying granulation tissue increased in mass that pushed up the superficial acellular grafts upward. The animals of groups II and III showed well-formed collagen and neovascularization with superficial epithelization but these processes were particularly faster and earlier in group III.

On day 21, the least score was observed in group III (14). The epithelialization was also more similar to the normal skin in group III. In the remodeling phase (around 21−30 days), these myofibroblast underwent apoptosis [52]. At 21 days of postimplantation, the acellular matrix group showed higher collagen synthesis as compared to the control group as appreciated by the Masson's Trichrome staining. Purohit [29] on day 21 found that the acellular dermal matrix throughout the width and length was replaced by mature collagenous connective tissue in experimentally created wounds in rabbits.

On day 28, postimplantation complete healing was observed in all the groups, but control group healed completely leaving abundant scar tissue. Similarly, in group II, the contraction and scar formation were present. Kim et al. [48] reported that full thickness skin wounds in rats treated with small intestinal submucosal sponges were completely covered with thin layer of epidermis by 4 weeks. In group III, the collagen fiber was best arranged and fibers oriented parallel to the skin surface suggesting a better repair of the damaged tissue. Dayamon et al. [23] observed best-arranged and parallel-oriented collagen fibers at day 28. Marked fibroblastic response associated with an abundant new collagenous fibrous tissue was observed in these groups. In acellular group, hair follicles and skin glands could be detected as in case of normal skin indicating the culmination of repair process. Here the healing was with minimum contraction and scar formation.

6.5 Conclusion

In this study, protocols for decellularizing the buffalo omasum were optimized. The omasum was made acellular using ionic detergent (sodium dodecyl sulfate), nonionic detergent (triton X-100) zwitterionic detergent (Tri (n-butyl) phosphate), and enzyme (trypsin) in 0.5% concentration and evaluated by gross, microscopic observations, DNA quantification, and SDS-PAGE analysis. Acellular matrix was also prepared from goat omasum using two different ionic biological detergents. Treatment with 1% sodium dodecyl sulfate for 48 hours resulted in completely acellular matrix as evaluated by histology, calorimetric protein estimation, DNA quantification, and SDS-PAGE analysis. Efficacy of acellular buffalo matrix was tested in reconstruction of full thickness skin wounds in rat model and was compared with commercially available bovine collagen sheet. The wounds were evaluated based on clinical, macroscopical, immunological, and histopathological parameters. The acellular buffalo omasum matrix showed better healing potential for the repair of full thickness skin wound defects in rats.

References

[1] Pandarinathan C, Sajithlal GB, Chandrakasan G. Influence of aloe vera on collagen characteristics in healing dermal wounds in rats. Mol Cell Biochem 1998;181:71—6.
[2] Shevchenko RV, James SL, James SE. A review of tissue-engineered skin bioconstructs available for the skin reconstruction. J R Soc Interface 2010;7:229.
[3] Zhong SP, Zhang YZ, Lim CT. Tissue scaffolds for skin wound healing and dermal reconstruction. Nanomed Nanobiotech 2010;2:510—25.
[4] Choi JS, Yang HJ, Kim BS, Kim JY, Yoo B. Human extracellular matrix (ECM) powder for injectable cell delivery and adipose tissue engineering. J Control Release 2009;139:2.
[5] Zhang X, Deng Z, Wang H, Guo W, Li Y. Expansion and delivery of human fibroblasts on micronized acellular dermal matrix for skin regeneration. Biomaterials 2009;30:26—66.
[6] Lun S, Irvine SM, Johnson KD, Fisher NJ, Floden EW, Negron L. A functional extracellular matrix biomaterial derived from ovine forestomach. Biomaterials 2010;31:4517—29.
[7] Irvine SM, Cayzer J, Lun S, Floden EW. Quantification of in-vitro and in-vivo angiogenesis stimulated by ovine forestomach matrix. Biomaterials 2011;32:6351—61.
[8] Ward B.R., Johnson K.D., May B.C.H. (2009) Tissue scaffolds derived from forestomach extracellular matrix. US Patent No. 12/512, 2009; 835.
[9] Rieder E, Kasimir MT, Silberhumer G, Seebacher G, Wolner E, Simon P. Decellularization protocols of porcine heart valves differ importantly in efficiency of cell removal and susceptibility of the matrix to recellularization with human vascular cells. J Thoracic CardiovascSurg 2004;127:399—405.
[10] Lin P, Chan WC, Badylak SF, Bhatia SN. Assessing porcine liver derived biomatrix for hepatic tissue engineering. Tissue Eng 2004;10:1046—53.
[11] Woods T, Gratzer PF. Effectiveness of three extraction techniques in the development of a decellularized bone-anterior cruciate ligament-bone graft. Biomaterials 2005;26:7339—49.
[12] Gamba PG, Conconi MT, Lo Piccolo R, Zara G, Spinazzi R, Parnigotto PP. Experimental abdominal wall defect repaired with acellular matrix. Pediatr Surg Int 2002;18:327—31.
[13] Singh P. Evaluation of omasum derived extracellular matrix for repair of full thickness skin wound in rats [MVSc thesis]. Submitted to Deemed University Indian Veterinary Research Institute; Izatnagar 243122, Uttar Pradesh, India; 2015.

[14] Singh P, Sharma AK, Kumar N, Tamil Mahan P, Sangeetha P, Singh AK, et al. Process development to prepare an acellular matrix from bovine omasum using biological detergents and enzymes. Trend Biomat. Artif Organs 2017;31(1):2−8.
[15] Seddon AM, Curnow P, Booth PJ. Membrane proteins, lipids and detergents: not just a soap opera. Biochem Biophys Acta 2004;1666:105−17.
[16] Gilbert TW, Sellaro TL, Badylak SF. Decellularization of tissues and organs. Biomaterials 2006;27:3675−83.
[17] Hasan A, Kumar N, Gopinathan A, Singh K, Sharma AK, Maiti SK, et al. Bovine reticulum derived extracellular matrix (b-REM) for reconstruction of full thickness skin wounds in rats. Wound Med 2016;12:19−31.
[18] Chen F, Yoo JJ, Atala A. Acellular collagen matrix as a possibet al.iomaterial for urethral repair. Urology 1999;54:407−10.
[19] Bader A, Schilling T, Teebken OE, Brandes G, Herden T, Steinhoff G. Tissue engineering of heart valves-human endothelial cell seeding of detergent acellularized porcine valves. Eur J Cardiothorac Surg 1998;14:279−84.
[20] Shakya P, Sharma AK, Kumar N, Remya V, Mathew DD, Dubey P, et al. Bubaline cholecyst derived extracellular matrix for reconstruction of full thickness skin wounds in rats. Scientifica 2016;2016. Available from: https://doi.org/10.1155/2016/2638371.
[21] Xu L, Shao A, Zhao Y. Quantification of residues DNA in animals-derived biological scaffold materials. J Biomed Engineering 2012;29(3):479−85.
[22] Laemmli UK. Cleavage of structural proteins during the assembly of the head of bacteriophage T4. Nature 1970;227:680−5.
[23] Mathew DD, Gangwar AK, Sharma AK, Khangembam SD, Shrivastava S, Remya V, et al. Bioengineered acellular dermal matrix for repair of full thickness skin wounds in rats. Trends Biomater. Artif Organs 2013;27(2):67−80.
[24] Raghuvanshi PDS, Kumar N, Maiti SK, Mohan D, Gautam D, Shrivastava S, et al. Preparation and characterization of caprine fore stomach matrix (CFM) for biomedical application. Trends Biomater Artific Organs 2019;33(1):9−14.
[25] Pellegata AF, Asnaghi MA, Stefani I, Maestroni A, Maestroni S, Dominioni T, et al. Detergent-enzymatic decellularization of swine blood vessels: insight on mechanical properties for vascular tissue engineering. BioMed Res Int 2013;918753. Available from: https://doi.org/10.1155/2013/918753.
[26] Singh P, Sharma AK, Kumar N, Tamil Mahan P, Singh R, Saxena S, et al. Bubaline omasal derived extracellular matrix for reconstruction of full thickness skin wounds in rats. BAO J Surgery 2017;3(Issue 2):021.
[27] Neagos D, Mitran V, Chiracu G, Ciubar R, Iancu C, Stan C, et al. Skin wound healing in a free-floating fibroblast populated collagen lattice model. Romanian J Biophys 2006;16:157−68.
[28] Schallberger SP, Stanley BJ, Hauptaman JG, Steficek BA. Effect of porcine small intestine submucosa on acute full-thickness wounds in dogs. Vet Surg 2008;37:515−24.
[29] Purohit S. Biocompatibility testing of acellular dermal grafts in a rabbit model: an invitro and in-vivo study [PhD thesis]. Submitted to Deemed University Indian Veterinary Research Institute, Izatnagar 243122, Uttar Pradesh India; 2008.
[30] Kaarthick DT, Sharma AK, Kumar N, Kumar V, Gangwar AK, Maiti SK, et al. Accelerating full-thickness dermal wound healing using primary mouse embryonic fibroblasts seeded bubaline acellular diaphragm matrix. Trends Biomater Artif Organs 2017;31(1):16−23.
[31] Ruszczak Z. Effect of collagen matrices on dermal wound healing. Adv Drug Deliv 2003;55:1595−611.

[32] Badylak SF, Gilbert TW. Immune response to biologic scaffold materials. Semin Immunol 2008;20:109−16.
[33] Rosenberg AS, Mizuochi T, Sharrow SO, Singer A. Phenotype, specificity and function of T cell subsets and T cell interactions involved in skin allograft rejection. J Exptl Med 1987;165:1296.
[34] Gulati AK, Cole GP. Immunogenicity and regenerative potential of acellular nerve allograft to repair peripheral nerve in rats and rabbits. Acta Neurochir Wein 1994;126:158−64.
[35] Townsend A, Bodmer H. Antigen recognition by class I restricted T lymphocytes. Ann Rev Immunol 1989;7:601−24.
[36] Zheng MH, Chen J, Kirilak Y. Porcine small intestine submucosa (SIS) is not an acellular collagenous matrix and contains porcine DNA: possible implications in human implantation. J Biomed Mater Res B Appl Biomater 2005;73:61−7.
[37] Otsuka H, Ikeya T, Ikano T, Karaoka K. Activation of lymphocyte proliferation by boronate containing polymer immobilized on substrate: The effect of boron content on lymphocyte proliferation. Europ Cells Mater 2006;12:36−43.
[38] Abbas AK, Murphy KM, Sher A. Functional diversity of helper T-lymphocytes. Nature 1996;383:787−93.
[39] Piccotti JR, Chan SY, Van Buskirk AM, Eichwald EJ, Bishop DK. Are Th2 helper T-lymphocytes beneficial, deleterious or irrelevant in promoting allograft survival. Transplantation 1997;63:619−24.
[40] Masson PJ. Some histological methods: trichrome staining and their preliminary techniques. J Tech Methods 1929;12:75−90.
[41] Borena BM, Pawde AM, Amarpal, Aithal HP, Kinjavdekar P, Singh R, et al. Autologous bone marrow-derived cells for healing excisional dermal wounds of rabbits. Vet Rec 2009;19:563−8. Available from: https://doi.org/10.1136/vr.165.19.563 165.
[42] Herndon DN, Barrow RE, Rutan RL, Desai MH, Abston S. A comparison of conservative versus early excision therapies in severely burned patients. Ann Surg 1989;209:547−52.
[43] Fosssum TW, Hedlund CS, Johnson AL, Schulz KS, Seim HB, Willard MD, et al. Surgery of integumentary system. Manual of Small Animal Surgery Mosby 2007;159−75.
[44] Swaim SF, Henderson RA. Wound management. Small Animal Wound Management, vol. 9. 2nd ed. Lea and Febiger; 1990. p. 33.
[45] Nithya M, Suguna L, Rose C. The effect of nerve growth factor on the early responses during the process of wound healing. Biochemica et Biophysica Acta 2003;1620:25−31.
[46] Judith R, Nithya M, Rose C, Mandal AB. Application of a PDGF containing novel gel for cutaneous wound healing. Life Sciences 2010;87:1−8.
[47] Midwood KS, Williams LV, Schwarzbauer JE. Tissue repair and the dynamics of the extracellular matrix. International J Biochem Cell Biol 2004;36:1031−7.
[48] Kim MS, Hong KD, Shin HW. Preparation of porcine small intestinal submucosa sponge and their application as a wound dressing in full thickness skin defect of rat. Int J Biol Macromol 2005;36:54−60.
[49] Perme H, Sharma AK, Kumar N, Singh H, Dewangan R, Maiti SK. In-vitro biocompatibility evaluation of cross-linked cellular and acellular bovine pericardium. Trends Biomater. Artif Organs 2009;23(2):66−75.
[50] Hinz B, White ES. The myofibroblast matrix: implications for tissue repair and fibrosis. J Pathol 2013;229:298−309.
[51] Boycee ST. Cultured skin substitutes: a review. Tissue Eng 1996;2:255−66.
[52] Li B, Wang JHC. Fibroblasts and myofibroblasts in wound healing: force generation and measurement. J Tissue Viability 2011;20:108−20.

Gall bladder-derived extracellular matrix scaffolds

7

Naveen Kumar[1,]*, Anil Kumar Gangwar[2], Sangeeta Devi Khangembam[3], Poonam Shakya[4], Ashok Kumar Sharma[1], Amit Kumar Sachan[4], Ravi Prakash Goyal[3], Parvez Ahmed[3], Kiranjeet Singh[1], Aswathy Gopinathan[1], Sonal Saxena[5], Sameer Shrivastava[5], Remya Vellachi[6], Dayamon David Mathew[7], Swapan Kumar Maiti[1] and Karam Pal Singh[8]

[1]Division of Surgery, ICAR-Indian Veterinary Research Institute, Izatnagar, Uttar Pradesh, India, [2]Department of Veterinary Surgery & Radiology, College of Veterinary Science & Animal Husbandry, Acharya Narendra Deva University of Agriculture and Technology, Ayodhya, Uttar Pradesh, India, [3]Department of Veterinary Surgery and Radiology, Acharya Narendra Deva University of Agriculture and Technology, Ayodhya, Uttar Pradesh, India, [4]Department of Surgery and Radiology, Acharya Narendra Deva University of Agriculture and Technology, Ayodhya, Uttar Pradesh, India, [5]Division of Veterinary Biotechnology, ICAR-Indian Veterinary Research Institute, Izatnagar, Uttar Pradesh, India, [6]Department of Veterinary Surgery and Radiology, College of Veterinary and Animal Sciences, Wayanad, Kerala, India, [7]Department of Veterinary Surgery & Radiology, Faculty of Veterinary and Animal Sciences, Banaras Hindu University, Rajiv Gandhi South Campus, Barkachha, Uttar Pradesh, India, [8]CADRAD, ICAR-Indian Veterinary Research Institute, Izatnagar, Uttar Pradesh, India

7.1 Introduction

The extracellular matrix (ECM) with adequate bioactive molecules, capable of supporting the growth of cells participating in regeneration, is an ideal graft suitable for wound healing application [1]. The ECMs isolated from certain mammalian organs and tissues have been found to have these essential biocomponents that support cell proliferation, migration, and differentiation [2]. These scaffolds are naturally rich in collagen, elastin, glycosaminoglycans (GAGs), laminin, and fibronectin on which the cells can migrate, attach, and grow. In addition, many of the bioactive degradation products released from the graft at the site of the grafting mimic growth factors required for healing [3]. The ECM is also known to aid angiogenesis by regulating the migration, proliferation, and sustenance of endothelial cells [4]. Intact decellularized allogenic/xenogenic ECM has the necessary

*Present affiliation: Veterinary Clinical Complex, Apollo College of Veterinary Medicine, Jaipur, Rajasthan, India.

requisites to provide for initial requirements of repair and subsequent remodeling. Hence, ECM is correctly termed as nature's ideal scaffold material [5]. The decellularization specifically removes cellular components that give rise to a residual immunological response. These decellularization techniques include chemical, enzymatic and mechanical means of removing cellular components, leaving a material composed essentially of ECM components. The decellularized tissues are expected to mimic closely the complex three-dimensional structure and mechanical properties of the native tissues from where it origins [6]. One of the major goals in using natural biodegradable materials is to induce the host, to replace the implanted construct with native tissue [7]. Cholecyst-derived ECM (CEM) recovered from ECM of porcine gall bladder had variable application in the field of regenerative medicine [8]. This CEM found to be a novel acellular proteinaceous biodegradable biomaterial and may have potential applications as scaffolds in heart valve tissue engineering. This matrix is rich in collagen and contains several other macromolecules useful in tissue remodeling [9].

7.2 Anatomy of gallbladder (cholecyst)

The cholecyst (gall bladder) is part of the digestive tract present within the abdominal cavity. It is accessory to the liver and empties into the duodenum through the common bile duct. Anatomically, it tucks between lobes of the liver. The gall bladder is a sac like organ. Histologically, the gallbladder is mainly composed of four layers, which are arranged from inside to out as follows: tunica mucosa, tunica submucosa, tunica muscularis, and tunica serosa. The mucosa is composed of lamina epithelialis, lamina propria, and lamina muscularis mucosa. In the buffalo gallbladder, lamina muscularis layer is absent. The cholecyst wall is lined on its luminal surface with a simple columnar epithelium. The mucosa rests on a basement membrane, which is continuous with the lamina propria. Lamina propria is rich in fenestrated capillaries and small venules. External to the lamina propria is a submucosa, having a thin muscularis layer with numerous smooth muscle bundles interlaced with collagenous fibers running in longitudinal and circumferential directions of the cholecyst wall. When the gall bladder is empty, mucosa layer is extremely folded. Muscularis layer is followed by a relatively acellular ECM made of loosely braided collagen bundles. The muscularis externa contains elastic and collagen fibers among the bands of irregularly arranged smooth muscle. Externally there is serosal layer that is made up of dense, irregularly arranged loose connective tissue and peritoneal mesothelium.

7.3 Cholecyst-derived extracellular matrix

Cholecyst-derived ECM is a fibro-porous decellularized perimuscular subserosal layer of gall bladder, which is composed of about 80% collagen [8]. Cholecyst-derived ECM is derived from a gallbladder from which the native cells have been removed to leave behind a mesh-like collagen structure. The nanotopography of the CEM makes it ideal for supporting new cell growth, while its mechanical properties ensure that it has the strength and elasticity for tissue repair. The CEM is completely reabsorbed by the body over time, making further surgery for its removal unnecessary. The surface of cholecyst-derived ECM is characterized by compact, parallel bundles of collagen fibers [10]. Its collagen fibers have consistent, wider range orientations, making it a suitable material for multiaxial loading applications [9]. This matrix has the necessary mechanical and regenerative properties to suit soft tissue regeneration application [8]. Compared to a commercially available skin-graft substitute made of porcine small intestinal submucosa, the cholecyst-derived scaffold was rich in natural biomolecules like elastin and glycosaminoglycans. When used as a xenograft, it promoted healing with excess cell proliferation at early phases and acceptable collagen deposition in the later remodeling phases [11].

The CEM is an acellular biomaterial that may address limitations associated with previously investigated scaffold materials. The CEM isolated from pig cholecyst has a high collagen, low muscle, and low cell content because cholecyst devoid of vasculature, submucosal gland, lymphatic, and nervous network and has a consistent thickness [9,12]. The cellular components have been shown to generate antigenicity in scaffold, but due to low cell content CEM superior than other scaffolds. Chan et al. [12] stated that CEM collagen comprises $75.5\% \pm 15.1\%$ and elastin comprises $13.3\% \pm 4.4\%$ of CEM dry weight.

The clue for proposing the use of CEM for skin-graft applications originated from its physical and biological properties as the biological and physical properties of any scaffold can influence wound healing [13]. Additionally, certain biomarkers that occur at the wound site can also predict the potential of a scaffold for cutaneous wound healing applications [14]. Revi et al. [11] studied some of these parameters like elastin and sulfated-GAG content of the scaffold. An important physical parameter often studied for evaluating healing potential is the water vapor transmission rate (WVTR) from the graft that indicates its ability to retain absorbed water [15,16]. High WVTR rate might cause the wound bed to become desiccated and consequently lead to loss of integrity. It may also modulate various tissue responses in the healing wound, for example, too much water loss increases the possibility of tissue necrosis and slowing of epithelial cell migration leading to impediment of reepithelization and decreased oxygen availability for bacterial killing leading to increased risk of infection and impaired nutrient flow [17]. These are all poor prognosis for wound healing. The novel scaffold CEM had lower WVTR compared to the other scaffold. Adequate moisture is also required for satisfactory activity of growth factors and proteolytic enzymes. Furthermore, moisture also enhances fibroblast/endothelial cell proliferation and is known to increase the immune defense of wound surface [18]. The epithelial cells need a moist ground to migrate and reepithelize faster [19]. The other hydration parameters such as percentage fluid uptake, moisture content of the graft and evaporative water loss of the CEM graft were similar with that of the commercially available graft.

The flexural rigidity and suture retention efficiency are two parameters that indicate the suitability of a scaffold as skin graft and for better wound healing [13]. The flexural rigidity corresponds to the ability of a dressing or graft to drape over the wound. It should be sufficiently low. On the other hand, high flexural rigidity means a rigid scaffold that will not be flexible to be in touch with the wound surface. In the present instance, the flexural rigidity of CEM was similar to that of the small intestinal submucosa (SIS). On the other hand, the CEM (2.3 ± 9 N) had lower suture retention ability than that of SIS. But the actual measure of the suture efficiency was higher than 1.2 N as expected for scaffold sheets commonly used in soft tissue engineering applications [20]. Hence, CEM can be considered for skin-graft applications.

The CEM had higher content of macromolecules like elastin and sulfated GAGs than SIS while the collagen content was similar. It was not sure whether the higher elastin and sulfated-GAG content reflected the natural content of these molecules in normal for CEM and whether the processing and extraction techniques have affected the values observed. Nevertheless, higher content of these biomolecules in CEM made it a better scaffold than SIS for skin-graft application. It is known that elastin enhances angiogenesis [21], promotes proliferation of endothelial cells [22] and also supports proliferation of dermal fibroblasts [23]. The GAGs facilitate specific interactions to cytokines [24] and chemotactic growth factors [25], which are important for wound healing regulate the release of growth factors in the healing environment. In addition, GAGs can also trap water in the form of gel and prevent loss of water. The presence of excess cellular content, especially nucleic acids, in a scaffold is known to cause inflammatory reactions that are not congenial for a good scaffold [26]. However, the CEM scaffold had contained lower DNA content compared to SIS probably because gall bladder is not a very cellular organ since its main function is storage and release of bile, unlike the small intestine which is involved in digestion, peristalsis, and secretion with wide absorptive surface organized into primary and secondary folding. Thus compared to the other material, the higher content of sulfated GAGs as well as elastin and lower content of DNA in CEM make it a preferred biomaterial for graft-assisted healing.

Pandit and Anilkumar [27] prepared cholecyst-derived ECM from porcine gallbladder. The cholecyst was removed from the pig gut and excised from the liver leaving cystic duct intact. The bile was drained through neck and cystic duct was removed. Subsequently the fundus and neck regions were removed and tissue was opened via a longitudinal incision to reveal the mucosal surface. The mucosa and lamina propria layers were peeled off. The serosal and adipose tissue layers on the outer side of gallbladder were removed by gentle abrasion. The isolated cholecyst-derived ECM was cleaned with 70% ethanol for 2−4 hours. Then it was washed thoroughly with distilled water for 24−48 hours. Extracted tissue was treated with ionic detergent 1% sodium dodecyl sulfate (in 10 mM Tris buffer) for 1−4 days and rinsed with Tris buffer at pH 8. This was followed by treatment with trypsin-EDTA (0.05% trypsin in 10 mM Tris buffer and 0.1% EDTA in pH 8) overnight. Burugapalli et al. [8] isolated CEM from porcine cholecyst and decellularized with peracetic acid solution in ethanol. Excess liver tissue was removed and bile fluid drained. The neck and fundus of the cholecyst were trimmed, followed by a longitudinal incision to obtain a flat sheet of tissue. The mucosa, lamina propria, and muscularis layers were peeled from the luminal side, followed by a similar process to remove the serosal mesothelium and its underlying connective tissue from the abluminal side. Any residual elements were removed by mechanical delamination on both sides. A solution of 0.15% peracetic acid and 4.8% ethanol solution in deionized water were used to decellularize the sheet of tissue for 30 minutes. Following decellularization, the samples were washed thoroughly in phosphate-buffered saline (PBS) and distilled water.

7.4 Preparation of acellular buffalo cholecyst-derived extracellular matrix

The buffalo gallbladder (BG) was collected from local abattoir. Immediately after collection, the gall bladder was kept in normal saline solution containing 0.02% EDTA and antibiotic (amikacin 1 mg/mL). The neck and fundus of the gallbladder were trimmed, followed by a longitudinal incision to obtain a flat sheet of tissue. The inner mucosal layer was peeled off and outer serosal layer was removed by mechanical delamination with a blunt edge. The tissue was cut into 2×2 cm^2 size pieces and treated with 70% ethanol for 2 hours and later on with distilled water for 24 hours (Fig. 7.1A−I) [28].

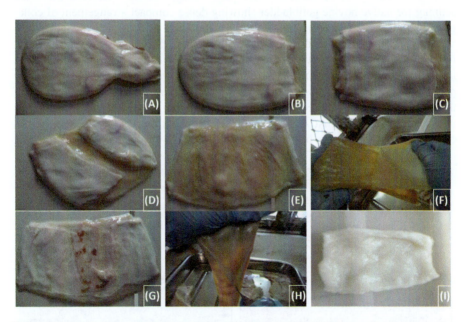

Figure 7.1 (A) Buffalo gallbladder collected from local abattoir, (B) removing the neck of the gallbladder, (C) removing the fundus of the gallbladder, (D) giving a longitudinal incision to obtain a flat sheet of tissue, (E) showing mucosal layer, (F) peeling of mucosal layer, (G) showing serosal layer, (H) removing serosal layer mechanically by forceps, (I) delaminated gall bladder.

The buffalo gallbladder tissue was treated with 0.5% and 1% sodium dodecyl sulfate (SDS) (ionic biological detergent) and 0.5% and 1% Triton X-100 (nonionic biological detergent) for 12, 24, 48, and 72 hours. The tissue was thoroughly washed in phosphate buffer saline solution after the completion of protocols. Acellularity of the b-CEM was evaluated by microscopic and scanning electron microscopic observations. The DNA concentration of native buffalo gallbladder and acellular buffalo cholecyst-derived extracellular matrices were evaluated.

7.4.1 Macroscopic observations

The mucosal layer was peeled off immediately after cleaning of the buffalo gallbladder. The serosal layer was separated mechanically with the help of forceps. The treated gallbladder with different detergents appeared soft, whiter, and slightly spongy in consistency than native tissue [28].

7.4.2 Microscopic observations

Microscopically, native buffalo gallbladder showed cellularity, mucosal and serosal layer, fatty tissue, and some liver portion. The delaminated submucosal layer showed cellularity, loose muscular layer and collagen fibers. Masson's trichrome staining for native bovine gallbladder showing dense compact arrangement of collagen fibers (Fig. 7.2A and B).

The delaminated gallbladder treated with 0.5% SDS ionic biological detergent for 12 hours under constant agitation showed loss of cellularity. The submucosal layer was completely acellular. The collagen fibers were compact with moderate porosity than the native tissue. Mild debris was observed between the spaces of thick collagen fibers. The collagen fibers became thin, compact and porosity was unchanged at 24 hours. At 48 hours, b-CEM showed thin, heavily loose arranged collagen fibers with very high porosity. No debris was evident. After 72 hours,

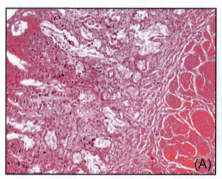
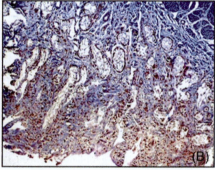

Figure 7.2 Microscopic appearance of native buffalo gall bladder showing cellularity, fatty tissue, mucosal and serosal layers: (A) H&E stain, ×100, (B) Masson's trichrome stain, ×100.

0.5% SDS-treated b-CEM was completely damaged with high porosity. No cellular debris was found (Fig. 7.3A−D) [28].

Delaminated 1% SDS-treated CEM at 12 hours showed thick collagen fibers with moderately loose arranged collagen bundles with moderate cellular debris. At 24 hours, collagen fibers were loosely arranged with moderate porosity. In 1% SDS, at 48 hours collagen fibers were heavily loose arranged with no cellular debris. At 72 hours heavily loose, thin damaged collagen fibers with high porosity were observed (Fig. 7.4A−D).

The delaminated CEM subjected to 0.5% Triton X-100 for 12 hours showed thick damaged collagen fibers with moderate cellular debris and porosity. At 24 hours the collagen fibers were thin and mildly loose with high porosity. The collagen fibers were thin, loosely arranged and damaged with heavily porosity at 48 and 72 hours (Fig. 7.5A−D) [28].

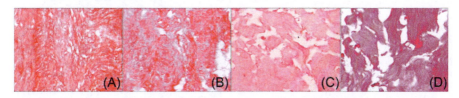

Figure 7.3 Delaminated buffalo gallbladder treated with 0.5% SDS biological detergent at different time intervals. (A) 12 h, (B) 24 h, (C) 48 h, (D) 72 h (H&E stain, ×100).

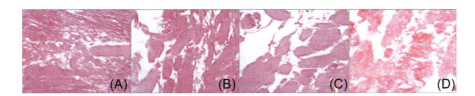

Figure 7.4 Delaminated buffalo gallbladder treated with 1% SDS biological detergent at different time intervals. (A) 12 h, (B) 24 h, (C) 48 h, (D) 7 h (H&E stain, ×100).

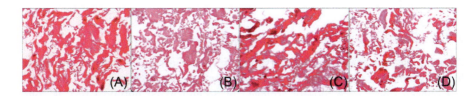

Figure 7.5 Delaminated buffalo gallbladder treated with 0.5% Triton X-100 biological detergent at different time intervals. (A) 12 h, (B) 24 h, (C) 48 h, (D) 72 h (H&E stain, ×100).

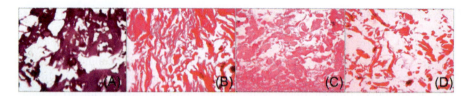

Figure 7.6 Delaminated buffalo gallbladder treated with 1% Triton X-100 biological detergent at different time intervals: (A) 12 h, (B) 24 h, (C) 48 h, (D) 72 h (H&E stain, × 100).

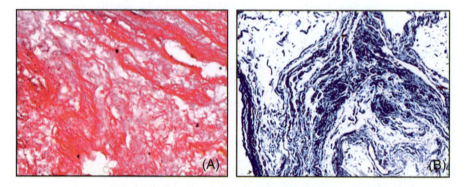

Figure 7.7 Microscopic appearance of acellular buffalo gall bladder, (A) H&E stain, × 100, (B) Masson's trichrome stain, × 100.

In 1% Triton X-100 treatment at 12, 24, 48, and 72 hours collagen fibers were loosely arranged and damaged from beginning with no cellular debris (Fig. 7.6A–D).

All the specimens treated for 12, 24, 48, and 72 hours were adequately decellularized and the best scaffold was prepared by treatment with 0.5% SDS. On Masson's trichrome staining thick, transversely and longitudinally arranged collagen fibers were observed (Fig. 7.7A and B). The specimens treated with 1% SDS and Triton X-100 (0.5% and 1%) were more or less similar and showed very loose arrangement with damage of collagen fibers [28].

7.4.3 Scanning electron microscopic observations

Scanning electron microscopic observations revealed loss of cellularity. The native buffalo gallbladder showed dense and closely packed collagen fibers (Fig. 7.8A–C).

Collagen fibers were loosely arranged in decellularized buffalo CEM as compared to native bovine gallbladder tissue. The collagen bundles were seen to be formed of many thick collagen fibrils (Fig. 7.9A–C) [28].

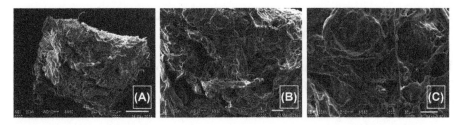

Figure 7.8 Scanning electron microscopic appearance of native buffalo gall bladder (A) 40×, (B) 100×, (C) 150×.

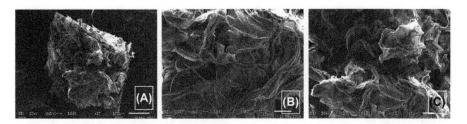

Figure 7.9 Scanning electron microscopic appearance of decellularized buffalo gall bladder: (A) 40×, (B) 100×, (C) 150×.

Table 7.1 Mean ± SE values of DNA content (ηg/μL) after decellularization.

Content	Native	0.5% SDS	1% SDS	0.5% TX-100	1% TX-100
BG	39.1 ± 0.06	8.87 ± 0.33[a]	9.09 ± 0.43[a]	10.67 ± 0.73[a]	10 ± 0.45[a]

[a]Differ significantly ($P < .05$) from native values.

7.4.4 DNA contents analysis

To measure the effectiveness of decellularization method, the DNA from the samples was isolated and the DNA concentration was measured using Nanodrop technique. This quantification was a direct measure to confirm the effectiveness of the decellularization processes, since the DNA is present in active nucleus of cell. In the native gallbladder, abundant cell components and nucleic acids were apparent. However, after the decellularization, cells and nucleic acids were hardly observed in ECM. The DNA contents (ηg/μL) before and after decellularization with different detergents at different concentration are presented in Table 7.1.

The concentration of native buffalo gallbladder DNA ($P < .05$) was 39.1 ηg/μL. After decellularization by different protocols, the DNA contents decreased and its ranges from 8.87 ± 0.33 ηg/μL to 10.67 ± 0.73 ηg/μL. Treatment with 0.5% SDS showed lowest values (8.87 ± 0.33 ηg/μL) of DNA contents [28].

7.5 Preparation of acellular pig cholecyst-derived extracellular matrix

The pig gallbladder (PG) was collected from local abattoir. Immediately after collection, the gall bladder was kept in normal saline solution containing 0.02% EDTA and antibiotic (amikacin 1 mg/mL). The detail procedure has been already described in the preparation of bovine gallbladder section. Acellularity of the p-CEM was evaluated by microscopic and scanning electron microscopic observations. The DNA concentration of native porcine gallbladder and acellular porcine cholecyst-derived extracellular matrices were evaluated.

The mucosal layer and serosal layer were removed mechanically. Microscopically, native porcine gallbladder showed cellularity, mucosal and serosal layer, fatty tissue and some liver portion. The delaminated submucosal layer showed cellularity, loose muscular layer, and collagen fibers (Fig. 7.10A and B) [28].

Masson's trichrome staining was done for native pig gallbladder and delaminated pig gallbladder showing dense compact arrangement of collagen fibers (Fig. 7.11A and B).

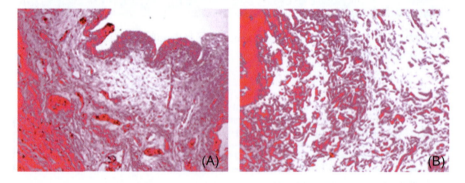

Figure 7.10 (A) Microscopic appearance of native porcine gall bladder, (B) delaminated gall bladder (H&E stain, ×100).

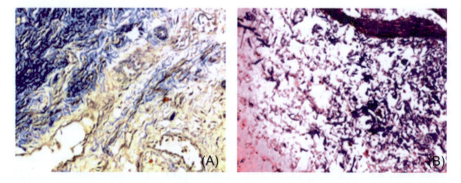

Figure 7.11 (A) Microscopic appearance of native pig gall bladder, (B) delaminated gall bladder (Masson's trichrome stain, ×100).

The delaminated pig gallbladder was subjected to macroscopic and microscopic observations.

7.5.1 Macroscopic observations

The pig gallbladder tissue treated with 0.5% and 1% sodium dodecyl sulfate and 0.5% and 1% Triton X-100 biological detergent was observed for 12, 24, 48, and 72 hours. The matrix appeared soft, whiter, and slightly spongy in consistency as compared to native delaminated tissue [28].

7.5.2 Microscopic observations

The delaminated gallbladder treated with 0.5% sodium dodecyl sulfate for 12 hours under constant agitation showed complete loss of cellularity. The submucosal layers were completely acellular. The collagen fibers were compact with moderate porosity than the native tissue. Mild debris was observed between the spaces of thick collagen fibers. The collagen fibers became thin and compact, and porosity was unchanged at 24 hours. At 48 hours, CEM showed thin, heavily loose arranged collagen fibers with very high porosity. No debris was evident, but after 72 hours, CEM was completely damaged with high porosity. No cellular debris was found (Fig. 7.12A–D).

The 1% SDS-treated CEM showed thick collagen fibers, moderately loose arranged collagen bundles and moderate cellular debris at 12h time interval. At 24 hours, collagen fibers were loosely arranged with moderate porosity. At 48 hours, collagen fibers were heavily loose arranged with no cellular debris. At 72 hours, heavily loose, thin damaged collagen fibers with high porosity were observed (Fig. 7.13A–D) [28].

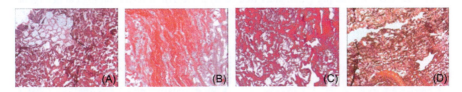

Figure 7.12 Delaminated porcine gallbladder treated with 0.5% SDS biological detergent at different time intervals: (A) 12 h, (B) 24 h, (C) 48 h, (D) 72 h (H&E stain, × 100).

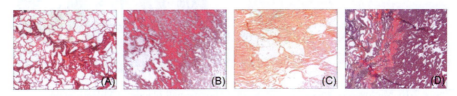

Figure 7.13 Delaminated pig gallbladder treated with 1% SDS biological detergent at different time intervals: (A) 12 h, (B) 24 h, (C) 48 h, (D) 72 h (H&E stain, × 100).

The delaminated CEM subjected to 0.5% Triton X-100 for 12 hours showed thick damaged collagen fibers with moderate cellular debris and porosity. At 24 hours, the collagen fibers were thin and mildly loose with high porosity. Collagen fibers were thin, loosely arranged, and damaged with heavily porosity at 48 and 72 hours (Fig. 7.14A–D) [28].

In 1% Triton X-100 at 12, 24, 48, and 72 hours, collagen fibers were loosely arranged and damaged from beginning with no cellular debris (Fig. 7.15A–D).

All the specimens treated for 12, 24, 48, and 72 hours were adequately decellularized, and the best scaffold was prepared by treatment with 0.5% sodium dodecyl sulfate. On Masson's trichrome staining, thick, transversely, and longitudinally arranged collagen fibers were observed (Fig. 7.16A and B). The specimens treated

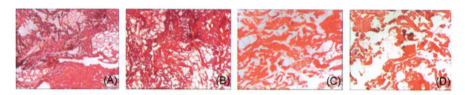

Figure 7.14 Delaminated pig gallbladder treated with 0.5% Triton X-100 biological detergent at different time intervals: (A) 12 h, (B) 24 h, (C) 48 h, (D) 72 h (H&E stain, × 100).

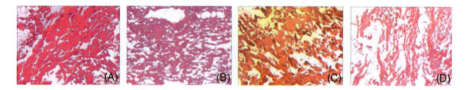

Figure 7.15 Delaminated pig gallbladder treated with 1% Triton X-100 biological detergent at different time intervals: (A) 12 h, (B) 24 h, (C) 48 h, (D) 72 h (H&E stain, × 100).

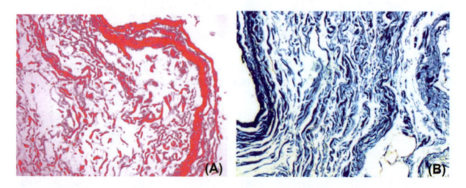

Figure 7.16 Microscopic appearance of acellular bovine gall bladder: (A) H&E stain, × 100, (B) Masson's trichrome stain, × 100.

with 1% sodium dodecyl sulfate and Triton X-100 (0.5% and 1%) were more or less similar [28].

7.5.3 Scanning electron microscopic observations

The native pig gall bladder (Fig. 7.17A−C) showed dense and closely packed collagen fibers.

Collagen fibers were loosely arranged in decellularized pig CEM comparison to native gallbladder tissue. The collagen bundles were seen to be formed of many fine threads like collagen fibrils. A meshwork of collagen bundles forms the bulk of the tissue, yet smaller fibers not associated with larger collagen bundles form a fine mesh-like fibrous component of the tissue. The fine filamentous noncollagenous content of the CEM was also clear, forming a fine weave between the collagen bundles (Fig. 7.18A−C) [28].

7.5.4 DNA contents analysis

The DNA contents ($\eta g/\mu L$) before and after decellularization with ionic and nonionic biological detergent at different concentration are presented in Table 7.2.

The concentration of native pig gallbladder DNA ($P < .05$) was 32.13 $\eta g/\mu L$. After decellularization by different protocols, the DNA contents decreased and its ranges from 9.74 ± 0.32 $\eta g/\mu L$ to 11.06 ± 0.55 $\eta g/\mu L$. There was significant

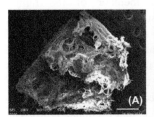
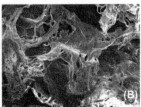
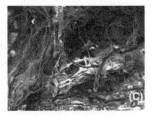

Figure 7.17 Scanning electron microscopic appearance of native pig gall bladder: (A) 40 ×, (B) 100 ×, (C) 150 ×.

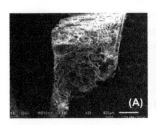
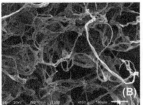
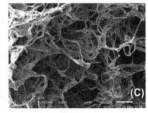

Figure 7.18 Scanning electron microscopic appearance of native pig gall bladder: (A) 40 ×, (B) 100 ×, (C) 150 ×.

Table 7.2 Mean ± SE values of DNA content (ηg/μL) after decellularization.

Content	Native	0.5% SDS	1% SDS	0.5% TX-100	1% TX-100
PG	32.13 ± 0.09	9.74 ± 0.32[a]	9.90 ± 0.30[a]	11.06 ± 0.55[a]	10.33 ± 0.75[a]

[a]Differ significantly ($P < .05$) from native values.

($P < .05$) decrease in DNA contents in acellular gallbladder as compared to native porcine gallbladder. Treatment with 0.5% SDS showed lowest values (9.74 ± 0.32 ηg/μL) of DNA contents [28].

7.6 Preparation of acellular goat cholecyst-derived extracellular matrix

The goat gall bladder is a simple muscular sac and the wall of the bladder does not have muscularis mucosae and submucosa. The muscularis externa (muscle layer) has bundles of smooth muscle cells, collagen, and elastic fibers. Various biological detergents like Triton X-100, Tween-20, sodium dodecyl sulfate, and sodium deoxycholate have been used for decellularization with variable results and certain cell cytotoxicity. There is a need to find out plant origin detergents that have low cell toxicity and have good biocompatibility. *Sapindus mukorossi*, a member of the family Sapindaceae, is commonly known by several names such as soapnut, soapberry, washnut, reetha, aritha, dodan, and doadni in India. The fruit of this plant is valued for the saponins (10.1%) present in the pericarp and forms up to 56.5% of the drupe. Plant-derived saponins are secondary metabolites mostly used in detergents due to its amphiphilic nature with the presence of a lipid-soluble aglycone and water-soluble chains in their structure. The major compounds are isolated from the soapnut pericarp are triterpenoidal saponins of mainly oleanane, dammarane, and tircullane types. The fruits are of much importance for their medicinal value for treating a number of diseases like eczema and psoriasis. George and Shanmugam (2014) observed antimicrobial activity of both ethanol and aqueous extracts of pericarp of *Sapindus mukorossi*. This plant has anticancer and antidiabetic properties. Other pharmaceutical properties of saponins are as follows: hemolytic, molluscicidal, antiinflammatory, antifungal, antibacterial or antimicrobial, antiparasitic, antitumor, and antiviral. The efficacy of 5% aqueous extract of the fruit pericarp of a plant, *Sapindus mukorossi*, for decellularization of gall bladder of caprine origin was studied.

7.6.1 Preparation of soapnut pericarp extract

Soapnut pericarp extract was prepared as per method described by Verma [29]. Briefly, commercially available well-dried ritha fruits (*Sapindus mukorossi*) were purchased from local market. Pericarp was separated from seed (Figs. 7.19) and 5 gm of pericarp was soaked overnight in 100 mL phosphate buffer saline (PBS) at 23°C. This mixture was further vortexed using magnetic stirrer for 2 hours at room temperature and then filtered using stainless steel sieve. The ritha concentration in this solution was determined on the basis of the ratio of weight of ritha pericarp to that of PBS volume. Thus the concentration of ritha in the above-mentioned solution becomes 5 wt.%. This solution was further subjected to centrifugation at 5000 rpm for 20 minutes and at room temperature to get suspended free supernatant solution (Fig. 7.20A—D), which was used for further studies.

Efficacy of the decellularization was evaluated using histology (Picro Sirius Red staining), biochemical assay (DNA assay), and DAPI (*4, 6-diamidino-2-phenylindole 2Hcl*) fluorescent staining. Crapo and Gilbert in 2011 proposed three criteria for satisfactory decellularization, namely, <50 ng of double stranded DNA (dsDNA) per mg of ECM (dry weight); a lack of visible nuclear material in tissue sections stained with 4,6-diamidino-2-phenylindole or hematoxylin and eosin (H&E); and a <200 bp DNA fragment length.

7.6.2 Preparation of acellular goat gall bladder matrix

The goat gall bladder (Fig. 7.21A) was procured from the local abattoir and stored in chilled sterile phosphate-buffered saline (PBS) containing (0.1% amikacin) and proteolytic inhibitor (0.25% EDTA). The samples were washed thoroughly with sterile phosphate-buffered saline (PBS, pH 7.4) to remove adherent blood and debris. The maximum time period between retrieval and the initiation of protocol was less than 4 hours. After initial washing, the native gall bladder tissues were cut into 1×1 cm^2 pieces and transferred into a sterilized plastic container having

Figure 7.19 Dried fruits, pericarp, and seeds of *Sapindus mukorossi*.

Figure 7.20 Preparation of 5% solution from pericarp of *Sapindus mukorossi*: (A) soapnut pericarp in grinded form, (B) loading of beaker on magnetic stirrer, (C) filtration of extract using stainless steel sieve, and (D) filtered extract.

Figure 7.21 (A) Goat gall bladder, (B) delaminated gall bladder, (C) decellularized gall bladder.

100 mL of 5% extract of soapnut pericarp and subsequently agitated on magnetic stirrer for 72 hours at room temperature. Tissue samples were collected at 6-, 12-, 24-, 48-, and 72-hour time intervals for confirmation of decellularization. The acellularity was confirmed by histological examination, DAPI staining, and scanning electron microscopy (SEM). All the samples were stored in PBS solution containing 0.1% amikacin at $-80°C$ until use [30].

The goat gall bladder was subjected to hypertonic saline treatment for delamination. Mechanical separation of the mucosa and tunica muscularis externa of the gall bladder was started at 6-hour time interval in broken pieces at places. Complete delamination was observed mechanically at 8-hour interval (Fig. 7.21B). At 12 hours, the gall bladder tissue treated with 5% solution of soapnut extract were soft, milky white and spongy in consistency (Fig. 7.21C). At 24 hours, the samples were slightly swollen. The samples were started dissolving in the solution at 72-hour interval [30].

7.6.3 Microscopic observations

For histological observations, the tissue samples were fixed in phosphate-buffered 10% formalin saline, dehydrated in ethanol, cleared in xylene, and embedded in paraffin to get 5 micron thick paraffin sections. The sections were stained with H&E and evaluated microscopically by using histological scoring system as per method of Gangwar et al. [31]. The histological parameters were graded as follows: cellular contents: normal cellularity (+ + +), moderate number of cells (+ +), mild number of cells (+), no cellular material (-); debris: more (+ + +), moderate (+ +), mild (+), no debris (-); collagen fibers arrangement: compact (+ + +), mildly loose (+ +), moderately loose (+), heavily loose (-); porosity: highly porous (+ + +), moderate (+ +), mild (+).

The microscopic results of the gall bladder scaffolds collected from the 5% solution of soapnut extract at different time intervals are presented in the Table 7.3 Native caprine gall bladder showed cellularity (Fig. 7.22A). At 6- and 12-hour intervals, the scaffolds showed 60%–70% decrease in the cellular contents. At 24 hours, cellularity decreased up to 80% and cellular debris was seen at places. At 48 hours, the collagen fibers were loosely arranged and thick. At 72 hours, the collagen fibers were heavily loose with more porosity and no debris was present (Fig. 7.22B) [30].

Table 7.3 Microscopic observations of goat gall bladder processed by 5% extract of soapnut pericarp at different time intervals.

S. No.	Parameters	Time interval					
		Native	6 h	12 h	24 h	48 h	72 h
1.	Cellular contents	+ + +	+ +	+	+	-	-
2.	Cellular debris	-	+ +	+ +	+	+	-
3.	Collagen fibers arrangement	+ + +	+ +	+	+	+	-
4.	Porosity	+	+ +	+ +	+ +	+ + +	+ + +

7.6.4 DAPI staining: (4, 6-diamidino-2-phenylindole 2Hcl)

Native and decellularized gall bladder tissues were fixed in 10% formalin. These tissues were processed and sectioned in standard manner. The sectioned tissue was loaded over the amino propyl tri-ethoxy silane (APTES)-coated slides. The sections were dewaxed in xylene (2 × 5 minutes), rehydrated in ethanol series (absolute, 95% for 5 minutes, 70%, 30% ethanol, dH_2O for 3 minutes). The slides were washed with 0.2% TBST (Tris buffered saline Tween-20) for 2–3 times. The slides were dried on paper towel, DAPI staining solution was applied on each slides (∼200 μL), and then the slides were incubated for 15 minutes in dark at room temperature. The slides were washed with 0.2% TBST (Tris buffered saline Tween-20) for 2–3 times, 3–5 minutes each to remove the unbind DAPI. Mount cover slips with Gel Mount. Nuclei emit cold blue fluorescence under fluorescent microscope [32].

DAPI staining of native gall bladder (Fig. 7.23A) showed cold blue fluorescence which depicts the nuclear DNA. The samples of gall bladder at 24- and 48-hour interval processed by soap nut extract showed few DNA remnants at places. However, at 72 hours, nearly complete removal of nuclear components in gall

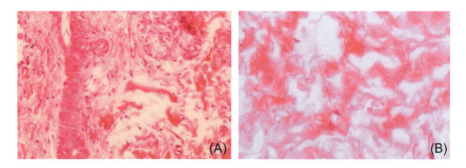

Figure 7.22 (A) Micrograph of goat native (normal) gall bladder. (B) Micrograph of goat gall bladder after processing in 5% extract of soapnut pericarp (72 h) (H&E stain, × 100).

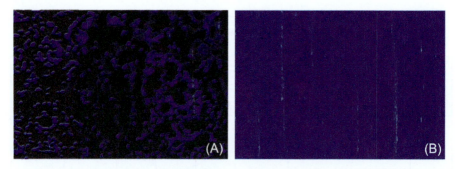

Figure 7.23 (A) 4',6-Diamidino-2-phynylindole (DAPI) staining of a native (normal) gall bladder matrix. (B) DAPI staining of a decellularized gall bladder matrix at 72 h.

bladder matrices was observed (Fig. 7.23B). Quantification of punctate nuclear staining in tissue sections demonstrated a reduction of approximately 99% in the cellularity, relative to the native tissues [30].

7.6.5 Scanning electron microscopic observations

The SEM examination of the native and decellularized gall bladder samples was performed. Scanning electron microscope (SEM), model Jeol JSM-840, was used for ultrastructure observations. Briefly, the gall bladder samples were cut into small pieces and fixed into freshly prepared Karnovsky fixative at 4°C. After fixation for 3—4 days, tissues were washed three times with phosphate buffer of pH 7.2. Specimens were dehydrated using serial gradient of alcohol in water. Finally, tissues were washed three times in absolute alcohol and treated with hexamethyldisilazane (HMDS). Specimens were mounted on metal stubs using adhesive carbon tape and sputtered with gold ions at 1.2 kV (7 mA current) for 5 minutes. Finally, specimens were observed under SEM at 5 kV using different magnifications as needed for the desired observations.

Native gall bladder tissue showed fibroplasia over the collagen fibers (Fig. 7.24A). Acellular gall bladder matrix showed collagen networks, composed of thin-fibrils in a mesh-like pattern. The collagen fibers were randomly oriented fibrillar structure with large interconnected pores and cells were absent (Fig. 7.24B) [30].

7.6.6 DNA contents analysis

To verify the extent of decellularization, tissue specimens for DNA quantification were taken from native gall bladder tissue (25 mg each) and placed on dry filter paper to remove excess fluid. Each sample was homogenized separately with a mortar and pestle. Total genomic DNA was isolated from native and decellularized scaffolds using the DNASure Tissue mini kit (Genetix Biotech Asia Pvt. Ltd.)

 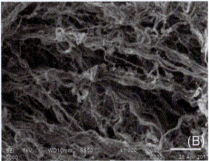

Figure 7.24 (A) Scanning electron microscopic (SEM) image of a normal gall bladder matrix. (B) SEM image of decellularized gall bladder matrix at 72 h.

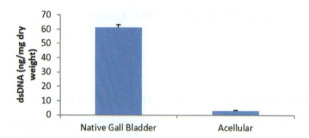

Figure 7.25 Mean ± SE of DNA quantification values (ng per mg dry weight) in the native and acellular gall bladder matrix.

following the manufacturer's instructions. The total amount of DNA was quantified using spectrophotometry (NanoDrop ND-1000, NanoDrop Technologies).

Quantitative evaluation of DNA in the decellularized tissues by spectrophotometry demonstrated a significant reduction of DNA content. Quantity of dsDNA in native and decellularized gall bladder was 61.5 ± 1.6 ng per mg dry weight and 3.1 ± 0.4 ng per mg dry weight, respectively (Fig. 7.25) [30].

7.7 Experimental evaluation of buffalo cholecyst-derived extracellular matrix in a rat model

7.7.1 Animals and ethics statement

Protocols for the study were approved by the Institute Animal Ethics Committee of the ICAR—Indian Veterinary Research Institute, Izatnagar, Uttar Pradesh, India, according to guidelines for the care and use of laboratory animals. Thirty-six clinically healthy Sprague Dawley rats of either sex, weighing from 250 to 300 g, and between 3 and 4 months of age were used in this study. Animals were housed individually in cages, provided with commercial diet and water ad libitum in a temperature and humidity-controlled environment. The animals were acclimatized to approaching and handling for a period of 10 days before start of the study [33].

7.7.2 Skin wound creation and implantation

Animals were anesthetized using xylazine (5 mg/kg body weight) and ketamine (50 mg/kg body weight) combination [34]. The animals were restrained in sternal recumbency and dorsal thoracic area was prepared for aseptic surgery. Using a sterile plastic template, the vertices of the experimental wounds of 20×20 mm^2 dimensions were outlined on the dorsothoracic region of the rats. A full thickness skin defect including the *panniculus carnosus* was excised with a #11 BP blade on each animal. Hemorrhage, if any, was controlled by applying pressure with sterile cotton gauze. The 36 animals were randomly divided into three equal groups of 12 animals each. The defect in group I (C) was kept as open wound and was taken as control (C). In group II, the defect was repaired with commercially available collagen sheet (b-CS) (Skin Temp II, absorbable bovine collagen, polyethylene nonwoven mesh, Human BioSciences India Limited, 142/143 P Vasna Chacharvadi, Ahmedabad 382213, India). In group III, the defect was repaired with cholecystic-derived ECM of buffalo origin (b-CEM). The scaffolds were secured to the edge of the skin wounds with eight simple interrupted 4/0 nylon sutures. The repaired wounds were covered with paraffin surgical dressings. After recovery from anesthesia, rats were housed individually in properly disinfected cages. Further, antimicrobial treatment (Enrofloxacin 10 mg/kg intramuscularly once a day) was continued for 5 days and meloxicam (0.3 mg/kg intramuscularly once a day) for 3 days [33].

7.7.3 Wound contraction

Wound area of all groups was measured by tracing its contour using transparent sheets with graph paper on postoperative days 0, 3, 7, 14, 21, and 28. The area (mm^2) within the boundaries of each tracing was determined and percent contraction was measured by a rate of wound reduction, as well as percent reduction in wound area. A gradual decrease in wound area (mm^2) was observed in all the groups during the entire observation period. The wound area decreased significantly ($P < .05$) at various time intervals in different groups. However, the rate of decrease in wound area was faster in group III as compared to other groups, but there was no significant difference ($P > .05$) between groups III and I in later stages (day 28).

A gradual increase in percent wound contraction was observed in all the groups during the observation period. Although the original wound area created was 20×20 mm^2 but the wounds were expanded to various extent in groups I and II, as in group I, no matrix was used and the wound was healed as open wound and in group II commercially available collagen sheet was applied, as it is very fragile in nature, suturing was not possible. In group III, prepared acellular cholecyst was sutured over the wound area to prevent drying and desiccation of the inner matrix. A significant increase ($P < .05$) in percent contraction was observed in all groups during the observation period [33].

The wound contraction is the centripetal displacement of the wound edges that facilitates its closure after trauma. This process is carried out by myofibroblasts that contain α-actin from smooth muscle and is mediated by contractile forces produced by granulation tissue from wound [35]. The wound healing rate defined as the gross epithelialization of the wound bed. The wound contraction was assessed by percent retention of the original wound area as reported by Schallaberger et al. [36]. The wound contraction has been used to monitor wound healing. The wound area decreased gradually as the healing progressed. Maximum percent contraction was recorded in group I as no biomaterials was applied and wound healed as open wound. The wound in animals of this group healed completely by 28 days leaving a large scar indicating the existence of severe contraction. In group III, the healing occurred with minimal contraction as the matrix was sutured with wound edges in this group [33].

7.7.4 Gross observations/planimetry

Color photographs were taken on postimplantation days 7, 14, 21, and 28 with the help of digital camera at a fixed distance. Analysis of shape, irregularity, and color of the lesion were determined. During the entire period of the study, none of the wounds showed any visible suppurative inflammation or any mortality. The representative images of wound of one animal in each group, at baseline (day zero), and end of healing are presented in Fig. 7.26. On day 3, in control group (I), wounds were covered with soft and fragile pinkish mass above the wound surface area with serous exudates oozing out. On day 7, the surface became more desiccated and necrosed with some amount of exudates. By day 14, the wound size decreased markedly and a thick crust developed and it starts detaching leaving a raw granular pink tissue. By day 28, wound healed up completely by severe contraction leaving a large scar. In b-CS(II), on day 3, the top layer of commercially collagen sheet implanted wound covered with soft and fragile pinkish mass necrosed margin with exudates. By day 7, the top layer of graft got dried and turned up brown. By day 14, the upper layer became more desiccated and shriveled. On day 28, the wound was not completely healed. In b-CEM on day 3, the top layer of acellular grafts appeared dark brown in color. By day 7, it got dried and turned up brown. On day 14, it was seen in a stage of detachment from the underlying tissue as only the ends sutured holding it in place. On day 21, the dried-up top layer was completely sloughed off and newly formed granulation tissue within the under lying acellular graft covered the entire surface of the wound. By day 28, size of the wound decreased markedly. The wound edges healed completely and the remaining granulation tissue in the center dried up indicating complete healing beneath it [33].

Graft-assisted healing is an important strategy for treating full thickness skin wounds. The cholecyst-derived scaffold was rich in natural biomolecules like elastin and glycosaminoglycans and when used as a xenograft, it promoted healing with excess cell proliferation at early phases and acceptable collagen deposition in the later remodeling phases [11]. After application of CEM matrix, the color of the implanted samples changed from white to dark brown and finally to dark revealed that on subsequent time intervals, decrease in vascularization and continuous loss of moisture contents of the matrix lead to change its color in to black. Kaarthick [37] also reported change in matrix color from white to brown and later on black color after the repair of full thickness skin defects in rats. Upper matrix layer detached from the wound in form of scar and the underlying layer of b-CEM matrix was completely absorbed and newly formed granulation tissue within the matrix covered the whole surface of the wound. In group I (open wound), the healing was completed at day 28 with severe contracture and scarring as no matrix was used in this group.

7.7.5 Immunological observations

7.7.5.1 Humoral response

For the ELISA procedure, serum was collected on days 0 and 28 from each rat. Harvested sera were assessed for the extent of antibodies titer generated toward the

Gall bladder-derived extracellular matrix scaffolds

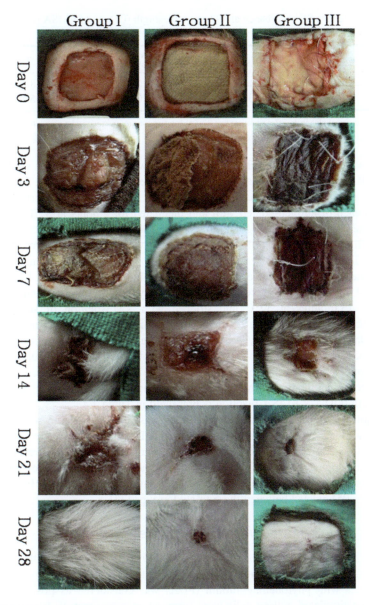

Figure 7.26 Digital color photographs of different groups at different time intervals.

implant. Briefly, 96-well flat bottom polystyrene plate was coated with 0.25 mg protein (derived from implant material) in 100 μL/well of 0.05 M sodium carbonate/sodium bicarbonate coating buffer (pH 9.6) per well. The plate was covered with aluminum foil and incubated at 4°C overnight and washed four times using washing buffer (PBS with 0.05% Tween-20; PBS-T; pH 7.4). The wells were

Table 7.4 Mean ± SE absorbance at 492 ηm wavelength (OD_{492}) of ELISA reaction on day 0 and 28.

Days	I (Control)	II (b-CS)	III (b-CEM)
Day 0	0.23 ± 0.01	0.24 ± 0.00	0.28 ± 0.03
Day 28	0.22 ± 0.03[a]	0.38 ± 0.01[a]	0.35 ± 0.01[a]

[a]Value with different alphabets differ significantly ($P < .05$) between the groups at particular time intervals.

blocked with 5% skimmed milk powder (200 μL/well) for 3 hours and washed four times. All plate washes and sample dilutions were made with PBS-T. Serum samples were run in triplicate. One hundred microliter of diluted (1:100) test serum was added to each of antigen-coated well and incubated for 2 hours at 37°C. The plate was washed and reincubated for 2 hours with 100 μL/well of the secondary antibody (goat antirat immunoglobulin G conjugated with horseradish peroxidase) in a 1:30,000 dilution. The plate was washed four times prior to the addition of 100 mL/well of the freshly prepared enzyme substrate solution (citric acid 0.1%, Na_3PO_4 0.1 M, 0.4 mg/mL o-phenylene diamine, 0.03% hydrogen peroxide) and was incubated for 15 minutes at 37°C and stopped with 3 M solution of sulfuric acid. Absorbance (OD) values were recorded at 492 nm wavelength using an automated ELISA plate reader (ECIL, Hyderabad, India). The antibodies titer present in serum samples harvested on day zero was taken as basal values [28]. The levels of antibodies present in serum prior to implantation were taken as basal values. The scaffolds specific antibodies were expressed as mean ± SE absorbance at 492 ηm wavelength (OD_{492}). ELISA reaction is presented in Table 7.4.

In control group, no rise in antibody titer was observed as the wound healed open without any matrix. In the treatment groups II and III, there were relative rise in antibody titer from day 0 to day 28 postimplantation. The B-cell response was significantly higher ($P < .05$) in the groups II and III on day 28 as evidenced by higher absorbance values when compared with day 0 values. No significant difference ($P > .05$) existed between these treated groups [33].

7.7.5.2 Cell-mediated immune response

The cell-mediated immune response was assessed by 3-(4,5-dimethylthiazol-2-yl)-2,5-diphenyltetrazolium bromide (MTT) colorimetric assay. The stimulation of rat lymphocytes with phytohemagglutinin (PHA) was considered as positive control, whereas unstimulated culture cells were taken as negative control. Blood collected from rats with implants (b-CS and b-CEM) on postimplantation days 0 and 28 was used for lymphocyte culture as per standard method. Briefly, 1 mL of blood was aseptically collected in a heparinized tube and mixed with equal volume of 1 × PBS. It was carefully layered over 2 mL of lymphocyte separation medium (Histopaque 1077) and centrifuged at 2000 rpm for 30 minutes. The buffy coat was harvested in a fresh tube and two washings were done with 1 × PBS at 1000 rpm for 5 minutes. Supernatant was discarded and pellet was resuspended in RPMI 1640 growth medium supplemented with 10% fetal bovine serum. The cells were adjusted to a concentration of 2×10^6 viable cells/mL in RPMI 1640 growth medium and seeded in 96-well tissue culture plate (100 μL/well). The cells were incubated at 37°C in 5% CO_2 environment. Cells from each rat were stimulated with antigens (b-CS and b-CEM) (10−20 μg/mL) and PHA (10 μg/mL). A T-cell mitogen in triplicates and three wells were left unstimulated for each sample. After 45 hours, 40 μL of MTT solution (5 mg/mL) was added to all the wells and incubated further for 4 hours. The plates were then centrifuged for 15 minutes in plate centrifuge at 2500 rpm. The supernatant was discarded, plates dried, and 150 mL dimethyl sulfoxide was added to each well and mixed thoroughly by repeated pipetting to dissolve the formazan crystals. The plates were immediately read at 570 nm with 620 nm as reference wavelength [28]. The stimulation index (SI) was calculated using formula:

SI = OD of stimulated cultures/OD of unstimulated cultures

The free protein contents of native bubaline gallbladder, decellularized bubaline gallbladder (b-CEM), and commercially available collagen sheet (b-CS) were estimated as per the methods of using bovine serum albumin (BSA) as a standard. The value of protein contents of native bubaline gallbladder, b-CEM, and b-CS are 3.76, 0.05, and 0.06 mg/mL, respectively. Mean ± SE stimulation index (SI) values of different groups on 28 days postimplantation are presented in Table 7.5.

Against the collagen sheet, native BG antigen the animals of group III showed significant amount of stimulation ($P < .05$). As compared to the SI values of Con A and PHA, all the groups exhibited significantly ($P < .05$) rise in SI values. The T-cell responses were lowest in the group III (implanted with acellular b-CEM) as evidenced by lower SI values [33].

The immunological response to biomaterial depends on the nature of processing and presence of potentially foreign antigen to various lymphocyte populations. The biomaterial could stimulate an antibody response, a cell-mediated sensitization, or minimal to no response. The immune response to xenogenic transplantation included both natural and induced humoral components [38]. The biological scaffolds are mainly composed of mammalian ECM and can be used for the reconstruction of various tissues and organs. These scaffolds are typically allogenic or xenogenic in origin and derived from tissues such as gallbladder, dermis, pericardium, diaphragm, small intestine submucosa, and so on. The cells in the ECM have

Table 7.5 Mean ± SE of Stimulation Index (SI) on stimulation with collagen sheet, native BG gallbladder antigen, PHA, and ConA in various groups.

Groups	Ag	PHA	ConA
Group I (C)	0.28 ± 0.07	1.13 ± 0.22[b]	1.22 ± 0.20[b]
Group II (b-CS)	0.22 ± 0.02	1.21 ± 0.21[b]	1.28 ± 0.19[b]
Group III (b-CEM)	0.29 ± 0.03[a]	1.35 ± 0.16[b]	1.02 ± 0.05[a,b]

[a]Value with different alphabets differ significantly ($P < .05$) between the groups at particular time intervals.
[b]Differ significantly ($P < .05$) from values of Ag.

class I and II histocompatibility antigens capable of eliciting rejection reactions. Also the cells have glycoproteins recognized by the immune system of hosts, which elicit rejection reactions. Therefore if these substances are eliminated from ECM, rejection reactions can be prevented. The animals of groups II and III showed slightly higher immune response on day 28 as compared to day 0. The least immune response was in animals of group I, where no graft was used. Slight immune response in groups II and III might be due to some immunogenic nature of decellularized ECM. The collagens are weak immunogenic as compared to other proteins. However, complete elimination of allo-antigens is considered difficult to perform and verify [39]. The tissues processed by decellularization and/or crosslinking to remove or mask antigenic epitopes and DNA [6]. The acellular grafts were less immunogenic having better tolerance by allogenic hosts and equally effective as isograft [40]. The biologic scaffold processing methods plays an important role in determining the host response.

7.7.6 Histopathological observations

The biopsy specimen from the implantation site was collected on 3, 7, 14, 21, and 28 days postimplantation days for histopathological evaluation. The sections were stained by H&E as per standard protocol. The host inflammatory response, neovascular tissue formation (fibroblasts, fine capillaries), deposition of neocollagen, and penetration of host inflammatory responses in the implanted matrix were evaluated as per Schallberger et al. [36]. Special staining for collagen fibers were done by using Masson's trichrome stain [41] to observe the deposition of collagen fibers and scoring was done as per the method of Borena et al. [42]. The histological parameters were graded as follows: epithelization: 1—present, 2—partially present, 3—absent; inflammation: 1—resembling normal skin, 2—mild, 3—moderate, 4—severe; fibroplasia: 1—resembling normal skin, 2—mild, 3—moderate, 4—severe; neovascularization: 1—resembling normal skin (0–1 new blood vessels), 2—mild (2–5), 3—moderate (6–10), 4—severe (>10); collagen fiber density: 1—denser, 2—dense, 3—less dense; collagen fiber thickness: 1—thicker, 2—thick, 3—thin; collagen fiber arrangement: 1—best arranged, 2—better arranged, 3—badly arranged, 4—worst arranged. The group having less histopathological score was considered the best [33].

The results of histopathological scores are presented in Table 7.6. The histopathological observations of different groups are presented in Figs. 7.27 and 7.28. In control group (C) on day 3, the wound was covered with necrotic debris and admixed with high degree inflammatory cell infiltration, edema, and congestion. The underlying stoma contained few proliferated fibroblasts and some neovascularization along with spilled neutrophils. The collagen fibers were less dense, thin, and worst arranged. The total histopathological score was 27. On day 7, proliferation of fibroblasts and angiogenesis became more, and there was no epithelialization at the margin of wound area covering the granulation tissue. The granulation tissue (fibroblast and neovascularization) was very wide, and had less dense, thin, and worst arranged collagen fibers and presence of moderate neutrophil and mononuclear cells (lymphocyte and macrophages). The histopathological score was 30 at this stage.

Table 7.6 Histopathological scores of various treatment groups at different time intervals.

Parameters	Group I (c)					Group II (b-CS)					Group III (b-CEM)				
	Days														
	3	7	14	21	28	3	7	14	21	28	3	7	14	21	28
Surface epithelium	4	4	3	2	2	3	2	2	2	2	3	2	2	2	1
Thickness of epithelium	1	2	4	3	2	2	4	4	4	3	2	4	3	2	2
Granulation tissue width	2	4	3	3	2	3	3	3	2	1	3	3	3	1	1
Inflammation	3	2	2	1	1	2	1	2	1	1	2	1	1	2	1
Fibroblast proliferation	2	4	3	4	2	4	3	4	3	2	4	4	3	2	2
Neovascularization	2	4	3	3	2	3	3	3	2	1	3	3	3	1	1
Collagen fiber density	3	3	3	2	1	3	2	2	2	1	3	1	2	1	1
Collagen fiber thickness	3	3	3	3	2	3	3	3	3	2	3	2	2	1	1
Collagen fiber arrangement	4	4	4	2	4	3	3	3	3	3	3	2	2	2	2
Total score	27	30	28	23	18	26	24	26	22	16	26	22	21	14	12

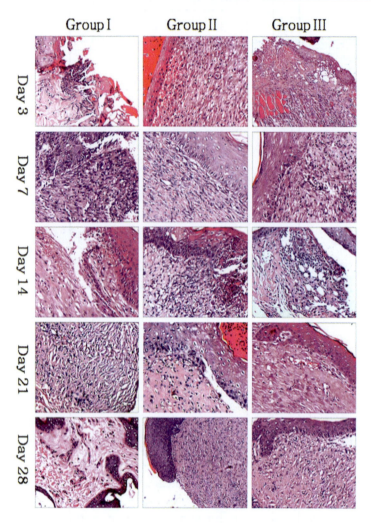

Figure 7.27 Microphotograph of different groups at different time intervals (H&E stain, 200×).

On day 14, severe proliferation of fibroblasts and neovascularization was observed along with inflammatory cell infiltration and the score was 28. On day 21, collagen formation was evident in some areas (immature collagen) and inflammation was greatly reduced and the histopathological score was reduced to 23. On day 28, a high degree of collagen deposition with worst arrangement was observed. The score was 18 at this stage [33].

In group II (b-CS), inflammatory changes were observed on day 3 postoperatively. The fibroblast proliferation was found in collagen sheet implanted tissue and some neovascularization was also observed. There was necrosis and sloughing of the

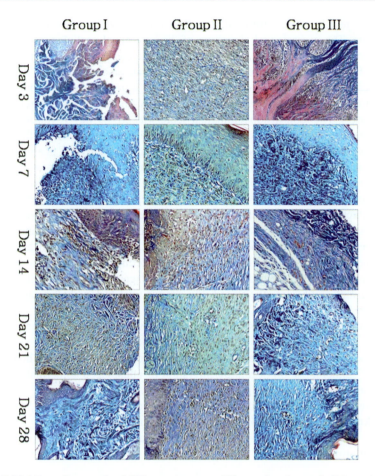

Figure 7.28 Microphotograph of different groups at different time intervals (Masson's trichrome stain, 200 ×).

superficial layer and epithelialization was observed at edges of wound. The histopathological score was 26. On day 7, the wounded tissue nearer to host tissue was severely infiltrated by proliferating fibroblasts. The surface epithelium was thicker than normal skin. The fibrous tissue also revealed numerous blood vessels. The total histopathological score was 24. On day 14, the deeper layer of dermis revealed deposition of new collagen fibers in which fibroblasts were dispersed and epithelialization was incomplete with reduced inflammation. The score was 26 at this stage. By day 21, epithelialization had started. The inflammatory changes and neovascularization were reduced as well as score was reduced to 22. On day 28, superficial epithelialization was observed. The histopathological score was 16 [33].

In group III (b-CEM), on day 3, CEM graft was present over the healing wound tissue, but the neoepithelium was not prominent. Mild inflammatory changes were observed. The fibroblasts proliferation and neovascularization were observed in the graft tissue. The necrosis and sloughing of the superficial acellular scaffold matrix were observed and epithelialization was started at the edges of wound area. The collagen fibers were less dense, thin with worst arrangement. However, connections of graft collagen fibers with wound collagen fibers were observed. At this stage, the granulation tissue included only scattered collagen fibers and some newly forming blood vessels. The total histopathological score was 26. By day 7, fibroblast proliferation became more prominent and surface epithelium was thicker than normal skin. The superficial graft showed sloughing and underlying acellular matrix showed severe neovascularization within the stroma. The total score was 22. By day 14, inflammation and neovascularization were less as compared to day 7. The total histopathological score was 14. By day 21, epithelialization observed in margin was thicker than normal skin. The new collagen deposition in matrix was with best arrangement and the score was 14 at this stage. By day 28, the granulation tissue was covered and the epithelization was almost complete. The hair follicles and skin glands were also seen. The process of remodeling was evident on these samples with reorganization and disappearance of collagen; densely packed thick bundles were replaced by loosely placed thin strands total histopathological score was reduced to 12 [33].

Full thickness skin wounds are characterized by a complete destruction of the epithelial regenerative elements that reside in the dermis. Full thickness skin wound healing occurs by granulation tissue formation, contraction and epithelialization [55]. The epithelialization occurs by migration of undamaged epidermal cells from the wound margins across the granulation bed [43].

Surface epithelium: 1—Complete, 2—Incomplete, 3—Initialization, 4—Absent

Thickness of surface epithelium: 1—Resembling normal skin, 2—Slightly thick, 3—Moderately thick, 4—Thicker than normal skin

Granulation tissue width: 1—Narrow, 2—Slightly wide, 3—Moderately wide, 4—Wide

Inflammation: 1—Absent, 2—Mild, 3—Moderate, 4—Severe

Fibroblast proliferation: 1—Resembling normal skin, 2—Mild, 3—Moderate, 4—Severe

Neovascularization: 1—Resembling normal skin, 2—Mild, 3—Moderate, 4—Severe

Collagen fiber density: 1—Denser, 2—Dense, 3—Less dense

Collagen fiber thickness: 1—Thicker, 2—Thick, 3—Thin

Collagen fiber arrangement: 1—Best, 2—Better, 3—Worse, 4—Worst

The exogenous collagen supplementation enabled faster migration of cells that are involved in cutaneous wound healing. Since the exogenous collagen is molecular in nature [44] and supplies endogenous collagen in vivo, it readily integrates with the wound tissue and facilitates the attachment, migration, and proliferation of cells on the wound site [45].

The double acellular matrix applied in group III showed necrosis of the superficial matrix, and underlying matrix showed severe fibroblastic proliferation with inflammatory cells. Group I exhibited severe proliferation of fibroblast with inflammatory cells in the wound area on day 3. Moderate degree of neovascularization was detected in all the groups. No new collagen formation was detected in any of the groups. But the graft implanted samples showed the existence of mature collagen of b-CEM origin. Although inflammation is necessary for healing by fighting infection and inducing the proliferation phase, healing proceeds only after inflammation is controlled [46]. During the initial stages of the wound healing, fibroblasts from the surrounding normal area are known to migrate and proliferate into the wound site and, within 3−4 days, get converted to myofibroblasts [47]. The main function of myofibroblasts in a healing wound is contraction by synthesizing ECM proteins, notably, collagen types I−VI and XVIII, laminin, thrombospondin, glycoproteins, and proteoglycans for the dermal repair. The myofibroblasts usually go on increasing from the inflammatory stage (3−4 days). During the onset of healing, the proliferating fibroblast starts to synthesize collagen, and the total collagen content increases preferentially.

On day 7, postimplantation, moderate-to-severe inflammation was present in all the groups and it was minimum in groups II and III. The early control of inflammation might be due to application of collagen matrix which facilitate the progress to the next phase of wound healing. The proliferation of fibroblasts was more in the group III. Similar findings were observed by Perme et al. [48]. There was an increase in the extent of collagen deposition by 7 days (about 20%) and 14 days (about 40%) [11].

On day 14, least scores were observed in groups III (21) followed by group II (26) may be due to early acceptance of matrix. The sloughing of the upper layer of matrix was observed in all the groups except control group where wound remained open. It may be either due to the desiccation of the graft in high environmental temperature or due to an impaired formation of new blood vessels [49]. The myofibroblasts proliferate till the end of proliferation phase (14−15 days). Meanwhile, the underlying granulation tissue increased in mass that pushed up the graft upwards. The animals of groups II and III showed well-formed collagen and neovascularization with superficial epithelialization. The epithelialization and neovascularization was faster in group III as compared to other groups.

On day 21, the least score was observed in group III (14) followed by group II (22). The epithelialization was more similar to the normal skin in the wounds of these groups. In the remodeling phase (around 21−30 days), these myofibroblasts undergo apoptosis [11]. At 21 days of postimplantation, the acellular matrix groups showed higher collagen synthesis as compared to the control group as appreciated by the Masson's trichrome staining. Purohit et al. [50] on day 21 found that the acellular dermal matrix throughout the width and length was replaced by mature collagenous connective tissue in experimentally created wounds in rabbits. On day 28, postimplantation the control group healed completely leaving abundant scar tissue but no complete healing was observed for other groups also in group II contraction and scar formation found. In group III, collagen fiber arrangement was almost similar to normal skin. Marked fibroblastic response associated with an abundant new collagenous fibrous tissue was observed in these groups.

7.8 Experimental evaluation of pig cholecyst-derived extracellular matrix in a rat model

7.8.1 Skin wound creation and implantation

Thirty-six clinically healthy Sprague Dawley rats of either sex, weighing from 250 to 300 g, and between 3 and 4 months of age were used in this study. The animals were randomly divided into three equal groups of 12 animals each. The surgical procedure and postoperative care was described in detail in experimental evaluation of b-CEM in a rat model section. The defect in group I (C) was kept as open wound and was taken as control (C). In group II, the defect was repaired with commercially available collagen sheet (b-CS). In group III, the defect was repaired with cholecyst-derived ECM of porcine origin (p-CEM). The healing was evaluated on the basis for the following parameters.

7.8.2 Wound contraction

Mean ± SE of the total wound area (mm^2) of the skin wounds at different time intervals are presented in Table 7.7. A gradual decrease in wound area (mm^2) was observed in all the groups during the entire observation period. The wound area decreased significantly ($P < .05$) at various time intervals in different groups. There was significant decrease ($P < .05$) in wound area on day 7 in group III as compared to groups I and II. On day 21, no significant difference in wound area was observed between all the three groups. However, the wound healed completely in group III on day 28.

Percent contraction (mean ± SE) of wound area at different time intervals in different groups is presented in Table 7.8. A gradual increase in percent wound contraction was observed in all the groups during the observation period. On day 7, severe wound contraction was observed in group I (C) as the wound healed denuded

Table 7.7 Mean ± SE of the wound area (mm^2) at different time intervals in various groups.

Groups	Time intervals (days)				
	0	7	14	21	28
Group I (C)	429.22 ± 17.71	341.44 ± 14.82[a,b]	157.22 ± 10.97[a,b]	45 ± 5.89[a,b]	5.56 ± 2.42[a,b]
Group II (b-CS)	417.78 ± 18.94	314.11 ± 22.50[a,b]	131.78 ± 25.61[a,b]	45.89 ± 8.44[a,b]	15.89 ± 6.55[a,b]
Group III (p-CEM)	447.44 ± 13.64	258.44 ± 17.32[a,b]	186 ± 18.18[a,b]	43.89 ± 7.917[a,b]	0.56 ± 1.98[a,b]

[a]Value with different alphabets differ significantly ($P < .05$) between the groups at particular time intervals.
[b]Differ significantly ($P < .05$) from day 0 values.

Table 7.8 Mean ± SE of percent contraction of wound (%) at different time intervals in various groups.

Groups	Time interval (days)			
	7	14	21	28
Group I (C)	70.76 ± 9.92[a]	62.96 ± 3.08[a]	89.46 ± 1.32[a]	98.77 ± 0.47[a]
Group II (b-CS)	49.18 ± 12.36[a]	72.75 ± 3.82[a]	88.68 ± 2.28[a]	95.89 ± 1.87[a]
Group III (p-CEM)	43.01 ± 3.41[a]	58.61 ± 3.84[a]	89.93 ± 2.17[a]	99.85 ± 0.15[a]

[a]Value with different alphabets differ significantly ($P < .05$) between the groups at particular time intervals.

and no scaffold matrix was used in this group. Wound contraction was significantly higher ($P < .05$) in control group at this stage when compared to groups II (b-CS) and III (p-CEM). However, no difference was observed on day 21 in all the three groups. On day 28, maximum wound contraction (99.85%) was recorded in group III animals where p-CEM was used.

The wound contraction is the centripetal displacement of the wound edges that facilitates its closure after trauma. This process is carried out by myofibroblasts that contain α-actin from smooth muscle and is mediated by contractile forces produced by granulation tissue from wound [35]. The wound contraction was assessed by percent retention of the original wound area as reported by Schallaberger et al. [36]. The wound contraction has been used to monitor wound healing. The wound area decreased gradually as the healing progressed. All wounds initially retracted and continued to retract because of viscoelastic properties of the skin. Delay in wound contraction was seen in groups II and III where matrices (b-CS, p-CEM) were used and in consistent with findings of a study by Prevel et al. [51], where the full thickness 20 mm diameter wounds in rodents were inflicted. They found average wound contraction rate of 33% of porcine small intestine submucosa-treated wounds and 56% for the control wounds. In the present study, on day 7, about 70% contraction was seen in control group whereas the contraction was 49% in b-CS and 43% in p-CEM-treated groups. There are numerous studies that report inhibition of wound contraction after implanting a collagenous ECM into full thickness wounds [36,52–54]. Maximum percent contraction was recorded in group I as no collagenous matrix was applied and wound healed as open wound. The wound in animals of this group healed completely by 28 days leaving a large scar indicating the existence of severe contraction. The p-CEM matrix implanted group healed with minimum contraction. The high rate of healing with minimum contraction was recorded in p-CEM-grafted group as compared to b-CS-treated group

7.8.3 Gross observations/planimetry

Color photographs were taken on postimplantation days 0, 7, 14, 21, and 28 with the help of digital camera at a fixed distance. The representative images are presented in Fig. 7.29.

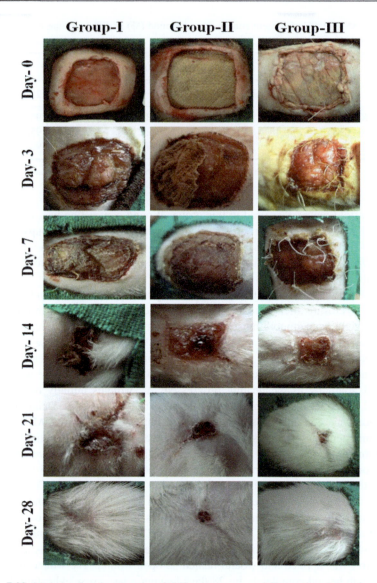

Figure 7.29 Digital color photographs of different groups at different time intervals

On day 7, in group I (C), the wound surface became more desiccated and necrosed with some amount of exudates. By day 14, the wound size decreased markedly and a thick crust developed. and it started detaching leaving a raw granular pink tissue. By day 28, wound healed up completely by severe contraction leaving a large scar. In group II (b-CS) on day 7, the top layer of graft got dried and turned up brown. By day 14, the upper layer became more desiccated and shriveled. On day 28, the wound was not completely healed. In group III (p-CEM), on day 7, the top layer became dried and brown in color. On day 14, the top layer appeared pinkish in color indicating upper layer sloughed off completely exposing the healing tissue. By day 21, wound margins healed leaving a small central area. By day 28, the wound healed up completely with no scar and on day 45, the implanted area appeared similar to the normal skin.

The color digital image processing was undertaken to study the healing of wounds in animals of different groups at different time intervals. After application of p-CEM matrix, the color of the implanted samples changed from white to dark brown and finally to dark. Upper matrix layer was detached from the wound in form of scab and the underlying layer of p-CEM matrix was completely absorbed and newly formed granulation tissue within the matrix covered the whole surface of the wound. Minimum scar formation was observed in p-CEM group followed by b-CS-implanted group.

7.8.4 Immunological observations

7.8.4.1 Humoral response

The value of protein contents of native porcine gallbladder, p-CEM, and b-CS are 2.13, 0.05, and 0.06 mg/mL, respectively. For the ELISA procedure, serum was collected on days 0 and 28 from each rat to assess the extent of antibodies titer generated toward the implant. The detail procedure has been already been described in evaluation of b-CEM section.

The scaffold-specific antibodies expressed as mean ± SE absorbance at 492 ηm wavelength (OD_{492}) are in presented Fig. 7.30. The levels of antibodies present in serum prior to implantation were taken as basal values. In control group, no rise in antibody titer was observed as the wound healed open without any matrix. In the treatment groups II and III, there were relative nonsignificant rise in antibody titer from day 0 to day 28 postimplantation. No significant difference ($P > .05$) existed between these treated groups.

7.8.4.2 Cell-mediated immune response

The cell-mediated immune response was assessed by 3-(4,5-dimethylthiazol-2-yl)-2,5-diphenyltetrazolium bromide (MTT) colorimetric assay. The stimulation of rat

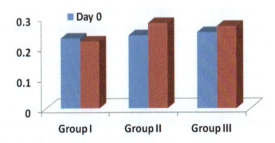

Figure 7.30 Scaffolds specific antibodies were expressed as mean ± SE absorbance at 492 ηm wavelength (OD_{492}).

lymphocytes with phytohemagglutinin (PHA) was considered as positive control, whereas unstimulated culture cells were taken as negative control. Blood collected from rats with implants (b-CS and b-CEM) on postimplantation days 0 and 28 was used for lymphocyte culture as per standard protocols. The plates were immediately read at 570 nm with 620 nm as reference wavelength. The stimulation index (SI) was calculated using formula:

SI = OD of stimulated cultures/OD of unstimulated cultures

No significant difference was observed between the treated groups for cell-mediated immune response. In humoral response, relative nonsignificant rise in antibody titer from day 0 to day 28 was observed. No significant difference existed between b-CS-treated animals and p-CEM-treated animals. It also shows that the developed p-CEM behaves more or less similar to commercially available bovine collagen sheath (b-CS). The nature and degree of immunological response to a foreign material is a crucial variable affecting the outcome and success or failure of implanted biomaterial. The biomaterial could stimulate an antibody response, a cell-mediated sensitization, or minimal to no response. The immune response to xenogenic transplantation included both natural and induced humoral components [38]. The biological scaffold processing methods play an important role in determining the host response, and in the present study processing methods remove all the cells; thus minimal or no host response was observed.

7.8.5 Histological observations

The biopsy specimens from the implantation site were collected on 7, 14, 21, and 28 days postimplantation for histopathological evaluation. The results of histopathological scores are presented in Table 7.9. The histopathological observations of different groups are presented in Figs. 7.31 and 7.32. The group having less histopathological score is considered the best.

In group I (C), on day 7, proliferation of fibroblasts and angiogenesis became more and there was no epithelialization at the margin of wound area covering the granulation tissue. The granulation tissue (fibroblast and neovascularization) was very wide having thin loose arranged collagen fibers with the presence of moderate neutrophil and mononuclear cells (lymphocyte and macrophages). The histopathological score was 30 at this stage. On day 14, severe proliferation of fibroblasts and neovascularization was observed along with inflammatory cell infiltration and the score was 28. On day 21, collagen formation was evident in some areas (immature collagen) and inflammation was greatly reduced and the histopathological score was reduced to 23. On day 28, a high degree of collagen deposition with worst arrangement was observed. The score was 18 at this stage.

In group II (b-CS), on day 7 postoperatively, the wounded tissue nearer to host tissue was severely infiltrated by proliferating fibroblasts. The surface epithelium was thicker than normal skin. The fibrous tissue also revealed numerous blood vessels. The total histopathological score was 24. On day 14, the deeper layer of dermis revealed deposition of new collagen fibers in which fibroblasts were dispersed, epithelialization was incomplete, and inflammation was found reduced. The score was

Table 7.9 Histopathological scores of various treatment groups at different time intervals.

Parameters	Group I (c)				Group II (b-CS)				Group III (p-CEM)			
	\multicolumn{12}{c}{Days}											
	7	14	21	28	7	14	21	28	7	14	21	28
Surface epithelium	4	3	2	2	2	2	2	2	2	2	2	1
Thickness of epithelium	2	4	3	2	4	4	4	3	4	3	2	2
Granulation tissue width	4	3	3	2	3	3	2	1	4	3	1	1
Inflammation	2	2	1	1	1	2	1	1	1	1	2	1
Fibroblast proliferation	4	3	4	2	3	4	3	2	4	3	2	2
Neovascularization	4	3	3	2	3	3	2	1	4	3	1	1
Collagen fiber density	3	3	2	1	2	2	2	1	2	1	1	1
Collagen fiber thickness	3	3	3	2	3	3	3	2	2	2	2	2
Collagen fiber arrangement	4	4	2	4	3	3	3	3	2	2	2	2
Total score	30	28	23	18	24	26	22	16	25	20	15	13

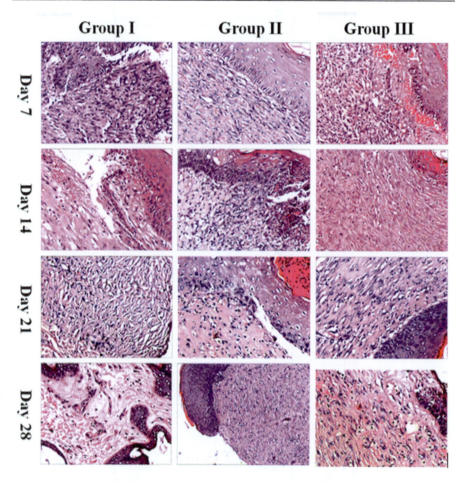

Figure 7.31 Histopathological observations of wound healing of different groups at various time intervals (H&E staining, ×100).

26 at this stage. By day 21, epithelialization had started. The inflammatory changes and neovascularization were reduced as well as score was reduced to 22. On day 28, superficial epithelialization was observed. The histopathological score was 16.

In group III (p-CEM) on day 7, cells started migrating inside the lower matrix and the sloughing changes of the superficial layer became more prominent. The stroma proliferation with fibroblasts and newer capillaries was very prominent. The neoepidermis was thicker and the granulation tissue was very prominent in the dermis. The newly formed keratinocyte was seen in both graft-induced healing reactions, and the collagen fibers were further organized into thick bundles. The collagen fibers were dense and thin with worst arrangement. The histopathological score was 25. On day 14, the reepithelialization was almost complete and marginal

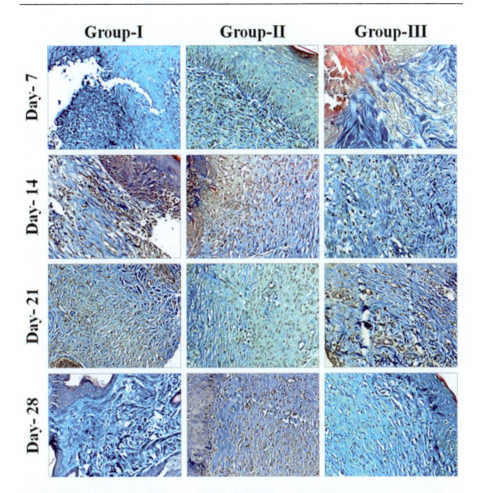

Figure 7.32 Histopathological observations of wound healing of different groups at various time intervals (Masson's trichrome staining, ×100).

epithelialization was thick than normal skin and denser collagen formation with better arrangement was observed and the histopathological score was 20. On day 21, the collagen formation was appreciable. The hair follicles and skin glands were seen and epithelium was like normal skin. The histopathological score was 15. By day 28, there was formation of complete epidermis. The dermis was well developed with newly formed blood vessels and well-organized collagen fibers. The process of remodeling was evident on these samples with reorganization and disappearance of collagen; densely packed thick bundles were replaced by loosely placed thin strands and the histopathological score was 13.

Full thickness skin wounds are characterized by a complete destruction of the epithelial regenerative elements that reside in the dermis. Full thickness skin wound healing occurs by granulation tissue formation, contraction, and epithelialization [55]. The epithelialization occurs by migration of undamaged epidermal cells from the wound margins across the granulation bed [43]. The exogenous collagen supplementation enabled faster migration of cells that are involved in cutaneous wound healing. Since the exogenous collagen is molecular in nature [44] and supplies endogenous collagen in vivo, it readily integrates with the wound tissue and facilitates the attachment, migration, and proliferation of cells on the wound site [45].

On day 7, postimplantation, moderate-to-severe inflammation was present in all the groups and it was minimum in groups II and III. The early control of inflammation might be due to application of collagen matrix as in case of groups II and III might facilitate the progress to the next phase of wound healing. The proliferation of fibroblasts was comparatively more in the group III. Remya et al. [54] also reported proliferation of fibroblasts in acellular fish-swim-treated rats. There was an increase in the extent of collagen deposition by 7 days (about 20%) and 14 days (about 40%) [11].

On day 14, least score observed in group III (20) followed by group II (26) may be due to early acceptance of matrix. In control group, the score was 28. The sloughing of the upper layer of matrix was observed in group III. It may be either due to the desiccation of the matrix in high environmental temperature or due to an impaired formation of new blood vessels [49]. The myofibroblasts proliferate till the end of proliferation phase (14–15 days). Meanwhile, the underlying granulation tissue increased in mass that pushed up the graft upward. The animals of groups II and III showed well-formed collagen and neovascularization with superficial epithelialization. The epithelialization and neovascularization were faster in group III as compared to other groups. Faster epithelialization and neovascularization were reported in acellular fish swim bladder construct seeded with mesenchymal stem cells for full thickness skin wounds in rats [54].

On day 21, the least score was observed in group III (15) followed by group II (22), and in control group, the score was 23. The epithelialization was more similar to the normal skin in the wounds of these groups. In the remodeling phase (around 21–30 days), these myofibroblasts undergo apoptosis [50]. At day 21 of postimplantation, the p-CEM-implanted group showed higher collagen synthesis as compared to the control group as appreciated by the Masson's trichrome staining. Purohit [56] on day 21 found that the acellular dermal matrix throughout the width and length was replaced by mature collagenous connective tissue in experimentally created wounds in rabbits.

On day 28 of postimplantation, the control group healed completely leaving abundant scar tissue and also in group II contraction and scar formation found. In group III, the collagen fiber arrangement was almost similar to normal skin. Marked fibroblastic response associated with an abundant new collagenous fibrous tissue was observed. The deposition of new collagen was oriented parallel to the skin surface. In p-CEM group, hair follicle and skin glands could be detected as in case of normal skin indicating the culmination of repair process. Here the healing was with minimum contraction and scar formation. Minimal histopathological score (13) was observed in this group.

7.9 Conclusion

We illustrate protocols for decellularization of the buffalo gall bladder with ionic biological detergent (0.5% and 1% SDS) and nonionic biological detergent (0.5% and 1% Triton X-100) for different time intervals. The efficacy of the decellularization method for the preservation of the ECM scaffold structure and integrity was confirmed using histology, scanning electron microscopy and DNA quantification. Treatment with 0.5% sodium dodecyl sulfate for 48 hours resulted in complete decellularization of buffalo gall bladder. Decellularization of pig gall bladder was done with ionic biological detergent (0.5% and 1% SDS) and nonionic biological detergent (0.5% and 1% Triton X-100) for different time intervals. Treatment with 0.5% sodium dodecyl sulfate for 48 hours resulted in complete decellularization of pig gall bladder. The soapnut pericarp extract (5%) for making native gall bladder tissue acellular was found suitable as complete acellularity of gall bladder was obtained at 48 hours with slight debris at places. The acellular matrices developed from goat gall bladder may be used without topical antibiotics or antifungal agents as the soapnut pericarp extract has the antimicrobial and fungicidal properties. The b-CEM and p-CEM both have healing potential and shown better healing response than commercially available collagen sheet to reconstruct full thickness skin wounds in rats. Histological observations showed that both b-CEM and p-CEM augmented wound healing activity by increasing cellular proliferation, neovascularization, synthesis of collagen, epithelialization, and early histological maturation in excisional wounds.

References

[1] Daley WP, Peters SB, Larsen M. Extracellular matrix dynamics in development and regenerative medicine. J Cell Sci 2008;121:255−64.
[2] Reing JE, Zhang L, Myers-Irvin J, Cordero KE, Freytes DO, Heber-Katz E, et al. Degradation products of extracellular matrix affect cell migration and proliferation. Tissue Eng Part A 2009;15:605−14.
[3] Kleinman HK, Philp D, Hoffman MP. Role of the extracellular matrix in morphogenesis. Curr Opin Biotech 2003;14:526−32.
[4] Raines EW. The extracellular matrix can regulate vascular cell migration, proliferation, and survival: relationships to vascular disease. Int J Exp Pathol 2000;81:173−82.
[5] Badylak SF, Freytes DO, Gilbert TW. Extracellular matrix as a biological scaffold material: structure and function. Acta Biomat 2009;5:1−13.
[6] Gilbert TW, Sellaro TL, Badylak SF. Decellularization of tissues and organs. Biomaterials 2006;27(36):75−83.
[7] Coburn JC, Pandit A. Development of naturally-derived biomaterials and optimization of their biomechanical properties. Biomaterials 2007;3:4−15.
[8] Burugapalli K, Thapasimuttu A, Chan JCY, Yao L, Brody S, Kelly JL, et al. Scaffold with a natural mesh-like architecture: isolation, structural, and in vitro characterization. Biomacromolecules 2007 2007;8:928−36.

[9] Coburn JC, Brody S, Billiar KL, Pandit A. Biaxial mechanical evaluation of cholecyst-derived extracellular matrix: a weakly anisotropic potential tissue engineering biomaterial. J Biomed Mater Res A 2007;81:250−6.
[10] Li B, Wang JHC. Fibroblasts and myofibroblasts in wound healing: force generation and measurement. J Tissue Viability 2011;20:108−20.
[11] Revi D, Vadavanath PV, Jaseer M, Akhila R, Anilkumar TV. Porcine cholecyst−derived scaffold promotes full-thickness wound healing in rabbit. J BiomaterTissue Eng 2013;3:261−72.
[12] Chen F, Yoo JJ, Atala A. Acellular collagen matrix as a possible "off the shelf" biomaterial for urethral repair. Urology 1999;54:407−10.
[13] Yannas IV, Burke JF. Design of an artificial skin. I. Basic design principles. J Biomed Mater Res 1980;14:65−81.
[14] Hahm G, Glaser JJ, Elster EA. Biomarkers to predict wound healing: the future of complex war wound management. Plast Reconstr Surg 2011;127(1):21−6.
[15] Palamand S, Reed AM, Weimann LJ. Testing intelligent wound dressings. J Biomater Appl 1992;6(3):198−215. Available from: https://doi.org/10.1177/088532829200600302.
[16] Vatankhah E, Prabhakaran MP, Jin G. Development of nanofibrous cellulose acetate/gelatin skin substitutes for variety wound treatment applications. J Biomater Appl 2013;2:1−10.
[17] Chang H, Wind S, Kerstein MD. Moist wound healing. Dermatol Nurs 1996;8:174−6.
[18] Bryan J. Moist wound healing: a concept that changed our practice. J Wound Care 2004;13:227−8.
[19] Szycher M, Lee SJ. Modern wound dressings: a systematic approach to wound healing. J Biomater Appl 1992. Available from: https://doi.org/10.1177/088532829200700204.
[20] Tran RT, Thevenot P, Zhang Y. Scaffold sheet design strategy for soft tissue engineering. Nat Mater 2010;3:1375−89.
[21] Robinet A, Arnaud A, Fahem A, Cauchard JH, Huet E, Vincent L, et al. Elastin-derived peptides enhance angiogenesis by promoting endothelial cell migration and tubulogenesis through upregulation of MT1-MMP. J Cell Sci 2005;118(2):343−56.
[22] Hafemann B, Ensslen S, Erdmann C, Niedballa R, Zühlke A, Ghofrani K, et al. Use of a collagen/elastin-membrane for the tissue engineering of dermis. Burns 1999;25(5):373−84.
[23] Rnjak J, Li Z, Maitz PKM, Wise SG, Anthony S, Weiss AS. Primary human dermal fibroblast interactions with open weave three-dimensional scaffolds prepared from synthetic human elastin. Biomaterials 2009;30(32):6469−77.
[24] Townsend A, Bodmer H. Antigen recognition by class I restricted T lymphocytes. Annu Rev Immunol 1989;7:601−24.
[25] Kirker R, Luo Y, Nielson JH, Shelby J, Glenn D, Prestwich GD. Glycosaminoglycan hydrogel films as bio-interactive dressings for wound healing. Biomaterials 2002;23(17):3661−71.
[26] Keane TJ, Londono R, Turner NJ. Consequences of ineffective decellularization of biologic scaffolds on the host response. Biomaterials 2012;33:1771−81.
[27] Pandit A., Anilkumar T.V. Tissue graft scaffold made from cholecyst-derived extracellular matrix. US patent publication 2009; US 2006/0159664 A1.
[28] Shakya P. Evaluation cholecyst derived-extracellular matrix reconstruction full thickness skwounds rats. MVSc thesis submitted to Deemed University Indian Veterinary Research Institute; Izatnagar 243122, Uttar Pradesh, India; 2015.
[29] Verma M.K. Study on development of protocol for decellularization and in-vitro determination of biocompatibility of acellular buffalo dermis and fish swim bladder. MVSc

thesis submitted to Narendra Deva University of Agriculture and Technology; Kumarganj, Faizabad, India; 2016.
[30] Sachan AK. Study on in-vitro determination of biocompatibility of cross-linked caprine urinary bladder and gall bladder acellular matrices decellularized by aqueous extract (5%) of Sapindus mukorossi. MVSc thesis submitted to Narendra Dev University of Agriculture and Technology, Kumarganj, Faizabad, Uttar Pradesh, India; 2015.
[31] Gangwar AK, Kumar N, Sharma AK, Kh Devi S, Negi M, Shrivastava S, et al. Bioengineered acellular dermal matrix for the repair of full thickness skin wounds in rats. Trends Biomat Artific Organs 2013;27(2):67−80.
[32] Tarnowski BI, Spinale FG, Nicholson JH. DAPI as a useful stain for nuclear quantification quantification. Biotechnic Histochemistry 1991;66(6):296−302. Available from: https://doi.org/10.3109/10520299109109990).
[33] Shakya P, Sharma AK, Kumar N, Remya V, Mathew DD, Dubey P, et al. Bubaline cholecyst derived extracellular matrix forreconstruction of full thickness skin wounds in rats. Scientifica 2016;2638371. Available from: https://doi.org/10.1155/2016/2638371.
[34] Stringer SK, Seligmann BE. Effects of two injectable anesthetic agents on coagulation assays in the rat. Lab Anim Sci 1996;46(4):430−3.
[35] Neagos D, Mitran V, Chiracu G, Ciubar R, Iancu C, Stan C, et al. Skin wound healing in a free floating fibroblast populated collagen latice model. Romanian J Biophys 2006;16(3):157−68.
[36] Schallberger SP, Stanley BJ, Hauptaman JG, Steficek BA. Effect of porcine small intestine submucosa on acute full-thickness wounds in dogs. Vet Surg 2008;37:515−24.
[37] Kaarthick D.T. Repair of cutaneous wounds using acellular diaphragm and pericardium of buffalo origin seeded with in-vitro cultured mouse embryonic fibroblasts cells in rat model. MVSc thesis submitted to Deemed University Indian Veterinary Research Institute, Izatnagar 243122, Uttar Pradesh, India; 2011.
[38] Ruszczak Z. Effect of collagen matrices on dermal wound healing. Adv Drug Deliv Rev 2003;55:1595−611.
[39] Malone JM, Brendel K, Duhamil RC, Reinert RL. Detergent-extracted small diameter vascular prostheses. J Vasc Surg 1984;1:181−91.
[40] Gulati AK, Cole GP. Immunogenicity and regenerative potential of acellular nerve allograft to repair peripheral nerve in rats and rabbits. Acta Neurochir Wien 1994; 126:158−64.
[41] Masson PJ. Some histological methods: trichrome staining and their preliminary techniques. J Tech Methods 1929;12:75−90.
[42] Borena BM, Pawde A, Amarpal, Aithal HP, Kinjavedekar P, Singh R, et al. Autologous bone marrow-derived cells for healing excisional dermal wounds of rabbits. Vet Rec 2009;165:563−8.
[43] Swaim SF, Henderson RA. Wound management. Small animal wound management, 9. Lea and Febiger; 1990. p. 33.
[44] Nithya M, Suguna L, Rose C. The effect of nerve growth factor on the early responses during the process of wound healing. Biochemica et Biophysica Acta 2003; 1620:25−31.
[45] Judith R, Nithya M, Rose C, Mandal AB. Application of a PDGF containing novel gel for cutaneous wound healing. Life Sci 2010;87:1−8.
[46] Midwood KS, Williams LV, Schwarzbauer JE. Tissue repair and the dynamics of the extracellular matrix. Int J Biochem Cell Biol 2004;36:1031−7.

[47] Hinz B, White ES. The myofibroblast matrix: implications for tissue repair and fibrosis. J Pathol 2013;229:298−309.
[48] Perme H, Sharma AK, Kumar N, Singh H, Maiti SK, Singh R. In-vivobiocompatibility evaluation of crosslinked cellular and acellular bovine pericardium. Indian J Anim Sci 2009;79:658−61.
[49] Boyce ST. Cultured skin substitutes: a review. Tissue Eng 1996;2:255−66.
[50] Purohit S, Kumar N, Maiti SK, Sharma SK, Shrivastava S, Saxena S, et al. In vivo biocompatibility determination of crosslinked acellular dermal matrix. Trends Biomat Artific Organs 2019;33(4):111−23.
[51] Prevel CD, Eppley BL, Summerlin DJ. Small intestinal submucosa: utilization as a wound dressing in full thickness rodent wounds. Ann Plast Surg 1995;35:382−8.
[52] Gangwar AK, Kumar N, Khangembam SD, Kumar V, Singh R. Primary chicken embryo fibroblasts seeded acellular dermal matrix (3-D ADM) improve regeneration of full thickness skin wounds in rats. Tissue Cell 2015;47:311−22. Available from: http://doi.org/10.1016/j.tice.2015.04.002.
[53] Remya V, Kumar N, Sharma AK, Sonal, Negi M, Maiti SK, et al. Acellular fish swim bladder biomaterial construct seeded with mesenchymal stem cells for full thickness skin wound healing in rats. Trends Biomat Artific Organs 2014;28(4):127−35.
[54] Kumar V, Kumar N, Gangwar AK, Singh H, Singh R. Comparative histologic and immunologic evaluation of 1,4-butanediol diglycidyl ether cross linked versus non cross linked acellular swim bladder matrix for healing of full-thickness skin wounds in rabbits. Wound Med 2015;7:24−33. Available from: http://doi.org/10.1016/j.jss.2015.04.080.
[55] Fossum TW, Hedlund CS, Johnson AL, Schulz KS, Seim HB, Willard MD, et al. Surgery of integumentary system. Manual of small animal surgery. Mosby; 2007. p. 159−75.
[56] Purohit S. Biocompatibility testing of acellular dermal grafts in a rabbit model. An in-vitro and in-vivo study. PhD thesis submitted to Deemed University, Indian Veterinary Research Institute, Izatnagar, Uttar Pradesh, India; 2008.

Aorta-derived extracellular matrix scaffolds and clinical application

Jetty Devarathnam[1], Ashok Kumar Sharma[2], Naveen Kumar[2,], Vineet Kumar[3], Shruti Vora[4], Kaarthick D.T.[5], Anil Kumar Gangwar[6], Rukmani Dewangan[7], Himani Singh[2], Sameer Shrivastava[8], Sonal Saxena[8], Kalaiselvan E.[2], Shivaraju S.[2] and Swapan Kumar Maiti[2]*

[1]Department of Surgery and Radiology, College of Veterinary Science, Sri Venkateswara Veterinary University, Proddatur, Andhra Pradesh, India, [2]Division of Surgery, ICAR-Indian Veterinary Research Institute, Izatnagar, Uttar Pradesh, India, [3]Department of Veterinary Surgery and Radiology, College of Veterinary and Animal Sciences, Bihar Animal Sciences University, Kishanganj, Bihar, India, [4]Department of Veterinary Surgery and Radiology, College of Veterinary Science and Animal Husbandry, Junagadh Agricultural University, Junagadh, Gujarat, India, [5]Veterinary Clinical Complex, Veterinary College and Research Institute, Thanjavur, Tamil Nadu, India, [6]Department of Veterinary Surgery & Radiology, College of Veterinary Science & Animal Husbandry, Acharya Narendra Deva University of Agriculture and Technology, Ayodhya, Uttar Pradesh, India, [7]Department of Veterinary Surgery and Radiology, College of Veterinary Science and Animal Husbandry, Dau Shri Vasudev Chandrakar Kamdhenu Vishwavidyalaya, Durg, Chhattisgarh, India, [8]Division of Veterinary Biotechnology, ICAR-Indian Veterinary Research Institute, Izatnagar, Uttar Pradesh, India

8.1 Introduction

Reconstructive surgery is an innovative field of science concerned with utilization of various synthetic and biological materials as implants and prostheses. Abdominal wall reconstruction is one of the challenging tasks in reconstructive surgery due to the limitations of available mesh materials. Even though much research has been carried out in this field, till date, there is no such material which can be used in abdominal wall reconstruction with promising success. The principal concept in management of abdominal wall defects is "tension-free" closure. Previously several synthetic materials like polypropylene and polyknitted mesh were used to achieve this objective. However, due to their suboptimal performance in clinical settings, many investigators have shown interest in the utilization of extracellular matrix (ECM) scaffolds for abdominal wall reconstruction because of their potential

---

*Present affiliation: Veterinary Clinical Complex, Apollo College of Veterinary Medicine, Jaipur, Rajasthan, India

Natural Biomaterials for Tissue Engineering. DOI: https://doi.org/10.1016/B978-0-443-26470-2.00008-9
© 2025 Elsevier Inc. All rights are reserved, including those for text and data mining, AI training, and similar technologies.

capacity to resist infection and induce a milder inflammatory response, angiogenesis, and host cell migration.

The biological matrices offer a new approach to the management of abdominal wall defects when the use of other foreign material is not ideal. Biological scaffolds composed of naturally occurring ECM have received significant attention for their potential therapeutic applications. The ECM is by definition nature's ideal biological scaffold material. The ECM is custom designed and manufactured by the resident cells of each tissue and organ and is in a state of dynamic equilibrium with its surrounding microenvironment [1]. The structural and functional molecules of the ECM provide the means by which adjacent cells communicate with each other and with the external environment [2–4]. The ECM is obviously biocompatible since host cells produce their own matrix and provide a supportive medium or conduit for blood vessels, nerves and lymphatics along with the diffusion of nutrients from the blood to the surrounding cells. For these reasons, the ECM has been developed as a biological scaffold for tissue engineering applications in virtually every body system.

The bioinductive properties of ECM scaffolds play a very important role in tissue remodeling. The viscoelastic behavior, biomechanical properties, and ability to support host cell attachment through collagen, fibronectin, and laminin are insufficient alone to explain the constructive remodeling events that are observed following in vivo implantation of ECM scaffolds. Component growth factors such as vascular endothelial cell growth factor (VEGF), basic fibroblast growth factor (b-FGF), and transforming growth factor beta (TGF-ß) are released during scaffold degradation and exert their biologic effects. These growth factors are dissociated from their binding proteins and activated. The processes of scaffold degradation and growth factor release continue until the scaffold is completely degraded. Therefore, sustained bioinductive properties are a hallmark of ECM scaffolds that are susceptible to in vivo degradation, that is, not chemically crosslinked. In contrast, ECM scaffolds that resist or retard the degradation process elicit a chronic inflammatory response and host fibrous connective tissue deposition [5]. The maintenance of the bioinductive properties of ECM scaffolds and the host response to such bioscaffolds can be critically dependent upon methods used to process these materials.

The mechanical properties of the ECM are largely a consequence of its collagen fiber architecture and kinematics. The tissue from which an ECM scaffold is harvested will define its structural characteristics and mechanical properties. An understanding of the collagen fiber alignment of ECM derived from each organ is obviously important for the design of tissue scaffolds if the intent is to closely match the scaffold mechanical properties to those of the target organ of its intended use. Since ECM scaffolds are typically degraded rapidly, it is important to remember that the mechanical properties of the scaffold material are only relevant for the time of surgical implantation. These properties will change immediately as a function of both the degradation rate and the remodeling that is facilitated by the bioinductive properties of the scaffold. The methods used to process tissues to create an ECM scaffold can affect the mechanical and biologic properties.

The main sources of currently available ECMs are skin (dermis), fascial structures (pericardium), and small intestine submucosa. Many of these ECM materials have been commercialized for a variety of therapeutic applications. The constructive remodeling induced by ECM scaffold materials and their widespread use across many clinical applications are a consequence of their bioinductive properties, mechanical and material properties, the host tissue response to naturally occurring ECM, and the degradation properties of the material. However, these presently available biological materials have significant limitations. Complications associated with these costly materials, such as rapid breakdown and loss of the graft material, especially in infected fields, and undesirable host foreign body reaction, have been reported in animal and human studies when used to repair abdominal wall defects [6−10]. As a consequence, the search for the optimal source of reconstruction material for clinical use continues.

The tissue engineered blood vessel has been used successfully as a xenogenic tissue graft in various urological and vascular applications. The decellularized vascular matrices retain their natural biological composition and three-dimensional architecture suitable for cell adhesion and proliferation. Therefore, these are used as scaffolds in cardiovascular tissue engineering. In bladder augmentation techniques, the blood vessel matrix demonstrated urodynamic and histologic properties superior to those resulting from augmentation with other biologic matrices in a rat model [11]. In vascular graft studies, blood vessel matrix seeded with autologous endothelial cells showed no long-term intimal hyperplasia, shrinkage, or fibrosis when used to replace the carotid artery in a sheep model [11,12]. The blood vessel matrix derived from porcine aorta served as a viable option in the repair of abdominal wall tissue defects [13]. Aortic grafts have been used to replace long segments of tracheal defects with promising results [14]. The ECM surrounding the vascular cells provided the biomechanical properties of the tissue [15]. The molecular network of blood vessel consists primarily of collagen (primarily types I and III), elastin in the form of fibers, proteoglycans (including versican, decorin, biglycan, lumican, and perlican), hyaluronan, and glycoproteins (laminin, fibronectin, thrombospondin, and tenascin). The collagens provide the tensile stiffness, the elastin the elastic properties and the proteoglycans contribute to the compressibility. The collagen, elastin, and proteoglycans together contribute to the viscoelastic properties. The collagen is the main protein constituent of muscular arteries where it serves a major structural role. Collagens types I and III are the major fibrillar collagens in blood vessels, representing 60% and 30% of vascular collagens respectively [16].

Unmodified natural materials, upon implantation are subjected to chemical and enzymatic degradation, thereby seriously decreasing the life of the prosthesis. The use of these natural biomaterials has typically required chemical or physical pretreatment aimed at preserving the tissue by enhancing the resistance of the material to enzymatic or chemical degradation, reducing the immunogenicity of the material and sterilizing the tissue. The decellularization specifically removes cellular components that are believed to promote calcification and to give rise to a residual immunological response. These decellularization techniques include chemical, enzymatic, and mechanical means of removing cellular components, leaving a material composed essentially of ECM components. For the most part, these acellular tissues retain natural mechanical properties and promote remodeling of the prosthesis by neovascularization and recellularization by the host. The decellularized tissues are expected to mimic closely the complex three-dimensional structure and mechanical properties of the native tissues from which they are derived [17]. The physical, chemical, and enzymatic treatments can have substantial effects on the composition, mechanical behavior, and host response to biologic scaffolds derived from the decellularization of native tissue and organs, and could have important implications for subsequent use for in vitro and in vivo applications. Thus, there is need of such decellularization protocols which can effectively remove cellular components and also maintain integrity of ECM.

8.2 Optimization of protocols for decellularization of buffalo aorta

Fresh posterior aorta of buffalo origin was collected from the local abattoir and immediately preserved in ice-cold sterile phosphate buffered saline (pH 7.4) containing antibiotic (amikacin 1 mg/mL) and 0.02% EDTA. The maximum time period between tissue procurement and processing was less than 4 hours. After transportation of aorta in ice-cold saline to the laboratory, it was removed from the solution and thoroughly washed with sterile phosphate buffered saline to remove all the adherent blood. After removing extraneous fat and fascia, aorta was cut into 2×2 cm^2 pieces to carry out decellularization protocols. Amikacin (1%) was used as preservative for tissue in nonionic biological detergent solution and 1% sodium azide was used for anionic biological detergent solution.

To make the aorta acellular, a total number of 12 protocols comprising of detergents (alone and in combination) and in combination with enzyme were tested. All the protocols were categorized into two groups (A and B). Aortic samples were subjected to continuous shaking in an orbital shaker at the rate of 180 rpm and at 37°C during the decellularization protocols to provide better contact of tissue with chemicals [18].

8.2.1 Group A

In this group, 6 protocols comprising of both Triton X-100 and sodium dodecyl sulfate detergents were tested, as follows (Table 8.1). Constant volume (20 mL) of detergent solution was used in each protocol.

Protocol A1: The aortic tissue was treated with 1% Triton X-100 detergent for 24 hours.

Protocol A2: The aortic tissue was treated with 1% sodium dodecyl sulfate detergent for 24 hours.

Protocol A3: The aortic tissue was treated with 1% Triton X-100 detergent for 48 hours.

Table 8.1 Group A (detergents alone and in combinations).

Protocol detergent duration of treatment
A1—1% Triton X-100 detergent (24 h)
A2—1% Sodium dodecyl sulfate detergent (24 h)
A3—1% Triton X-100 detergent (48 h)
A4—1% Sodium dodecyl sulfate detergent (48 h)
A5—1% Triton X-100 detergent + 1% sodium dodecyl sulfate (24 h + 24 h)
A6—1% Sodium dodecyl sulfate + 1% Triton X-100 detergent (24 h + 24 h)

Protocol A4: The aortic tissue was treated with 1% sodium dodecyl sulfate for 48 hours.

Protocol A5: The aortic tissue was treated with 1% Triton X-100 detergent for 24 hours followed by treatment with 1% sodium dodecyl sulfate for next 24 hours.

Protocol A6: The aortic tissue was treated with 1% anionic biological detergent for 24 hours followed by treatment with 1% Triton X-100 detergent for next 24 hours.

8.2.2 Group B

In this group, 6 different protocols using Triton X-100, sodium dodecyl sulfate detergents, and trypsin enzyme were used alone and in combinations as follows (Table 8.2). Constant volume (20 mL) of detergents and enzyme solutions were used in each protocol.

Protocol B1: The aortic tissue was treated with 1% Triton X-100 detergent for 24 hours followed by treatment with 0.25% trypsin enzyme solution for 2 hours.

Protocol B2: The aortic tissue was treated with 1% sodium dodecyl sulfate for 24 hours followed by treatment with 0.25% enzyme solution for 2 hours.

Protocol B3: The aortic tissue was treated with 1% Triton X-100 detergent for 24 hours followed by treatment with 0.25% trypsin enzyme solution for 2 hours. This step was followed by treatment with same 1% Triton X-100 detergent again for next 24 hours.

Protocol B4: The aortic tissue was treated with 1% sodium dodecyl sulfate detergent treatment for 24 hours followed by treatment with 0.25% trypsin enzyme solution for 2 hours. This step was followed by treatment with same 1% sodium dodecyl sulfate detergent again for next 24 hours.

Protocol B5: The aortic tissue was treated with Triton X-100 detergent for 24 hours followed by treatment with 0.25% trypsin enzyme solution for 2 hours. This step was followed by treatment with 1% sodium dodecyl sulfate detergent again for 24 hours.

Protocol B6: The aortic tissue was treated with 1% sodium dodecyl sulfate detergent treatment for 24 hours followed by treatment with 0.25% trypsin enzyme solution for 2 hours. This step was followed by treatment with 1% Triton X-100 detergent for next 24 hours.

Table 8.2 Group B (detergents alone and in combinations with enzymatic treatment).

Protocol detergent duration of treatment
B1—1% Triton X-100 detergent + 0.25% trypsin enzyme (24 h + 2 h)
B2—1% Sodium dodecyl sulfate detergent + 0.25% trypsin enzyme (24 h + 2 h)
B3—1% Triton X-100 detergent + 0.25% trypsin enzyme + 1% Triton X-100 detergent (24 h + 2 h + 24 h)
B4—1% Sodium dodecyl sulfate detergent + 0.25% trypsin enzyme + 1% sodium dodecyl sulfate detergent (24 h + 2 h + 24 h)
B5—1% Triton X-100 detergent + 0.25% rrypsin enzyme + 1% sodium dodecyl sulfate detergent (24 h + 2 h + 24 h)
B6—1% Sodium dodecyl sulfate detergent + 0.25 trypsin enzyme + 1% Triton X-100 detergent (24 h + 2 h + 24 h)

Each procedure was performed under aseptic condition. After completion of protocols, biomaterials were preserved in 10% formalin for histopathological examination to judge the efficacy of different protocols used for acellularity of tissues.

The acellular aortic matrix prepared by different aforesaid protocols was evaluated on the basis of the histological scores recorded in following four parameters.

 1. Cellularity: $0 =$ complete acellular; $1 = 1\%-30\%$; $2 = 30\%-50\%$; $3 = 50\%-70\%$; $4 \geq 70\%$.

 2. Collagen fiber arrangement: $0 =$ compact; $1 =$ mildly loose; $2 =$ moderately loose; $3 =$ extensively loose.

 3. Collagen fiber bundle thickness: $0 =$ normal; $1 =$ mildly thin; $2 =$ moderately thin; $3 =$ extremely thin.

8.2.3 Optimization of protocols for preparation of acellular aortic matrix

Native posterior aortic tissue specimens of buffalo origin were subjected to anionic and/ or nonionic biological detergents with and without enzyme solution for decellularization. A total of 12 protocols were tested to make the aorta acellular. All the 12 protocols were categorized into two groups.

8.2.4 Group A

In this group of protocols, aortic tissue was treated with both sodium dodecyl sulfate and Triton X-100 detergents, alone as well as in combination. The concentration of detergents used was 1%. The tissues were treated with detergents for a period of 24 and 48 hours.

8.2.5 Protocol A1

In this protocol, aortic tissue was treated with Triton X-100 detergent for 24 hours. The color and consistency of the aortic tissue treated with Triton X-100 for

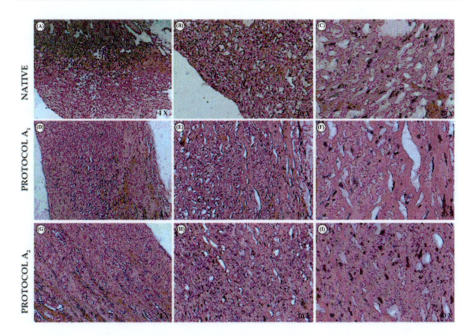

Figure 8.1 (A–I) Optimization of protocols for decellularization of aortic matrix (H&E stain, ×200) (Native, Protocol A1, Protocol A2).

24 hours were comparable to that of native tissue. The cellularity of the tissue was more than 70% (Fig. 8.1D). All the layers of aorta were intact. The arrangement of collagen fiber bundles was compact (Fig. 8.1E) and comparable to that of native tissue. The thickness of the collagen fibers was normal (Fig. 8.1F) and comparable to that of native tissue.

8.2.6 Protocol A2

In this protocol, the aortic tissue was treated with sodium dodecyl sulfate detergent for 24 hours. The aortic tissue treated with sodium dodecyl sulfate for 24 hours was white in color and soft in consistency when compared to the tissue treated with Triton X-100 alone. The tunica intima and adventitia were completely acellular. However, nuclei were evident in tunica media (Fig. 8.1G). The cellularity of the tissue was less than 30%. The collagen fibers were swollen (Fig. 8.1H) and compactly arranged when compared to that of native tissue (Fig. 8.1I). All the layers of aorta were intact.

8.2.7 Protocol A3

In this protocol, the aortic tissue was treated with Triton X-100 detergent for 48 hours. There was not much difference in the color and consistency of the aortic tissue treated with Triton X-100 for 48 hours when compared to native tissue. The cellularity of the tissue was 50%−70% (Fig. 8.2A). Increase in treatment time with

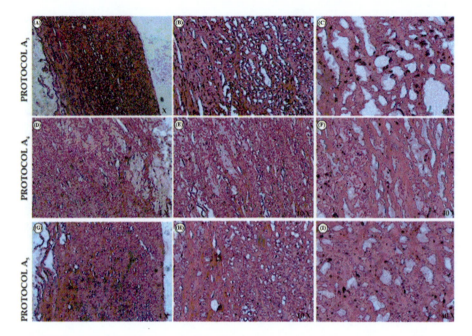

Figure 8.2 (A–I) Optimization of protocols for decellularization of aortic matrix (H&E stain, ×200) (Protocol A3, Protocol A4, Protocol A5).

detergent resulted in decreased cellularity. All the layers of aorta were intact. The collagen fibers were loosely arranged than the native tissue (Fig. 8.2B). The collagen fiber thickness was normal and comparable to that of native tissue (Fig. 8.2C).

8.2.8 Protocol A4

In this protocol, the aortic tissue was treated with sodium dodecyl sulfate detergent for 48 hours. Increase in treatment time with sodium dodecyl sulfate resulted in whiter and softer tissue than the tissue treated with sodium dodecyl sulfate alone for 24 hours. There was no evidence of cellularity of the tissue (Fig. 8.2D). All the layers of aorta were intact. The collagen fibers were compactly arranged than the native tissue (Fig. 8.2E). The collagen fibers were homogenized (Fig. 8.2F).

8.2.9 Protocol A5

In this protocol, the aortic tissue was treated with Triton X-100 detergent for 24 hours followed by treatment with sodium dodecyl sulfate for next 24 hours. The treated tissue was white in color and its consistency was not much different from the native tissue. There was no evidence of cellularity of the tissue (Fig. 8.2G). All the layers of aorta were intact. The arrangement of collagen fibers was mildly loose when compared to native tissue (Fig. 8.2H). The collagen fibers were partially digested resulting in mild to moderately thin fibers (Fig. 8.2I).

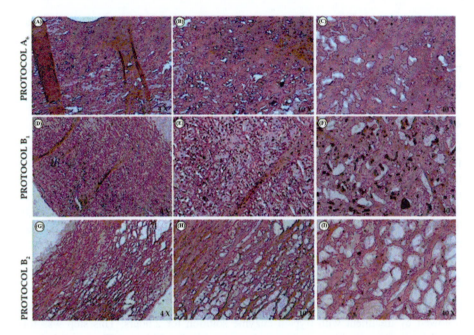

Figure 8.3 (A−I) Optimization of protocols for decellularization of aortic matrix (H&E stain, ×200) (Protocol A6, Protocol B1, Protocol B2).

8.2.10 Protocol A6

In this protocol, the aortic tissue was treated with sodium dodecyl sulfate detergent for 24 hours followed by treatment with Triton X-100 detergent for next 24 hours. The treated tissue was white in color and soft in consistency in comparison to the native tissue. There was no evidence of cellularity of the tissue (Fig. 8.3A). All the layers of aorta were intact. The arrangement of collagen fibers was compact in comparison to the native tissue (Fig. 8.3B). The collagen fibers were moderately thin (Fig. 8.3C) [18,19].

8.2.11 Group B

In this group of protocols, sodium dodecyl sulfate and Triton X-100 either alone and in combinations were used followed by treatment with trypsin enzyme solution. The concentrations of enzyme and detergents used were 0.25% and 1% respectively.

8.2.12 Protocol B1

In this protocol, the aortic tissue was treated with Triton X-100 detergent for 24 hours followed by treatment with trypsin enzyme solution for 2 hours. The treated tissue was soft and white in color in comparison to the tissue treated with nonionic detergent alone for 24 hours. The cellularity of the tissue was more than 70% (Fig. 8.3D). All the layers of aorta were intact. The arrangement of collagen fibers was compact and comparable to that of native tissue (Fig. 8.3E). The thickness of collagen fibers was normal and comparable to that of native tissue (Fig. 8.3F).

8.2.13 Protocol B2

In this protocol, the aortic tissue was treated with sodium dodecyl sulfate detergent for 24 hours followed by treatment with trypsin enzyme solution for 2 hours. The treated tissue was soft in consistency and white in color when compared to the tissue treated with anionic biological detergent alone for 24 hours. There was no evidence of cellularity of tissue. The excessive digestion of the tissue was evident (Fig. 8.3G). The tunica intima and media were disrupted (Fig. 8.3H). The collagen fibers were moderately thin and their arrangement was moderately loose (Fig. 8.3I).

8.2.14 Protocol B3

In this protocol, the aortic tissue was treated with Triton X-100 detergent for 24 hours followed by treatment with trypsin enzyme solution for 2 hours. This step was followed by treatment with same Triton X-100 again for 24 hours. The treated tissue was soft and white in color in comparison to the tissue treated with nonionic detergent alone for 48 hours. The tissue cellularity was about 30%–50% (Fig. 8.4A). The thickness of the collagen fibers (Fig. 8.4B) was normal and comparable to that of the native tissue. All the layers of aorta were intact. The arrangement of collagen fibers was mildly loose in comparison to native tissue (Fig. 8.4C).

8.2.15 Protocol B4

In this protocol, the aortic tissue was treated with sodium dodecyl sulfate detergent for 24 hours followed by treatment with trypsin enzyme solution for 2 hours. This treated tissue was again exposed to same sodium dodecyl sulfate for next 24 hours. The treated tissue was soft and white in color in comparison to the tissue treated with anionic detergent alone for 48 hours. There was no evidence of tissue cellularity (Fig. 8.4D). All the layers of aorta were intact. The collagen fibers were compactly arranged than the native tissue (Fig. 8.4E). The collagen fibers were homogenized (Fig. 8.4F).

8.2.16 Protocol B5

In this protocol, the aortic tissue was treated with Triton X-100 detergent for 24 hours followed by treatment with the trypsin enzyme solution for 2 hours. This step was followed by treatment with the sodium dodecyl sulfate detergent for next 24 hours. The treated tissue was soft and white in color in comparison to the tissue treated with nonionic and anionic detergent combination. There was no evidence of tissue cellularity

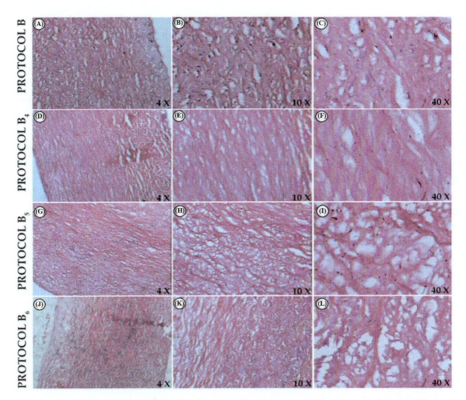

Figure 8.4 (A—I) Optimization of protocols for decellularization of aortic matrix (H&E stain, ×200) (Protocol B3, Protocol B4, Protocol B5, Protocol B6).

(Fig. 8.4G). All the layers of aorta were intact. The collagen fibers were slightly thin (Fig. 8.4H) and loosely arranged when compared to the native tissue (Fig. 8.4I).

8.2.17 Protocol B6

In this protocol, the aortic tissue was treated with sodium dodecyl sulfate detergent for 24 hours followed by treatment with trypsin enzyme solution for 2 hours. This step was followed by treatment with Triton X-100 detergent for next 24 hours. The treated tissue was soft and white in color in comparison to the tissue treated with nonionic and anionic detergent combination. There was no evidence of tissue cellularity (Fig. 8.4J). In tunica media, the collagen fibers were homogenized and loosely arranged in comparison to the native tissue (Fig. 8.4K). All the layers of aorta were intact. The collagen fibers were excessively digested in tunica adventitia (Fig. 8.4L).

This study revealed that protocol B4 in which the aorta was treated with 1% sodium dodecyl sulfate detergent for 24 hours followed by treatment with 0.25% trypsin enzyme solution for 2 hours and then again with same 1% sodium dodecyl sulfate for again next 24 hours showed complete acellularity with normal thickness and arrangement of collagen fibers. Therefore, protocol B4 was found best among all the protocols tested for decellularization of aorta [18,19].

The aorta is mainly composed of three main layers, or tunicae, that are from the inside out: the intima, the media, and the adventitia. The intima is composed of a confluent monolayer of endothelial cells arranged longitudinally, which form the endothelium. The media is the major component of muscular arteries. It is composed of smooth muscle cells aligned concentrically along with collagen, elastin fibers, and proteoglycans. The media confers the majority of the mechanical properties. The adventitia is mainly composed of collagen and fibroblast cells both arranged longitudinally. Each layer contributes uniquely to the overall mechanical properties of the vessel. Collagens types I and III are major collagens in blood vessels. The great strength of collagen fibers, however originates from the stable intermolecular covalent bonds between adjacent tropocollagen molecules [20].

The goal of any decellularization protocol is to effectively remove immunogenic cellular material while maintaining the biological activity and mechanical integrity of the ECM. Through decellularization procedures, vascular tissue can be reduced to a sterilized scaffold and implanted with a low risk of rejection. The treatments effectively reduce antigenicity while creating free volume spaces upon which the host's native cells are able to proliferate [21]. Several methods of decellularization have been cited in the literature, including physical, chemical, and enzymatic treatments. Studies have shown that the efficiency of a particular decellularization protocol is largely dependent upon the tissue of interest [17]. Because there have been no previous studies to determine which method is most suitable for decellularizing the aorta of buffalo origin, the present investigation analyzed the effects of detergent and enzyme combination on aorta.

In general, nonionic surfactants are used extensively in decellularization protocols due to their mild effects on tissue structure. These surfactants disrupt lipid–protein and lipid–lipid interactions, but generally leave protein–protein reactions intact with the result of maintaining their functional conformations [17]. However, mixed results on the efficacy of this chemical treatment have been reported. Regarding ECM components, Triton X-100 has led to a significant loss of glycosaminoglycans (GAGs) in heart valve tissues [22], whereas it has left GAG components in the anterior cruciate ligament unchanged [23]. The results of the present study indicate that nonionic detergent may not be the most effective decellularization method for the aorta, as the cellularity was evident in the aortic tissues even up to 48 hours of treatment [19].

Anionic detergents are generally effective for solubilizing both nuclear and cytoplasmic cellular membranes. However, results in the literature have also been discrepant regarding the efficiency of this class of chemical agents. A study by Seddon et al. [24] reported that ionic detergents tend to denature proteins by disrupting protein–protein interactions. In contrast, a study by Schaner et al. [25] concluded that sodium dodecyl sulfate (SDS) treatment is a feasible option for vascular tissue engineering efforts among several detergents studied. SDS is a commonly used anionic detergent because of its ease of use, success in a published model of vascular tissue decellularization [26] and ability to lyse cells uniformly within all layers of the tissue. Furthermore, its mechanism of action may make it less damaging than other enzymatic, mechanical, or tissue fixation methods.

In the present study, the cell extraction using anionic detergent was effectively achieved without significant disturbances in ECM morphology and strength. Decellularization of the aorta using anionic detergent in combination with enzyme solution resulted in a complete loss of cellular structures from both the media and adventitia. The tunica media became more compact with smaller interlamellar spaces after decellularization. The decellularization in aortic valve conduits has been observed with 0.03% SDS, along with disintegration of matrix fibers during SEM analysis [27]. The hydrophobic−hydrophilic and ionic properties of SDS can alter protein−protein interactions by opening the molecular structure of elastin and disrupting the hydrogen bonding of collagen [28]. The SDS treatment was also reported to destabilize the collagen triple helical domain and to swell the elastin network [29,30]. Although it has been reported that 1% SDS was able to decellularize porcine heart valve cusps in 24 hours [31], decellularization of aortic tissues was incomplete at 24 hours of 1% anionic detergent treatment. These differences may occur because of higher tissue thickness and density of the aorta compared to cusp tissue. Incomplete decellularization at short incubation times [32] and damage to the ECM of vascular tissue at longer incubation times [33] was reported with trypsin treatment. The combination of nonionic detergent and enzyme solution was ineffective in complete cell removal. Successful decellularization of aorta was attained after 2 days extraction in 1% anionic detergent followed by 2 hours in enzyme solution. Similar results were reported with porcine aorta by Lu et al. [34].

8.3 Preparation and characterization of the buffalo aortic matrix

8.3.1 Histological observations

The buffalo aortic matrix (BAM) was prepared using the published protocol of Kumar et al. [35]. Briefly, native aorta was cut into desired size and decellularized using 1% sodium dodecyl sulfate (SDS) for 24 hours followed by 0.25% trypsin for 2 hours and again with 1% SDS for 24 hours at room temperature under continuous agitation. Prepared BAM was thoroughly washed and stored in phosphate buffer saline (PBS) supplemented with 0.048% gentamicin at $-20°C$ until implantation. Native and decellularized aortae were fixed in 10% neutral buffered formalin, serially dehydrated with ethanol, cleared in xylene, and embedded in paraffin wax. Tissue sections (5 μm) were cut on a semiautomated rotary microtome (RM2245, Leica Microsystems, Wetzlar, Germany) and stained with hematoxylin and eosin (H&E). Further, Masson's trichrome (MTS) and Weigert's resorcin fuchsin (WRF) staining was used to identify collagen and elastic fibers, respectively. Slides were examined under $100 \times$ magnifications utilizing a light microscope (Axiocam ERc5 s, Primo star, Carl Zeiss, Germany). Microscopic images of native and decellularized aortae are shown in Fig. 8.2 [36].

Intima and media layers of the native aorta were composed of endothelial cells and smooth muscle cells (SMCs) with connective tissues, respectively. Masson's trichrome-stained section of the native aorta revealed compactly arranged collagen fibers (Fig. 8.5A). WRF-stained section of the native aorta revealed the presence of distinct wavy elastic fibers (Fig. 8.5B). Treatment of the native aorta with 1% SDS for 24 hours followed by 0.25% trypsin for 2 hours and again with 1% SDS for 24 hours resulted in complete loss of cells and nuclei. MTS-stained sections of BAM revealed preserved collagen structure and integrity (Fig. 8.5C). WRF-stained sections of BAM revealed abundant elastic fibers within matrix (Fig. 8.5D). After decellularization, all the cells were completely removed. Moreover, elastic and collagen fibers were preserved without structural distortion. Treatment with these chemicals resulted in complete decellularization, and retention of intact collagen structure within developed matrix. SDS is more effective than Triton X-100 for removing nuclei from dense tissues and organs while preserving tissue mechanics. SDS is typically more effective for removing cell residues from tissue compared to other detergents [23]. Further, SEM examination also confirmed effective decellularization of the bubaline aorta and the preservation of collagen structure and integrity within BAM [36].

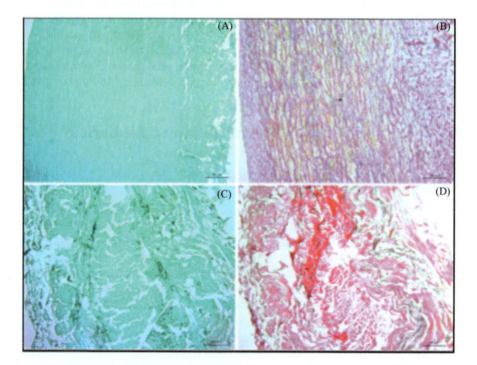

Figure 8.5 (A–D) Masson's trichrome-stained images of native aorta (A) and bubaline aortic matrix (C) ($\times 100$ magnification, 200 μm scale bar). Weigert's resorcin fuchsin-stained images of native aorta (B) and bubaline aortic matrix (D) ($\times 100$ magnification, 200 μm scale bar).

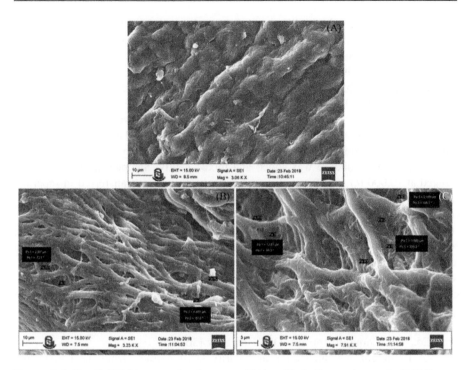

Figure 8.6 (A–C) SEM images of native aorta (NA) and bubaline aortic matrix (BAM).

8.3.2 Scanning electron microscopic (SEM) observations

Native and decellularized aortic samples were cut in minute pieces and washed in chilled 0.1 M phosphate buffer saline (PBS) (pH 7.2) and subjected to fixation in 2.5% glutaraldehyde in 0.1 M PBS for 6 hours. Fixed tissues were washed in 0.1 M PBS with three changes of 15 minutes each at 4°C. Then, tissues were dehydrated in ascending grade of acetone solution, namely, 30%, 50%, 70%, 80%, 90%, 95%, and 100%, at 4°C for 15 minutes in each solution. Subsequently, samples were mounted on aluminum stubs, coated with gold palladium in Emitech SC620 sputter coater. Backscattered electrons were used for ultrastructure observations, and topographical images of processed tissues were acquired using a scanning electron microscope (EVO-18, Carl Zeiss, Germany). SEM examination of native aorta demonstrated intact ultrastructure of endothelial and adventitial surfaces (Fig. 8.6A). The BAM showed intact collagen fiber structure and integrity. Thickness of collagen fibers varied between 1.191 and 4.489 μm (Fig. 8.6B and C) [36].

8.3.3 DNA extraction, quantification, and purity

DNA was extracted from the native and decellularized aortae by phenol/chloroform/isoamyl alcohol as per method described by Green and Sambrook [37] with modifications. Two hundred milligrams of the native and decellularized aortae were separately triturated in 1 mL lysis buffer (tris 50 mM; EDTA 0.1 M; SDS 1%) using glass Teflon tissue homogenizer, and lysates were incubated for 2 hours at 37°C in a water bath. To each tissue homogenate, 50 lL proteinase K solution (20 mg/mL) was added and incubated overnight at 56°C. Thereafter, tissue homogenates were centrifuged at 10,000 rcf for 10 minutes at room temperature. Supernatants were collected, and equal volume of tris-saturated phenol (pH 8.0) was added and mixed gently. Samples were again centrifuged at 10,000 rcf for 10 minutes, and upper aqueous phase was transferred to the new tubes. Similar volume of chloroform/isoamyl alcohol (24:1) mixture was added and mixed gently. Samples were again centrifuged at 10,000 rcf for 10 minutes, and aqueous phase was transferred in separate tube. The DNA was precipitated with 2 volumes of absolute ethanol and 0.2 volumes of 3 M sodium acetates (pH 5.2). Tubes were centrifuged at 10,000 rcf for 10 minutes, and supernatant was discarded. Pellets were then washed two times in 70% ethanol and air-dried, dissolved in nuclease-free water, and kept in $-20°C$ till further use. After extraction, DNA concentration and purity was quantified using Nanodrop 000 spectrophotometer (Thermo Scientific, United States). Subsequently, samples were separated by electrophoresis on a 0.8% low electroendosmosis (EEO) agarose gel with ethidium bromide (0.2 μg/mL) at 60 V for 1 hour stained and visualized with ultraviolet trans illumination (Fluor Shot PRO II SC850, Shanghai Bio-Tech Co., Ltd, China) to confirm DNA integrity [36].

DNA Extraction, Quantification and Purity DNA extracts from the native and decellularized aortae were of high purity as evident by the 260/280 ratios. DNA content of the native aorta was 488.11 ± 49.12 ng/mg. However, the DNA content significantly $(P < .001)$ reduced to 6.47 ± 1.26 ng/mg after decellularization (Fig. 8.7). Treatment of the native aorta with 1% SDS for 24 hours followed by 0.25% trypsin for 2 hours and again with 1% SDS for 24 hours resulted in 98.68% reduction in DNA content of the aorta. DNA fragmentation was confirmed through DNA agarose gel electrophoresis [36]. DNA extracts from native aorta show a large band indicative of large DNA fragment. Extracts from decellularized aortic tissue

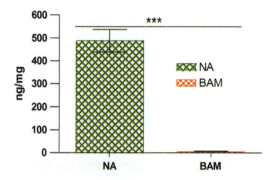

Figure 8.7 DNA content (ng/mg of tissue) (mean ± SE) in native aorta (NA) and bubaline aortic matrix (BAM). ***$P < .001$ versus other.

show significant removal of DNA material, with the absence of DNA bands. DNA retained in decellularized tissues may induce an immune response and foreign body reaction by the host. Therefore, the following minimum criteria for acceptable amounts of residual DNA after decellularization have been established: (a) <50 ng dsDNA per mg dry weight and (b) <200 base pair DNA fragment length [38]. These values have been approached in the present study. The absence of visible nuclei based on H&E staining complemented the spectrophotometric detection of DNA [36].

8.3.4 Fourier transform infrared (FTIR) spectroscopy

One milligram of each freeze-dried native and decellularized aortic tissues were mixed with pure dry KBr powder in 1:10 ratio, and pelleted. The FTIR spectra were recorded by an infrared spectrophotometer (FTIR 8400 s Shimadzu Corporation, Tokyo, Japan) in the 500–4000 cm^{-1} wave number spectral range with a spectral resolution of 2 cm^{-1} and 45 scans. The FTIR spectra of NA and BAM are shown in Fig. 8.8 [36].

The spectrum of BAM matched well with the FTIR spectrum of NA collagen and other collagens [39–42]. The amide A band (3294 cm^{-1}) is associated with H-bonded N–H stretching [40,43,44] and was found at 3282.95 cm^{-1} for NA and 3280 cm^{-1} for BAM. The amide B band (2953 and 2928 cm^{-1}) is related to CH2 asymmetric stretching [39,43,44] and was observed at 2958.9 cm^{-1} for NA and 2954.08 cm^{-1} for BAM. The amide I band (1641–1658 cm^{-1}) is associated with C=O hydrogen bonded stretching [42–44] as recorded at 1658.84 cm^{-1} for NA and 1658.84 cm^{-1} for BAM. The amide II (1539–1546 cm^{-1}) is associated with C–N stretching and N–H in plane bending from amide linkages, including wagging vibrations of CH2 groups from the glycine backbone and proline side chains [41,43,44] in NA and BAM appeared at 1526.71 and 1529.60 cm^{-1}, respectively. The amide III (NH bend) band was found at 1282.55 cm^{-1} for NA and

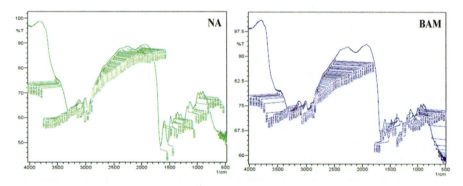

Figure 8.8 FTIR spectra showing transmittance peaks of native aorta (NA) at 1282.55, 1526.71, 1658.84, 2958.9, and 3282.95 cm^{-1}; and bubaline aortic matrix (BAM) at 1230.69, 1529.60, 1658.84, 2954.08, and 3280 cm^{-1}.

1230.69 cm^{-1} for BAM [43–45]. Based on the location of amide A, amide B, amide I, amide II, and amide III peaks, it seems that prepared bubaline aortic matrix was composed of collagen [36].

8.4 In vivo biocompatibility determination of acellular aortic matrix

In vivo biocompatibility determination was done bysubcutaneous implantation of native, acellular, and crosslinked scaffolds. Sixteen clinically healthy adult albino guinea pigs of either sex were randomly divided into four groups having four animals in each group. In group I native, in group II acellular, in group III BDDGE crosslinked, and in group IV EDC crosslinked grafts were implanted. Animals were maintained under uniform conditions of feeding, management and environment throughout experimental period [46].

8.4.1 Surgical procedure

Food for 12 hours and water for 6 hours were withheld before the operation. Xylazine was administered at the dose rate of 10 mg/kg bodyweight and it was immediately followed by ketamine HCl at the dose rate of 50 mg/kg bodyweight. Both the drugs were given intraperitoneally for anesthesia. The either side of spine of all the animals was properly clipped, shaved, and scrubbed with 5% cetrimide and chlorhexidine solution and painted with povidone iodine solution. Four subcutaneous pouches were prepared on either side of the spine. The biomaterials of 20 × 10 mm^2 in size were taken and implanted in each pouch. The biomaterial kept in each pouch was anchored to the subcutaneous tissue using nylon. The native, acellular, and crosslinked aortic tissues were implanted in separate guinea pigs. Skin incision was closed with nylon using horizontal interrupted suture pattern. These grafts were retrieved back on 15, 30, and 60 postoperative implantation days and subjected to following observations (Fig. 8.9A–I).

8.4.2 Macroscopic observations

Macroscopic assessment of the retrieved implant was done as per the procedure described by Lu et al. [47]. On day 15, all the aortic grafts were easily retrieved from the surrounding host tissue. The grafts were covered with thin white fibrous connective tissue. Complications like infection or pus formation was not seen in the vicinity of the implanted biomaterials. The BDDGE crosslinked materials were light brown in color and stiff in consistency. No change in color and consistency was observed in native, acellular, and EDC crosslinked aortic grafts. No cellular reaction was observed in acellular and crosslinked grafts. The cellular reaction was more prominent in animals implanted with native tissue [46].

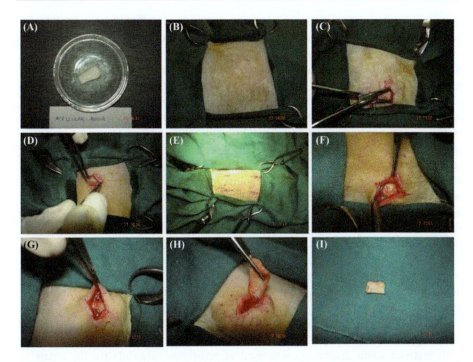

Figure 8.9 (A–I) Subcutaneous implantation and retrieval of aortic grafts in a guinea pig model.

On day 30, all the grafts were covered with fibrous connective tissue, which was denser than that observed at 15th day. The implanted biomaterials were more deeply seated within the fibrous connective tissue. The grafted tissue was difficult to retrieve as compared to day 15 tissue samples. No change was observed in color and consistency in all grafts. Cellular reaction was comparatively less in animals implanted with native tissue as compared to 15-day graft. Complications like infection or pus formation was not seen in the vicinity of the implanted biomaterials [46].

On day 60, the aortic grafts were covered with dense fibrous connective tissue, which was denser than that seen at day 30. No complications like infection or pus formation was observed at the implantation site. The implanted biomaterials were more deeply seated within the fibrous connective tissue and were difficult to retrieve as compared to day 30 postimplantation. No change was observed in color and consistency in any of the retrieved grafts. Any abnormal cellular reaction was not seen at the host−graft junction in any animal [46].

The acellular aortic matrix grafts were subcutaneously implanted after treatment with different crosslinking agents and the resorption rate and other changes were compared with acellular uncrosslinked grafts (control). The degree of immune response to graft varies with type of graft. Xenografts exhibit the greatest genetic disparity and therefore engender a vigorous graft rejection. Before biomaterials can be applied for its clinical use, the tissue response to these biomaterials had to be evaluated in vivo. This approach is to identify a suitable xenogenic tissue and modify the structure to give a material that will be immunologically inert, mechanically robust, and will support cell attachment and proliferation [48]. The preparation of natural matrices commonly involves a combination of physical methods to delaminate layers of tissue, followed by chemical and enzymatic methods to remove cell bodies from the remaining ECM [17] and such decellularization strategies, designed to limit the immunogenicity of the matrix. Decellularization process may attenuate severe xenogenic immune response [49], but the removal of cellular components may not be sufficient to eliminate inflammation, and fixation techniques may still be necessary to prevent degradation [50]. The chemical crosslinking of collagen had been used for several years to improve scaffold stability [51]. The control resorption of biological biomaterials is essential where it is to be used for tissue regeneration.

In this study, a uniform layer of white connective tissue was found covering all the implanted biomaterials at days 15 and 30 postimplantation. However, it was dense at day 30. Shoukry et al. [52] also observed similar observations where commercial polyester fabric was used to repair the abdominal hernias in horse. At day 60, the implanted biomaterials were present beneath the fibrous connective tissue. Similarly, Kanade et al. [53] found uniform layer of connective tissue covering the graft material when diaphragm was used as prosthetic materials for the repair of ventral abdominal wall defects in bovines. Deokiouliyar et al. [54] also reported similar findings when glycerol-treated pericardium was used for hernioplasty in bovines.

8.4.3 Microscopic observations

The retrieved implants were preserved in 10% formalin saline solution. The tissues were processed by routine paraffin embedding technique and the sections were cut at 5-micron thickness. The sections were examined for inflammatory reaction around the implant material, degenerative changes of the graft, neovascularization, lymphocytes infiltration, and fibroblastic proliferation. Special staining for collagen fibers (Masson's trichrome stain) was also done. The histopathological picture of retrieved native, acellular EDC and BDDGE crosslinked acellular aortic matrix grafts are depicted in Fig. 8.10. The collagen contents were evaluated by Masson's Trichome stain [46].

Native aortic matrix graft: On day 15, native aortic matrix showed extensive infiltration of mononuclear cells comprising of macrophages and epithelioid cells indicating chronic inflammatory response. The collagen fibers were moderately degraded. Formation of a delimiting membrane around the layer of cellular infiltration was also observed. However, by day 30, the inflammatory reaction was remarkably reduced and there was proliferation of fibrous tissue on the outer layer. On day 60, the graft was surrounded on one side by proliferating connective tissue, with infiltration of mononuclear cells and fibroblasts indicating chronic inflammatory response. There

Aorta-derived extracellular matrix scaffolds and clinical application 223

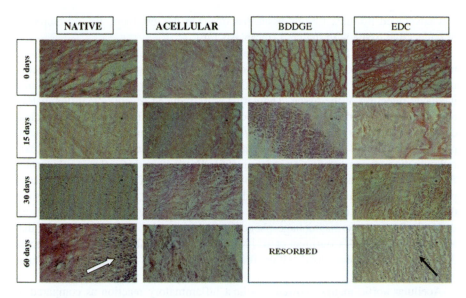

Figure 8.10 Photomicrographs of native, acellular, BDDGE, and EDC crosslinked acellular aortic grafts retrieved at 15, 30, and 60 days after subcutaneous implantation in guinea pig model (H&E stain, ×40). Native grafts showing chronic inflammatory response (*white arrow*) at day 60. BDDGE crosslinked grafts got resorbed by day 60. EDC crosslinked grafts showing development of connective tissue with mature fibroblasts (*black arrow*) at day 60.

was marked ingrowth of host tissue in the graft with infiltration of mononuclear cells and fibroblasts at different stages of maturation [46].

Acellular aortic matrix graft: On day 15, the acellular aortic matrix showed mild chronic inflammatory response with less infiltration of mononuclear cells when compared to the native aortic matrix graft. Cellular infiltration was limited only to the periphery of graft. Degradation of collagen fibers was mild and confined to the periphery. On day 30, severe inflammatory response was observed at both the interfaces. Moderate degradation of collagen fibers was observed. There was extensive proliferation of fibrous cellular tissue. On day 60, the graft was degraded and covered with connective tissue with fibroblasts at different stages of maturation [46].

The BDDGE crosslinked acellular aortic matrix graft: On day 15, there was severe mononuclear cell infiltration, indicating chronic inflammatory response which was mostly confined to the periphery of the graft. Degradation of collagen fibers was observed only at the surface. On day 30, the cellular infiltration was observed inside the graft. But the inflammatory response was reduced when compared to day 15. Moderate degradation of collagen fibers was observed owing to infiltrating mononuclear cells. On day 60, the graft was resorbed [46].

The EDC crosslinked acellular aortic matrix graft: On day 15, there was mild chronic inflammatory response, confined to the periphery of the graft. The interface was covered with thin band of connective tissue. On day 30, the inflammatory response was severe with mild to moderate degradation of collagen fibers. The graft was enveloped by thick fibrous tissue reaction. On day 60, there was development of connective tissue with matured fibroblasts [46].

The intensity of the inflammatory response to an implanted material is critical toward acceptance or rejection of the material. The xenogenic tissues are recognized as foreign by the host and induce an inflammatory response or an immune-mediated rejection of the tissue. The native aorta induced severe host inflammatory reaction as compared to acellular aorta, characterized by infiltration of mononuclear cells and fibroblasts which persisted up to 60 days postimplantation. The macrophages differentiate toward a phenotype that is associated with either cytotoxic inflammation or constructive remodeling [55]. The factors that influence the proinflammatory versus antiinflammatory polarization profile of a mononuclear macrophage population are largely unknown. It appears, however, that acellular materials that are resistant to degradation elicit a proinflammatory type of response whereas the antiinflammatory macrophage phenotype predominates with native tissues that are readily degraded. Courtman et al. [28] hypothesized that cell extraction process decreased the antigenic load within the tissue due to the elimination of cellular antigens.

Acellular aortic matrix showed less host inflammatory reaction as compared to the native tissue for the first 15 days postimplantation suggesting the decreased antigenicity of these matrices due to decellularization. Similar results were obtained using the acellular bovine pericardium by Gilberto and Pereira [56]. Petter-Puchner et al. [57] found a concerning degree of inflammation in a porcine-derived biologic matrix obtained from small intestinal submucosa when it was implanted in a rodent model of abdominal wall hernia repair. An acute inflammatory response consisted mostly of polymorphonucleocyte infiltration and subsequent loss of the graft material following implantation. This early remodeling process, which involved reestablishment of cell infiltrate and progressive deposition of organized connective tissue in a manner that is consistent with natural wound healing, is essential for effective regeneration and repair of abdominal wall defects. At 30 days postimplantation, it was found that inflammatory cells and fibroblasts were able to infiltrate into acellular tissues. Penetration of cells into the acellular tissue may be caused by the extraction of soluble proteins, lipids, nucleic acids, salts, and carbohydrates, leading the tissue more permeable to cellular infiltrates. Enzymatic degradation is necessary for efficient migration of cells into the acellular scaffolds [58]. The depth of cell infiltration into the acellular tissue decreased with increase in crosslinking degree.

In BDDGE and EDC crosslinked grafts, host inflammatory reaction was limited to the surface of the graft and may be attributed to the fact that crosslinking within the acellular tissue may produce a physical barrier for cell infiltration [46]. Additionally, crosslinking of the acellular tissue increased its resistance against enzymatic attack, which is necessary for cell migration into scaffolds. Infiltration of inflammatory cells was accompanied by degradation of collagen fibers. Among the various inflammatory cells such as polymorphonuclear leukocytes, macrophages, and fibroblasts that infiltrate implanted materials, macrophages are known to be able to secrete collagenase among other proteases [59]. This allows the fibroblasts from the host tissue to migrate into implanted grafts. The acellular graft was covered with connective tissue with fibroblasts at different stages of maturation at day 60, suggesting synthesis of neo collagen fibrils. In BDDGE crosslinked grafts, moderate degradation of collagen fibers was observed from day 30 postimplantation. By day 60, the grafts were completely absorbed suggesting decreased resistance of graft toward enzymatic attack in vivo. The EDC crosslinked grafts showed less inflammatory response when compared to native and acellular grafts [46]. Host reaction was limited to the periphery of the graft. EDC increases collagen biostability and reduces antigenicity while preserving compatibility [60]. Moreover, by day 60, there was proliferating connective tissue with fibroblasts at end stage of maturation suggesting host in growth into the graft.

8.4.4 Immunological studies

For immunological study lymphocyte proliferation assay and ELISA tests were performed.

8.4.5 Lymphocyte proliferation assay

The cell-mediated immune response toward xenogenic acellular aortic matrix graft was studied by lymphocyte proliferation assay.

8.4.5.1 Preparation of antigen

The crosslinked acellular aortic matrix grafts were cut into small pieces and extract was made by grinding with sterile glass powder in sterile normal saline solution containing penicillin and streptomycin at a concentration of 100 IU/mL and 1 μg/mL, respectively. The samples were centrifuged at 2000 rpm for 30 minutes and supernatant was filtered through 0.22 μm syringe filter and used in the assay to stimulate lymphocytes in vitro. The uncrosslinked acellular and native aortic tissues were also processed similarly and used in the assay to stimulate T cells so as to compare the stimulation index with crosslinked graft [18].

8.4.5.2 Peripheral blood lymphocytes

Blood (2 mL) was aseptically collected from anterior vena cava of guinea pig in heparinized tubes on 0, 15, and 60 days postimplantation. Sterile PBS (2 mL) was added to the 2 mL of blood and properly mixed. It was layered carefully over 2 mL of lymphocyte separation medium (Histopaque 1077) and centrifuged at 2200 rpm for 30 minutes. The buffy coat was collected in a fresh tube and two washings were done with sterile PBS at 1800 rpm for 10 minutes. Supernatant was discarded and pellet was resuspended in RPMI1640 growth medium. The cells were adjusted to a concentration of 2×10^6 viable cells/mL in RPMI1640 growth medium and seeded in 96-well tissue culture plate @100 μL/well. The cells were incubated at 37°C in 5% CO_2 environment. Cells from each guinea pig were stimulated with antigen (10−20 μg/mL) and PHA (10 μg/mL) in triplicates and three wells were left unstimulated for each sample. After 45 hours, 40 μL of MTT solution (5 mg/mL) was added to all the wells and incubated further for 4 hours. The plates were then centrifuged for 15 minutes in plate centrifuge at 2500 rpm. The supernatant was discarded, plates were dried, and 150 μL of DMSO was added to each well and mixed thoroughly by repeated pipetting to dissolve the formazan crystals. The plates were immediately read at 570 nm with 620 nm as reference wavelength. The stimulation index (SI) was calculated using the following formula.

$$\text{Stimulation Index (SI)} = \frac{\text{OD of stimulated cultures}}{\text{OD of unstimulated cultures}}.$$

The cell-mediated immune response toward the subcutaneously implanted native, acellular, and crosslinked acellular aortic matrix grafts in all the guinea pigs was assessed by MTT colorimetric assay. The mean ± SE of stimulation index (SI) values of guinea pig groups implanted with native, acellular, and crosslinked acellular aortic matrix grafts at 0, 15, and 60 days postimplantation, stimulated with PHA, native, and acellular aortic antigens are presented in Fig. 8.11A−C [46].

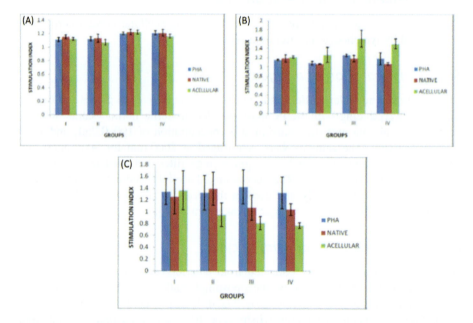

Figure 8.11 (A) Mean ± SE of stimulation index (SI) of guinea pigs (peripheral blood lymphocytes) subcutaneously implanted with native (I), acellular (II), BDDGE (III), and EDC (IV) crosslinked aortic grafts at day 0. (b) Mean ± SE of stimulation index (SI) of guinea pigs (peripheral blood lymphocytes) subcutaneously implanted with native (I), acellular (II), BDDGE (III), and EDC (IV) crosslinked aortic grafts at days 15. (C) Mean ± SE of stimulation index (SI) of guinea pigs (peripheral blood lymphocytes) subcutaneously implanted with native (I), acellular (II), BDDGE (III), and EDC (IV) crosslinked aortic grafts at days 60.

At day 15, the SI values of groups I and III (1.18 ± 0.08) stimulated with native antigen were higher when compared to group II (1.06 ± 0.01) and group IV (1.07 ± 0.03) stimulated with the same antigen. Stimulation of group III with acellular antigen resulted in higher SI value (1.61 ± 0.18) when compared to groups I (1.21 ± 0.03), II (1.26 ± 0.16), and IV (1.50 ± 0.11) (Fig. 8.11B).

At day 60, the SI value of group II (1.39 ± 0.28) stimulated with native antigen was higher than the SI values of groups I (1.25 ± 0.29), IV (1.04 ± 0.21), and III (1.07 ± 0.10), stimulated with same antigen whereas the group I stimulated with acellular antigen showed higher SI value (1.36 ± 0.33) when compared to groups II (0.95 ± 0.20), IV (0.77 ± 0.11), and III (0.81 ± 0.05) stimulated with the same antigen. The stimulation index values were lower in group IV at 15 and 60 days postimplantation when stimulated with both acellular and native antigens (Fig. 8.11C) [46].

8.4.5.3 Splenocytes culture

Sixty days postimplantation, guinea pigs were sacrificed and spleen of each animal was collected in sterile PBS (pH 7.4). Individual spleens were macerated with steel mesh and rinsed with PBS to obtain single cell suspension. Splenocyte suspension was centrifuged at 1800 rpm for 10 minutes. Supernatant was discarded and pellet was resuspended in 2 mL PBS. RBC lysis was done by adding 2 mL of RBC lysis buffer and keeping in dark for 5–10 minutes. Immediately cells were diluted with 10 times volume of PBS, and again centrifuged at 1800 rpm for 10 minutes. Supernatant was discarded and 2–3 more washings were given with PBS at 1800 rpm for 10 minutes. Supernatant was discarded and pellet was resuspended in RPMI1640 growth medium. The cells were adjusted to a concentration of 2×10^6 viable cells/mL in RPMI1640 growth medium and seeded in 96-well tissue culture plate @ 100 µL/well. The cells were incubated at 37°C in 5% CO_2 environment. Cells from each guinea pig were stimulated with antigen (10–20 µg/mL) and PHA (10 µg/mL) in triplicates and three wells were left unstimulated for each sample. After 45 hours, 40 µL of MTT solution (5 mg/mL) was added to all the wells and incubated further for 4 hours. The plates were then centrifuged for 15 minutes in plate centrifuge at 2500 rpm. The supernatant was discarded, plates dried, and 150 µL DMSO was added to each well and mixed thoroughly by repeated pipetting to dissolve the formazan crystals. The plates were immediately read at 570 nm with 620 nm as reference wavelength. The stimulation index (SI) was calculated using the following formula [18].

$$\text{Stimulation Index (SI)} = \frac{\text{OD of stimulated cultures}}{\text{OD of unstimulated cultures}}.$$

At sixty days postimplantation, guinea pigs from the four groups were sacrificed and spleen of each animal was collected for lymphocyte proliferation using MTT colorimetric assay in the same way as described for peripheral blood lymphocytes. The mean ± SE of stimulation index (SI) values of native, acellular, and crosslinked

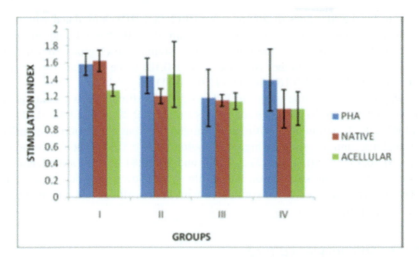

Figure 8.12 Mean ± SE of stimulation index (SI) of guinea pigs (splenocytes) subcutaneously implanted with native (I), acellular (II), BDDGE (III), and EDC (IV) crosslinked aortic grafts.

acellular aortic matrix grafts at 60 days postimplantation, stimulated with PHA, native, and acellular aortic antigens are presented in Fig. 8.12. The SI value of group I (1.62 ± 0.13) stimulated with native antigen was higher than groups II (1.20 ± 0.09), III (1.15 ± 0.07), and IV (1.05 ± 0.23) stimulated with the same antigen. The SI value of group II (1.46 ± 0.39) stimulated with acellular antigen was higher than the groups I (1.27 ± 0.07), III (1.14 ± 0.10), and IV (1.05 ± 0.20). Among all the groups, stimulation index was lowest in guinea pigs subcutaneously implanted with EDC crosslinked acellular aortic matrix graft when stimulated with both native and acellular antigens [46].

Lymphocyte proliferation assay (LPA) measures the ability of lymphocytes placed in short-term tissue culture to undergo a clonal proliferation when stimulated in vitro by a foreign molecule (antigen/mitogen). $CD4^+$ lymphocytes proliferate in response to antigenic peptides in association with class II major histocompatibility complex (MHC) molecules on antigen presenting cells (APCs). Antigen-specific T-cell proliferation is a major technique for assessing the functional capacity of $CD4^+$ lymphocytes to respond to various stimuli.

An immune response against nonself and self-antigens is initiated by presentation of the antigen in a suitable form to T cells. Antigen can only be presented to T cells in the context of molecules of MHC [61]. Hence, practically each nucleated cell of the body is able to present antigen, first by virtue of a constitutive MHC class I expression, and second by a de novo expression of MHC class II molecules on the surface of the cell. It is well known that when an antigen enters the body, two different types of immune response may occur. The first type, known as the "humoral immune response," involves the synthesis and release of free antibody into the blood and other body fluids. The second type of immune response, known as the "cell-mediated immune response," involves the production of sensitized lymphocytes that are themselves the effectors of this type of immunity. Lymphocytes consist of various subpopulations with distinctive functions, which play important roles in immune responses [62]. Activation and proliferation of these subpopulations can be achieved by treating them with antigens.

Graft rejection is usually mediated by activity of T cells, especially cytotoxic T cells. The T-cell subsets (Th1 and Th2) generated by naïve T cells on MHC antigen stimulation play a major role in the graft rejection through activity of different sets of cytokines that activate macrophages and B cells. Cells in ECMs have Class I and II histocompatibility antigens capable of eliciting rejection reactions. Also, the cells have glycoproteins recognized by the immune system of hosts, which elicit rejection reactions. Therefore, if these substances are eliminated from ECMs, rejection reactions can be prevented. However, complete elimination of all antigens is considerably difficult to perform and verify [63]. In this study, the antigen prepared from acellular tissue showed highest SI in MTT assay. The SI recorded for crosslinked samples was lower in comparison to the values of uncrosslinked samples. The greater ability of this antigen to stimulate the lymphocytes in vitro may be attributed to the fact that on treatment with biological detergent, the bonds between protein molecules are broken and results into a change from quaternary and tertiary structure to primary and secondary structures. Therefore, the acellular antigen had greater ability to trigger cell-mediated immune response in host because of presence of shorter peptide fragments which can be presented to the immune system by MHC class II pathway and stimulate the $CD4^+$ lymphocytes. Whereas, the crosslinked tissue is not processed in the body to form shorter immunogenic fragments that can elicit the cell-mediated immune response in host. This may also be because of the fact that on crosslinking tissues with different chemicals, the site where biological enzymes act in vivo, are masked and the crosslinked tissue is no longer broken down into smaller peptide fragments to elicit immune response. Crosslinking of the proteins on treatment with EDC might have masked immunogenic epitopes, and therefore, there is either delayed cellular immune response in host body.

8.4.5.4 ELISA

To evaluate the immune compatibility of the crosslinked biomaterials, ELISA was performed. Serum samples from the guinea pigs were collected on 15, 30, and 60 days postimplantation for ELISA. The test was done as per standard protocol. Microtiter ELISA plate was coated with 0.25 μg of protein (derived from grafted material) in 100 μL of 0.05 M sodium carbonate buffer (pH 9.6) per well. The plate was incubated at 4°C overnight. After incubation plate was washed with PBS-T [0.15 M sodium chloride 0.02 M phosphate buffer (pH 7.2) containing 0.005% Tween 20]. Subsequently, blocking was done with 1% bovine serum albumin in PBS-T and further incubated at 37°C for 2 hours. Plate was washed with PBS-T followed by adding 1:100 dilution of sera obtained from different guinea pigs grafted with various graft materials. Plate was incubated again for 2 hours at 370 C, followed by washing with PBS-T. Peroxidase-labeled anti guinea pig conjugate having 1:20,000 dilutions was made in PBS-T and instilled 100 μL in each well and then incubated at 37°C for 2 hours. Finally, plate was washed as before and peroxidase substrate was added [100 μL of 17 mM Na citrate buffer, pH 6.3 containing 0.2% (wt/vol.) ophenylene diamine and 0.015% (wt/vol.) hydrogen peroxide] per well. Substrate was allowed to act for 30 minutes at 37°C, keeping the plate in dark. Absorbance was recorded at 492 nm using ELISA reader. The values of antibodies titer (absorbance) were expressed as absorbance (OD492) and are presented in Fig. 8.13 [18].

The levels of antibodies present in serum samples collected prior to implantation were taken as basal values. Hyperimmune sera raised against native aorta were used as standard positive control (1.5 ± 0.18). The graft-specific antibody levels started increasing on 15th postimplantation day in all the groups. The graft-specific antibody levels showed an increasing trend till 30th postimplantation day and then onward showed a decreasing trend in all groups except group II, which showed increasing trend up to day 60postimplantation. At 15 days postimplantation, the graft-specific antibody levels were higher in group IV (0.414 ± 0.03) and lower in group I (0.316 ± 0.001). At 30 days postimplantation, the antigraft-specific antibody levels were higher in group II (0.729 ± 0.003) and lower in group IV (0.414 ± 0.04). At 60 days postimplantation, the graft-specific antibody levels were higher in group II (0.767 ± 0.005) and lower in group IV (0.402 ± 0.03). Among all the groups, group IV showed minimal graft-specific antibody levels when compared to native, acellular, and BDDGE groups and the levels remained constant and more or less equal to basal value (0.306 ± 0.01).

The immune response to xenogenic transplantation included both natural and induced humoral components. The presence of antibodies to xenogenic collagen was an epiphenomenon and not an indicator for rejection of the implant [64].

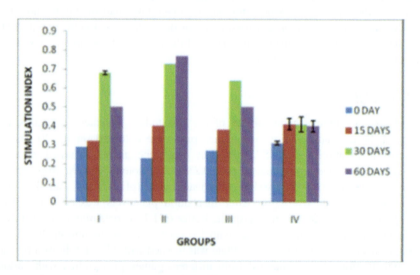

Figure 8.13 Mean ± SE of absorbance values (ELISA) of guinea pigs subcutaneously implanted with native (I), acellular (II), BDDGE (III), and EDC (IV) crosslinked aortic grafts.

The ELISA was performed to check the extent of antibody generated toward the graft components. The absorbance values were taken as a measure to compare the magnitude of immune response. Seddon et al. [24] reported that ionic detergents like 1% SDS are effective for solubilizing both cytoplasmic and nuclear cellular membranes, but tend to denature proteins by disrupting protein−protein interactions. Collagens are weakly immunogenic as compared to other proteins. The major antigenic determinants are situated in the telopeptide regions of the molecule. The other two types of determinants are composed of the triple helix and of the amino acid sequence of the alpha chains. The latter type is accessible only when the collagen is denatured [65]. When the antigenic determinants are exposed due to collagen degeneration, it results in severe immune response.

Acellular group showed higher immune response when compared to crosslinked and native groups. The antigenicity of a collagen biomaterial can be reduced by the process of crosslinking [66]. Therefore, the immune response was less in BDDGE and EDC crosslinked groups when compared to acellular group. It is a well-known fact that production of antibodies requires about 21 days after antigen administration. As a result, the antibody levels increased in all groups at 30 days postimplantation. As the EDC resulted in efficient masking of antigenic determinant sites, low levels of antibodies were detected at 15, 30, and 60 days postimplantation in the group that received EDC implants. The same finding was reported in EDC crosslinked BAMG by Dewangan et al. [67].

8.4.6 Molecular weight analysis

Molecular weight analysis of the biomaterials crosslinked with various reagents was done by sodium dodecyl sulfate polyacrylamide gel electrophoresis (SDS-PAGE) as per the standard method. It was performed by 10% sodium dodecyl sulfate polyacrylamide gel electrophoresis (SDS-PAGE). Samples were prepared by triturating 100 mg of crosslinked and native biomaterials with 10% SDS (1 mL) and supernatant was obtained after centrifugation at 10,000 rpm for 10 minutes. Sample buffer was added to the supernatant in the ratio 1:1 and heated for 10 minutes. The solution was allowed to cool at room temperature. Then the sample was loaded onto 14 slot applicators. The finished gel was stained with Coomassie blue 2 and destained with 20% methanol and 10% acetic acid. Known molecular weight marker was used to calibrate the gel. The values of high molecular weight protein bands were expressed in kDa.

SDS-PAGE was performed to determine the crosslinking ability of different chemicals. Crosslinking resulted in the formation of high molecular weight protein, which was determined by the expression of protein bands (Fig. 8.14). The typical aortic collagen pattern is represented in the native aorta (Lane 2). In the SDS resolving gel, the collagen bands showed molecular weight of about 25, 30, 45, 50, 60, 66, and 200 kDa. Native collagen molecules remained in the stacking gel. After decellularization process, the soluble protein decreased as revealed in SDS-PAGE of acellular aorta (Lane 3). The protein band pattern of collagen in acellular aorta in SDS-PAGE did not show any high protein band. The GA-treated acellular aortic matrix

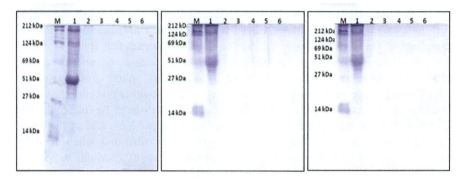

Figure 8.14 Analysis of protein crosslinking with different chemicals by performing SDS-PAGE using 10% resolving gel and 3% stacking gel.

graft did not show any higher protein band pattern in SDS-PAGE gel. The GA crosslinked collagen resulted in the formation of large covalently crosslinked complex. This complex did not disassociate by chemical treatment with sample buffer and was too large to enter the stacking gel. Therefore, all the GA-treated acellular aortic matrix graft did not show any band pattern in SDS-PAGE gel. The EDC- and BDDGE-treated tissues produced a characteristic crosslinking of the proteins which was observed in all crosslinked samples, suggesting that the chemical treatment had effectively crosslinked the different chains of collagen proteins resulting into formation of high mass which did not find entry even in the stacking gel. Therefore, all the EDC- and BDDGE-treated acellular aortic matrix grafts did not show any band pattern in SDS-PAGE gel. The aortic grafts crosslinked for 24 hours with EDC and BDDGE were ideal for subcutaneous implantation in guinea pig model.

Once the protein is crosslinked in the acellular graft, it will delay the degradation of transplanted tissue, thereby, providing sufficient line for the host body to replace the damaged tissue. The aortic tissues treated with GA, EDC, and BDDGE showed fewer amounts of low molecular weight proteins, which indicated efficient crosslinking. This was evidenced by absence of specified bands in the SDS-PAGE gel. Therefore, the maximum ability to crosslink the aortic tissues was seen with GA, followed by EDC and BDDGE. Similar results were reported with GA and EDC crosslinked bladder acellular matrix graft by Dewangan et al. [67].

8.5 Clinical applications in different species of animals

Kumar et al. [68] repaired umbilical hernias in nine client-owned calves using acellular buffalo matrix. The calves were female and either crossbred Holstein–Friesian ($n = 7$) or undefined breed ($n = 2$). The age of the calves ranged from 1 to 11 months (mean, 4.56 months). Physical examination in each case revealed a painless, reducible soft swelling with a discernible ring at the umbilicus (Fig. 8.15A) diagnosed as congenital umbilical hernia. The aortic

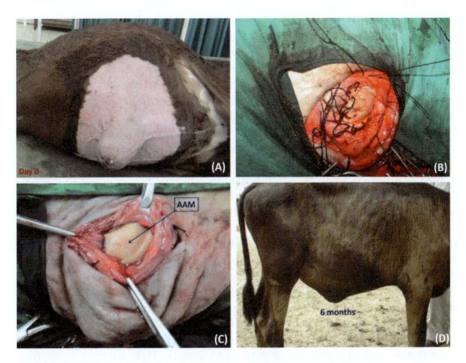

Figure 8.15 (A) Preoperative appearance of an umbilical hernia in a female Holstein–Friesian calf, (B) intraoperative image of placement of acellular aortic matrix (AAM) graft between parietal peritoneum and internal rectus sheath with suture ends retrieved through the abdominal muscles, (C) intraoperative image of placement of AAM, (D) postoperative appearance of umbilical hernia 6 months after hernioplasty.

matrix was prepared by decellularized using 1% SDS for 24 hours followed by treatment with 0.25% trypsin for 2 hours and 1% SDS for a further 24 hours, during continuous agitation in a horizontal orbital shaker (180 rotations/minute). The acellularity of the prepared matrix was confirmed microscopically using H&E staining. The calves were restrained in lateral recumbency. To expose the hernial sac, a fusiform skin incision, spanned the length of the hernia and extended 2 cm beyond the cranial and caudal margins of the hernial ring. The hernial sac was dissected from overlying skin and the dissection was continued laterally to expose the hernial ring and the external sheath of the rectus abdominis muscle (aponeurosis of the external and internal abdominal oblique muscles) abaxial to it. Manual pressure was used to control any subcutaneous hemorrhage. The unopened, isolated hernial sac was inverted into the abdomen. A space was created by separating the parietal peritoneum from the internal sheath of the rectus abdominis muscle (aponeurosis of the transverse abdominis muscle). An appropriately sized AAM graft with preplaced horizontal mattress sutures of number 2 surgical silk with long ends attached to its cranial, caudal, and mid-lateral edges was introduced between the parietal peritoneum and internal sheath of the rectus abdominis muscle. After orientation of the AAM graft, the suture ends were retrieved using a nontraumatic needle (Fig. 8.15B). Each of the sutures was tied with the knots resting on the external sheath of the rectus abdominis muscle, thus provisionally securing the AAM (Fig. 8.15C).

While the graft was being implanted, the surgical site was lavage periodically with sterile PBS containing 0.1% amikacin. Excess skin was excised and the subcutaneous tissues were closed in two layers, using number 1 chromic catgut placed in a simple continuous suture pattern. The skin incision was then closed using number 1 surgical silk suture material in a horizontal mattress suture pattern.

To assess the integrity of the repair, clinical evaluation of calves was performed at 4-weekly intervals up to 6 months. Owners were advised to maintain the bandage up to 30 days, at which time healing was assessed by physical examination. The technique of hernioplasty using AAM graft appears to be a satisfactory treatment regimen for umbilical hernia repair in calves (Fig. 8.10D). Evidence from a small number of treated animals suggests that AAM of buffalo origin has adequate strength to be used for the surgical repair of umbilical hernias in calves [68]. Umbilical hernias are fairly common in calves. The condition is considered to be hereditary and most commonly occurs in the Holstein−Friesian breed. It is the result of incomplete closure of the umbilicus at birth, because of mal development or hypoplasia of the abdominal muscles and the only effective treatment is surgery to restore the integrity of the abdominal wall and prevent incarceration and strangulation of herniated contents. Tight suturing to approximate and close the defect can lead to wound dehiscence, recurrent hernias, and nonhealing of the wound. The use of nonabsorbable, synthetic mesh material has been reported to cause complications such as mesh extrusion, bowel adherence, fistula formation, wound infection, skin erosion, and seroma development. To overcome the disadvantages of synthetic meshes, biomaterials may be preferable for the surgical repair of hernias [68].

Successful tracheal segment replacement using acellular aortic graft of buffalo was done in a 5-year old, crossbred Holstein−Friesian cow with postanastomotic tracheal stenosis [69]. Under xylazine sedation and local analgesia, through midline cervical incision, the stenotic tracheal segment was exposed and resected. Defect was repaired with acellular aortic graft, supported by a plastic stent to prevent airway collapse. Postoperative complications were not observed in this animal during a 2.5 years follow-up examination. The acellular aortic graft of buffalo origin can be used as a tracheal substitute for the tracheal segment replacement in Holstein−Friesian cow [69].

Successful umbilical hernia repair was done in four crossbred Holstein−Friesian calves with an average age of 7 months and average weight of 60 kg. Average hernial ring diameter was 8 cm [70]. Acellular matrix from buffalo aorta was prepared using 1% anionic biological detergent for 24 hours followed by 0.25% enzyme solution for 2 hours and then again with same 1% anionic biological detergent for next 24 hours [18]. After proper anesthesia, an elliptical incision was made over the hernial sac and fascia and muscles were separated from the hernial ring. All animals were having slight to strong adhesions. The hernial contents were pushed back in cases where no adhesions or slight adhesions were present. The adhesions were removed by blunt dissection. The hernial ring was freed and repaired by acellular aortic graft using inlay technique. The graft was anchored in position by black braided silk No. 2 using horizontal mattress sutures in all the animals. Finally, the skin was closed by black braided silk No. 2. All the animals recovered completely and no complication of wound healing was observed up to 10th–14th postoperative days [70].

Acellular aortic matrix was used for the repair of large ventral hernia in a buck [71]. A Jamuna Pari breed buck aged 3 years having bodyweight of 35 kg was presented with the history of swelling in the right ventral abdomen (Fig. 8.16A). Clinical examination revealed large ventral hernia of abdominal wall (Fig. 8.16B). The size of hernial ring was 10–12 cm in diameter. After proper anesthesia, an elliptical incision was made over the hernial sac and fascia and muscles were separated from the hernial ring (Fig. 8.16C) [71].

The hernial contents were pushed back in the abdominal cavity. The hernial ring was repaired by acellular aortic matrix using inlay technique. The graft was anchored in position by black braided silk No. 2 using horizontal mattress suture pattern (Fig. 8.16D and E). Finally the skin was closed by silk sutures (Fig. 8.16F). Acellular matrices were prepared as per method described by Devasthanam [18]. The buck recovered completely without clinical signs of wound dehiscence or infection. In this study, we demonstrated that ventral hernia repaired with AAM result in no herniation up to 3 months after reconstruction [71].

Vora et al. [36] prepared acellular aortic matrix with 1% SDS for 24 hours followed by 0.25% trypsin for 2 hours and again with 1% SDS for 24 hours.

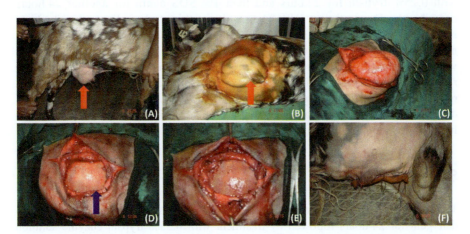

Figure 8.16 (A) Preoperative appearance of an umbilical hernia in a buck (red arrow), (B) showing large umbilical hernia(red arrow), (C) intraoperative image of hernia, (D) intraoperative image of placement of acellular aortic matrix between parietal peritoneum and internal rectus sheath (blue arrow), (E) intraoperative image of placement of acellular aortic matrix after completion of suturing, (F) postoperative appearance.

Clinical, hematobiochemical and antioxidant parameters were evaluated to assess biocompatibility of xenogenic BAM. Histologically, the absence of cells and orderly arranged collagen fibers were observed in treated aorta. SEM confirmed the preservation of collagen structure and integrity. They were clinically applied for the repair of umbilical hernias in six cattle. The study was conducted on six cattle (two males and four females) clinically affected with abdominal hernias. Mean weight and age of animals were 132.50 ± 21.86 kg and 6.50 ± 1.50 months. The bubaline aorta, upon treatment with 1% SDS for 24 hours followed by 0.25% trypsin for 2 hours and again with 1% SDS for 24 hours, results in complete removal of cells and cellular components, with the preservation of collagen structure and integrity. Xenogenic bubaline aortic matrix shows excellent repair efficiency and biocompatibility for abdominal hernia repair in cattle without complications. Cattle with the BAM implant recovered uneventfully and remained sound. Hematobiochemical and antioxidant findings were unremarkable. Bubaline aortic matrix shows excellent repair efficiency and biocompatibility for abdominal hernia repair in cattle without complications [36]. Nine crossbred Murrah buffaloes (seven males and two females) with an average age of 7.44 months and with congenital umbilical hernias were repaired with acellular aortic matrices [72]. On physical examination, a hernial ring was observed with average diameter of 6.89 cm. In another 11-month-old, female crossbred buffalo, there was a swelling behind the 10th rib which had resulted from a gore injury. Clinical examination revealed a hernial ring of about 12×10 cm and fracture of the costal cartilages but an intact diaphragm. This case was diagnosed as a ventral hernia. Native aorta of buffalo origin was decellularized using 1% SDS for 24 hours followed by treatment with 0.25% trypsin for 2 hours and then 1% SDS again for another 24 hours described by Devarathnam [18]. The results of this study indicate that AAM of buffalo origin can be used safely in buffaloes for the reconstruction of abdominal hernias with adequate strength and minimal foreign body reaction [72]. In vivo biocompatibility determination of acellular aortic matrix of buffalo origin revealed that the acellular matrices were biocompatible and produced least reaction when implanted in laboratory animals [46]. In vitro biocompatibility determination of acellular aortic matrix of buffalo origin revealed that the acellular matrices were biocompatible [73].

8.6 Conclusion

The study revealed that protocol B4 in which the aorta was treated with 1% SDS detergent for 24 hours followed by treatment with 0.25% trypsin enzyme solution for 2 hours and then again with same 1% SDS for again next 24 hours showed complete acellularity with normal thickness and arrangement of collagen fibers. The developed acellular aortic matrices were tested for the repair of abdominal wall defects in guinea pigs, clinical cases of hernia of buffalo, cattle, buck, calves, and repair of trachea in cow.

References

[1] Bissell MJ, Aggeler J. Dynamic reciprocity: how do extracellular matrix and hormones direct gene expression? Prog Clin Biol Res 1987;249:251−62.
[2] Brown E, Dejana E. Cell-to-cell contact and extracellular matrix editorial overview: cell-cell and cell-matrix interactions-running, jumping, standing still. Curr Opin Cell Biol 2003;15:1−4.
[3] Kleinman HK, Philp D, Hoffman MP. Role of the extracellular matrix in morphogenesis. Curr Opin Biotechnol 2003;14:526−32.
[4] Rosso F, Giordano A, Barbarisi M, Barbarisi A. From cell-ECM interactions to tissue engineering. J Cell Physiol 2004;199:174−80.
[5] Valentin JE, Badylak JS, McCabe GP, Badylak SF. Extracellular matrix bioscaffolds for orthopaedic applications: a comparative histologic study. J Bone Jt Surg Am 2006;88:2673−86.
[6] Cobb GA, Shaffer J. Cross-linked acellular porcine dermal collagen implant in laparoscopic ventral hernia repair: case-controlled study of operative variables and early complications. Int Surg 2005;90:24−9.
[7] Helton WS, Fisichella PM, Berger R. Short-term outcomes with small intestinal submucosa for ventral abdominal hernia. Arch Surg 2005;140:549−60.
[8] Ko R, Kazacos EA, Snyder S. Tensile strength comparison of small intestinal submucosa in body wall repair. J Surg Res 2006;135:9−17.
[9] Gaertner WB, Bonsack ME, Delaney JP. Experimental evaluation of four biologic prostheses for ventral hernia repair. J Gastrointest Surg 2007;11:1275−85.
[10] Nahabedian MY. Does AlloDerm stretch? Plast Reconstr Surg 2007;120:1276−80.
[11] Amiel G.E., Yu R.N., Shu T. Head-to-head comparison of urodynamics and histology of rat bladders augmented with blood vessel matrix versus small intestine submucosa. Society of Biomaterials 35th Annual Meeting, Pittsburgh; 2006.
[12] Kaushal S, Amiel GE, Guleserian KJ. Functional small-diameter neo vessels created using endothelial progenitor cells expanded ex- vivo. Nat Med 2001;7:1035−40.
[13] Bellows CF, Jian W, McHale MK, Cardenas D, West JL, Lerner SP, et al. Blood vessel matrix: a new alternative for abdominal wall reconstruction. Hernia 2008;12:351−8.
[14] Anoosh F, Hodjati H, Dehghani S, Tanideh N, Kumar PV. Tracheal replacement by autogenous aorta. J Cardio Thorac Surg 2009;4:23−7.
[15] Wight TN. Arterial wall. In: Comper W, editor. Extracellular Matrix., Volume I. The Netherland: Harwood Academic Publishers; 1996, p. 175−202.
[16] Jacob MP, Badier-Commander C, Fontaine V, Benazzoug Y, Feldman L, Michel JB. Extracellular matrix remodeling in the vascular wall. Pathol Biol 2001;49:326−32.
[17] Gilbert TW, Sellaro TL, Badylak SF. Decellularization of tissues and organs. Biomaterials 2006;27:3675−83.
[18] Devarathnam J. Biocompatibility determination of acellular aortic matrix of buffalo origin. [MVSc thesis]. Submitted to Deemed University Indian Veterinary Research Institute, Izatnagar 243122, Uttar Pradesh, India; 2010.
[19] Devarathnam J, Sharma AK, Kumar N, Rai RB. Optimization of protocols for decellularization of buffalo aorta, J Biomat. Tissue Eng 2014;4(10):778−85.
[20] Chang MC, Tanaka J. FTIR study for hydroxyapatite/collagen nanocomposite crosslinked by glutaraldehyde. Biomaterials 2002;23:4811−18.
[21] Roy S, Silacci P, Stergiopulos N. Biomechanical proprieties of decellularized porcine common carotid arteries. Am J Physiol Heart Circ Physiol 2005;289:1567−76.

[22] Grauss RW, Hazekamp MG, Oppenhuizen F, van Munsteren CJ, Gittenberger-de Groot AC, DeRuiter MC. Histological evaluation of decellularized porcine aortic valves: matrix changes due to different decellularization methods. Eur J Cardio Thorac Surg 2005;27(4):566−71. Available from: https://doi.org/10.1016/j.ejcts.2004.12.052.
[23] Woods T, Gratzer PF. Effectiveness of three extraction techniques in the development of a decellularized bone−anterior cruciate ligament− bone graft. Biomaterials 2005;26:7339−49.
[24] Seddon AM, Curnow P, Booth PJ. Membrane proteins, lipids and detergents: not just a soap opera. Biochem Biophys Acta 2004;1666:105−17.
[25] Schaner P, Martin ND, Tulenko TN, Shapiro IM, Tarola NA, Leichter RF, et al. Decellularized vein as a potential scaffold for vascular tissue engineering. J Vasc Surg 2004;40:146−53.
[26] Allaire E, Bruneval P, Mandet C, Becquemin JP, Michel JB. The immunogenicity of the extracellular matrix in arterial xenografts. Surgery 1997;122:73−81.
[27] Kasimir MT, Rieder E, Seebacher G, Silberhumer G, Wolner E, Weigel G, et al. Comparison of different decellularization procedures of porcine heart valves. Int J Artif Organs 2003;26:421−7.
[28] Courtman DW, Pereira CA, Kashef V, McComb D, Lee JM, Wilson GJ. Development of a pericardial acellular matrix biomaterial: Biochemical and mechanical effects of cell extraction. J Biomed Mater Res 1994;28:655−66.
[29] Bodnar E, Olsen EG, Florio R, Dobrin J. Damage of porcine aortic valve tissue caused by the surfactant sodium dodecyl sulphate. Thorac Cardiovasc Surg 1986;34:82−5.
[30] Samouillan V, Dandurand-Lods J, Lamure A, Maurel A, Lacabanne C, Gerosa G, et al. Thermal anlaysis characterization of aortic tissues for cardiac valve prostheses. J Biomed Mater Res 1999;46:531−8.
[31] Booth C, Korossis SA, Wilcox HE, Watterson KG, Kearney JN, Fisher J, et al. Tissue engineering of cardiac valve prostheses I: Development and histological characterization of an acellular porcine scaffold. J Heart Valve Dis 2002;11:457−62.
[32] Rieder E, Kasimir M, Silberhumer G, Seebacher G, Wolner E, Simon P, et al. Decellularization protocols of porcine heart valves differ importantly in efficiency of cell removal and susceptibility of the matrix to recellularization with human vascular cells. J Thorac Cardiovasc Surg 2004;127:399−405.
[33] Bader A, Steinhoff G, Strobl K, Schilling T, Brandes G, Mertsching H, et al. Engineering of human vascular aortic tissue based on a xenogeneic starter matrix. Transplantation 2000;70:7−14.
[34] Lu Q, Kavitha G, Dan T, Simionescu S, Vyavahare NH. Novel porous aortic elastin and collagen scaffolds for tissue engineering. Biomaterials 2004;25:5227−37.
[35] Kumar N, Devarathnam J, Sharma AK, Singh H, Kumar V, Gangwar AK, et al. Optimization of protocols for decellularization of aortic matrix of buffalo origin. Presented in XXIV Annual Congress of ISVS and International symposium on Newer Concepts in Surgical Techniques for Farm and Companion Animal Practice, Puduchery; 2010, P. 117−18.
[36] Vora SD, Kumar V, Singh VK, Fefar DT, Gajera HP. Bubaline aortic matrix: Histologic, imaging, Fourier transform infrared spectroscopic characterization and application into cattle abdominal hernia repair. Proc Natl Acad Sci, India Sect B: Biol Sci 2020;90(1):161−70. Available from: https://doi.org/10.1007/s40011-019-01094-w.
[37] Green MR, Sambrook J. Isolation of high-molecular weight DNA using organic solvents. Cold Spring Harbor Laboratory Press; 2017. https://doi.org/10.1101/pdb.prot093450.

[38] Crapo PM, Gilbert TW, Badylak SF. An overview of tissue and whole organ decellularisation processes. Biomaterials 2011;32:3233−43.
[39] Abe Y, Krimm S. Normal vibrations of crystalline polyglycine II. Biopolymers 1972;11:1817−39.
[40] Doyle BB, Bendit E, Blout ER. Infrared spectroscopy of collagen and collagen-like polypeptides. Biopolymers 1975;14:937−57.
[41] Krimm S, Bandekar J. Vibrational spectroscopy and conformation of peptides, polypeptides, and proteins. Adv Protein Chem 1986;38:181−364.
[42] Payne KJ, Veis A. Fourier transform IR spectroscopy of collagen and gelatin solutions: deconvolution of the Amide I band for conformational studies. Biopolymers 1988;387:1949−60.
[43] Asodiya FA. Clinical assessment of caprine acellular dermal matrix (CADM) for abdominal hernioplasty in buffaloes. [MVSc thesis]. Submitted to Junagadh Agricultural University, Gujrat, India; 2018.
[44] Vora SD, Kumar V, Asodiya FA, Singh VK, Fefar DT. Bubaline diaphragm matrix: development and clinical assessment into cattle abdominal hernia repair. In: 42nd Annual Congress of Indian Society forVeterinary Surgery, Navsari, India; 2018, p. 187−88.
[45] Muyonga JH, Cole CGB, Duodu KG. Characterisation of acid soluble collagen from skins of young and adult Nile perch (Lates niloticus). Food Chem 2004;85:81−9.
[46] Devarathnam J, Sharma AK, Kumar N, Shrivastava S, Sonal, Rai RB. In vivo biocompatibility determination of acellular aortic matrix of buffalo origin. Progr Biomat 2014;3:115−22. Available from: https://doi.org/10.1007/s40204-014-0027-6.
[47] Lu HJ, Chang Y, Sung HW, Chiu YT, Yang PC, Hwang B. Heparinization on pericardial substitutes can reduce adhesion and pericardial inflammation in the dog. J Thorac Cardiovasc Surg 1998;115:1111−20.
[48] Schmidt CE, Baier JM. Acellular vascular tissue: natural biomaterials for tissue repair and tissue engineering. Biomaterials 2000;21:2215−31.
[49] Goldstein S, Black K, Clark D, Orton EC, O'Brien MF. Inflammatory responses to uncross-linked xenogenic heart valve matrix. World Symposium on Heart Valve Diseases. London, England; 1999, p. 205.
[50] Courtman DW, Wilson GJ. Development of an acellular matrix vascular xenograft: Modification of the in-vivo immune response in rats. In: Proceedings of the society for biomaterials 25th annual meeting; 1999, p. 20.
[51] Mckegney M, Taggart I, Grent MH. The influence of cross-linking agents and diamines in the pore size, morphology and biological stability of collagen sponges and their effect on cells penetration through the sponge matrix. J Mater Sci Mater Med 2001;23:833−44.
[52] Shoukry M, El-Keiey M, Hamouda M, Gadallah S. Commercial polyester fabric repair of abdominal hernias and defects. Vet Rec 1997;140:606−7.
[53] Kanade MG, Kumar A, Sharma SN. Diaphragm as a prosthetic material for the repair of ventral body defects in bovines. Indian J Vet Surg 1986;7:8−14.
[54] Deokiouliyar UK, Khan AA, Sahay PN, Prasad R. Evaluation of preserved homologous pericardium for hernioplasty in buffalo calves. J Vet Med 1988;35:391−4.
[55] Mantovani A, Sica A, Locati M. Macrophage polarization comes of age. Immunity 2005;23:344−6.
[56] Gilberto G, Pereira DR. A study on biocompatibility and integration of acellular polyanionic collagen: elastin matrices by soft tissue. Rev Bras Eng Biomed 2003;19:167−73.

[57] Petter-Puchner AH, Fortelny RH, Mittermayr R, Walder N, Ohlinger W, Redl H. Adverse effects of porcine small intestine submucosa implants in experimental ventral hernia repair. Surg Endosc 2006;20:942–6.
[58] Courtman DW, Errett BF, Wilson GJ. The role of cross-linking in modification of the immune response elicited against xenogenic vascular acellular matrices. J Biomed Mater Res 2001;55:576–86.
[59] Silver IA, Murills RJ, Etherington DJ. Microelectrode studies on the acid microenvironment beneath macrophages and osteoclasts. Exp Cell Res 1988;175:266–76.
[60] Hardin-Young J, Carr RM, Downing GJ, Condon KD, Termin PL. Modification of native collagen reduces antigenicity but preserves cell compatibility. Biotechnol Bioeng 1996;49:675–82.
[61] Townsend A, Bodmer H. Antigen recognition by class I restricted T lymphocytes. Annu Rev Immunol 1989;7:601–24.
[62] Otsuka H, Ikeya T, Okano T, Kataoka K. Activation of lymphocyte proliferation by boronate containing polymer immobilised on substrate: the effect of boron content on lymphocyte proliferation. Eur Cell Mater 2006;12:36–43.
[63] Malone JM, Brendel K, Duhamil RC, Reinert RL. Detergent-extracted small diameter vascular prosthesis. J Vasc Surg 1984;1:181–91.
[64] Ruszezak Z. Effect of collagen matrices on dermal wound healing. Adv Drug Delivery Rev 2003;55:1595–611.
[65] Chevallay B, Herbage D. Collagen-based biomaterials as 3-D scaffold for cell cultures: applications for tissue engineering and gene therapy. J Med Biol Eng Comp 2000;38:211–18.
[66] O'Brien TK, Gabbay S, Parkes AC, Knight RA, Zalesky PJ. Immunological reactivity to a new tanned bovine pericardial heart valve. Trans Am Soc Artif Intern Organs 1984;30:440–4.
[67] Dewangan R, Sharma AK, Kumar N, Maiti SK, Singh H, Kumar A, et al. *In-vivo* determination of biocompatibility of bladder acellular matrix in a rabbit model. Trends Biomat Artific Organs 2012;26:43–55.
[68] Kumar V, Kumar N, Gangwar AK, Saxena AC. Using acellular aortic matrix to repair umbilical hernias of calves. Aust Vet J 2013;91(6):252–3. Available from: https://doi.org/10.1111/avj.12058.
[69] Kumar V, Kumar N, Devarathnam J, Pawde AM, Gangwar AK, Singh H. Successful tracheal replacement with acellular aortic graft in a cow. J Vet Adv 2012;2(11):552–6.
[70] Devarathnum J, Sharma AK, Gangwar AK, Kumar V, Kumar V, Singh H, et al. Acellular aortic grafts for the reconstruction of umbilical hernias in Holstein-Friesian calves. Vet Pract 2012;13(2):307–8.
[71] Gangwar AK, Kumar V, Devarathnum J, Kumar N, Sharma AK, Singh H, et al. Acellular aortic matrix for the repair of large ventral hernia in a buck. Vet Pract 2012;13(2):293–4.
[72] Kumar V, Devarathnam J, Gangwar AK, Kumar N, Sharma AK, Pawde AM, et al. Use of acellular aortic matrix for reconstruction of abdominal hernias in buffaloes. Vet Rec 2012;170:392–4.
[73] Devarathnam J, Sharma AK, Rai RB, Maiti SK, Shrivastava SK, Sonal, et al. In vitro biocompatibility determination of acellular aortic matrix of buffalo origin. Trends Biomat Artific Organs 2014;23(3):92–8.

Pericardium-derived extracellular matrix scaffolds

Naveen Kumar[1],*, Honjon Perme[1], Ashok Kumar Sharma[1], Himani Singh[1], Rukmani Dewangan[2] and Swapan Kumar Maiti[1]
[1]Division of Surgery, ICAR-Indian Veterinary Research Institute, Izatnagar, Uttar Pradesh, India, [2]Department of Veterinary Surgery and Radiology, College of Veterinary Science and Animal Husbandry, Dau Shri Vasudev Chandrakar Kamdhenu Vishwavidyalaya, Durg, Chhattisgarh, India

9.1 Introduction

The pericardium is the outermost covering of the heart. It is composed of an outer layer, the fibrous pericardium, and an inner layer, the serous pericardium. The serous pericardium consists of a parietal and visceral layer. The parietal layer is contiguous with the fibrous pericardium and is separated from the visceral layer, which covers the muscular wall of the heart, by the pericardial cavity. The material used to make the leaflets of pericardial heterografts consists of the fibrous pericardium and the parietal layer of the serous pericardium. In vivo, most of the functions of the pericardium are mechanical in nature [1].

The primary constituent of pericardium is collagen, which is essentially polymer of amino acids. The collagen molecule consists of three chains of poly amino acids or polypeptides arranged in a trihelical configurations ending in a nonhelical carboxyl and amino terminals one at each end. These nonhelical ends are believed to contribute to most of the antigenic properties of collagen. In natural state, the collagen trihelical configurations are held in place by direct chemical bonds, hydrogen bonds, and water bridged crosslinks. The elastin, proteoglycans, and mucopolysaccharides are associated with collagen in pericardium. These are believed to modulate collagen fibrillo genesis, fill space, bind, and organize water and repel negatively charged molecules in these tissues. The amino acids in collagen contain pendent groups such as amines (NH_2), acids (COOH), and hydroxyls (OH). Additionally, water molecules surrounding the collagen molecules form another source of entry for reaction, since they can be displaced upon dehydration exposing previously concealed groups for potential crosslinking.

The extracellular matrix (ECM) is a secreted product of cells that populate a given tissue or organ. The ECM plays a central role in mammalian development and physiology and the amino acid sequence. The quaternary structure of many components of ECM such as collagen is highly conserved across species lines [2].

---

*Present affiliation: Veterinary Clinical Complex, Apollo College of Veterinary Medicine, Jaipur, Rajasthan, India.

Natural Biomaterials for Tissue Engineering. DOI: https://doi.org/10.1016/B978-0-443-26470-2.00009-0
© 2025 Elsevier Inc. All rights are reserved, including those for text and data mining, AI training, and similar technologies.

The most commonly used naturally occurring scaffold material has been the structural protein collagen. Collagen is a naturally occurring, highly conserved protein that is ubiquitous among mammalian species and accounts for approximately 30% of all body proteins. Bovine and procaine type 1 collagen provide a readily available source of scaffold material for numerous applications and have been shown to be very compatible with human systems. The collagen can be stabilized by various methods of crosslinking and must be sterilized prior to surgical use.

Collagen is generally treated as a "self" tissue by recipients into whom it is placed and is subjected to the fundamental biological processes of tissue degradation and integration into adjacent host tissues when left in its native ultrastructure. Certain treatment methods, however, significantly alter the mechanical and physical properties of collagen-based materials and may negatively affect the natural physiological processes of cell attachment and proliferation and tissue remodeling.

Collagenous tissues obtained from abattoir, cadaver, or patient being degraded immediately. Therefore, in the exploitation of tissue as clinical material this deterioration must be arrested and deferred preferably beyond the recipient's natural life. Resorption control of collagen membrane is essential where it is used for tissue regeneration controlled by varying the crosslinks density of the biomaterials. Crosslinking is an effective method to control resorption rate of collagen-based biomaterials and to prevent a rapid elution of the material into wound fluids. This is an important for synchronizing degradation of biomaterial with wound healing as well as for increasing tensile properties and flexibility of the membrane to the level necessary for application as a synthetic derma for biosynthetic skin substitute. However crosslinked biomaterials are more resistant to enzymes, with a higher elasticity modulus and a lower degree of swelling. The aim of crosslinking is to prolong the material's original structural and mechanical integrity. Another aim of crosslinking is to remove or at least neutralize the antigenic properties attributed to these materials. Natural collagenous materials are being investigated for surgical repair because of inherent low antigenicity and their ability to integrate with surrounding tissue [3]. Degradable collagenous materials have shown the potential but may lose strength in vivo if they are not crosslinked [3,4].

Methods of crosslinking, concentrate mainly to create new additional chemical bonds between the collagen molecules. These supplementary links (crosslinking) reinforce the tissue to give a tough and strong but nonviable material that maintains the original shape of the tissue. The process of crosslinking involves the chemical agents initiating ideally, irreversible and stable intra- and intermolecular chemical bonds between collagen molecules. Preferably, the agent promotes bonds between the groups of the amino acids. The efficiency and extent of crosslinking reactions depend upon the thickness of the layers of the collagenous tissue and defines the magnitude of the penetration. The other parameters like concentration of the crosslinker, the time and the temperature of exposure affects the crosslinking [5].

Ideally, the crosslinking treatment should also maintain much of the original character of the tissue, such as the flexible mechanical properties of the tissue (biomaterial) without its shrinkage. Hence, it is necessary to keep the tissue near to neutral pH, ensuring an aqueous media environment and minimizing denaturation of the collagen for optimizing crosslinking. Therefore, a balance must be achieved for attaining enough reliable crosslinks for the biomaterials to last lifetime of the recipients, yet permit the biomaterial to perform as it would be in its natural state. The methods that have been developed do not and probably cannot satisfy the dual requirements.

9.2 Preparation of acellular goat pericardium matrix

The fresh pericardium of goat after collection from abattoir was rinsed with normal saline to remove the adhered blood. The maximum time period between the retrieval and initiation of protocols was less than 4 hours. The tissue was cut into 2×2 cm^2 in size. The tissue samples were placed in 4% sodium deoxycholate (anionic biological detergent) which was continuously agitated for 24 hours on shaker for acellularity. The tissue was then thoroughly washed in phosphate buffer saline (PBS) solution. The prepared acellular biomaterials were stored in PBS solution containing 1% amikacin at 4°C.

Anionic biological detergent was used for preparing acellular pericardium of goat. Microscopic observations of acellular pericardium revealed that the protocol for making them acellular was effective. The treatment procedure resulted in complete removal of cellular material including the nucleus. The acellular tissue was primarily composed of ECM. The prepared acellular biomaterials were stored in PBS solution containing 1% amikacin at 4°C [6].

The decellularization techniques resulted in the removal of nucleus and cytoplasmic cellular components, lipids, and its membranes along with soluble proteins and basement membrane components of cellular material, while preserving the components of ECM, consisted of primarily the elastin, insoluble collagen, and tightly bound glycosaminoglycans [7]. The acellular tissue matrices possessed the appropriate mechanical properties [8] and induced appropriate interaction with the host cells that resulted in the regeneration of functional tissues [9]. No inflammation, aneurysm, or dystrophic calcification was noted for the acellular matrix vascular prostheses [10] and complete reendothelialization of the graft in canines had been reported [11]. The acellular tissue matrices were biocompatible, slowly degraded upon implantation and were replaced and remodeled by the ECM proteins synthesized and secreted by ingrowing host cells, which reduced the inflammatory response [12]. The acellular matrices supported the regeneration of tissues with no evidence of immunogenic reaction [13]; however, it was important to keep in mind that even after the removal of cells and cell debris the intact ECM of the acellular tissue itself might have elicited an immune response [14].

9.3 Crosslinking of native and acellular goat pericardium matrix

In vitro evaluation of cellular and acellular goat pericardium following crosslinking with glutaraldehyde (GA), glyoxal (GO) diphenyl phosphoryl azide (DPPA), and ethylene glycol diglycidyl ether (EGDGE) for 12, 24, 48, and 72 hours, respectively, was carried out. The uncrosslinked native and acellular pericardium was used as control. Fresh pericardium was divided into two equal halves. One part was crosslinked as such while another part was made acellular and then crosslinked. In vitro studies included the gross observations, nonenzymatic degradation, enzymatic degradation (1% cyanogen bromide, pepsin and papain), free protein determination, free amino group determination, fixation index, moisture percentage, and SDS-PAGE analysis. Present study revealed that 0.65% GA and 1% GO, DPPA, and EGDGE with a duration of 24 hours at room temperature showed better crosslinking as compared to other combinations. GO treatment showed highest resistance to nonenzymatic and enzymatic degradation at different time interval. Significant reduction ($P < .05$) of free amino groups was observed with GA and DPPA at different time intervals as compared to control. Moisture percentage after GA treatment showed significant reduction ($P < .01$) at different time intervals as compared to control. GO treatment resulted in better crosslinking and formation of high molecular weight protein as evaluated by SDS-PAGE. Pericardium crosslinked for 24 hours with all the four crosslinking agents at room temperature were found best for in vivo studies [15].

9.4 Preparation of acellular buffalo pericardium matrix

Fresh buffalo pericardium was procured from the local abattoir. Immediately after the collection, the biomaterial was preserved in cold physiological saline solution. The tissue was gently rinsed with fresh saline to remove the adhered blood. Excess fatty layer was carefully removed from the pericardial surface. The biomaterial was cut into 2×2 cm^2 size pieces and was placed in a 20 mL of distilled water and shaken for 12 hours to lyse the cells and to release the intracellular contents. It was followed by suspension of tissues in 4% sodium deoxycholate for 2 to 4 hours, then it was treated with 2000 Kunitz units DNase—I suspended in 1 M sodium chloride solution and continuously stirred for 2 hours. The process was repeated twice. Finally, the tissue was thoroughly washed in phosphate buffer saline (PBS) solution. The prepared acellular biomaterials were stored in PBS solution containing 1% amikacin at 4°C (Fig. 9.1A and B). The tissue was also preserved in 10% formal saline solution for histological examination to confirm the acellularity [16].

Microscopic observation of acellular pericardium revealed that the protocol for making them acellular was effective. It resulted in complete removal of cells. No nuclear bodies were seen. The tissue was primarily composed of ECM (Fig. 9.2A and B).

Cellular antigens are predominantly responsible for the immunological reaction associated with allograft [17]. Remnants of cell components in xenograft may contribute to calcification and/or immunogenic reaction [7]. In particular, lipids and DNA fragments are likely to play a role in calcification [18]. Decellularization techniques resulted in the removal of nucleus and cytoplasmic cellular components, lipids, and its membranes along with soluble proteins and basement membrane components of cellular material, while preserving the components of ECM, which consist of primarily of elastin, insoluble collagen and tightly bound glycosaminoglycans [7]. Acellular tissue matrices are biocompatible, slowly degraded upon

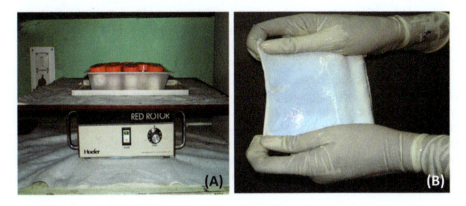

Figure 9.1 (A) Continuous shaking of biomaterial on orbital shaker for making acellular, (B) prepared acellular buffalo pericardium matrix.

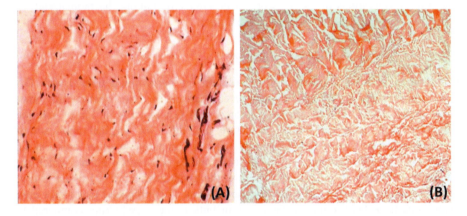

Figure 9.2 (A) Native buffalo pericardium, (B) acellular buffalo pericardium (H&E stain, × 100).

implantation and are replaced and remodeled by the ECM proteins synthesized and secreted by ingrowing host cells, which reduce the inflammatory response [12]. After the removal of cells and cell debris, the intact ECM of the acellular tissue itself may elicit an immune response [19].

9.5 Crosslinking of native and acellular buffalo pericardium matrix

Native buffalo pericardium and acellular buffalo pericardium matrix was crosslinked with different crosslinking agents. The following chemicals were used to crosslink the biomaterials:

- Glutaraldehye (GA):[$C_5H_8O_2$], molecular weight 100.12 (Spectro Chem. Pvt. Ltd., Mumbai).
- Formaldehyde (FA): [HCHO], molecular weight 30.03 (Merck Ltd., Mumbai-400018).
- 1-ethyl-3-(3-dimethylaminopropyl)−carbodiimide (EDC): [$C_8H_{17}N_3$] molecular weight 155.2 (Sigma-Aldrich, New Delhi 110029).

The biomaterials were cut into 2×2 cm^2 size pieces and fixed with 0.5% and 1% concentration of crosslinking agents. The solutions were prepared in phosphate buffer saline. The amount of solution used to crosslink each sample was 20 mL. Tissues of each study group was kept for 6, 12, 24, 48, 72, and 144 hours in chemical for crosslinking. The crosslinking was done at 4°C and room temperature. Normal saline solution (NSS) preserved biomaterials were treated as control. The suitability/biocompatibility of biomaterials was evaluated on the basis of the following parameters [16].

9.5.1 Gross observations

The gross observations of tissues were made after crosslinking the biomaterials at different concentration, duration and temperature.

9.5.1.1 Concentration of solution

Crosslinking in 0.5% concentration of solution the pericardium treated with GA was comparatively stiffer and tougher in comparison to those treated with other crosslinking agents. The pericardium treated with GA and FA in 1% of solution was very stiff, hard, and brittle in nature and unfit for surgical applications. Pericardium treated with GA was more stiff as compared to other crosslinking agents [20]. In glutaraldehyde, there are only one carbon-carbon bonds, which are known to be relatively inflexible; therefore, the glutaraldehyde fixed tissue is usually comparatively stiffer [21]. Cheung and Nimmi [22] proposed that lower concentrations have been found to be better in bulk tissue crosslinking compared to higher concentration. 0.5% solution was found to be better over 1% solution. In 0.5% concentration solution, the properties of biomaterials were more suitable and ideal for surgical implantation. High concentration of glutaraldehyde promoted rapid surface crosslinking of tissue, generating a barrier that impeded or prevented the further diffusion of glutaraldehyde into the tissue bulk. Similarly, Santillan-Doherty et al. [23] recommended the use of lower concentration of GA for the preservation of biological tissue.

9.5.1.2 Duration of treatment

The 6-hour treatment did not produce much difference in the physical nature of the graft and resembled more or less with the native tissue. The biomaterials treated for 12, 24, 48, and 72 hours with different chemicals were stiffer and harder as compared to those treated for 6-hour time interval. The treatment with EDC for 144 hours showed degradation and putrefaction of biomaterials. However, the biomaterials treated with GA and FA did not show any sign of degradation and putrefaction up to 144 hours of crosslinking. At 6-hour duration, crosslinking the pericardium did not differ much from the native pericardium [20]. Prolonged contact time of 24 hours was suggested for biological fixation by glutaraldehyde [21].

9.5.1.3 Temperature

There was no change in gross appearance of biomaterials at 4°C including the color and consistency. The biomaterials crosslinked at room temperature (25°C) appeared stiffer and stronger as compared to biomaterials treated at 4°C. The GA-treated biomaterials were more strong as compared to other crosslinking agents. The EDC-treated biomaterials were slightly softer as compared to FA- and GA-treated biomaterials. At 4°C no change in the physical nature was seen even after prolonged period of crosslinking. At room temperature, the rate of crosslinking reaction was better [20]. Higher temperature resulted in lower free amino group content, and in addition, the tissues crosslinked at 4°C showed slightly higher moisture percentage as compared to 25°C, 37°C, and 45°C [21].

9.5.2 In vitro enzymatic degradation

The degradation of the samples was studied by exposing the materials to collagenase, elastase and trypsin solution.

9.5.2.1 In vitro collagenase enzymatic degradation

It was performed as per the procedure of Connolly et al. [24]. Collagenase type I from *Clostridium histolyticum* (Sigma-Aldrich Co., St Louis, MO, USA, C0130), 20 U/mL in phosphate buffered saline was used for enzymatic degradation at 12, 48 and 72 hours. The tissues were blotted and the mass was recorded. Weight loss of biomaterial was then calculated that of the original tissue. The rates of weight loss (in %) of native and acellular pericardium due to collagenase degradation are presented in Table 9.1. The native pericardium treated at 12, 48, and 72 hours with GA, FA, and EDC showed higher percentage of weight loss ($P < .01$) as compared to NSS-treated samples. The GA-treated acellular pericardium showed significant ($P < .01$) rate of weight loss at 12-, 48-, and 72-hour intervals, whereas, in other groups, no significant change ($P > .01$) in weight loss was observed at different time intervals [20].

9.5.2.2 In vitro elastase enzymatic degradation

Elastase enzymatic degradation was performed as per the method described by Leach et al. (2005). Elastase (pancreatic solution, type I: from porcine pancreas, Sigma-Aldrich Co., St Louis, MO, USA, E1250), 0.1 U/mL in PBS was used for enzymatic degradation. The rates of weight loss of native and acellular pericardium due to elastase degradation are presented in Table 9.1. The native pericardium showed significantly lesser ($P < .05$) percentage of weight loss in the samples

Table 9.1 Rate of weight loss (percentage) (mean ± SE) due to collagenase, elastase, and trypsin degradation of native and acellular pericardium after crosslinking with different chemical agents.

Test	Biomaterials	Treatment	12 h	Time (hours) 48 h	72 h
Collagenase degradation	Native pericardium	GA	35.81^b ± 0.60	34.12^b ± 0.27	36.06^b ± 0.57
		FA	37.57^b ± 0.29	39.39^b ± 0.33	39.50^b ± 0.28
		EDC	37.59^b ± 0.07	38.89^b ± 0.25	39.06^b ± 0.07
		NSS	40.06 ± 0.15	41.77 ± 0.60	41.73 ± 0.05
	Acellular pericardium	GA	37.28^b ± 0.07	37.12^b ± 0.72	38.06^a + 0.57
		FA	38.79 ± 0.43	39.29 ± 0.16	39.50 ± 0.28
		EDC	38.57 ± 0.08	39.07 ± 0.56	38.87 ± 0.75
		NSS	40.26 ± 0.35	39.79 ± 0.87	40.23 ± 1.05
Elastase degradation	Native pericardium	GA	25.94^b ± 0.41	25.14^b ± 0.84	25.45^b ± 0.27
		FA	29.99^b ± 0.78	29.31^b ± 1.06	29.87^b ± 0.71
		EDC	29.37^b ± 0.56	28.07^b ± 0.92	29.37^b ± 0.56
		NSS	35.13 ± 0.63	32.90 ± 0.55	33.50 ± 0.76
	Acellular pericardium	GA	28.13^b ± 0.06	28.11^b ± 0.72	29.56^b ± 0.20
		FA	32.06 ± 078	32.29^b ± 0.27	32.39^b ± 0.24
		EDC	32.22^b ± 0.57	32.03^b ± 0.62	33.96^b ± 0.77
		NSS	35.72 ± 0.57	36.12 ± 0.43	36.45 ± 0.11
Trypsin degradation	Native pericardium	GA	23.87^b ± 0.01	23.20^b ± 0.73	23.17^b ± 0.35
		FA	25.27^a ± 0.39	24.29^a ± 0.41	24.06^a ± 0.45
		EDC	24.08^b ± 0.14	24.16^b ± 0.71	23.61^b ± 0.04
		NSS	28.18 ± 0.49	26.60 ± 0.07	26.64 ± 0.68
	Acellular pericardium	GA	22.76^b ± 0.13	21.48^b ± 0.00	22.65^b ± 0.17
		FA	27.20 ± 1.64	23.78 ± 0.04	24.20 ± 0.37
		EDC	24.38 ± 0.34	23.66 ± 0.56	24.55 ± 0.89
		NSS	26.57 ± 0.54	25.07 ± 0.54	26.14 ± 0.81

GA: glutaraldehyde; FA: formaldehyde; EDC: 1-ethyl-3-(3-dimethyl aminopropyl)-carbodiimide; NSS: normal saline solution.
^{a}Differ significantly ($P < .05$) compared to control.
^{b}Differ significantly ($P < .01$) compared to control.

treated with GA, FA, and EDC at 12, 48-, and 72-hour intervals as compared to control. The acellular pericardium treated with GA, FA, and EDC showed significantly lesser percentage of weight loss ($P < .01$) as compared to control [20].

9.5.2.3 In vitro trypsin enzymatic degradation

The procedure of trypsin enzymatic degradation was similar to elastase enzymatic degradation. The 0.006 Anson U/mL of trypsin (HiMedia Laboratories Pvt. Ltd., India, RM 6216_0) in PBS was used for enzymatic degradation. The rates of weight loss of native and acellular pericardium following trypsin degradation are presented in Table 9.1. The native pericardium treated with GA, FA, and EDC showed significantly less percentage of weight loss at 12-, 48-, and 72-hour intervals. The glutaraldehyde-treated acellular pericardium showed significantly less ($P < .01$) weight loss at 12-, 48-, and 72-hour intervals as compared to control [20].

The crosslinking with different chemicals retards resorption of collagen-based biomaterials in tissues. Therefore, the study of enzymatic degradation can be a good model for evaluation of the resorption rate in tissue. In vitro degradation studies (collagenase, elastase, and trypsin degradation) revealed that uncrosslinked tissues were more prone to enzymatic degradation as compared to crosslinked tissue. Collagen-based biomaterials are biocompatible and nontoxic to tissues and have well-documented structural, physical, chemical, biological, and immunological properties. Additionally, mechanical and, to some extent, immunologic properties of collagen scaffolds can be influenced by modification of matrix properties (porosity, density) or by different chemical treatment affecting its degradation rate [5].

The native buffalo pericardium as well as acellular buffalo pericardium crosslinked with GA and FA showed greater resistance to collagenase degradation as compared to EDC [20]. The GA and FA reacted with the amino group of lysyl residues in protein (collagen) and induced the formation of interchain crosslinks [25], which stabilized the tissue against chemical and enzymatic degradation depending upon the extent of crosslinking [26]. The GA and FA crosslinked native and acellular pericardium tissue has shown significantly increased resistance to elastase degradation at different time intervals. Native and acellular pericardium crosslinked with EDC also showed increased resistance to elastase degradation. However, noncrosslinked tissue samples, whether native or acellular, degraded earlier as compared to crosslinked tissue samples. The GA- and FA-treated samples also showed increased resistance to trypsin degradation as compared to other groups. Fixing of samples with crosslinking agents resulted in increased resistance against enzymatic degradation [20]. This increased resistance against enzymatic degradation (collagenase, elastase, and trypsin) probably results from the cleavage sites of collagen being hidden or altered by the action of crosslinking agents, resulting in inhibition of enzyme substrate interaction [27].

9.5.3 Free amino group contents determination

Each tissue (native and acellular buffalo pericardium) of 1 g was weighed, minced with scissors, and washed thrice with cold normal saline and was homogenized in

3 mL 10% sodium dodecyl sulfate with pastel mortar. The homogenate was centrifuged at 3000 rpm for 10 minutes, and supernatant was separated [20]. Ninhydrin assay was used to determine the free amino group contents of each test sample after enzymatic degradation, as per the procedure of Sung et al. [21]. The free amino group contents determination for native and acellular pericardium is presented in Table 9.2. The native pericardium treated with GA showed significant ($P < .01$) reduction in amino group content at 12-, 48-, and 7-hour intervals as compared to control. The FA and EDC crosslinked samples also showed significantly less ($P < .01$) free amino group content as compared to control at 12-, 48-, and 72-hour intervals. The acellular pericardium treated with GA showed significant ($P < .01$) decrease in amino acid content when compared to control. The FA- and EDC-treated samples also showed significantly less ($P < .01$) free amino group contents as compared to control at different time intervals.

9.5.4 Moisture content analysis

The moisture content was analyzed as per the method Sung et al. [21]. The moisture content of the test tissue was calculated as follows:

$$\text{Moisture content}(\%) = [(\text{Wet tissue weight} - \text{Dry tissue weight})/\text{Wet tissue weight}] \times 100.$$

Moisture content analysis/swelling ratio and moisture percentage of native and acellular pericardium are presented in Table 9.2. The GA- and EDC-treated native pericardium revealed significantly lower ($P < .01$) moisture percentage at 12- and 48-hour intervals as compared to control. Significantly, lower ($P < .01$) moisture percentage was observed at 48 hours in FA-treated samples. The other crosslinked samples also showed significantly lower ($P < .05$) moisture percentage at different time intervals. The GA and FA crosslinked acellular pericardium samples showed significantly ($P < .01$) lower moisture percentage at 12-, 48-, and 72-hour time intervals. EDC-treated samples revealed significantly lower ($P < .05$) moisture percentage at 12-, 48-, and 72-hour time intervals [20].

The free amino group contents analysis indicates that the GA has the greatest ability to crosslink insoluble collagen fibrils. Lastowka and Maffia [28] reported that the GA-induced crosslinking resulted in the least number of free amines, as also observed in the present study. It is also known that the reduction of free amino groups (determined by ninhydrin assay in the present study) in biological tissue diminishes its antigenicity [29]. However, ninhydrin assay cannot directly predict the crosslinking density of the fixed tissue because an agent that simply reacts with free amino groups (but does not crosslink) may similarly reduce the amount of free amino groups in the tissue. A rapid initial drop at 12 hours of the free amino group contents after fixation with GA indicated that the fixation was completed earlier as compared to other fixative agents (crosslinkers). Rapid fixation of biological tissue by GA has been reported by Sabatini et al. [30]. Raymond et al. [31] also reported decrease in amine group contents of GA-treated dermal collagen.

Table 9.2 Mean ± SE of amino acid concentration (g/mL) and moisture percentage of native and acellular pericardium treated with different chemical agents.

Test	Biomaterials	Treatment	Time (h) 12 h	48 h	72 h
Amino acid concentration (g/mL)	Native pericardium	GA	22.70[b] ± 0.75	19.69[b] ± 0.76	24.21[a] ± 0.75
		FA	39.29[b] ± 0.75	31.30 ± 1.20	34.77[b] ± 0.75
		EDC	39.30[b] ± 0.76	35.52[b] ± 1.51	40.81[b] ± 0.76
		NSS	95.86 ± 1.52	92.85 ± 1.51	95.11 ± 0.76
	Acellular pericardium	GA	24.71[b] ± 1.25	22.20[b] ± 0.24	22.70[b] ± 0.74
		FA	35.37[b] ± 1.16	33.52[b] ± 1.01	36.29[b] ± 0.74
		EDC	37.99[b] ± 0.56	34.32[b] ± 0.71	39.29[b] ± 0.75
		NSS	83.55 ± 1.25	81.55 ± 0.75	81.54 ± 0.75
Moisture percentage	Native pericardium	GA	74.72[b] ± 0.40	73.67[b] ± 0.21	74.64[b] ± 0.57
		FA	76.34[a] ± 0.00	77.70[b] ± 0.15	78.40[a] ± 0.54
		EDC	76.11[b] ± 0.89	77.47[b] ± 0.41	78.14[a] ± 0.85
		NSS	78.76 ± 0.27	80.34 ± 0.40	80.74 ± 0.87
	Acellular pericardium	GA	69.72[b] ± 0.09	63.60[b] ± 0.02	70.08[b] ± 0.040
		FA	73.19[b] ± 0.37	73.69[b] ± 1.27	74.48[b] ± 0.67
		EDC	73.76 ± 1.22	74.50[b] ± 0.96	75.98 ± 1.00
		NSS	77.05 ± 0.67	78.42 ± 0.17	79.22 ± 0.67

GA, glutaraldehyde; FA, formaldehyde; EDC, 1-ethyl-3-(3-dimethyl aminopropyl)-carbodiimide; NSS, normal saline solution.
[a]Differ significantly ($P < .05$) compared to control.
[b]Differ significantly ($P < .01$) compared to control.

The moisture contents of the samples showed significant reduction in moisture percentage of crosslinked samples as compared to control samples. The degree of moisture percentage decreases nonlinearly with increasing crosslinking density. Leach et al. [32] observed that the swelling ratio of the single step neutral crosslinking with EGDE (ethylene glycol di glycidyl ether) was greater than the swelling ratio of either the single step alkaline or the double step alkaline procedures. Sung et al. [21] also reported that the moisture content of glutaraldehyde and genipin crosslinked tissues was significantly lower than the fresh tissue. The EDC-treated tissues were found to be soft in consistency, which may be attributed to their more absorption of moisture or less crosslinking as compared to GA-treated tissue, which showed less swelling. The crosslinked samples revealed lower moisture percentage as compared to control. This may be attributed to the shrinkage of tissue during fixation, which reduces the free volume in tissue and thus expels some water molecules out of the fixed tissue. GA and FA crosslinked samples of native and acellular pericardium revealed significantly ($P < .05$) lower moisture contents, which indicated that the fixation with aldehydes caused more shrinkage of tissue as compared to EDC, expelling more number of water molecules out of the fixed tissue [20].

9.5.5 Molecular weight analysis

It was performed by 10% sodium dodecyl sulfate polyacrylamide gel electrophoresis (SDS-PAGE). One gram of each tissue samples was weighed out and minced with scissors. The tissue samples were washed thrice with cold normal saline and homogenized in 5 mL of 10% SDS with pastel mortar. The homogenate was centrifuged at 3000 rpm for 10 minutes and supernatant was separated [20]. The tissues samples were mixed with an equal volume of $5 \times$ nonreducing sample buffer and using 10% polyacrylamide gel under nonreducing condition as per method described by Laemmli (1970). After electrophoresis, the gels were removed and stained with 0.5% Coomassie brilliant blue R-250 dye in 30% methanol and 10% acetic acid for three hours. After staining, it was destained with three changes of 30% methanol and 10% acetic acid for 15, 30, and 60 minutes, respectively, for each change.

Crosslinking of matrices resulted in the formation of high molecular weight protein, which determines the expression of protein bands. The SDS-PAGE of native and acellular pericardium crosslinked for 12 hours are presented in Figs. 9.3 and 9.4.

Native pericardium treated with GA for 12 hours showed 15 KDa bands. Whereas acellular pericardium showed 18 KDa protein bands. The EDC-treated native pericardium showed 18 KDa bands but no bands were visible in acellular pericardium. FA-treated native and acellular pericardium both showed no bands in SDS-PAGE. Native pericardium without any crosslinking showed 150, 120, 50, and 20 KDa protein bands. Acellular pericardium without any crosslinking showed only one protein band between 60 and 50 KDa. SDS-PAGE of native and acellular pericardium crosslinked for 72 hours is presented in Figs. 9.5 and 9.6.

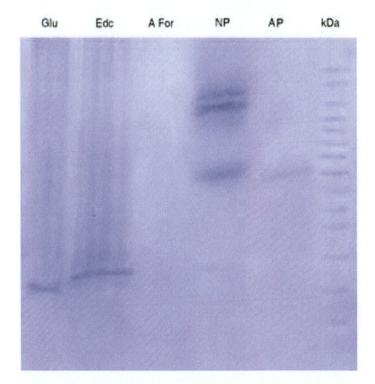

Figure 9.3 Native pericardium in SDS-PAGE treated with different chemicals for 12 h: Lane 1—treated with GA, Lane 2—treated with EDC, Lane 3—treated with FA.

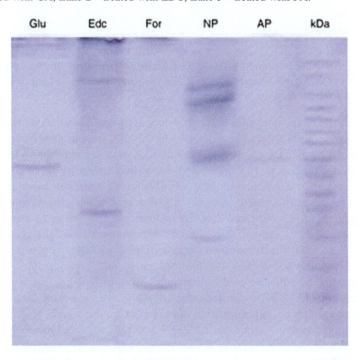

Figure 9.4 Acellular pericardium in SDS-PAGE treated with different chemicals for 12 h: Lane 1—treated with GA, Lane 2—treated with EDC, Lane 3—treated with FA, Lane 4—native pericardium, Lane 5—acellular pericardium.

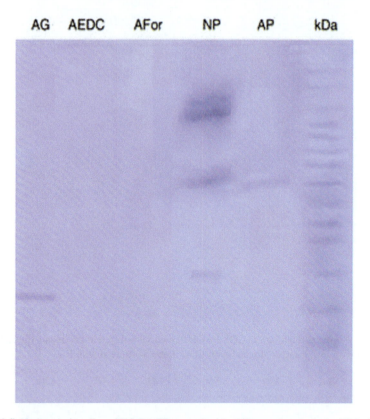

Figure 9.5 Native pericardium SDS-PAGE treated with different chemicals for 72 h: Lane 1—treated with GA, Lane 2—treated with EDC, Lane 3—treated with FA, Lane 4—native pericardium, Lane 5—acellular pericardium.

GA-treated native pericardium for 72 hours showed only one 50 KDa band. Acellular pericardium showed very lightly one 85 KDa protein bands. Native pericardium treated with EDC showed 150 and 25 KDa protein bands but acellular pericardium shows no bands. The FA-treated native tissue showed only one 12 KDa bands but acellular pericardium showed 200 KDa higher molecular bands [20].

Native pericardium treated with GA showed less amount of low molecular weight proteins, which indicated that GA had the greatest ability to crosslink the biomaterials. In EDC-exposed biomaterial, high molecular weight protein (116 kDA) was expressed, which indicated the lesser ability of EDC to crosslink the biomaterials in comparison to GA and FA. Similar to native pericardium, acellular pericardium treated with GA exhibited greater ability to crosslink the biomaterials in comparison to EDC. FA treatment exhibited the least crosslinking ability among all the chemical agents [20]. Formation of higher molecular weight protein following the GA treatment has been reported [28]. Further, these higher molecular weight proteins were found maximum in tissues exposed to GA as compared to dehydrothermal drying and microbial transglutaminase method of crosslinking.

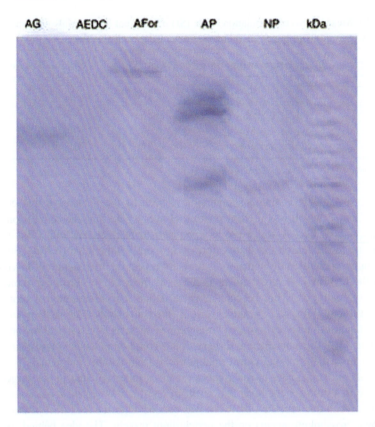

Figure 9.6 Acellular pericardium in SDS-PAGE treated with different chemicals for 72 h: Lane 1—treated with GA, Lane 2—treated with EDC, Lane 3—treated with FA, Lane 4—native pericardium, Lane 5—acellular pericardium.

9.5.6 In vitro cell cytotoxicity

In vitro cell cytotoxicity of crosslinked biomaterials was done as per the method described by Goswami et al. [33]. The cytotoxicity of native pericardium and acellular pericardium was tested in peripheral blood mononuclear cell (PBMC) culture. Stimulation index (SI) of different biomaterials is shown in Table 9.3. The SI of Con A (1.276 ± 0.0791) was used as control to evaluate the immunosuppression of biomaterials with and without Con A. The SI revealed that all the biomaterials when used with Con A showed moderate suppression of blastogenic effect of Con A, while suppression was greater when biomaterial was used alone without Con A. As there is ethical hindrance for indiscriminate use of animal models to test the cytotoxic effect biomaterials, the alternative approach is in vitro cytotoxicity examination before in vivo implantation study [20].

The peripheral blood mononuclear cell (PBMC) was used to detect the immunosuppressive effect of biomaterials. The result indicated that the biomaterials with and

Table 9.3 Mean ± SE of stimulation index (SI) of different biomaterials with and without Con A cultured in peripheral blood mononuclear cell (PBMC).

Sl. No.	Biomaterials (with and without Con A)	Stimulation index (SI)
1	Con A (control)	1.27 ± 0.00
2	Con A + Native pericardium	1.06 ± 0.00[a]
3	Con A + Acellular pericardium	1.05 ± 0.01[a]
6	Native pericardium	0.64 ± 0.01[a]
7	Acellular pericardium	0.63 ± 0.01[a]

[a]Differ significantly ($P < .05$) as compared to control Con A.

without Con A showed immunosuppression as seen by the SI when it was compared with the Con A, which was used as control. However, the immunosuppression of biomaterials without Con A was more as compared to biomaterials with Con A, which may be due to the reason that native form was more cytotoxic when used alone [20]. The immunosuppression of biomaterials may not reflect the true picture of in vivo findings as the in vivo host environment is different from the in vitro environment. The preliminary finding may not be conclusive to define the unsuitability of biomaterials for reconstructive surgery, because various inert materials like rubber and nylon although failed to satisfy cytotoxicity tests yet has an acceptable clinical history Wallin and Upman [19]. In reconstructive surgery, it is a common practice to transplant the tissue either from the same individual (autograft) or from another individual of the same species (allograft). In order to make these practices more efficient in the surgery, it is imperative to search for methods that result in rapid and complete replacement of the host tissue. With this very aim, the work was conducted to see the effect of three different known crosslinking agents on the pericardium protein. The idea behind this study was that once the protein is crosslinked in the acellular graft, it will delay the degradation of transplanted tissue, thereby, providing sufficient line for the host body to replace the damaged tissue. Moreover, in the crosslinked protein, there is masking of immunogenic epitopes, and therefore, there is either delayed or altogether no immune response in host body resulting into successful tissue regeneration [20].

9.6 In vivo evaluation in a rabbit model

New Zealand white rabbits were kept off fed for 12 hours and water for 6 hours before the operation. The back of the animal was properly clipped, shaved, and scrubbed with 5% cetrimide and chlorhexidine solution and painted with betadine. Four subcutaneous pouches were prepared on either side of the spine. The biomaterials of 2×0.5 cm^2 in size were taken and implanted in each pouch. The crosslinked biomaterials and acellular crosslinked biomaterials using one chemical were implanted in one rabbit. On one side of the spine, crosslinked biomaterials and, on other side, the acellular crosslinked biomaterials were implanted. The implants were retrieved back on 14, 30, and 90 days (Fig. 9.7).

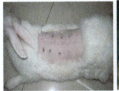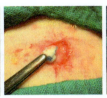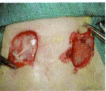

Figure 9.7 For in vivo study, the crosslinked biomaterials were subcutaneously implanted in the back muscle of rabbits and retrieved after at different time intervals.

9.6.1 Macroscopic observations

After subcutaneous implantation of treated biomaterials with different crosslinking agents, the site will be examined at days 14, 30, and 90 to evaluate the reaction, if any. Macroscopic assessment of the retrieved implant was done as per the procedure described by Lu et al. [34].

At day 14: All treated or untreated biomaterials were covered with white fibrous connective tissue which was separated from the surrounding host tissue as shown in Fig. 9.7. The biomaterials could be easily retrieved from the host tissue. There was no sign of infection or pus formation in the vicinity of the implanted biomaterials treated with GA, FA, PEG, EDC, and HMDC. In native FA-treated group, increased cellular reaction was observed after 14 days of implantation, which had reduced at 30 days. During the period of increased cellular reaction, the animals showed the increased sensitivity at the site. The implanted biomaterials at day 14 did not show any loss of mass when compared with original tissue.

At day 30: The biomaterials were covered with fibrous connective tissue, which was dense than that seen at 14th day. Therefore, the implanted biomaterials were more deeply seated within the fibrous connective tissue than it was on 14th day and the implanted tissues were difficult to retrieve as compared to 14th day. There was slight decrease in the mass of the implanted biomaterials as compared to 14th day.

At day 90: The implanted tissues were quite deeply seated within the host tissue and the implanted biomaterials were difficult to separate from the host tissue. By day 90, the native and acellular pericardium treated with NSS, FA, and HMDC were resorbed. Whereas GA- and EDC-treated grafts were partially resorbed. There was complete resorption of native diaphragm treated with NSS, FA, PEG, and HMDC, whereas, GA- and EDC-treated grafts were present deep beneath the fibrous connective tissue; however, they had significantly reduced in size due to resorption. The acellular diaphragm treated with NSS, PEG, EDC, and HMDC showed complete resorption, while the biomaterials treated with GA and FA were deeply seated within the host tissue.

The scaffold used for tissue regeneration must provide the necessary support until the new tissue achieved its biological function. In this study, a uniform layer of white connective tissue was found covering all the implanted biomaterials at 14- and 30-day intervals postimplantation; similar observations were reported by Shoukry et al. [35]. On 90th day, the implanted biomaterials were covered with white dense connective tissue and revealed the significant resorption. Similar findings were also observed when diaphragm was used as prosthetic materials for the repair of ventral abdominal wall defects in bovines [36] and hernioplasty of bovines with glycerol-treated pericardium [37]. Since there was no crosslinking with NSS-treated biomaterials, these untreated biomaterials were completely absorbed by 90th day postimplantation. HMDC-treated biomaterials were also completely absorbed by 90th day postimplantation. Liang et al. [27] found out that fresh and the 30% crosslinked acellular tissue with genipin were degraded significantly at 1 month postoperatively and completely disappeared by 90th day. In contrast, the 60% and 90% crosslinked acellular tissues were still present at 1 year postoperatively. The biomaterials treated with GA underwent slight degradation at 90 days postimplantation, which indicated that degradation was less as compared to other crosslinking agents. Native and acellular pericardium treated with PEG and EDC were significantly resorbed at 90 days postimplantation, while diaphragm (native and acellular) treated with PEG and EDC was completely resorbed by 90 days postimplantation. During tissue regeneration, the cells begin to secrete their own ECM, the scaffold degrades, and is eventually eliminated from the body [38].

9.6.2 Microscopic observations

The grafted pericardium (native and acellular) were retrieved from the grafted site of the rabbits and fixed in 10% formalin saline. After fixation of the graft, thin pieces were cut and processed for paraffin embedding technique to get 4- to 5-μm-thick paraffin sections. The sections were stained by hematoxylin and eosin as per the standard protocol. The sections were also subjected to Masson's Trichrome staining to see the deposition of collagen and by von Kossa staining to demonstrate the deposition of calcium salts in the grafted tissue, if any.

Analysis of the implanted grafts of six treatment groups, including normal saline treatment (NSS) group, was carried out microscopically by evaluating histopathological changes in the graft. The host inflammatory response, neovascular tissue formation (fibroblasts, fine capillaries), deposition of neocollagen, and penetration of host inflammatory responses in the grafted matrix were evaluated. The microscopic pictures were taken for each treatment with respect to pericardium and the diaphragm for comparison [39].

Native pericardium: The histology of host reaction toward native pericardium treated with different crosslinking agents is presented in Fig. 9.8.

The NSS-treated native pericardium implant retrieved at day 14 showed severe inflammatory reaction around the graft, which persisted till day 30, characterized by heavy infiltration of mononuclear cells consisted of lymphocytes, macrophages, and fibroblasts. The degenerated/necrosed collagen of the graft showed lesser fibrous tissue reaction with little neocollagen. By day 90, the grafted tissue got completely resorbed.

On the contrary, the native glutaraldehyde graft by day 14 showed severe host reaction characterized by infiltration of lymphocytes, macrophages, eosinophils, and fibroblasts. The fibroplasia was pronounced with thick neocollagen deposition. The reaction became chronic by day 30 with formation of mature collagen fibers

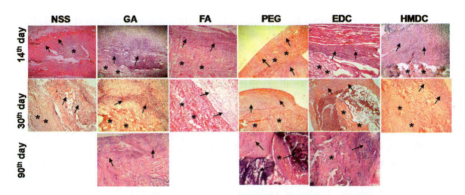

Figure 9.8 Histology of native pericardium treated with different crosslinking agents including NSS. Arrows indicate host reaction and asterisks (*) indicate implanted biomaterials. NSS-treated grafts showed more severe host reaction followed by HMDC and GA. EDC-treated graft showed invasion of graft with inflammatory cell with dissolution of graft in EDC treated, while FA and PEG showed persistent thick capsule of inflammatory reaction up to day 30. At day 90, GA and EDC showed incorporation of host tissue within the graft. PEG-treated acellular pericardium showed necrosis of graft at day 90. *NSS*, normal saline solution; *GA*, glutaraldehyde; *FA*, formaldehyde; *PEG*, polyethylene glycol; *EDC*, 1-ethyl-3-(3-dimethylaminopropyl)-carbodiimide; *HMDC*, hexamethylene diisocyanate (H&E, 10 ×).

within the graft. There was deposition of calcium by day 30. By day 90, the graft was found completely organized with focal calcification within the graft [39].

The formaldehyde-treated native pericardium graft on days 14 and 30 showed thin fibrous tissue mantle around the graft. The mononuclear cells and the fibroblasts with neocollagen were lesser than the GA. However, by day 90, no graft was traceable at the site.

By day 14, the native pericardium treated with PEG showed mature collagen formation around the graft and infiltration with large number of inflammatory cells invading the graft. By day 30, it got thicker with fibrocellular (lymphocytes, macrophages, and eosinophils) reaction showing fibroplasia and fragmented necrosed collagen fibers confined to the surface of the graft. By day 90 there was formation of focal areas of necrocalcification within the graft [39].

On day 14, the EDC-treated native pericardium showed extensive heavy reaction confined to the surface of the graft. By day 30, the graft showed thick covering of mature fibrous connective tissue with infiltration of lymphocytes and macrophages, which were found in pool of eroded graft collagen with necrosis and degradation. By day 90, there was extensive host reaction with infiltration of lymphocytes and eosinophils within the graft with necrosis and calcification of graft.

On day 14, the graft treated with HMDC showed thick fibrous connective tissue reaction with heavy infiltration of macrophage and fibroblast invading the graft. However, on day 30, thick band of granulation tissue was formed over the surface of the graft characterized by presence of numerous fine capillaries and presence of plump shaped fibroblasts showing disorderly arranged neocollagen and infiltrated with mild number of eosinophils. There was calcification of graft on day 30. Few cells were found invading into the graft collagen at places. By 90 days, the graft was resorbed [39].

Acellular pericardium: The histology of host reaction toward acellular pericardium treated with different crosslinking agents are presented in Figs. 9.9 and 9.10.

The NSS-treated acellular pericardium showed thick fibroblast covering with many eosinophils and few mononuclear cells confined to periphery of the graft. By day 30, the graft showed extensive reaction but it was confined to the periphery of the graft with few numbers of inflammatory cells invading the graft.

Compared to its native counterparts, the acellular pericardium showed less host inflammatory reaction. By day 90, the graft was completely resorbed. Compared to native pericardium, the acellular pericardium treated with GA showed fibrocellular reaction confined to the periphery with less infiltration within the graft on days 14 and 30. On day 90, the graft showed thick fibrous reaction with necrocalcification of the graft [39].

On day 14, FA-treated acellular pericardium showed host reaction confined to the periphery of the host with invasion of inflammatory cells within the graft. On day 30, very mild inflammatory reaction was found at the periphery of the graft and the collagen were loosely arranged. By day 90, the tissue was resorbed.

PEG-treated acellular pericardium on day 14 showed fibrous tissue reaction with orderly arranged fibroblast and fibrocyte running inside the graft. Few eosinophils were also present inside the graft. On day 30, thick layer of fibrous tissue reaction was found enveloping the graft, which was similar to the reaction observed in

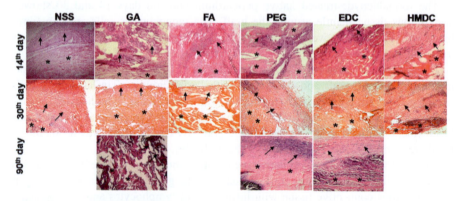

Figure 9.9 Histology of acellular pericardium treated with different crosslinking agents including NSS. Arrows indicate host reaction and asterisks (*) indicate implanted biomaterials. All graft at day 14 showed heavy infiltration of inflammatory cells which subsides at day 30 in GA- and FA-treated graft. PEG and EDC showed invasion within the graft; however, host inflammatory reaction was persistent in NSS- and HMDC-treated graft up to day 30. At day 90, GA, PEG, and EDC showed incorporation of host tissue within the graft with focal areas of necrosis. NSS: normal saline solution; *GA*, glutaraldehyde; *FA*, formaldehyde; *PEG*, polyethylene glycol; *EDC*, 1-ethyl-3-(3-dimethylaminopropyl)-carbodiimide; *HMDC*, hexamethylene diisocyanate, (H&E, 10×).

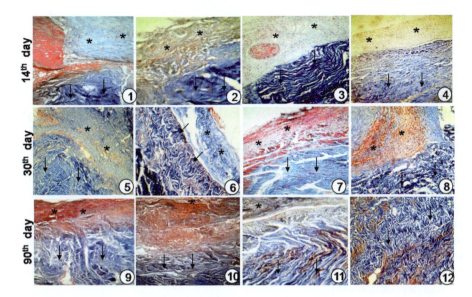

Figure 9.10 Histology of acellular pericardium treated with different crosslinking agents including NSS. Arrows indicate host reaction and asterisks (*) indicate implanted biomaterials (Massons trichrome stain, ×400).

native pericardium. On day 90, the acellular pericardium showed thick fibrocellular reaction but degradation of graft was less compared to native pericardium. There was invasion of the graft by mononuclear cells and few eosinophils with focal areas of calcification [39].

On day 14, EDC-treated acellular pericardium showed less fibrocellular reaction confined to the periphery of the graft without infiltration within the graft. On day 30, the reaction was confined to the graft with focal areas of necrosis but less degradation of graft compared to native pericardium with invasion of graft by few mononuclear cells. By day 90, there was formation of thick connective tissue around the graft. The graft was invaded with necrotic cell debris and presence of calcification within the graft.

On day 14, HMDC-treated acellular pericardium showed necrotic reaction around the disintegrated graft, which was similar to HMDC-treated native pericardium. On day 30, the inflammatory reaction was similar to HMDC-treated native pericardium with extensive host reaction with focal areas of necrosis around the graft. By day 90, the graft was absorbed hence could not be retrieved [39].

Before biomaterials can be applied for clinical use, the tissue response to these biomaterials had to be evaluated. Chemical crosslinking of collagen had been used for several years to improve scaffold stability [40]. Resorption control of biological biomaterials is essential where it is used for tissue regeneration controlled by varying the crosslinking density of biomaterials. Crosslinking is an effective method to control resorption rate of collagen-based biomaterials and to prevent a rapid elution of the materials into the wound fluid [41].

In this study, bovine pericardium (native and acellular) was evaluated after treatment with different crosslinking agents including NSS as control and implanted subcutaneously in rabbits. Biocompatibility of crosslinked biomaterials was concluded from the induction of a transitional inflammatory response and the acceptance of these implanted biomaterials. Graft rejection is mediated by T cells, including cytotoxic T cells. T-cell subsets (Th1 and Th2) generated by naïve T cell on MHC antigen stimulation play a major role in the graft rejection through different sets of cytokines that activate macrophages and B cells in the effector phases of cell mediated immunity (CMI) and humoral immunity. Cells in extracellular matrices have Class I and II histocompatibility antigens capable of eliciting rejection reactions. Also, the cells have glycoproteins recognized by the immune system of hosts, which elicit rejection reactions. Therefore, if these substances are eliminated from extracellular matrices, rejection reactions can be prevented. However, complete elimination of all antigens is considerably difficult to perform and verify [11]. Decellularization process may attenuate severe xenogenic immune response [42], but the removal of cellular components may not be sufficient to eliminate inflammation, and fixation techniques may still be necessary to prevent degradation [43].

NSS-treated native pericardium was found to induce more host inflammatory reaction as compared to its acellular counterpart, which was characterized by infiltration of mononuclear cells consisted of macrophage, lymphocytes, eosinophils, and fibroblast, which persisted up to 30 days, and ultimately, there was absorption of graft by 90 days postimplantation. Similar findings were observed by Liang et al. [27] where fresh pericardia were filled with inflammatory cells, which were severely degraded. Courtman et al. [43] hypothesized that cell extraction process decreased the antigenic load within the tissue due to the elimination of cellular antigens. The native pericardium treated with GA showed more inflammatory reaction up to 30 days postimplantation but at day 90, the native pericardium was more organized as compared to acellular pericardium, which showed necrosis and calcification of graft. Native and acellular pericardium treated with FA showed similar type of inflammatory reaction, which was confined to the periphery of the graft without infiltration within the graft. For tissue regeneration to occur, there should be some enzymatic degradation for efficient cellular migration for tissue regeneration [44] which allows fibroblast from the host tissue to secrete neocollagen fibrils [45]. Native pericardium treated with PEG showed fibrocellular reaction at day 14 at the periphery while its acellular counterpart shows infiltration of host reaction within the graft. By day 30, the reaction was comparable in both native and acellular pericardium. By day 90, native and acellular pericardium showed necrosis of graft with more necrosis and degeneration seen with native pericardium, which indicated that native pericardium was degraded earlier as compared to its acellular counterpart, which was due to its more antigenicity. Native and acellular pericardium treated with showed similar type of reaction with infiltration and degradation of graft more severe in native pericardium. By 90 days, both treatment showed focal areas of necrocalcification. HMDC-treated native and acellular pericardium showed similar reaction with heavy infiltration of graft with host inflammatory cells, which resulted in degradation of graft as early as 30 days, and by day 90, the graft was completely reabsorbed. Biomaterials must remain in the body for desired period of time so as to serve as a scaffold for tissue regeneration till host replaced it with its own tissue.

Calcification is the deposition of calcium salts in the implanted biomaterials (Fig. 9.11). Calcification of biological tissue had been described as one of the major causes of failure of bioprosthetic heart valves derived from glutaraldehyde-treated bovine pericardium or porcine aortic valves [46–48]. Primary tissue degeneration due to calcification necessitates reoperation or causes death in 20%–25% of adult recipients of porcine aortic bioprosthesis by 7–10 years after operation [49]. In this study, at day 14, calcification was observed with native pericardium treated with GA and HMDC while no calcification was observed with acellular pericardium. On the other hand, at day 30, calcification was observed in GA and EDC treated with native diaphragm as well as with acellular diaphragm treated with PEG and EDC. All the retrieved samples were showing calcification with necrosis of the graft at 90 days postimplantation (Fig. 9.12). Calcification limits the durability of implanted biomaterials as it decreased the function and resulted in early degradation of implants. Various methods have been used to reduce the calcification such as making biological tissue acellular as cellular remnant may act as site of deposition, treatment with ethanol, heparin with a considerable success but the exact mechanism for calcification remains obscure.

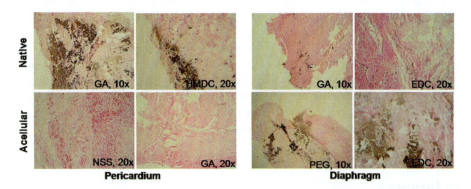

Figure 9.11 Von Kossa's staining of implanted tissue showing calcium salts at the site of necrosis in different treatment groups at 30 days. NSS- and GA-treated acellular pericardium graft did not show any calcium deposition. *GA*, glutaraldehyde; *FA*, formaldehyde; *PEG*, polyethylene glycol; *EDC*, 1-ethyl-3-(3-dimethylaminopropyl)-carbodiimide; *HMDC*, hexamethylene diisocyanate.

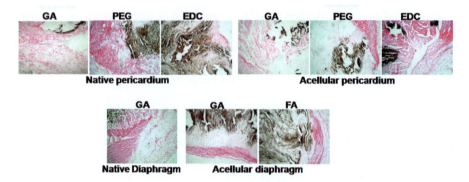

Figure 9.12 Calcification of retrieved samples after 90 days of implantation. *GA*, glutaraldehyde; *FA*, formaldehyde; *PEG*, polyethylene glycol; *EDC*, 1-ethyl-3-(3- dimethylaminopropyl)-carbodiimide; *HMDC*, hexamethylene diisocyanate (von Kossa staining, 10×).

9.7 Conclusion

In the present study, the technique for making biomaterials acellular was a success as complete acellularity of pericardium was obtained. Crosslinking of biomaterials was found better at concentration of 0.5% at room temperature. In vitro study revealed that concentration of 0.5% with a duration of 48 hours at room temperature showed better crosslinking as compared to other combination. The GA treatment showed highest resistance to enzymatic degradation at different time interval. Significant reduction ($P < .01$) of free amino group was observed with GA treatment at different time interval as compared to control. Moisture percentage after GA treatment showed significant reduction ($P < .01$) at different time intervals when it was compared to control. The GA treatment resulted in better crosslinking and formation of high molecular weight protein as evaluated by SDS-PAGE. In vivo study revealed all the biomaterials showed increase host response at day 14, which decreased at days 30 and 90. NSS- and HMDC-treated biomaterials showed more host inflammatory reaction as compared to GA-, FA-, and EDC-treated biomaterials. The acellular biomaterials treated with different crosslinking agents were less immunogenic compared to its native counterpart. The biomaterials retrieved on day 90 showed reorganization of graft with host tissue.

References

[1] Crofts CE, Trowbridge EA. The tensile strength of natural and chemical modified bovine pericardium. J Biomed Mater Res 1988;22:89–98.
[2] Vanderrest M, Garrone R. Collagen family of proteins. FASEB J 1991;47:2814–23.
[3] Van-Der Laan JS, Lopez GP, Van Wachem PB, Nieuwenhuis P, Ratner BD, Bleichrodt RP, et al. TFE-plasma polymerized dermal sheep collagen for the repair of abdominal wall defects. Int J Artif Org 1991;14:661–6.

[4] Van-Wachem PB, Van-Luyn MJA, Olde-Damink LHH, Dijkstra PJ, Feijen J, Nieuwenhuis P. Tissue regenerating capacity of carbodiimide cross-linked dermal sheep collagen during repair of the abdominal wall. Int J Artif Organs 1994;17(4):230−9.
[5] Khor E, Wee A, Tan BL, Chew TY. Methods for the treatment of collagenous tissues for bio prosthesis. Biomaterials 1997;18(2):95−105.
[6] Kumar A. Biocompatibility of crosslinked pericardium and diaphragm of caprine origin. MVSc thesis submitted to Deemed University Indian Veterinary Research Institute, Izatnagar 243122, Uttar Pradesh, India; 2009.
[7] Courtman DW, Pereira CA, Kashef V, McComb D, Lee JM, Wilson GJ. Development of a pericardial acellular matrix biomaterial-biochemical and mechanical effects of cell extraction. J Biomed Mater Res 1994;28:655−66.
[8] Sacks MS, Gloeckner DC. Quantification of the fiber architecture a biaxial mechanical behaviour of porcine intestinal submucosa. J Biomed Mater Res 1999;46:1−10.
[9] Voytik-harbin SL, Brightman AO, Waisner BZ, Robinson JP, Lamar CH. A tissue-derived extracellular matrix that promote tissue growth and differentiation of cells in-vitro. Tissue Eng 1998;4:157−74.
[10] Wilson GJ, Courtman DW, Klement P, Lee JM, Yeger H. Acellular matrix allograft of small caliber vascular prostheses. ASAIO Trans 1990;36:340−3.
[11] Malone JM, Brendel K, Duhamil RC, Reinert RL. Detergent-extracted small diameter vascular prosthesis. J Vasc Surg 1984;1:181−91.
[12] Parien JL, Kim BS, Atala A. In-vitro and in-vivo biocompatibility assessment of natural derived and polymeric biomaterials using normal human urothelial cells. J Biomed Mat Res 2001;55:33−9.
[13] Yoo JJ, Meng J, Oberpenning F, Atala A. Bladder augmentation using allogenic bladder submucosa seeded with cells. Urology 1998;51:221−5.
[14] Coito AJ, Kupiec-Weglinsky JW. Extracellular matrix protein-bystanders or active participants in the allograft rejection cascade. Ann Transpl 1996;11:14−18.
[15] Kumar A, Sharma AK, Kumar N, Maiti SK, Dewangan R, Kumar V, et al. In vitro evaluation of cross-linked native and acellular caprine pericardium. Trends Biomat Artific Organs 2015;29(2):96−105.
[16] Perme, H. In-vitro and in-vivo biocompatibility of crosslinked bovine pericardium and diaphragm. MVSc thesis submitted to Deemed University Indian Veterinary Research Institute, Izatnagar 243122, Uttar Pradesh, India; 2006.
[17] Weber RA, Proctor WH, Warner MR, Verheyden CN. Autonomy and the sciatic functional index. Microsurgery 1993;14:323−8.
[18] Jorge-Herrero E, Getierrez MP, Castillo-Olivares JL. Calcification of soft tissue employed in the construction of heart valve prostheses: Study of different chemical treatments. Biomaterials 1991;12:249−55.
[19] Wallin RF, Upman PJ. Evaluating the biological effects of medical devices and materials. In: Wise DL, editor. Encyclopedic Hand Book of Biomaterials and Bioengineering, Part B. Applications, Volume I. New York: Marcel Dekker Inc; 1995. p. 415.
[20] Perme H, Sharma AK, Kumar N, Singh H, Dewangan R, Maiti SK. In-vitro biocompatibility evaluation of cross-linked cellular and acellular bovine pericardium. Trends Biomat Artific Organs 2009;23:65−75.
[21] Sung HW, Chang Y, Liang IL, Chang WH, Chen YC. Fixation of biological tissues with a naturally occurring cross-linking agent: Fixation rate and effects of pH, temperature, initial fixative concentration. J Biomed Mat Res 2000;52:77−82.
[22] Cheung DT, Nimni ME. Mechanism of crosslinking of protein by glutaraldehyde-I: reaction with monomeric and polymeric and polymeric collagen. Connect Tissue Res 1982;10:655−64.

[23] Santillan-Doherty, Victoria RJ, Sotres-Vega A, Olmos R, Arrelo JL, Garcia D, et al. Thoraco abdominal wall repair with preserved bovine pericardium. J Invest Surg 1996;9:45−55.
[24] Connolly MJ, Alferiev I, Eidelman N, Sacks M, Palmatory E, Kronsteiner A, et al. Tri glycidyl amine crosslinking of porcine aortic valve cusps or bovine pericardium results in improved biocompatibility, biomechanics, and calcification resistance: chemical and biological mechanisms. Am J Pathol 2005;166(1):1−13.
[25] Yannas IV. Natural materials. In: Ratner BD, Hoffman AS, Schoen FJ, Lemons JE, editors. Biomaterial Science. Academic Press: San Diego; 1996. p. 84−94.
[26] Golomb G, Schoen FJ, Smith MS, Linden J, Dixon M, Levy RJ. The role of glutaraldehyde-induced cross-links in calcification of bovine pericardium used in cardiac valve bio prosthesis. Am J Pathol 1987;127:122−30.
[27] Liang HC, Chang Y, Hsu CK, Lee MH, Sung HW. Effect of cross-linking degree on an acellular biological tissue on its tissue regeneration pattern. Biomaterials 2004;25:3541−52.
[28] Lastowka A, Maffia GJ, Brown EM. A comparison of chemical, physical and enzymatic cross-linking of bovine type I collagen fibrils. JALCA 2004;100:196−202.
[29] Imamura E, Sawatani O, Koyanagi H, Noishiki Y, Miyata T. Epoxy compounds as a new crosslinking agent for porcine aortic leaflets: subcutaneous implants studies in rats. J Card Surg 1989;4:50−9.
[30] Sabatini DD, Bensch K, Barnett R. Cytochemistry and electron microscopy. J Cell Biol 1963;17:19−23.
[31] Raymond Z, Dijkstra PJ, Pauline B, van Wachem, Marja JA, van Luyn, et al. Successive epoxy and carbodiimide crosslinking of dermal sheep collagen. Biomaterials 1999;20:92−7.
[32] Leach JB, Jesse AB, Wolinsky A, Phillip J, Stone B, Joyce Y, et al. Crosslinked á-elastin biomaterials: towards a processable elastin mimetic scaffold. Acta Biomaterialia 2005;1:155−60.
[33] Goswami TK, Kumar N, Gupta OP, Sharma AK, Pawde AM. In vitro determination of biocompatibility of different grafts/implants used for reconstructive surgery of tendon. Indian J Ani Sci 2002;72:957−61.
[34] Lu HJ, Chang Y, Sung HW, Chiu YT, Yang PC, Hwang B. Heparinization on pericardial substitutes can reduce adhesion and pericardial inflammation in the dog. J Thorac Cardiovasc Surg 1998;115:1111−20.
[35] Shoukry M, El-Keiey M, Hamouda M, Gadallah S. Commercial polyester fabric repair of abdominal hernias and defects. Vet Rec 1997;140:606−7.
[36] Kanade MG, Kumar A, Sharma SN. Diaphragm as a prosthetic material for the repair of ventral body defects in bovines. Indian J Vet Surg 1986;7:8−14.
[37] Deokiouliyar UK, Khan AA, Sahay PN, Prasad R. Evaluation of preserved homologus pericardium for hernioplasty in buffalo calves. J Vet Med 1988;35:391−4.
[38] Pieper JS, Hafman T, Veerkamp JH. Development of tailor-made collagen glycosaminoglycan matrices: EDC/NHS crosslinking and ultrastructural aspects. Biomateials 2000;21:581−93.
[39] Perme H, Sharma AK, Kumar N, Singh H, Maiti SK, Singh R. In-vivo biocompatibility evaluation of crosslinked cellular and acellular bovine pericardium. Indian J Ani Sci 2009;79:658−61.
[40] Mckegney M, Taggart I, Grent MH. The influence of cross-linking agents and diamines in the pore size, morphology and biological stability of collagen sponges and their effect on cells penetration through the sponge matrix. J Mater Sci: Mater Med 2001;23:833−44.
[41] Bakos D, Koniarová D. Collagen and collagen/hyaluronan complex modifications. Chem Pap 1999;53(6):431−5.

[42] Goldstein S, Black K, Clark D, Orton EC, O'Brien MF. Inflammatory responses to uncross-linked xenogenic heart valve matrix. World Symposium on Heart Valve Diseases. 1999;205 London, England.
[43] Courtman DW, Wilson GJ. Development of an acellular matrix vascular xenograft: modification of the in-vivo immune response in rats. Proceedings of the Soceity for Biomaterials, 25th Annual Meeting. 1999;20.
[44] Courtman DW, Errett B, Wilson GJ. The role of cross-linking in modification of the immune response elicited against xenogenic vascular acellular matrices. J Biomed Mater Res 2001;55:576—86.
[45] Chang Y, Tsai CC, Liang HC, Sung HW. In vivo evaluation of cellular and acellular bovine pericardium fixed with naturally occurring cross-linking agent. Biomaterials 2002;23:2447—57.
[46] Schoen FJ, Harasaki H, Kim KM, Anderson HC, Levy RJ. Biomaterial-associated calcification: pathology, mechanisms, and strategies for prevention. J Biomed Mater Res 1988;22:11—36.
[47] Potkins B, Mcintosh C, Cannon R, Roberts W. Bio prostheses in tricuspid and mitral valve position or 95 months with heavier calcific deposit on the right-sided valve. Am J Cardiol 1988;61:949—56.
[48] Pelletier L, Carrier M, Leclere Y, Lepage G, Deguise P, Dyrda I. Porcine versus pericardial bio prostheses: a comparison of late results in 1,593 patients. Ann Thorac Surg 1989;47:352—61.
[49] Ionescu MJ, Smith DR, Hansan SS, Chidambaran AP, Tandom AP. Clinical durability of the pericardial xenograft valve: ten years' experience with mitral replacement. Ann Thorac Surg 1982;34:265—77.

Diaphragm-derived extracellular matrix scaffolds and clinical application

Vineet Kumar[1], Naveen Kumar[2,*], Anil Kumar Gangwar[3], Kaarthick D.T.[4], Harendra Rathore[2], Swapan Kumar Maiti[2], Ashok Kumar Sharma[2], Dayamon David Mathew[5], Jetty Devarathnam[6], Sameer Shrivastava[7], Sonal Saxena[7], Apra Shahi[8], Himani Singh[2] and Karam Pal Singh[9]

[1]Department of Veterinary Surgery and Radiology, College of Veterinary and Animal Sciences, Bihar Animal Sciences University, Kishanganj, Bihar, India, [2]Division of Surgery, ICAR-Indian Veterinary Research Institute, Izatnagar, Uttar Pradesh, India, [3]Department of Veterinary Surgery & Radiology, College of Veterinary Science & Animal Husbandry, Acharya Narendra Deva University of Agriculture and Technology, Ayodhya, Uttar Pradesh, India, [4]Veterinary Clinical Complex, Veterinary College and Research Institute, Thanjavur, Tamil Nadu, India, [5]Department of Veterinary Surgery & Radiology, Faculty of Veterinary and Animal Sciences, Banaras Hindu University, Rajiv Gandhi South Campus, Barkachha, Uttar Pradesh, India, [6]Department of Surgery and Radiology, College of Veterinary Science, Sri Venkateswara Veterinary University, Proddatur, Andhra Pradesh, India, [7]Division of Veterinary Biotechnology, ICAR-Indian Veterinary Research Institute, Izatnagar, Uttar Pradesh, India, [8]Department of Surgery and Radiology, College of Veterinary Science and Animal Husbandry, Nanaji Deshmukh Veterinary Science University, Jabalpur, Madhya Pradesh, India, [9]CADRAD, ICAR-Indian Veterinary Research Institute, Izatnagar, Uttar Pradesh, India

10.1 Introduction

Collagen is regarded as one of the most useful biomaterial due to its excellent biocompatibility, biodegradability, and weak antigenicity. In cellular grafts, the histocompatibility antigens of the cells cause immunological reaction phenomenon. Less immunogenicity and better tolerance of acellular grafts were observed in rats and rabbits [1]. The need of the readily availability of a nonimmunogenic and nonprosthetic biomaterial that could guide the regeneration of normal tissue is a fascinating possibility. Acellular biological tissues have been proposed to be used as natural biomaterials for tissue repair [2]. Natural biomaterials are composed of extracellular matrix (ECM) proteins that are conserved and can be served as

*Present affiliation: Veterinary Clinical Complex, Apollo College of Veterinary Medicine, Jaipur, Rajasthan, India.

scaffolds for cell attachment, migration, and proliferation. The acellular matrix can stimulate exact regeneration of missing tissue.

In the surgical repair of congenital abdominal wall defects, the easy availability of a nonimmunogenic and nonprosthetic biomaterial that could guide the regeneration of normal tissue is a fascinating possibility. Biomaterials are already in use, but an acellular matrix (ACM) can arouse exact regeneration of the mislaid tissue. Decellularized scaffolds can be prepared from animal tissues and represent a promising biomaterial for exploration in tissue regeneration studies [3]. The natural bioscaffolds have advantages over synthetic materials that they impersonate natural ECM formation and composition, imitate natural stimulatory effects of ECM on cells, and permit the merger of growth factors and other matrix proteins to further enhance cell functions. The ultimate goal of any decellularization protocols is to remove all cellular material without adversely affecting the composition, mechanical integrity, and eventual biological activity of the remaining ECM. A modified method for decellularization of buffalo diaphragm was developed to prepare an acellular biological scaffold, which can be utilized for repair of varied muscular defect in both large and small animals of different species.

The biomaterials are used in close or direct contact with the body to augment or replace faulty materials. Biomaterial has been proven not to be completely inert after implantation and does generate an inflammatory response as a foreign body reaction that differs between individuals and depends on the amount of material and the structure of the mesh [4,5]. Enzyme digestion of hydrolysable bonds of implanted polymers is usually not observed [6]. Williams and Mort [7] demonstrated in vitro that enzymes could increase the rate of degradation of several nominally stable polymers as, for example, PETP, PMMA, nylon 66, or a poly (ether urethane). In general, two classes of enzymes are of interest to be studied in the immediate surroundings of the implantation site. First, the hydrolases, hydrolytic enzymes like phosphatases, esterases, and amino peptidases. These enzymes are predominantly lysosomal and are mostly contained within macrophages and giant cells. The second class of enzymes is represented by the oxidoreductases, providing a way for further hydrolytic breakdown. It is interesting to know to what extent the cells in the immediate environment of an implanted biomaterial influence the in vivo degradation, for example, by the production of specific enzymes [8]. The concentrations of some intracellular enzymes as well as that of protein increase in the lymph draining from a rabbit hindlimb after the limb has been subjected to thermal or chemical injury, but the nature of the enzyme pattern depends upon the degree of cellular injury [9].

Skin defects often arise following acute trauma, chronic wounds, and burns. Current strategies to repair such defects were allogenic, xenogenic transplantation, and autologous skin flaps, which have their limitations in the availability of sufficient donor sites, donor site morbidity and immunological rejections [10]. Although various materials have been described to restore the integrity of full-thickness skin defects, no material has yet proved to be optimal. Recently, tissue engineering (TE) and regenerative medicine have been used as an alternative to restore or replace lost tissues through the use of cells and ECM scaffolds [11]. In TE approach, the cells isolated from autogenic, allogenic, or xenogenic sources are cultured in an ECM to regenerate new tissues [12]. These ECMs provide necessary support for cell adhesion, proliferation, and phenotypic differentiation and offer a biochemical and biophysical cue to modulate the neotissue formation [13,14]. Acellular diaphragm matrix (ADM) has been used as a bioscaffold for the repair of abdominal defects in swine without complications [15]. Fibroblasts play an important role in skin regeneration and enhance wounds healing [16]. Hence, in this study, advantage of both ADM as a bioscaffold and allogenic fibroblasts that has been well tolerated by the host [13] were combined to prepare a bioengineered scaffold. Moreover, fibroblasts lack major histocompatibility complex (MHC) class II antigen, which is responsible for graft rejection and hence are relatively immunologically inert [17].

10.2 Preparation and characterization of buffalo diaphragm matrix

10.2.1 Materials and methods (I)

Fresh diaphragm (tendinous portion) of buffalo was collected from a local abattoir in chilled (4°C) sterile phosphate buffered saline (PBS, pH 7.4), containing 0.048% gentamicin, 0.0205% EDTA, and 0.1% sodium azide and processed immediately. After initial washing, buffalo diaphragm matrix (BDM) was cut into 20×20 mm^2 size pieces and subjected to 0.5%, 1%, 2%, 3%, and 4% sodium dodecyl sulfate (SDS) solution, respectively. The tissue was continuously agitated at the rate of 250 rpm in SDS solutions on magnetic stirrer (C-MAG HS7, IKA, USA) for 12, 24, 48, and 72 hours at room temperature. Prepared BDM was extensively rinsed with sterile phosphate buffer saline solution (PBS) to remove residual detergent and stored in sterile PBS containing 0.048% gentamicin at $-20°C$ till clinical use. Sterility test of stored sterile PBS was performed to check any bacterial or fungal growth [18].

10.2.2 Histological observations

The native and SDS-treated diaphragm tissues were fixed in 10% neutral buffered formalin, serially dehydrated with ethanol, cleared in xylene and embedded in paraffin wax. Tissue sections (6 μm) were cut on a semiautomated rotary microtome (RM2245, Leica Microsystems, Wetzlar, Germany) and stained with hematoxylin and eosin (H&E). Special staining using Masson's trichrome stain (MTS) was performed to

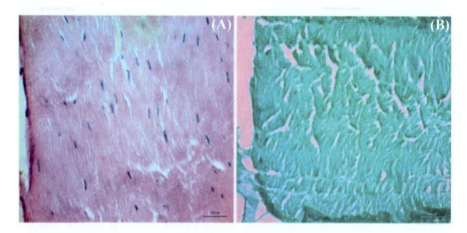

Figure 10.1 (A) Microscopic images of native diaphragm after hematoxylin-eosin staining. (B) Masson's trichrome staining (×40; scale bar 200 μm).

identify collagen fibers. Slides were examined under 10 magnifications using an upright compound microscope (Axiocam ERc5s, Primo star, Carl Zeiss, Germany). Histologically, native diaphragm (ND) showed cell nuclei (Fig. 10.1A) and acellular diaphragm showing no nuclei after Masson's trichrome staining (Fig. 10.1B). Microscopic images of SDS-treated diaphragm are shown in Fig. 10.2A–T. The diaphragm treated with 0.5% SDS for 12, 24, and 48 hours showed compactly arranged collagen fibers and cellular debris (Fig. 10.2A–C). At 72 hours, mild porous collagen lattice with presence of cell nuclei was observed (Fig. 10.2D). Treatment with 1% SDS solution for 12, 24, and 48 hours resulted in decrease in cellular components and debris, moderate porosity, and preserved collagen structure (Fig. 10.2E–G). At 72 hours, collagen fibers were thin, moderately loose with moderate porosity (Fig. 10.2E–H). Treatment with 2% SDS solution for 12 and 24 hours showed well-preserved collagen architecture with sporadic presence of cells (Fig. 10.2I and J). At 48 hours, complete loss of cell nuclei and well-preserved collagen fibers were observed (Fig. 10.2K). At 72 hours, absence of cell nuclei and thin, highly loose collagen fibers were observed (Fig. 10.2L). Diaphragm treated with 3% and 4% SDS for 12, 24, 48, and 72 hours showed complete loss of cell nuclei and extensive damage to collagen structure (Fig. 10.2M–T). Masson's trichrome-stained images of SDS-treated diaphragm are shown in Fig. 10.3A–T. Treatment of a diaphragm with 2% SDS solution for 48 hours revealed orderly arranged collagen fibers within BDM (Fig. 10.3A–H). Treatment of a diaphragm with 3% and 4% SDS solutions at all incubation period resulted in extensive damage to collagen fibers with increased porosity (Fig. 10.3M–T) [18].

10.2.3 Scanning electron microscopy

For scanning electron microscopy (SEM) the native and decellularized diaphragm tissues were cut in small pieces and washed in chilled PBS (0.1 M; pH 7.2) and

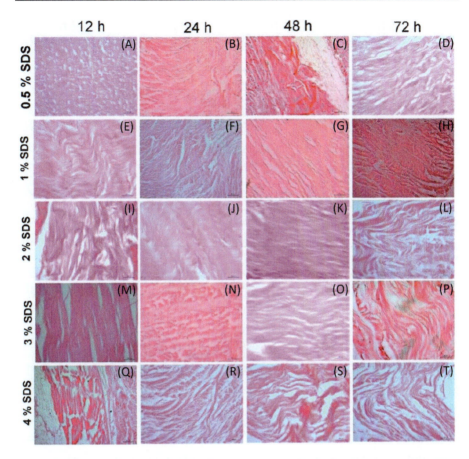

Figure 10.2 (A−T) Microscopic images of bubaline diaphragm after treatment with aqueous SDS solutions for 12, 24, 48, and 72 h (H&E; ×40; scale bar 200 μm).

subjected to fixation in 2.5% glutaraldehyde in 0.1 M PBS (pH 7.2) for 6 hours. Fixed samples were washed in 0.1 M PBS with three times for 15 minutes each at 4°C. Thereafter, samples were dehydrated in ascending grade of acetone solution, namely, 30%, 50%, 70%, 80%, 90%, 95%, and 100%, respectively at 4°C for 15 minutes. Thereafter, samples were mounted on aluminum stubs, coated with gold palladium in Emitech SC620 sputter coater and topographical imaging of processed tissues was performed using a scanning electron microscope (EVO-18, Carl Zeiss, Germany). SEM images of ND and BDM are shown in Fig. 10.4. Ultrastructure of the native diaphragm revealed compact collagen fibers with low porosity Fig. 10.4A and B. Thickness of collagen fibers varied between 2.074 and 4.818 μm. The BDM (treated with 2% SDS for 48 hours) showed intact collagen structure and integrity. Thickness of collagen fibers was slightly lesser as compared to the native diaphragm and varied between 1.454 and 4.154 μm (Fig. 10.4C) [18].

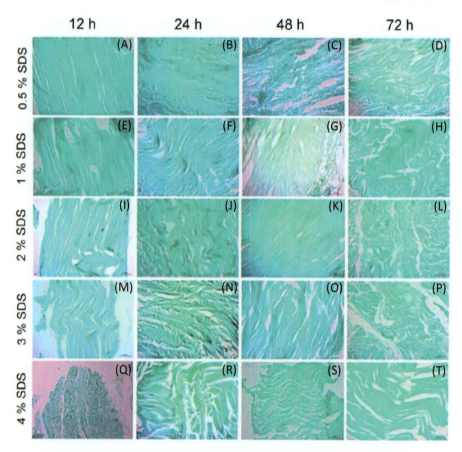

Figure 10.3 (A−T) Masson's trichrome-stained images of bubaline diaphragm after treatment with aqueous SDS solutions for 12, 24, 48, and 72 h (×40; scale bar 200 μm).

10.2.4 DNA extraction

DNA was extracted from native and decellularized diaphragm tissues by phenol: chloroform: isoamyl alcohol method described by Green et al. [19] with slight modifications. Two hundred milligrams of native and decellularized diaphragm tissues were separately triturated in 1 mL lysis buffer (Tris: 50 mM; EDTA:0.1 M; SDS:1%) using glass Teflon tissue homogenizer, and lysates were incubated for 2 hours at 37°C in a water bath. To each tissue homogenate, 50 μL proteinase K solutions (20 mg/mL) was added and incubated overnight at 56°C. Thereafter, tissue homogenates were centrifuged at 10,000 rcf for 10 minutes at room temperature. Supernatants were collected, and equal volume of Tris-saturated phenol (pH 8.0) was added and mixed gently. Samples were again centrifuged at 10,000 rcf for 10 minutes and upper aqueous phase was transferred to new tube. Equal volume of chloroform:isoamyl alcohol (24:1) mixture was added and mixed gently. Samples were again centrifuged at 10,000 rcf for 10 minutes and aqueous phase was

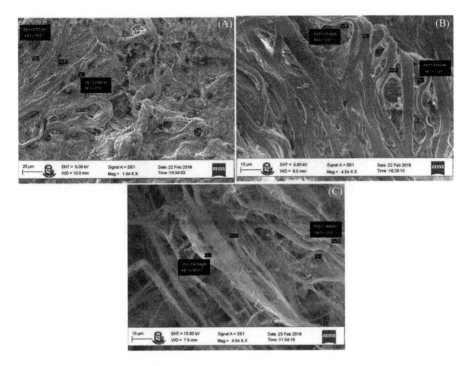

Figure 10.4 (A) SEM image of native diaphragm (ND) (1940 ×, scale bar 20 μm). (B) SEM image of native diaphragm (ND) (4540 ×, scale bar 10 μm). (C) Acellular buffalo diaphragm matrix (BDM) (4540 ×, scale bar 10 μm).

transferred in separate tube. The DNA was precipitated with 2 volumes of absolute ethanol and 0.2 volume of 3 M sodium acetate (pH 5.2). Tubes were centrifuged at 10,000 rcf for 10 minutes and supernatant was discarded. Pellets were then washed two times in 70% ethanol and air dried, dissolved in nuclease-free water, and kept in −20°C till further use. After extraction, DNA concentration and purity was quantified using Nanodrop 2000 spectrophotometer (Thermo Scientific, USA). Further, samples were separated by electrophoresis on a 0.8% low electroendosmosis (EEO) agarose gel with ethidium bromide (0.2 μg/mL) at 60 V for 1 hour stained and visualized with ultraviolet transillumination (Fluor Shot PRO II SC850, Shanghai Bio-Tech Co., Ltd, China) to confirm DNA integrity.

10.2.5 DNA quantification

Extracted DNA from ND and BDM was of high purity as evident by the 260/280 ratios. DNA content was 433.96 ± 162.60 ng/mg and 33.12 ± 5.40 ng/mg in ND and BDM, respectively. DNA content was significantly ($P = .045$) decreased in BDM as compared to ND (Fig. 10.5A). Treatment with 2% SDS for 48 hours resulted in 92.54% reduction in DNA contents of the diaphragm. DNA fragmentation was

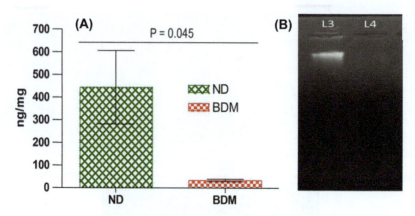

Figure 10.5 (A) DNA content (ng/mg of tissue) (mean ± SE) in native diaphragm (ND) and bubaline diaphragm matrix (BDM). (B) Ethidium bromide-stained agarose gel image showing DNA band (L-3 = ND, L-4 = BDM).

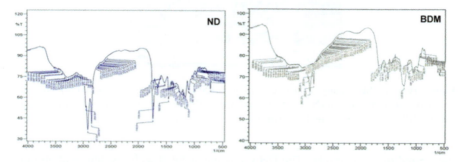

Figure 10.6 FTIR spectra showing peaks of the native diaphragm (ND) at 1238.34, 1535.39, 1657.87, 2955.04, and 3386.15 cm^{-1}; bubaline diaphragm matrix (BDM) at 1220.02, 1534.11, 1649.19, 2954.08, and 3343.71 cm^{-1}.

confirmed through agarose gel electrophoresis. DNA extracts from native diaphragm show broadband indicative of large DNA fragments (Fig. 10.6, Lane 3). Extracts from decellularized diaphragm tissue show considerable removal of DNA material, with absence of DNA band (Fig. 10.5B, Lane 4). This is indicative of effective removal of remnant DNA from developed BDM [18].

10.2.6 Fourier transform infrared spectroscopy

One milligram of each freeze-dried native and decellularized diaphragm tissues were mixed with pure dry KBr powder in 1:10 ratio, and pelleted. The Fourier transform infrared spectra (FTIR) were recorded by an infrared spectrophotometer (FTIR 8400 s Shimadzu Corporation, Tokyo, Japan) in the 500–4000 cm^{-1} wave number spectral range with a spectral resolution of 2 cm^{-1} and 45 scans. The FTIR spectra of

ND and BDM are shown in the Fig. 10.6. The amide A band (3294 cm^{-1}) is associated with H-bonded N-H stretching[14] and was found at 3386.15 cm^{-1} for ND and 3343.71 cm^{-1} for BDM [18]. The amide B band (2953 and 2928 cm^{-1}) is related to CH_2 asymmetric stretching [20] and was observed at 2955.04 cm^{-1} for ND and 2954.08 cm^{-1} for BDM. The amide I band (1641−1658 cm^{-1}) is associated with C = O hydrogen bonded stretching [21] as recorded at 1657.87 cm^{-1} for ND and 1649.19 cm^{-1} for BDM. The amide II (1539−1546 cm^{-1}) is associated with C-N stretching and N-H in plane bending from amide linkages, including wagging vibrations of CH_2 groups from the glycine backbone and proline side chains [22] in ND and BDM appeared at 1535.39 and 1534.11 cm^{-1}, respectively. The amide III band was found at 1238.34 cm^{-1} for ND and 1220.02 cm^{-1} for BDM confirming the presence of hydrogen bonds [23].

The increasing use of bioscaffolds for tissue repair has prompted its development. They are typically prepared after decellularization of source tissues. No visible nuclei per histologic evaluation via H&E is a part of the minimum criteria for effective decellularization in terms of the DNA content remaining in decellularized biological scaffolds [24]. In the present study, xenogenic BDM scaffold was developed from fresh diaphragm of buffalo origin using aqueous SDS solution. SDS was used in the present study because it is readily available, cost effective and has been used for decellularization of other biological tissues such as aorta [25] and skin [26,27]. Histologically, treatment of fresh diaphragm with 2% SDS solution for 48 hours resulted in complete loss of cell nuclei, and retention of the distinctive, natural, three-dimensional structures of the collagen within developed matrix. SDS, an ionic detergent, is effective for solubilizing both cytoplasmic and nuclear cellular membranes [28]. It is typically more effective for removing cell residues from tissue compared to other detergents [29]. Further, SEM examination also confirmed effective decellularization of the bubaline diaphragm and preservation of collagen structure and integrity within BDM.

A recent method to assess effective decellularization of xenogenic tissue is by quantification of remnant DNA [30]. DNA retained in decellularized tissues may induce an immune response and foreign body reaction by the host to scaffold implants. Carpo et al. [24] have suggested that 50 ng/mg dry weights is a part of the minimum criteria for effective decellularization in terms of the DNA content remaining in decellularized biological matrices. In the present study, DNA content was significantly ($P = .045$) lesser in BDM as compared to the native diaphragm. Treatment with 2% SDS solution for 48 hours resulted in almost complete reduction in DNA content, indicating effective acellularity. Absence of DNA band on ethidium bromide-stained agarose gel further indicated loss of DNA from the prepared matrix. FTIR spectroscopy is used to characterize functional groups of chemical compounds [31]. In the present study, FTIR spectrum of BDM shows all characteristic transmittance peaks of collagen [20−23,32]. The amide A (NH stretch coupled with hydrogen bonding) and amide B (CH_2 asymmetrical stretch) peaks of the BDM (3343.71 and 2954.08 cm^{-1}, respectively) were similar to ND (3386.15 and 2955.04 cm^{-1}, respectively). The amide I (C = O stretch, and hydrogen bonding coupled with COO$^-$), amide II (NH bend coupled CN stretch), and amide III (NH bend) peaks of the BDM (1649.19, 1534.11, and 1220.02 cm^{-1}, respectively) were similar to ND (1657.87, 1535.39 and 1535.39 cm^{-1}, respectively). Based on the location of amide A, amide B, amide I, amide II and amide III peaks, it seems that developed BDM was composed of collagen.

10.3 Preparation of acellular diaphragmatic scaffold of buffalo origin

10.3.1 Materials and methods (II)

Ten pieces of fresh buffalo diaphragm were collected in phosphate buffer saline (PBS) supplemented with antibiotics, that is, gentamicin @80 µL/mL, from the local slaughter house. All the blood clots and visible contaminants were removed by thorough cleaning in sterile PBS. The diaphragmatic tissues were cut to the size of 2×2 cm^2. These were decellularized by using method of Kumar et al. [33]. After cutting, these were placed in sterile PBS approximately for 2 hours at room temperature. The diaphragmatic tissues were placed in 1% sodium dodecyl sulfate (SDS) solution and kept in an orbital shaker at a rotation speed of 50 rotations/min at 21°C for 12 hours. Later on, tissues were washed three times in sterile PBS and again placed in freshly prepared 1% SDS solution. Tissues were placed in orbital shaker for next 12 hours at 50 rpm and 21°C for next 12 hours. This procedure was repeated for next 48 hours (total three days). Thereafter, tissues were washed four times in sterile PBS and placed in high-grade 70% ethanol in orbital shaker for 3 hours at 21°C, 20 rpm. Tissues were taken out and washed in sterile PBS three times and again placed in orbital shaker at 21°C, 20 rpm, for 3 hours to remove residual ethanol. Tissues were finally rinsed in fresh PBS and stored in their flat position with minimum quantity of sterile PBS at −20°C till use.

10.3.2 Gross observations

The color of diaphragmatic tissue changed from pale yellow (before treatment) to milky white after treatment (Fig. 10.7).

10.3.3 Histological observations

Diaphragmatic tissue was embedded in paraffin wax after being fixed in 10% buffered neutral formalin. Embedded tissues were sectioned at a thickness of 5 µm along the perpendicular direction of the fibers. H&E staining along with Masson's trichome staining (for collagen fibers) was done to visualize the cellularity and orientation of collagen fibers.

Histological evaluation of tendinous portion of buffalo diaphragm revealed that cellularity was intact in diaphragm before treatment as suggested by H&E and Masson's trichome staining (Figs. 10.8 and 10.9). The chemical process of decellularization

Figure 10.7 Diaphragmatic tissue before and after decellularization.

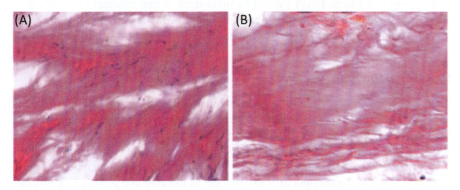

Figure 10.8 Section of diaphragmatic tissue before decellularization. (A) Nuclei visible, (B) nuclei absent (H&E, ×40).

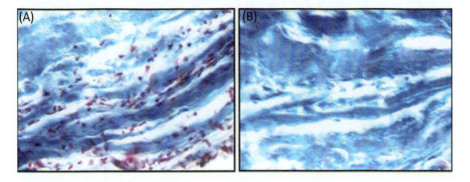

Figure 10.9 Section of decellularized diaphragmatic tissue showing collagen fibers. (A) Nuclei are visible. (B) Nuclei absent (Masson's trichome, ×40).

resulted in a sheet of homogenous ECM consisting mainly of collagen and also elastin and reticulin and removed all soluble proteins in the matrix. These findings are in consonance with the study of Lin et al. [34] and Kumar et al. [33]. The clearance of nuclear content was further evident in DAPI staining. These findings are in accordance with the study of Gilbert et al. [35] who find SDS as an effective agent for decellularization of biological tissue. However, Kumar et al. [15] observed that 2% sodium deoxycholate was effective for decellularization of bubaline diaphragm.

10.3.4 (4,6-Diamino-2-phenylindole) dihydrochloride staining

For staining of nucleus 4,6-diamino-2-phenylindole dihydrochloride (DAPI) staining was done as per standard protocol in diaphragmatic tissues before and after decellularization. Working DAPI stain (1 μg/mL) was prepared by adding double sterile PBS to stock solution (1 mg/mL). The slides having thin deparaffinized samples were smeared with working DAPI solution and incubated for 15 minutes in a lightproof room. Excess solution was drained and slides were washed many times with sterile PBS. The stained slides were viewed using a fluorescent microscope with appropriate filters. DAPI stained sections showed that treatment with 1% SDS for 72 hours resulted in loss of cellularity without affecting the three-dimensional structure of collagen fibers (Fig. 10.10).

10.3.5 DNA quantification

DNA quantification was done as per method described by John et al. [36] with some modifications. One gram of diaphragmatic tissue (before and after decellularization) from each diaphragm was weighed and triturated using pestle and mortar in liquid nitrogen till it became fine powder. The powdered tissue was transferred in a 15 mL sterile tube and mixed in 2 mL of solution 1 by continuous shaking for 10 minutes. It was centrifuged at 5000 rpm for 10 minutes and supernatant was discarded. The sediment was mixed in 400 μL of solution 2 and 2 μL of proteinase K

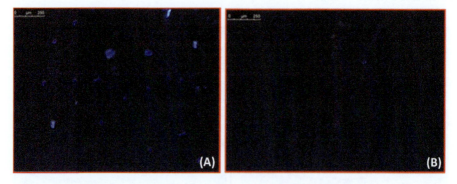

Figure 10.10 (A) Section of decellularized diaphragmatic scaffold giving fluorescence indicating presence of nuclei. (B) No fluorescence indicating absence of nuclei (DAPI, × 40).

in a 2 mL sterile tube and maintained in a warm water bath at 56°C for 12 hours. Then 1 mL mixture of phenol, chloroform, and isoamyl alcohol at a ratio of 25:24:1 was mixed in this and again centrifuged at 8000 rpm for 15 minutes. By using supernatant, it was mixed in 400 μL of solution containing chloroform and isoamyl alcohol at a ratio of 24:1 and centrifuged at 11,000 rpm for 2 minutes. Supernatant was separated and equal volume of chilled isoamyl alcohol was added in it and it was kept at −20°C for 10 minutes.

The solution was centrifuged at 10,000 rpm for 15 minutes and supernatant was discarded. The remaining pellet at the bottom of the tube was mixed 1 mL of high-grade 70% ethanol and centrifuged at 10,000 rpm. This procedure was repeated for two more times. The resulting pellet was dried by keeping the tube open in the environment till all the alcohol was evaporated. The pellet was dissolved in 200 μL of 0.3 × TE buffer by mixing it properly. Then the optical density was measured against the blank (TE buffer) at a wavelength of 260 nm and DNA was quantified using Nanodrop ND-1000 spectrophotometer V3.5 and expressed in terms of ng/μL. Average of DNA content before and after decellularization was calculated.

Modifications in the original protocol given by Johns et al. [36] for DNA isolation from blood were found appropriate for isolation of DNA from diaphragmatic tissue. The tissue was triturated in liquid nitrogen which provided a homogenous mixture of tissue and dissolving solution without any solid tissue remaining at the bottom of the tube. DNA isolation from this method resulted in intact DNA without RNA contamination and final DNA preparation was translucent to slightly whitish in color. The isolated DNA was easily dissolved in standard TE buffer. The mean quantity of DNA in untreated samples ($n = 12$) was 952.20 ± 25.43 ng/μL, which was significantly ($P < .01$) reduced to 51.31 ± 4.24 ng/μL in decellularized diaphragmatic tissue samples, indicating 94.61% efficacy of decellularization protocol. Lin et al. [34] obtained 86.4% DNA clearance by using SDS along with trypsin and DNase. As it is not possible to remove all immunogenic materials (here, DNA) but the residual cellular antigens may be insufficient to elicit the type of proinflammatory or immune response that could adversely affect biologic scaffold remodeling [37]. It is also possible that the chemical decellularization has altered the protein of reactive antigens in such a manner that it can no longer stimulate the adverse reaction.

10.3.6 Agarose gel electrophoresis

The 0.5 g agarose powder was mixed in 70 mL of TAE buffer and melted. Melted agarose was poured into sealed gel casting tray with the comb positioned appropriately. Once the gel got solidified, it was transferred to electrophoresis tank filled with 1 × TAE buffer. The level of buffer was kept at least one cm above the gel. The wells were carefully charged with combination of 2 μL of DNA loading dye (ethidium bromide) and 8 μL of DNA sample in each well. Electrophoresis was carried out first at 100 V for 5 minutes followed by 70 V for next one hour. On completion of process, the gel was visualized under UV light and documented by Geldoc gel documentation system and presence or absence of DNA in samples after and before decellularization was judged. Agarose gel electrophoresis revealed a thick white band over gel indicating presence of nuclear content in untreated tissue samples. However, gel images of DNA samples from decellularized diaphragmatic tissues revealed only a smear without any detectable bands indicating complete degradation of DNA in the tissue (Fig. 10.11). Similar findings are also reported by Andrea et al. [38] during decellularization of skeletal muscles of rat, rabbit and human.

Recipients of an allogenic and more specifically xenogenic scaffolds based on ECM may encounter adverse immunogenic response due to foreign antigenic epitopes related to intracellular organelles and cell membranes. This has to be eliminated for their acceptance in the biological systems [39]. This antigenicity of xenogenic biomaterials can be reduced by a variety of chemical and physical methods. Chemicals, such as acid solutions, hypotonic/hypertonic solutions, detergents, organic solvents, and enzyme solutions, tend to disrupt the cellular and extracellular components and their antigenic epitopes. Gamma radiations, sonication, and freezing thawing are commonly used physical modalities to decellularize the biological tissues and these are most of the times used in conjunction with chemical treatment [40].

The tendinous portion of an adult buffalo diaphragm is a thick piece of collagen fibers oriented in multiple direction covered with peritoneum from abdominal side and with pleura from thoracic side. This provides a rigid configuration and greater weight bearing ability. It is also elastic in nature and can be stretched up a certain extent due to elastin and reticulin content. These advantages of buffalo diaphragm were utilized for preparation of a decellularized scaffold which on application can withhold its mechanical properties and is resistant to tearing due to over weight of large animals where conventional meshes (nylon, polypropylene, pogalactin, proline, etc.) cannot withstand.

Detergents are molecules with exclusive properties of alteration, in form of either disintegration or synthesis, of hydrophobic—hydrophilic interactions among molecules biological tissues. Here we utilized cell lysing and protein solubilization property of

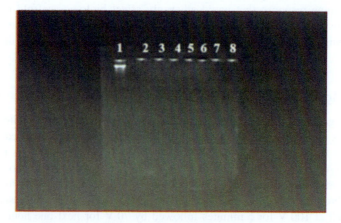

Figure 10.11 Imaging of DNA samples by gel documentation systems after agarose gel electrophoresis of untreated diaphragmatic tissue (1), decellularized diaphragmatic scaffolds (2—8).

SDS. The SDS is an anionic detergent which acts on membrane proteins (hydrophobic) as well as nonmembrane (hydrophilic) proteins. At its higher concentration (more than 2%), increased pH (alkaline), and temperature (above 50°C) of solution, it tends to denature the conformation of collagen. Here we kept the pH of the solution toward neutral and maintained 21°C temperature in orbital shaker to avoid undue damage to ultrastructure of collagen. Also, the concentration of solution was maintained at 1%, for this solution was replaced with freshly prepared solution of 1% SDS after each 12 hours up to 3 days. Zhou et al. [41] also used 1% SDS for decellularization of bovine pericardium and reported that it preserved the ECM components and architecture. Youngstorm et al. [42] reported no alteration in ultimate tensile stress and maximum stress in decellularized equine superficial digital flexor tendon using 2% SDS. However, glycosaminoglycan (GAG) content was reduced.

10.4 Biocompatibility evaluation of pig diaphragm

10.4.1 Materials and methods (III)

10.4.1.1 Cell extraction

orcine diaphragm procured from the abattoir was used as raw material. After proper washing with sterile normal saline solution, the diaphragm was cut in the pieces of 2×2 cm^2 in size. The diaphragm was made acellular by using the method of Meezan et al. [43] with slight modifications. Briefly, the graft pieces were placed in distilled water and agitated for 1 hour using a magnetic bar to lyse the cells and to release the intracellular contents. Tissues were then suspended in 4% sodium deoxycholate for 3 hours under continuous agitation followed by the treatment with deoxyribonuclease-1 (2000 Kunitz units) suspended in 1 M sodium chloride solution. This process was repeated thrice to extract all cells from the tissue. Samples were then stored until grafting in 0.9% phosphate buffered saline containing 10% gentamycin at 4°C. The method used for acellularity of diaphragm tissue proved ideal. The diaphragm was found completely acellular on histological examination (Fig. 10.12).

Acellular tissue matrix in the present study was produced by extracting cells through a multistep chemical and enzymatic process. The extraction leaves behind a sheet of homogenous ECM consisting mainly of collagen and elastin and removes all soluble proteins in the matrix [44]. This ECM can organize the regeneration of abdominal wall/muscles.

10.4.1.2 Cell cytotoxicity

The in vitro cells cytotoxicity studies were done in two types of cell systems i.e. peripheral blood leukocytes of rabbits and chicken embryo fibroblast (CEF) cell culture. Saline extracts of graft material was made and different dilutions of biomaterials were used for cell cytotoxicity studies. The highest concentration of biomaterials (1:20 dilution) did not produce any appreciable change in cell morphology at 24 hours postincubation in rabbit leukocytes. About 20% of cell death was recorded

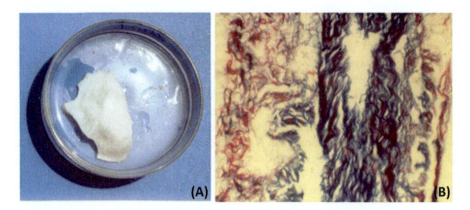

Figure 10.12 (A) Gross photograph of acellular porcine matrix; (B) microphotograph of acellular porcine diaphragm matrix (Masson's trichome stain, ×150).

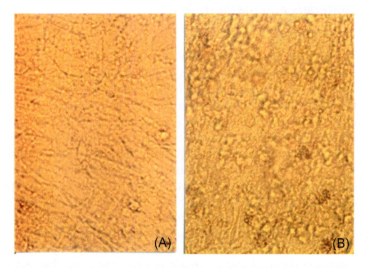

Figure 10.13 (A) Primary chick embryo fibroblasts (CEF) grown in MEM 199 (×400); (B) CEF showing rounding of cells 24 h postincubation with 1:10 dilution of acellular porcine diaphragm extract (×400).

when tissue culture fluid was replaced with 50% biomaterials. Whereas, in the control culture, cell death to a level of 10% was recorded at 96 hours postincubation (Fig. 10.13A), which normally occurs in vitro leukocyte culture without any mitogen or antigen. In CEF cell culture studies, cytopathic effects with rounding of cells were observed with acellular diaphragm (Fig. 10.13B), at 1:10 dilution after 24 hours postincubation.

Two types of cell systems were used in the present study due to its easy availability and suitability for immunocompatibility test. Similar cell systems (lymphocytes, PMN, monocytes) have been used in earlier reports [45]. In presence of highest concentration of biomaterial (1:20 dilution), no appreciable changes in cell morphology at 24 hours postincubation were recorded in rabbit leukocytes. In the control culture, cell death to a level of 10% was recorded at 96 hours postincubation, which normally occur in vitro leukocyte culture without any mitogen or antigen. In a different approach using CEF cell culture, the cell cytotoxicity was tested once the monolayer was completed. Simultaneous addition of biomaterial extract during seeding of CEF cells was not practiced as this can hamper the growth of mitotic cells to differentiate into fibroblastic appearance. Instead of established cell line (BHK-21, Vero cells and NDBK cells), CEF cell culture was preferred due to its most susceptibility to toxic effects being a primary culture and of vertebrate origin. Moderate cell cytotoxicity was observed with acellular diaphragm and acellular dermis at 1:10 dilution within 24 hours postincubation. The exact cause of cytotoxicity could not be defined, as individual constituents in the tissue extract was not probed. Possibly the treatment of above biomaterials in sodium deoxycholate and DNase along with high concentration of salt (1 M NaCl) might have contributed in cytotoxicity in spite of repeated washings before the preparation of tissue extract. Previous reports also have advocated in favor of extraction technique using high salt solution involving detergent like Triton X-100, DNase, RNase, which enhance the patency [46] but no discussion has been made by them regarding the residual effect of these substances. The observed in vitro cell cytotoxicity in the present findings may not reflect the true picture of in vivo findings as the detoxification ability of the host resulting in graft acceptability. Attachment of fibroblastic cells to the surface layer of culture tube usually a healthy sign of biocompatibility test, yet failure to adherence of cell type does not necessarily indicate toxicity of the biomaterials [47]. As there is ethical hindrance for indiscriminate use of animal model to test the cytotoxic effects of biomaterial implants, thus the only alternative is the in vitro cytotoxicity test before any in vivo application.

10.5 Biochemical changes in rabbit organs after subcutaneous implantation of bovine diaphragm

10.5.1 Materials and methods (IV)

Fresh bovine diaphragm was procured from local abattoir. Diaphragm was divided into two equal halves and one portion was used as native and another portion was made acellular as per the technique of Kumar [48]. Both native and acellular tissues matrices were crosslinked using with 5% glutaraldehyde (GA) and hexamethylene diisocyanate (HMDC) for 72 hours.

10.5.2 Subcutaneous implantation of buffalo diaphragm in rabbits

Adult New Zealand white rabbits (18) of either sex were utilized for evaluation of diaphragm biomaterial. The animals were kept off feed and water for 6 hours and 12 hours respectively, before the implantation. The back of the animals was properly clipped, shaved, and scrubbed with 5% cetrimide and chlorohexidine and painted with povidone iodine solution. The biomaterials were cut in 10×20 mm^2 size and implanted in two pouches created on either side of the back. The animals (18) were randomly divided into different groups/subgroups as shown in the Table 10.1.

10.5.3 Collection of organs

Immediately, after euthanasia at day 90 postoperatively, the organs (kidney, liver, lung and heart) were removed and chilled in crushed ice. The organs were thoroughly washed with cold deionized distilled water. The organs were then blotted on filter paper, cut into small pieces, weighed, and homogenized using a glass mortar and pestle in extraction buffer (50 mM Tris, pH 7.4 containing 0.25% Triton X-100 and 0.5% sodium dodecyl sulfate). The material was centrifuged at 3000 rpm for 10 minutes. After centrifugation, the supernatants were separated with the help of pasture pipette and extracts were stored at $-20°C$ until used. Free amino group, protein, acid phosphatase (ACP), and alkaline phosphatase (ALP) analysis were performed in the different organs.

10.5.4 Free amino group contents analysis

Ninhydrin assay was used to determine the free amino group contents of each test sample, as per the procedure of Sung et al. [49]. Uncrosslinked groups (A1 and B1) showed significantly ($P < .05$) increased values of free amino group contents as compared to crosslinked groups (A2, A3, B2, B3) in different organs (kidney, liver, lung, and heart) on day 90 postimplantation.

Kidney: HMDC-crosslinked groups (A3, B3) showed significant ($P < .05$) decrease in free amino group content as compared to GA-crosslinked groups (A2, B2). The highest free amino group content (36.33 ± 1.45 mg/mL) was observed in native diaphragm group (A1) and lowest free amino group content (21.00 ± 1.53 mg/mL) was seen in

Table 10.1 Subcutaneous implantation of biomaterials before and after crosslinking in different groups/subgroups.

Groups	Subgroups	No. of animals	Treatment given
ND (A)	A1	3	ND
	A2	3	ND crosslinked with GA
	A3	3	ND crosslinked with HMDC
AD (B)	B1	3	AD
	B2	3	AD crosslinked with GA
	B3	3	Acellular diaphragm crosslinked with HMDC

AD, Acellular diaphragm; GA, glutaraldehyde; HMDC, hexamethylene diisocyanate; ND, native diaphragm.

acellular diaphragm group (B3). The significant ($P < .05$) decrease in free amino group content in A3, B1, B2, and B3 groups was observed as compared to control.

Liver: HMDC-crosslinked groups (A3, B3) showed significant ($P < .05$) decrease in free amino group contents as compared to GA-crosslinked groups (A2, B2). The highest free amino acid content (46.66 ± 1.33 mg/mL) was observed in uncrosslinked native diaphragm group (A1) and lowest value (16.00 ± 0.58 mg/mL) was observed in HMDC-crosslinked acellular diaphragm group (B3). The significant ($P < .05$) decrease in free amino group content was observed in A3, B1, B2, and B3 groups as compared to control.

Lung: GA-crosslinked group (A2, B2) showed significant ($P < .05$) increased in free amino group content as compared to HMDC-crosslinked groups (A3,B3). The highest value (40.33 ± 0.33 mg/mL) was observed in uncrosslinked native diaphragm group (A1) and lowest value (13.00 ± 1.00 mg/mL) in HMDC-crosslinked acellular diaphragm group (B3). The significant ($P < .05$) decreased in free amino group content was observed in all implanted groups except uncrosslinked native diaphragm group (A1) as compared to control.

Heart: GA-crosslinked groups (A2, B2) showed significant ($P < .05$) increased of free amino group content as compared to HMDC-crosslinked groups (A3, B3). The highest free amino group (30.00 ± 0.57 mg/mL) content was observed in uncrosslinked native diaphragm group (A1) and lowest (12.66 ± 0.33 mg/mL) in HMDC-crosslinked acellular diaphragm group (B3). Uncrosslinked and crosslinked acellular diaphragm groups (B1, B2, and B3) showed significantly ($P < .05$) decrease in amino group as compared to control.

10.5.5 Protein contents analysis

The protein contents in test sample homogenates were estimated by the method of Lowry et al. [50] using bovine serum albumin (BSA) as a standard.

Kidney: The protein content was found significantly ($P < .05$) higher in uncrosslinked groups (A1, B1) on day 90 as compared to crosslinked groups (A2, A3 and B2, B3). HMDC-crosslinked groups (A3, B3) showed significant ($P < .05$) decrease in protein as compared to GA-crosslinked groups (A2, B2). The protein content was significantly ($P < .05$) decreased in all diaphragm-implanted groups as compared to control.

Liver: The protein content was significantly ($P < .05$) lower in uncrosslinked groups (A1, B1) as compared to crosslinked groups (A2, A3 and B2, B3). HMDC-crosslinked groups (A3, B3) showed significantly ($P < .05$) decrease value as compared to GA-crosslinked groups (A2, B2). The protein content remained significantly ($P < .05$) increase in all implanted groups in comparison to control.

Lung: Protein content showed significant ($P < .05$) decrease in uncrosslinked groups (A1, B1) as compared to crosslinked groups (A2, A3 and B2, B3). GA-crosslinked groups (A2, B2) showed significant ($P < .05$) increase in protein content as compared to HMDC-crosslinked groups (A3, B3). Protein content remained significantly ($P < .05$) increased in diaphragm-implanted groups as compared to control.

Heart: Uncrosslinked groups (A1, B1) showed significant ($P < .05$) decrease in protein content as compared to crosslinked groups (A2, A3 and B2, B3). HMDC-crosslinked groups (A3, B3) showed significant ($P < .05$) increase in values of protein as compared to GA-crosslinked groups (A2, B2). Protein content was significantly ($P < .05$) increased in diaphragm-implanted groups as compared to control group.

10.5.6 Acid phosphatase and alkaline phosphatase analysis

The determination of acid phosphatase (ACP) and alkaline phosphatase (ALP) activity in each test sample homogenates was based on the method of King and Armstrong [51]. In all diaphragm-implanted groups, uncrosslinked groups (A1,B1) showed significant ($P < .05$) increase as compared to crosslinked groups (A2, A3, B2, B3) in different organs (kidney, liver, lung, and heart). HMDC-crosslinked groups (A3, B3) showed significant ($P < .05$) decrease in acid phosphatase and alkaline phosphatase activity in all the four organs (kidney, liver, lung, and heart) of diaphragm- and pericardium-implanted rabbit groups as compared to GA-crosslinked groups (A2, B2). The enzyme activity was significantly ($P < .05$) increased in all implanted groups as compared to control. Highest activity was observed in native diaphragm-implanted group.

A postoperative fall of plasma amino acids has been reported. However, the mechanism by which surgical operation results in a lowered concentration of plasma amino acids has not been known [52,53]. It has been suggested that postoperative malnutrition is the main causative factor [54]. Free amino group (mg/mL) significantly decreased in all implanted groups as compared to control in different organs (kidney, liver, lung, and heart). In diaphragm- and pericardium-implanted groups, uncrosslinked groups (native and acellular) showed significantly increase as compared to crosslinked groups. In all implanted groups, free amino group was significantly lower in HMDC-crosslinked groups as compared to GA-crosslinked groups in liver, lung, heart. Whereas, only kidney extract of pericardium-implanted groups exhibited significantly increase value of amino acid concentration in HMDC-crosslinked groups as compared to GA-crosslinked groups. The reduction of free amino groups (determined by ninhydrin assay in the present study) in biological tissue diminishes it antigenicity [55]. A relatively large increase in the plasma free amino acid levels to accompany tumor growth [56].

Norten et al. [57] have shown that cancer patients with no or minimum body weight loss maintained their plasma free amino acid levels within normal range [19]. Protein concentration in kidney, liver, and lung was found significantly increased in GA-crosslinked diaphragm-implanted groups as compared to HMDC-crosslinked groups. Whereas heart showed significantly increase in HMDC-crosslinked group as compared to GA-crosslinked group. All pericardium-implanted groups showed significantly increased protein contents in GA-crosslinked groups as compared to HMDC-crosslinked groups in liver, lung, and heart but low protein contents in kidney of GA-crosslinked groups was observed. Hypoalbuminemia could have a profound effect on the response of patients to surgery. This effect has been investigated in experimental animals and in man. Impaired wound healing [58] increased susceptibility to hemorrhagic shock [59] and increased incidence of postoperative infection [60] have long been known to be associated with hypoalbuminemia. More recently, impaired immune function has been reported in surgical patients with low plasma albumin levels [61].

The enzymes have been differentiated in various tissues and in serum on the basis of their inhibition by different physical and chemical treatments. Decreased serum alkaline phosphatase activity has been observed in infants who are undergoing cardiac surgery with profound hypothermia, circulatory arrest, and limited cardiopulmonary bypass [62]. Total alkaline phosphatase activity in sera reflects a number of alkaline phosphohydrolases of several tissue origins. Tissue-specific alkaline phosphatases have been identified in bone, liver, intestine, and placenta [63,64]. The increase levels of alkaline phosphatase in traumatic area as a result of tissue injuries have been reported to help in the proliferation of fibroblasts [65]. Alkaline phosphatase appeared to be associated with the metabolic process concerning collagen formation [66]. Acid phosphatase and alkaline phosphatase activity was significantly decreased in HMDC-crosslinked groups as compared to GA-crosslinked groups in different organs (kidney, liver, lung, and heart) of diaphragm- and pericardium-implanted groups. In the healing process of wound, various tissue enzymes play crucial role. The alkaline phosphatase (a zinc containing enzymes) is located in cell membrane of various cells (neutrophils, macrophages giant cells, fibroblasts) of the body. Alkaline phosphatase activity was markedly affected by the tumor and associated with various metabolic processes of bone, and its blood level provides valuable means in the diagnosis of bone disease [67]. Alkaline phosphatase activity in kidney, bone, intestine, and liver provides a sensitive biochemical index for monitoring any change in the cellular activities of these organs.

10.6 Dermal wound healing using primary mouse embryonic fibroblasts seeded buffalo acellular diaphragm matrix in a rat model

10.6.1 Materials and methods (V)

Acellular diaphragm matrices were prepared as per method described by Kaarthick [68]. The prepared tissue matrices were divided into two groups. One group was used for seeding of primary mouse embryo fibroblasts (p-MEF) and another group was taken as such. Microscopic observation of native diaphragm treated with 1% sodium deoxycholate at 48 hours revealed complete removal of cells. No nuclear bodies were seen. The tissue was primarily composed of ECM (Fig. 10.14A–D). Present findings were similar to the report described earlier [15]. Sodium

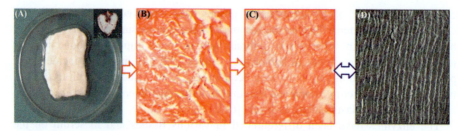

Figure 10.14 Photograph showing preparation of acellular diaphragm matrix: (A) gross photograph of acellular bovine diaphragm (in inset whole buffalo diaphragm); (B) microphotograph showing native diaphragm (H&E staining, ×100); (C) microphotograph of acellular bovine diaphragm; (D) scanning electron microscope image of acellular bovine diaphragm (×2500).

deoxycholate is effective for solubilizing both cytoplasmic and nuclear cellular membranes [28]. It is used extensively in decellularization due to their mild effects on tissue structure. It disrupts lipid−protein and lipid−lipid interactions but generally leaves protein−protein reactions intact with the result of maintaining their functional conformations [35].

10.6.2 Isolation, culture, and seeding of p-MEF over ADM

Isolation and culture of p-MEF were done as per earlier report [69]. Prepared ADM of 20×20 mm^2 sizes were washed 4−5 times in Dulbecco's Modified Eagle Medium containing antibiotic and placed in 6-well tissue culture plates. Further, p-MEF cells were seeded on ADM at the rate of 2×10^6/mm^2 as per Zhang et al. [70] and seeded matrices were incubated under 5% CO_2 tension in CO_2 incubator. The morphological assessment of cellular growth on ADM was done by the histologic and scanning electron microscopic (SEM) examinations.

The p-MEF cells showed characteristic growth and adherence pattern in vitro and proliferated rapidly to complete the monolayer in 72 hours. The morphology of in vitro cultured cells clearly indicated the presence of embryonic fibroblast in the culture. On 48 hours postseeding, p-MEF cells proliferated throughout the ADM (Fig. 10.15A). Further, SEM examination also confirmed proliferation and penetration of p-MEF cells into the ADM scaffold (Fig. 10.15B) [71].

10.6.3 Wound creation and implantation

Forty-eight clinically healthy Sprague-Dawley rats of either sex, weighing from 250 to 300 g, and 3−4 months of age were used in this study and were randomly divided into four groups (I to IV) of 12 animals each. The study was approved by Institute Animal Ethics Committee, Indian Veterinary Research Institute, Izatnagar, India. Animals were housed individually in cages, provided with commercial diet and water ad libitum, and maintained under uniform conditions. Animals were acclimatized to approaching and

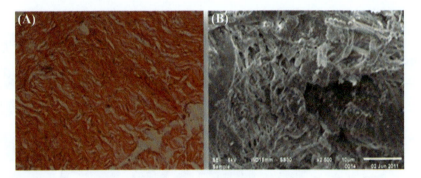

Figure 10.15 (A) Photomicrograph (H&E staining, $\times 100$) showing growth and penetration of the p-MEF in ADM at 48 h. (B) SEM image ($\times 2500$) showing growth and penetration of the p-MEF in ADM at 48 h.

handling for a period of 10—15 days before start of the study. Wound was created aseptically under anesthesia using xylazine (5 mg/kg intramuscularly) and ketamine (50 mg/kg intramuscularly). The animals were restrained in sterna recumbency and dorsal thoracic area was prepared for aseptic surgery. Using a sterile plastic template, the vertices of the experimental wounds of 20×20 mm^2 dimensions were outlined on the dorsothoracic region of rats. A full-thickness skin defect including the panniculus carnosus was excised on each animal. Hemorrhage, if any, was controlled by applying pressure with sterile cotton gauze. The defect in group I was left open and taken as control. In group II, the defect was repaired with autograft. In group III, the defect was covered with ADM and in group IV with p-MEF-seeded ADM [71].

10.6.4 Wound healing evaluation
10.6.4.1 Macroscopic observation and planimetry

Color photographs were taken on days 0, 3, 7, 14, 21, 28, or till completion of healing with the help of digital camera. The shape, irregularity, and color of the lesion were determined. On day 3, the graft was brown in color in group III. In group IV, hyperemia was observed around the graft. The upper layer that was used as the anchor for the graft had come out as scab on day 14 in group IV. In group III, the upper layer came out as scab on day 21 onward. Reduction in wound area was observed in all groups from day 7 onward. Group IV exhibited remarkable decrease in wound size from day 14 onward and complete healing was observed on day 19. Group I healed on postimplantation days 28. The color of the implants changed from white to brown and finally dark brown (Fig. 10.16) [71].

It might be due to decrease in vascularization and continuous loss of moisture from the implants. The contraction was lesser in autograft-treated wounds due to early acceptance and neovascularization. This is consistent with the findings of a recent study [33]. Change in color from dark to dark brown in animals treated with

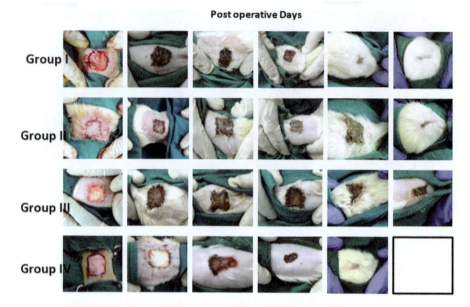

Figure 10.16 Image of skin wounds of control (I), autograft (II), ADM (III), p-MEF-seeded ADM (IV) groups on days 0, 3, 7, 14, 21, and 28.

acellular intestinal matrix was also observed in earlier report [72]. Gangwar et al. [69] reported the color of implanted skin graft changed from white to brown and finally dark brown on the subsequent time intervals as healing progressed.

10.6.4.2 Wound area and contraction

The wound area and percent contraction was measured at day 0, 3, 7, 14, 21, 28, or till completion of healing as per Kumar et al. [72]. The results are presented in Figs. 10.17 and 10.18.

As the healing progressed, there was significant ($P < .05$) decrease in the wound area at different time intervals. On day 3, significantly ($P < .05$) decreased values were recorded in group IV as compared to other groups. On day 14 and 21, the wound area decreased significantly ($P < .05$) in group IV as compared to others, indicating faster healing rate in this group. Complete healing was recorded on day 28 and the entire wound area was covered with hair. On day 3, maximum wound contraction was recorded in group III ($26.58 \pm 5.27\%$). On day 14 and 21, percent contraction in group IV was 92.65 ± 0.40 and 94.25 ± 0.58, respectively. Wound area decreased gradually as healing progressed. The percent contraction on days 14 and 21 in group IV was significantly ($P < .05$) higher as compared to others. The present study revealed significantly decreased wound area on days 14 and 21 in p-MEF-seeded ADM-implanted group as compared to others. This may be due to enhanced healing due to proliferation of mouse embryo fibroblasts in the ADM [71]. There are numerous studies that report inhibition of wound contraction after implanting a collagenous ECM into full-thickness wounds [33,69,72].

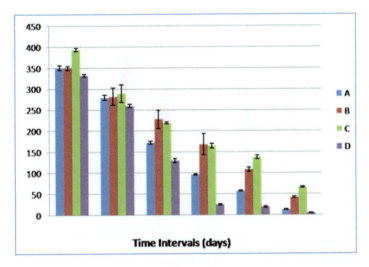

Figure 10.17 Wound area in control (I), autograft (II), ADM (III), p-MEF-seeded ADM (IV) groups on days 3, 7, 14, 21, and 28. Data are expressed as mean ± standard error of the mean.

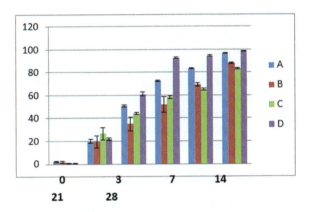

Figure 10.18 Wound contraction (%) in control (I), autograft (II), ADM (III), p-MEF-seeded ADM (IV) groups on days 3, 7, 14, 21, and 28. Data are expressed as mean ± standard error of the mean.

10.6.5 Immunologic observations

10.6.5.1 Indirect ELISA

The humoral response elicited by the implants in animals of different groups was determined by indirect ELISA using rat monoclonal anti IgG antibodies. The blood samples were collected on days 0 and 45 postsurgery. Absorbance was recorded at 492 nm using ELISA reader. The levels of antibodies present in serum samples collected prior to implantation were taken as basal values. A hyperimmune serum raised against native diaphragm antigen was used as positive control and the sera of acellular diaphragm and

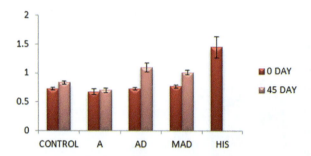

Figure 10.19 ELISA reading (absorbance at 492 nm) of sera from rats of control (I), autograft (II), ADM (III), p-MEF-seeded ADM (IV) groups on days 0 and 45. Data are expressed as mean ± standard error of the mean. A, autograft antigen; AD, acellular dermal matrix antigen; MAD, p-MEF-seeded acellular dermal matrix antigen; HIS, hyperimmune sera.

p-MEF-seeded ADM diaphragm were compared with positive control. In all the groups, there was increase in antibody titer from 0 day to 45 day (Fig. 10.19). There was higher B-cell response in animals with ADM and p-MEF-seeded ADM implants as compared to negative control and autograft-implanted animals [71].

Reduced humoral response was observed in p-MEF-seeded ADM-implanted group as compared to group with ADM implant. It might be due to presence of MEF cells in ADM. Moreover, mouse and rats were genetically similar hence less humoral response against the bioengineered matrices [71]. The nature and degree of the immunological response to a foreign material is a crucial variable affecting the outcome and success or failure of that biomaterial. The biomaterial could stimulate an antibody response, a cell-mediated sensitization, or minimal to no response. The presence of antibodies to xenogenic collagen was an epiphenomenon and not an indicator for rejection of the implant [73].

10.6.5.2 Lymphocyte proliferation evaluation

The cell-mediated immune (CMI) response toward implants in all experimental rats was assessed by MTT colorimetric assay. On day 45 postimplantation, rats were euthanized and spleens were harvested. The spleenocytes were cultured in vitro. The spleen collected from a normal rat, which was not exposed to any antigen, was used as negative control. Spleenocytes culture and the assay were performed as per the procedure described by Kruisbeek et al. [74]. The stimulation of rat spleenocytes with concanavalin A (Con A) and phytohemagglutinin (PHA) was considered as positive control, whereas unstimulated culture cells were taken as negative control. Against the diaphragm antigen, the group IV showed maximum stimulation (1.68 ± 0.10), whereas group II exhibited less stimulation (0.79 ± 0.08) (Fig. 10.20). As compared to the SI values of Con A and PHA, all the groups exhibited considerable rise in SI values except group II. The T-cell response was higher in the groups implanted with either ADM or p-MEF-seeded ADM as evidenced by higher SI values as compare to autograft or control [71].

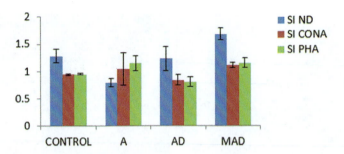

Figure 10.20 Stimulation index values (mean ± standard error of the mean) of rats splenocytes in control (I), autograft (II), ADM (III), p-MEF-seeded ADM (IV) groups on day 45. A, autograft; AD, acellular dermal matrix; MAD, p-MEF-seeded acellular dermal matrix; ND, native diaphragm antigen; CONA, concanavalin A; PHA, phytohemagglutinin.

LPA measures T-lymphocyte functions in vitro. An immune response against nonself-antigens is initiated by presentation of the antigen in a suitable form to T cells. Antigen can only be presented to T cells in the context of molecules of MHC [75]. Practically each nucleated cells is able to present antigen, first by virtue of a constitutive MHC class I expression, and second by a de novo expression of MHC class II molecules on the surface of the cell [76]. Graft rejection is usually mediated by activity of T cells, especially cytotoxic T cells. The T-cell subsets (Th1 and Th2) generated by native T cell on MHC antigen stimulation play a major role in the graft rejection through activity of different sets of cytokines that activate macrophages and B cells. Allman et al. [77] reported that the immune responses to xenogenic acellular tissue are of Th2 class, not involving fixation of complement or graft rejection, but rather inducing tolerance. The SI recorded for bioengineered matrices implanted animals was lower in comparison to the values of acellular matrices implanted animal's samples. The greater ability of this antigen to stimulate the lymphocytes in vitro may be attributed to the fact that on treatment with detergent, the bonds between protein molecules are broken and result into a change from quaternary and tertiary structure to primary and secondary structures. Therefore, the acellular antigen had greater ability to trigger CMI response in host [71].

10.6.6 Histologic observations

The biopsy specimen from the implantation site was collected on 7, 14, 21, and 28 days. The host inflammatory response, neovascularization, fibroplasia, deposition of neocollagen, and penetration of host inflammatory cells in the implanted matrices were evaluated.

In group I, on day 7, fibroblasts and granulation cells were observed at the wound area and there were evidence of collagen formation, infiltration of inflammatory cells, and neovascularization (Fig. 10.21A). On day 14, the inflammatory cells, especially, the polymorphic cells, were more, and there was proliferation of fibroblasts and neovascularization. Persistence of granulation tissue was also observed. On day 21, the

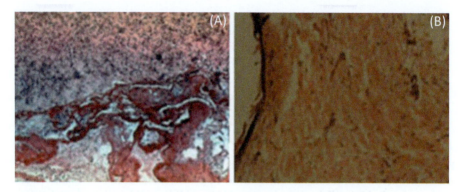

Figure 10.21 (A) Image showing epithelial proliferation under necrosed tissue, collagen formation, infiltration of inflammatory cells and neovascularization. (B) Formation of skin glands and keratin layer (H&E stain, × 100).

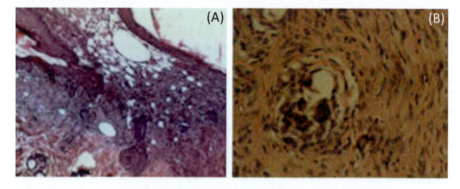

Figure 10.22 (A) Image showing excessive epithelial proliferation and delayed formation of granulation tissue, infiltration of inflammatory cells especially polymorphs, proliferation of fibroblasts and neovascularization. (B) More fibroblasts and collagen synthesis along with giant cells (H&E stain, × 100).

inflammation was greatly reduced and more formation of collagen was observed. On day 28, collagen formation was similar to that of normal skin and few inflammatory cells were present. Formation of skin glands and keratinous layer was also observed (Fig. 10.21B) [71].

In group II, on day 7, there was excessive epithelial proliferation and delayed formation of granulation tissue. Infiltration of inflammatory cells especially polymorphs, proliferation of fibroblasts, and neovascularization were observed at the site (Fig. 10.22A). On day 14, inflammatory reaction was decreased as compared to day 7. Complete proliferation of epidermis was seen. On day 21, inflammation reduced further as compared to day 14. Necrosis of the superficial layer and detachment of the scab was seen. On day 28, more number of fibroblasts and collagen synthesis was observed (Fig. 10.22B) [71].

In group III, on day 7, proliferation of fibroblasts was observed. Growth of connective tissue into the upper layer was also observed (Fig. 10.23A). On day 14, a greater number of inflammatory cells was present. The epithelial proliferation was seen under the superficial layer (Fig. 10.23B). On day 21, presence of mononuclear cells was appreciated. There was less collagen formation at the site. The superficial layer was removed. On day 28, inflammation was reduced and more collagen formation was observed (Fig. 10.23C) [71].

In group IV, on day 7, mild inflammation and more proliferation of fibroblasts and neovascularization were seen (Fig. 10.24A). On day 14, inflammation was reduced and restricted to the peripheral area of the wound. Fibroblasts proliferation still increased as compared to day 7. Excessive collagen formation was observed. On day 21, inflammation reduced further, and excessive collagen formation was seen all over the wound area. Epithelial proliferation was also observed (Fig. 10.24B) [71].

Overall in all the groups, there was inflammatory cells infiltration with fibroblasts proliferation. Excessive epithelial proliferation was observed on day 7 in group I, whereas others showed epithelialization after day 14. Group I also showed formation of granulation tissue faster due to earlier contraction and scarring of the

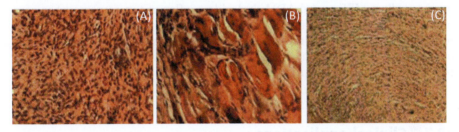

Figure 10.23 (A) Image showing proliferation of fibroblasts and growth of connective tissue into the upper layer. (B) Necrosis of superficial area and epithelial proliferation under necrosed area. (C) Reduced inflammation and more collagen formation (H&E stain, ×100).

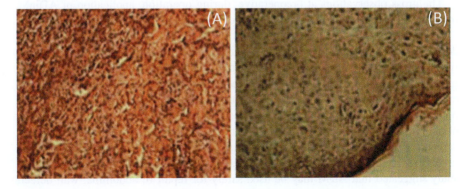

Figure 10.24 (A) Image showing mild inflammation, more proliferation of fibroblasts and neovascularization. (B) Reduced inflammation and excessive collagen formation with epithelial proliferation all over the wound area (H&E stain, ×100).

wound as the wound left open and no scaffold was applied. Group IV showed lesser inflammatory response from day 14 onward and more collagen synthesis were observed. At various postimplantation days, minimum host tissue reaction, less fibroplasia, and minimum collagen density were observed in wounds treated with autograft (group II) as compared with wounds of groups I, III, and IV [71]. On day 7, all the groups showed infiltration of inflammatory cells especially the neutrophils. Proliferation of fibroblasts was seen more in the acellular matrices and the bioengineered acellular matrices [69]. Courtman et al. [2] hypothesized that cell extraction process decreased the antigenic load within the tissue due to the elimination of cellular antigens. On day 14, more proliferation of fibroblasts was observed along with the inflammatory cells. Starting of epithelialization was observed in all the groups. Massive fibroblasts proliferation along with angiogenesis was observed. These findings were consistent with earlier report [78]. On day 21, p-MEF-seeded ADM group showed higher collagen synthesis as compared to others, which can be appreciated by the Masson's trichrome staining. This may be due to the presence of the fibroblasts that have the major functions of collagen secretion collagen and ECM deposition. On day 28, the control group showed less inflammatory cells than autograft and ADM groups and complete epithelialization was also observed. In the present study, p-MEF-seeded ADM showed less inflammatory response when compared to others. Host reaction was limited to the periphery of the graft. The p-MEF-seeded ADM implantation leads to increased collagen synthesis and ECM deposition. Although ADM has healing potential, seeding with p-MEF increases their potential for repair of full-thickness skin wounds in rats.

10.7 Clinical applications

Acellular diaphragm matrix (ADiaM) was used to reconstruct the abdominal wall defects in different animals. The detail of each case treated is described in Table 10.2.

The gross observations at preoperative, during operation, application of ADiaM, immediate postoperative and 1 month postoperatively are presented in Fig. 10.25.

The ADiaM matrices were applied in eight clinical cases of abdominal wall defects of different species of animals. Animals with ADiaMs recovered uneventfully and remained sound for at least up to 3 months. Hematological and immunological findings were unremarkable. Bubaline diaphragm showed excellent repair efficiency and biocompatibility for abdominal wall defects repair in animals without complications [15]. In another study, prepared ADiaMs were used for the repair of abdominal wall defects in four different species of animals. The abdominal wall defects repaired with ADM remained sound over a period of 3 months. All the defects repaired with ADM healed completely without graft rejections. The present study suggested that ADM may be used safely for repair of abdominal wall defects in different animal species (buffalo, cattle, goat, and pig) [79,80].

Acellular diaphragm matrix was successfully used in pigs and dogs for the repair of umbilical and perineal hernia, respectively [81]. The buffalo diaphragm after

Table 10.2 Showing repair of abdominal wall defects (hernia) cases using ADiaM.

S. no	Species	Age (month)	Sex	Ring diameter (RD) (cm)	Type of hernia	Hernial contents	Outcome
1	Cattle	3	Male	6	Umbilical	Large intestine	Repaired
2	Cattle	6	Female	8	Lateral	Small intestine	Repaired
3	Buffalo	5	Male	5	Umbilical	Omentum/omental fat	Repaired
4	Buffalo	12	Male	10	Ventral	Small intestine/mesentery	Repaired
5	Goat	5	Male	8	Ventral	Small intestine/mesentery	Repaired
6	Goat	2	Female	14	Ventral	Small intestine/mesentery	Repaired
7	Goat	6	Female	7	Lateral	Intestine/omentum	Repaired
8	Pig	3	Male	6	Umbilical	Omentum/fat	Repaired

ADiaM, Acellular diaphragm matrix.

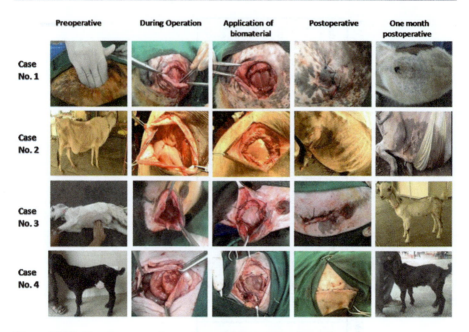

Figure 10.25 Showing gross observations at different time intervals.

decellularization with 2% sodium deoxycholate for 48 hours was used for umbilical hernioplasty in 12 crossbred Landrace pigs. Treatment with 2% sodium deoxycholate leads to complete decellularization at 48 hours. All the hernias repaired with acellular diaphragm matrix healed completely without graft rejection and remained sound over a period of 3 months [81]. Remya et al. [82] performed reconstruction of umbilical hernia with acellular diaphragm matrix in a 7-month-old crossbred calf using the inlay technique. The animal completely recovered after 30 days without postoperative complications. Hysterocele in a Dachshund dog was surgically corrected with decellularized diaphragm matrix [83]. Acellular diaphragm matrix was found to be a promising and immunotolerable prosthetic for hernioplasty [84].

10.8 Conclusion

The bubaline diaphragm, upon treatment with 2% SDS for 48 hours, results in complete removal of cells and cellular components, with preservation of collagen structure and integrity. Treatment with 1% sodium SDS for 72 hours also results in complete decellularization of bubaline diaphragm matrix. Xenogenic bubaline diaphragm matrix shows excellent repair efficiency and biocompatibility. The acellular matrix, which is produced, can be used for repair of various soft tissue defects in small and large animals without any fear of life threatening immunological reaction and graft rejection. Free amino acid concentration, protein concentration, acid phosphatase, and alkaline phosphatase activity were observed significantly decreased in HMDC-crosslinked groups as compared to GA-crosslinked and native groups in different organs (kidney, liver, lung, and heart) of diaphragm-implanted groups. Xenogenic nature of scaffold also reduces enslavement on availability of autologous and allogenic tissue materials.

References

[1] Gulati AK, Cole GP. Immunogenicity and regenerative potential of acellular nerve allograft to repair peripheral nerve in rats and rabbits. Acta Neurochir Wien 1994;126:158−64.
[2] Courtman DW, Pereira CA, Kashef V, McComb D, Lee JM, Wilson GJ. Development of a pericardial acellular matrix biomaterial-biochemical and mechanical effects of cell extraction. J Biomed Mater Res 1994;28:655−66.
[3] Wu Z, Tang Y, Fang H, Su Z, Xu B, Lin Y. Decellularized scaffolds containing hyaluronic acid and EGF for promoting the recovery of skin wounds. J Mat Sci Mat Med 2015;26:532−40.
[4] Klosterhalfen B, Klinge U, Hermanns B. Pathology of traditional surgical nets for hernia repair after long-term implantation in humans. Chirurg 2000;71:43−51.
[5] Coda A, Bendavid R, Botto-Micca F. Structural alterations of prosthetic meshes in humans. Hernia 2003;7:29−34.
[6] Visser GE, Robison RL, Maulding HV, Fong JW, Pearon JE, Argentieri GJ. Note: Biodegradation of and tissue reaction to poly-(DLlactide) microcapsules. J Biomed Mater Res 1986;20:667−76.
[7] Williams DF, Mort E. Enzyme-accelerated hydrolysis of polyglycolic acid. J Bioeng 1977;1:231−8.
[8] Williams DF. Some observations on the role of cellular enzymes in the in-vivo degradation of some polymers. In: Syvett B, Achyara A, editors. Corrosion of Implant Materials. Philadelphia: ASTM; 1979. p. 61.
[9] Lewis GP. Changes in the composition of rabbit hind limb lymph after thermal injury. J Physiol 1969;205:619−34.
[10] Schallberger SP, Stanley BJ, Hauptman JG, Steficek BA. Effect of porcine small intestinal submucosa on acute full thickness wounds in dogs. Vet Surg 2008;37:515−24.
[11] Shin H, Jo S, Mikos AG. Biomimetic materials for tissue engineering. Biomaterials 2003;24:4353−64.
[12] Hollister SJ. Porous scaffold design for tissue engineering. Nat Mater 2005;4:518−24.
[13] Clark RA, Ghosh K, Tonnesen MG. Tissue engineering for cutaneous wounds. J Invest Dermatol 2007;127:1018−29.
[14] Tabata Y. Biomaterial technology for tissue engineering applications. J R Soc Interface 2009;6(Suppl):311−24.
[15] Kumar V, Gangwar AK, Kumar N, Singh H. Use of the bubaline acellular diaphragm matrix for umbilical hernioplasty in pigs. Vet Arh 2015;85:49−58.
[16] van der Veen VC, van der Wal MB, van Leeuwen MC, Ulrich MM, Middelkoop E. Biological background of dermal substitutes. Burns 2010;36:305−21.
[17] Dickman S, Strobel G. Life in the tissue factory. N Scientists 1995;145:32−7.

[18] Vora SD, Kumar V, Asodiya FA, Singh VK, Fefar DT, Gajera HP. Bubaline diaphragm matrix: development and clinical assessment into cattle abdominal hernia repair. Braz Arch Biol Technol 2019;62:e19180442. Available from: https://doi.org/10.1590/1678-4324-2019180442.
[19] Green MR, Sambrook J, Inglis J. A Laboratory Manual to Accompany Quantitative Chemistry and InstrumentalAnalysis, Natural Science Division. Pepperdine University; 2012.
[20] Abe Y, Krimm S. Normal vibrations of crystalline polyglycine II. Biopolymers 1972;1(9):1817–39.
[21] Payne K, Veis A. Fourier transform IR spectroscopy of collagen and gelatin solutions: deconvolution of the amide I band for conformational studies. Biopolymers 1988;27(11):1749–60.
[22] Krimm S, Bandekar J. Vibrational spectroscopy and conformation of peptides, polypeptides and proteins. Adv Protein Chem 1986;38:181–364.
[23] Muyonga JH, Cole CGB, Duodu KG. Characterisation of acid soluble collagen from skins of young and adult Nile perch (*Latesniloticus*). Food Chem 2004;85:81–9.
[24] Crapo PM, Gilbert TW, Badylak SF. An overview of tissue and whole organ decellularisation processes. Biomaterials 2011;32:3233–43.
[25] Kumar V, Devarathnam J, Gangwar AK, Kumar N, Sharma AK, Pawde AM. Use of acellular aortic matrix for reconstruction of abdominal hernias in buffaloes. Vet Rec 2012;70(15):392–6.
[26] Kumar V, Gangwar AK, Mathew DD, Ahmad RA, Saxena AC, Kumar N. Acellular dermal matrix for surgical repair of ventral hernia in horses. J Equine Vet Sci 2013;33(4):238–43.
[27] Kumar V, Gangwar AK, Kumar N. Evaluation of the murine dermal matrix as a biological mesh in dogs. Proc Natl Acad Sci India Sect B: Biol Sci 2016;86(4):953–60.
[28] Seddon AM, Curnow P, Booth PJ. Membrane proteins, lipids and detergents: not just a soap opera. Biochem Biophys Acta 2004;1666:105–17.
[29] Woods T, Gratzer PF. Effectiveness of three extraction techniques in the development of a decellularized bone-anterior cruciate ligament-bone graft. Biomaterials 2005;26:7339–49.
[30] Londono R, Dziki JL, Haljasmaa E, Turner NJ, Leifer CA, Badylak SF. The effect of cell debris within biologic scaffolds upon the macrophage response. J Biomed Mater Res A 2017;105:2109–18.
[31] Baker MJ, Trevisan J, Bassan P, Bhargava R, Butler HJ, Dorling KM. Using Fourier Transform IR spectroscopy to analyze biological materials. Nat Protoc 2014;8:1771–91.
[32] Doyle BB, Bendit E, Blout ER. Infrared spectroscopy of collagen and collagen-like polypeptides. Biopolymers 1975;14:937–57.
[33] Kumar V, Kumar N, Gangwar AK, Singh H, Singh R. Comparative histologic and immunologic evaluation of 1,4-butanediol diglycidyl ether crosslinked versus noncrosslinked acellular swim bladder matrix for healing of full-thickness skin wounds in rabbits. J Surg Res 2015;197:436–46.
[34] Lin CH, Yang JR, Chiang NJ, Ma H, Tsay RY. Evaluation of decellularized extracellular matrix of skeletal muscle for tissue engineering. Inter J Art Org 2014;37(7):546–55.
[35] Gilbert TW, Sellaroa TL, Badylak SF. Decellularization of tissues and organs. Biomaterials 2006;27:3675–83.
[36] John SW, Weitzner G, Rozen R, Scriver CR. A rapid procedure for extracting genomic DNA from leukocytes. Nucl A Res 1991;19:408–15.
[37] Thomas W, Gilbert JF, Stephen FB. Quantification of DNA in biologic scaffold materials. J Surg Res 2009;152:135–9.

[38] Andrea P, Maria MS, Alex P, Veronica M, Lucia P, Piero GP. Decellularized human skeletal muscle as biologic scaffold for reconstructive surgery. Inter J Mol Sci 2015;16:14808—31.
[39] Trivedi HL. Immunobiology of rejection and adaptation. Transplan Proc 2009;39:647—52.
[40] Derek DC, Jerry CH, Leigh GG, Kyriacos AA. Antigen removal for the production of biomechanically functional, xenogeneic tissue grafts. J Biomech 2013;. Available from: https://doi.org/10.1016/j.jbiomech.2013.10.041i Online.
[41] Zhou J, Schleicher OM, Wendel HP, Schenke-Layland K, Harasztosi C, Hu S, et al. Impact of heart valve decellularization on 3-D ultrastructure, immunogenicity and thrombogenicity. Biomaterials 2014;31:2549—54.
[42] Youngstrom DW, Barrett JG, Rose RR, Kaplan DL. Functional characterization of detergent decellularized equine tendon extracellular matrix for tissue engineering applications. PLoS One 2013;27:41—51.
[43] Meezan E, Hjelle JT, Brendel K. A simple, versatile, non disruptive method for isolation of morphologically and chemically pure basement membranes from several tissues. Life Sci 1975;17:1721—32.
[44] Kumar N, Sharma AK, Singh GR, Gupta OP. Carbon fiber and plasma preserved tendon allografts for the gap repair of flexor tendon in bovines: clinical, radiological and angiographical observations. J Vet Med Ser A 2002;49:161—8.
[45] Pizzorferrato A, Ciapetti G, Stea S, Cenni E, Aricola CR, Granchi D, et al. Cell culture methods for testing biocompatibility. Clin Mater 1994;15:173—90.
[46] Wilson GJ, Courtman DW, Klement P, Lee JM, Yegar H. Acellular matrix-a biomaterial approach for coronary artery bye pass and heart valve replacement. Ann Thorac Surg 1995;60:353—8.
[47] Tateishi T., Ushida T., Aoki H., Ikada Y., Nakamura M., Williams D.F., et al. Round robin test forstandardization of biocompatibility test procedure by cell culture method. In: Doherty PJ, Williams RL, Williams DF, Lee AIJ, editors, Biomaterials—Tissue Interfaces. Amsterdam: Elsevier Science Publishers; 1992. p. 89—97.
[48] Kumar V. Acellular buffalo small intestinal submucosa and fish swim bladder for the repair of full thickness skin wound in rabbits. MVSc thesis submitted to Deemed University Indian Veterinary Research Institute, Izatnagar 243122, Uttar Pradesh India; 2010.
[49] Sung HW, Chang Y, Liang IL, Chang WH, Chen U. Fixation of biological tissues with a naturally occurring crosslinking agent: fixation rate and effects of pH, temperature, and initial fixative concentration. Biomed Mater Res 2000;52:77—87.
[50] Lowry OH, Rosenbrough NJ, Farr AL, Randall RJ. Protein measurement with the Folin phenol reagent. J Biol Chem 1951;93:265—75.
[51] King EJ, Armstrong AR. Plasma alkaline phosphtase in disease. Can Med Assoc J 1934;31:376—9.
[52] Woolf LI, Groves AC, Moore JP, Duff JM, Finley RJ, Loomer RL. Arterial plasma amino acids in patients with serious postoperative infection and in patients with major fractures. Surgery 1976;79:283—92.
[53] Dale G, Young F, Latner AL, Goode A, Tweedle D, Johnston ID. The effect of surgical operation on venous plasma free amino acids. Surgery 1977;81:295—301.
[54] Schonheyder F, Bon J, Skjoldborg H. Variations in plasma amino acid concentrations after abdominal surgical procedures. Acta Chir Scand 1974;140:271—5.
[55] Imamura E, Sawatani O, Koyanagi H, Noishiki Y, Miyata T. Epoxy compounds as a new crosslinking agent for porcine aortic leaflets: subcutaneous implants studies in rats. J Card Surg 1989;4:50—7.
[56] Wu C, Bauer JM. A study of free amino acids and of glutamine synthesis in tumour bearing rats. Cancer Res 1960;20:848—56.

[57] Norton JA, Gorschboth CM, Wesley RA, Burt ME, Brennan MF. Fasting plasma amino acid levels in cancer patients. Cancer 1985;56:1181−6.
[58] Thompson WD, Ravdin IS, Frank IL. Effect of hypoproteinemia on wound disruption. Arch Surg 1938;36:500−18.
[59] Ravdin IS, Mcnamee HG, Kamholz JH, Rhoads JE. Effect of hypoproteinemia on susceptibility to shock resulting from hemorrhage. Arch Surg 1944;48:491−7.
[60] Rhoads JE, Alexander CF. Nutritional problems of surgical patients. Ann NY Acad Sci 1955;63:268−75.
[61] Law DK, Dudrick SJ, Abdou NI. The effects of protein calorie malnutrition on immune competence of the surgical patient. Surg Gynecol Obstet 1974;139:257−66.
[62] Neutze JM, Drakeley MJ, Barratt-Boyes BG. Serum enzymes after cardiac surgery under profound hypothermia with circulatory arrest and limited cardiopulmonary bypass. Am Heart J 1974;88:553−6.
[63] Fishman WH, Ghosh NK. Isozymes of human alkaline phosphatase. Advan Clin Chem 1967;10:255−62.
[64] Posen S. Alkaline phosphatase. Ann Intern Med 1967;67:183−9.
[65] Soni NK, Patel MR, Shrivastava RK. Alkaline phosphatase, total protein, protein bound hexoses and hexosamine changes in serum and wound fluids in surgically induced subcutaneous wounds in buffalo calves. Acta Vet AcadSci Hung 1976;26(3):263−70.
[66] French JE, Beneditt EP. Observations on the localization of alkaline phosphates in healing wound. Arch Path 1954;57:352−6.
[67] King EJ, Moss DW. Enzymes in metabolic bone disease. In: Sissons HA, editor. Bone Metabolism in Relation to Clinical Medicine. London: Pitman Medical Publishing Co; 1962. p. 42−54.
[68] Kaarthick D.T. Repair of cutaneous wound using acellular pericardium and diaphragm of buffalo origin seeded with mouse embryo fibroblast cells in a rat model. MVSc thesis submitted to Deemed University, Indian Veterinary Research Institute, Izatnagar, Bareilly, Uttar Pradesh, India; 2012.
[69] Gangwar AK, Kumar N, Devi KS, Kumar V, Singh R. Primary chicken embryo fibroblasts seeded 3-D acellular dermal matrix (3-D ADM) improve regeneration of full thickness skin wounds in rats. Tissue Cell 2015;47:311−22.
[70] Zhang X, Deng Z, Wang H, Yang Z, Guo W, Li Y, et al. Expansion of delivery of human fibroblasts on micronized acellular dermal matrix for skin regeneration. Biomaterials 2009;30:2666−74.
[71] Kaarthick DT, Sharma AK, Kumar N, Kumar V, Gangwar AK, Maiti SK, et al. Accelerating full-thickness dermal wound healing using primary mouse embryonic fibroblasts seeded bubaline acellular diaphragm matrix. Trends Biomat Artific Organs 2017;31(1):16−23.
[72] Kumar V, Kumar N, Gangwar AK, Singh H. Comparison of acellular small intestinal matrix (ASIM) and 1-ethyl-3-(3-dimethylaminopropyl) carbodiimide crosslinked ASIM (ASIMEDC) for repair of full-thickness skin wounds in rabbits. Wound Med 2014;7:24−33.
[73] Ruszezak Z. Effect of collagen matrices on dermal wound healing. Adv Drug Deliv Rev 2003;55:1595−611.
[74] Kruisbeek AM, Shevach E, Thornton AM. Proliferative assays for T cell function. Curr Protoc Immunol 2004; Unit 3.12.
[75] Townsend A, Bodmer H. Antigen recognition by class I restricted T lymphocytes. Annu Rev Immunol 1989;197:601−24.

[76] Otsuka H, Ikeya T, Okano T, Kataoka K. Activation of lymphocyte proliferation by boronate containing polymer immobilised on substrate: The effect of boron content on lymphocyte proliferation. Eur Cell Mater 2006;12:36—43.
[77] Allman AJ, McPherson TB, Badylak SF, Merril LC, Kallakury B, Sheenan C, et al. Xenogenic extracellular matrix grafts elicit a Th2 restricted immune response. Transplantation 2001;71:1631—40.
[78] Kim MS, Hong KD, Shin HW, Kim SH, Kim SH, Lee MS, et al. Preparation of porcine small intestinal submucosa sponge and their application as a wound dressing in full-thickness skin defect of rat. Int J Biol Macromol 2005;36:54—60.
[79] Rathore HS, Raghuvansi PDS, Gautam D, Mohan D, Singh AK, Singh AP, et al. Use of the bubaline acellular diaphragm matrix (ADM) for repair of abdominal wall defects in four different species of animals. Inter J Chem Stud 2018;6(4):157—61.
[80] Rathore HS, Kumar N, Singh K, Maiti SK, Shrivastava S, Shivaraju S, et al. Clinical application of acellular matrix derived from the bubaline diaphragm and caprine rumen for the repair of abdominal wall defects in animals. Aceh J Ani Sci 2019;4(2):50—60. Available from: https://doi.org/10.13170/ajas4.2.13071.
[81] Raghuvanshi PDS, Mohan D, Gautam D, Shivaraju S, Maiti SK, Kumar N. Clinical application of animal based extracellular matrix in hernioplasty. MOJ Immunology 2018;6(4):115—18.
[82] Remya V, Shakya P, Sivanarayanan TB, Dubey P, Pawde AM, Kumar N. Umbilical hernioplasty using acellular bovine diaphragm matrix in a calf. Rumin Sci 2013;2(2):229—31.
[83] Mathew D.D., Kumar N., Ahmad R.A., Remya V., Vijaykumar H., Shakya P., et al. Hysterocele in a Dachshund dog and its surgical correction with decellularised diaphragm matrix. Paper presented in XXXVII Annual Congress of Indian Society for Veterinary Surgery and National Symposium on Need for specilaization and Super-specialization in Veterinary Surgery and Imaging techniques for Professional Efficiency development held at Mannuthy, Thrissur, Kerala; 2013. p 244.
[84] Remya V., Kumar N., Singh K., Gopinathan A., Shakya P., Sangeetha P., et al. Acellular Diaphragm Matrix: A promising and immuno-tolerable prosthetic for hernioplasty. Paper presented in XXXVIII Annual Congress of Indian Society for Veterinary Surgery and International Symposium on New Horizons of Camel Surgery and Large Ruminant Surgery held at COVS, RUVAS, Bikaner (Rajasthan); 2014. p 360.

Fish swim bladder−derived tissue scaffolds

Remya Vellachi[1], Naveen Kumar[2,*], Ashok Kumar Sharma[2], Sonal Saxena[3], Swapan Kumar Maiti[2], Vineet Kumar[4], Dayamon David Mathew[5] and Sameer Shrivastava[3]

[1]Department of Veterinary Surgery and Radiology, College of Veterinary and Animal Sciences, Wayanad, Kerala, India, [2]Division of Surgery, ICAR-Indian Veterinary Research Institute, Izatnagar, Uttar Pradesh, India, [3]Division of Veterinary Biotechnology, ICAR-Indian Veterinary Research Institute, Izatnagar, Uttar Pradesh, India, [4]Department of Veterinary Surgery and Radiology, College of Veterinary and Animal Sciences, Bihar Animal Sciences University, Kishanganj, Bihar, India, [5]Department of Veterinary Surgery & Radiology, Faculty of Veterinary and Animal Sciences, Banaras Hindu University, Rajiv Gandhi South Campus, Barkachha, Uttar Pradesh, India

11.1 Introduction

Organ damage or loss can occur from congenital disorders, cancer, trauma, infection, inflammation, iatrogenic injuries, or other conditions and often necessitates reconstruction or replacement. Depending on the organ and severity of damage, autologous tissues can be used for reconstruction. However, there is unavailability of sufficient tissue and there is a degree of morbidity associated with the harvest procedure. For functional replacement, organ transplants are used for damaged tissues. However, there is a severe shortage of donor organs, which is worsening with the ageing of the population. Both aforementioned approaches rarely replace the entire function of the original organ. Tissues used for reconstruction can lead to complications because of their inherent divergent functional parameters. The replacement of deficient tissues with functionally equivalent tissues would improve the outcome for these patients. Therefore, engineered biological substitutes that can restore and maintain normal tissue function would be useful in tissue and organ replacement applications [1].

Tissue engineering, one of the major components of regenerative medicine, follows the principles of cell transplantation, materials science, and engineering toward the development of biological substitutes that can restore and maintain normal function. Tissue engineering strategies generally fall into two categories: the use of acellular matrices, which depend on the body's natural ability to regenerate for proper orientation and direction of new tissue growth and the use of matrices

---

*Present affiliation: Veterinary Clinical Complex, Apollo College of Veterinary Medicine, Jaipur, Rajasthan, India.

Natural Biomaterials for Tissue Engineering. DOI: https://doi.org/10.1016/B978-0-443-26470-2.00011-9
© 2025 Elsevier Inc. All rights reserved, including those for text and data mining, AI training, and similar technologies.

with cells. Acellular tissue matrices are usually prepared by manufacturing artificial scaffolds or by removing cellular components from tissues by mechanical and chemical manipulation to produce collagen-rich matrices [2–5]. These matrices tend to slowly degrade on implantation and are generally replaced by the extracellular matrix (ECM) proteins that are secreted by the in growing cells. Biomaterials have become critical components in the development of effective new medical therapies for wound care. Natural collagenous materials are being investigated for surgical repair because of inherent low antigenicity and their ability to integrate with surrounding tissue [6].

The outbreak of bovine spongiform encephalopathy and the foot-and-mouth disease have caused restrictions on use of collagen from bovine or porcine origin. Collagens from fish swim bladders may be good substitutes, because of their safety. Absorbable suture material prepared from fish intestine collagen and used in the gastric wounds revealed good healing in rabbit model [7].

The use of natural biomaterials has typically pretreatment aimed at (1) preserving the tissue by enhancing the resistance of material to enzymatic or chemical degradation, (2) reducing the immunogenicity of the material, and (3) sterilizing the tissue [8]. Biological materials composed of ECM are typically processed by methods that include decellularization and/or chemical crosslinking to remove or mask antigenic epitopes, DNA, and damage associated molecular pattern (DAMP) molecules [9–11]. Biomaterials composed of decellularized ECM have been shown to promote the healing process via modulation of the host immune response, resistance to bacterial infections, allowing reinnervation and reestablishing homeostasis in the healing region [12–15].

When native cells are used for tissue engineering, a small piece of donor tissue is dissociated into individual cells. These cells are expanded in culture, attached to a support matrix, and then reimplanted into the host after expansion. Cells can also be used for therapy through injection, either with carriers such as hydrogels or alone. The source of donor tissue can be heterogenic (divergent species), allogeneic (same species, divergent individual), or autogenic (same individual). Ideally, both structural and functional tissue replacement will occur with minimal complications. The preferred cells to use are autologous cells, in which a biopsy of tissue is obtained from the host, the cells are dissociated and expanded in culture, and the expanded cells are implanted into the same host [4,16–19]. The use of autologous cells, although it may cause an inflammatory response, avoids rejection, and thus the deleterious side effects of immunosuppressive medications can be avoided.

Most current strategies for tissue engineering depend on a sample of autologous cells from the diseased organ of the host. However, for many patients with extensive end-stage organ failure, a tissue biopsy may not yield enough normal cells for expansion and transplantation. In other instances, primary autologous human cells cannot be expanded from a particular organ, such as the pancreas. In these situations, embryonic and adult stem cells are an alternative source of cells from which the desired tissue can be derived. Embryonic stem cells can be derived from discarded human embryos or from fetal tissues. Adult stem cells can be harvested from adult tissues including bone marrow, fat, muscle, and skin. These cells can be differentiated into the desired cell type in culture and then used for bioengineering.

Groups	No. of animals	Treatment
I	8	Control (open wounds)
II	8	Acellular fish swim bladder (A-FSB)
III	8	Rat bone marrow–derived mesenchymal stem cell-seeded acellular FSB matrix (R-BMSC)
IV	8	Goat bone marrow–derived mesenchymal stem cell-seeded acellular FSB matrix (G-BMSC)

To complete the list of possible cell sources for bioengineering of tissues and organs, therapeutic cloning must be mentioned. Therapeutic cloning, which has also been called nuclear transplantation and nuclear transfer, involves the introduction of a nucleus from a donor cell into an enucleated oocyte to generate an embryo with a genetic makeup identical to that of the donor. Stem cells can be derived from this source, which may have the potential to be used therapeutically [1].

The use of native cells and adult stem cells is ethically sound and accepted by all major religions and governments. On the other hand, the use of embryonic stem cells and therapeutical cloning are more controversial because the same methods could theoretically be used to clone human beings. Major advances have been achieved in engineering of tissues in the past decade. Regenerative medicine may extend the treatment options for various diseases. However, like every new evolving field, regenerative medicine and tissue engineering are expensive. Several of the clinical trials involving bioengineered products have been placed on hold because of costs involved with the specific technology. With a bioengineered product, costs are usually high because of the biological nature of the therapies involved. As with any therapy, the cost that the medical health care system can allow for a specific technology is limited. Therefore, the costs of bioengineered products have to be reduced for them to have an impact clinically. This is currently being addressed for multiple tissue-engineered technologies. As the technologies advance over time and the volume of the application is considered, costs will naturally decrease.

11.2 Preparation of acellular matrix from fish swim bladder

Swim bladders of the fresh water fish (*Labeo rohita*) (Fig. 11.1A) were treated with 0.5% sodium deoxycholate (ionic biologic detergent) for 24 hours for decellularization. The tissues were washed with phosphate buffer saline (PBS) and stored at $-20°C$ with antibiotics. Acellularity and collagen fiber arrangements were assessed by light microscopic and scanning electron microscopic examination. After the decellularization, the tissue became slightly soft and spongy in consistency and whiter than native tissue (Fig. 11.1B) [20].

Figure 11.1 (A) Fish swim bladder and. (B) decellularized fish swim bladder.

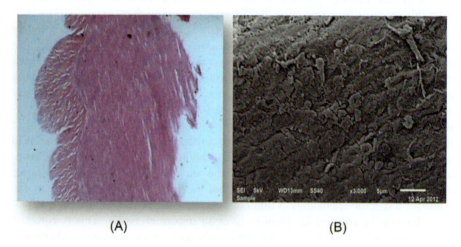

Figure 11.2 (A) Microphotograph of native fish swim bladder (H&E stain, ×100) and (B) SEM photograph of native fish swim bladder (×3000).

Light microscopic observations of native fish swim bladder showed cellularity dense and closely packed fibers (Fig. 11.2A). Scanning electron microscopic examination also revealed dense and closely packed fibers (Fig. 11.2B). After decellularization there was complete loss of cellularity. The tunica externa and interna were completely acellular. The collagen fibers were loosely arranged than the native tissue (Fig. 11.3A). Scanning electron microscopic observations revealed loss of cellularity. Collagen fibers were loosely arranged (Fig. 11.3B) [20].

The goal of decellularization is to efficiently remove all cellular and nuclear material while minimizing any adverse effect on the composition, biological activity, and mechanical integrity of the ECM [9]. Decellularization can be brought by physical,

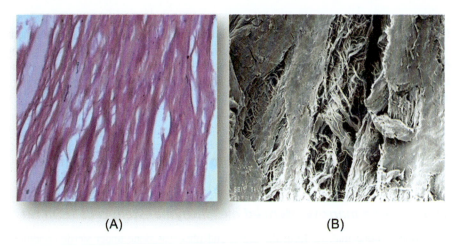

Figure 11.3 (A) Microphotograph of acellular fish swim bladder (H&E stain, ×200) and (B) SEM photograph of acellular fish swim bladder (×5000).

chemical, and enzymatic methods which leave a material composed of ECM components. These acellular tissues retained their natural mechanical properties and promote remodeling of the prosthesis by neovascularization and recellularization by the host [8]. Cellular antigens are predominantly responsible for the immunological reaction associated with allograft [21]. Remnants of cell components in xenograft may contribute to calcification and/or immunogenic reaction [22]. In particular, lipids and DNA fragments are likely to play a role in calcification [23]. The acellular tissue matrices possess the appropriate mechanical properties [24] and induced appropriate interaction with the host cells that resulted in the regeneration of functional tissues [25]. No inflammation, aneurysm or dystrophic calcification was noted for the acellular matrix vascular prostheses [26] and complete reendothelialization of the graft in canines had been reported [27]. Acellular tissue matrices are biocompatible, slowly degraded upon implantation and are replaced and remodeled by the ECM proteins synthesized and secreted by ingrowing host cells, which reduce the inflammatory response [28]. Acellular matrices support the regeneration of tissues with no evidence of immunogenic reaction [4]. However, it is important to keep in mind that even after the removal of cells and cell debris the intact ECM of the acellular tissue itself may elicit an immune response [29].

Ionic detergents are effective for solubilizing both cytoplasmic and nuclear cellular membranes, but tend to denature proteins by disrupting protein–protein interactions [30]. In general, ionic detergents are used extensively in decellularization protocols due to their mild effects on tissue structure. These surfactants disrupt lipid–protein and lipid–lipid interactions, but generally leave protein–protein reactions intact with the result of maintaining their functional conformations [9]. Sodium deoxycholate is very effective for removing cellular remnants but tends to cause greater disruption to the native tissue architecture when compared to sodium dodecyl sulfate [9]. The cell extraction was effectively achieved without significant disturbances in ECM morphology and strength. Decellularization of the fish swim bladder using 0.5% ionic biologic detergent revealed complete acellularity as per the findings of Kumar [31]. Results of the present study also supports that treatment with 0.5% sodium deoxycholate for a period of 24 hours is a viable option for decellularizing fish swim bladder. This resulted in a complete loss of cellular structures. The tunica externa and interna were completely acellular. The collagen fibers were loosely arranged as compared to the native tissue.

11.3 Collection and isolation of bone marrow−derived mesenchymal stem cells

11.3.1 Bone marrow−derived mesenchymal stem cells from rat

Collection of bone marrow from the femur and tibia was done under sterile condition. The material collected was washed with HBSS containing antibiotic and centrifuged at 1500 rpm for 15 minutes. The supernatant was discarded and the pellet was resuspended in RPMI medium. To separate the mononuclear fraction of the bone marrow, the resuspended material was submitted to the Histopaque protocol [32]. After centrifugation, an interface formed between the Histopaque and the RPMI medium containing the mononuclear cells was selected and resuspended in RPMI medium, supplemented with 10% fetal bovine serum and antibiotic (Fig. 11.4). The cells were seeded into a tissue culture

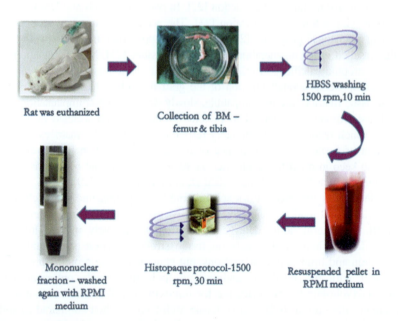

Figure 11.4 Showing procedure for collection and isolation of bone marrow−derived mesenchymal stem cells from rat.

Figure 11.5 Showing procedure for seeding and subculturing of mesenchymal stem cells.

flask and kept in humidified 5% CO_2 incubator at 37°C. Adherent cells, when reached a subconfluence, were detached with 0.25% trypsin-EDTA solution and then seeded on acellular fish swim bladder matrix (Fig. 11.5) [32,33].

11.3.2 Bone marrow–derived mesenchymal stem cells from goat

Under epidural anesthesia bone marrow was aspirated with bone marrow biopsy needle from the iliac crest under sterile condition. The material collected was washed with HBSS containing antibiotic and centrifuged at 1500 rpm for 15 minutes. The detail procedure for separation of mesenchymal stem cells has been already described in rat section. Adherent cells when reached a subconfluence, they were detached with 0.25% trypsin-EDTA solution and then seeded on acellular fish swim bladder matrix (Fig. 11.5) [32,33].

11.4 Seeding of stem cells on acellular fish swim bladder matrix

Seeding procedure was done under sterile conditions in biosafety cabinet type II. The scaffolds were transferred to a 60 mm dish containing sterile phosphate buffer saline (PBS) and were allowed to remain in PBS while preparing the cell solution. Bone marrow–derived mesenchymal stem cells were trypsinized, centrifuged, and resuspended in RPMI with fish swim bladder scaffolds (FBS) and antibiotics. Each matrix was then placed in separate wells of a 6-well cell culture plate. The cells were seeded in a drop-wise fashion at a density of 2.7×10^5 cells/cm^2 on the 2×2 cm^2 size FSB scaffolds kept in each well of 6-well cell culture plate. Culture was maintained at 37°C in a humidified atmosphere of 5% CO_2 in a CO_2 incubator. The cells were seeded in a drop-wise fashion at a density of 2.7×10^5 cells/cm^2 on the 2×2 cm^2 size FSB scaffolds kept in each well of 6 well cell culture plate.

The cells were seeded on culture plates and were observed daily under the inverted phase contrast microscope to assess the viability and proliferation of cells. The cells showed characteristic growth and adherence pattern in vitro and proliferated rapidly to complete the monolayer in about 12–15 days in case of rat cells and 7–9 days in case of goat cells. The morphology of in vitro cultured cells clearly indicated the presence of mesenchymal stem like cells. After 2—3 subcultures, when adherent cells reached a subconfluency, they were detached with 0.25% trypsin-EDTA solution and then seeded on acellular fish swim bladder matrix. The flushing of femur and tibia of young rats resulted in good amount of bone marrow collection. The aspiration from iliac crest of goats also gave a good amount of bone marrow for culturing of bone marrow—derived mesenchymal stem like cells. The cell separation technique as per Histopaque protocol isolated the mononuclear fraction of bone marrow very clearly [32,33].

11.5 Attachment and growth of mesenchymal stem cells

The seeded FSB matrix was processed for morphological assessment at different time intervals. For the morphological assessment of cellular growth in the matrix, inverted phase contrast microscopic examination, light microscopic examination, and scanning electron microscopic examination were done. The seeded matrices were preserved in 10% formalin for histological examination and in 2% glutaraldehyde for scanning electron microscopic (SEM) examination.

Under phase contrast microscope, the proliferation and viability of cell was observed. Histological observations prior to seeding did not reveal any nuclear bodies in the acellular matrices. By day 7 of seeding marked growth and proliferation of goat bone marrow, derived mesenchymal stem cells (G-BMSC) were observed (Fig. 11.6A). Whereas confluency of rat bone marrow—derived mesenchymal stem cells (R-BMSC) was observed on day 14 (Fig. 11.6B) [32,33].

Under scanning electron microscopy, it was observed that by day 3 of seeding G-BMSC started growing over acellular fish swim bladder (Fig. 11.7A) and growth became more evident by day 7 of seeding (Fig. 11.7B). Whereas in case of R-BMSC, the growth and proliferation of cells over the matrix were in a slower rate as compared to G-BMSC. Microphotographs showed proliferation of stem cells over acellular fish swim bladder matrix (Fig. 11.8A and B).

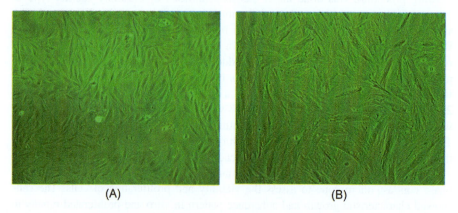

Figure 11.6 (A) Goat bone marrow—derived mesenchymal stem cells showing confluency on day 7. (B) Rabbit bone marrow—derived mesenchymal stem cells showing confluency on day 14.

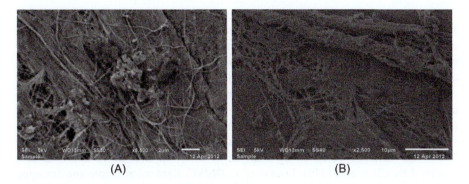

Figure 11.7 SEM photographs showing proliferation of goat stem cells over acellular fish swim bladder matrix on (A) day 3 (× 2500) and (B) day 7 (× 2500).

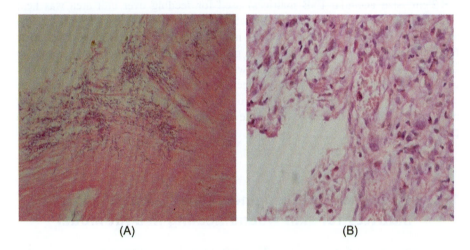

Figure 11.8 (A) Microphotographs showing proliferation of stem cells over acellular fish swim bladder matrix (H & E stain, × 100) and (B) in higher magnification (H & E stain, × 400).

The flushing out method of collection from femur and tibia of young rats yielded a good amount of bone marrow. The aspiration technique from iliac crests of goats also gives a good amount of bone marrow for culturing of bone marrow–derived mesenchymal stem like cells. The cell separation technique as per Histopaque protocol resulted in isolation of the mononuclear fraction of bone marrow very clearly. The G-BMSC and R-BMSC were cultured over RPMI medium in cell culture flask as per the technique described by Lovitt et al. [34]. The methods used for isolation and culture of G-BMSC and R-BMSC were found suitable. Using the same protocol of cell culture, R-GMSC took more time to reach at confluency (12–15 days) than G-BMSC (where it took 7–9 days). These cells were subjected to a minimum of 3 passaging/subculturing before seeding over acellular matrix in order to get maximum pure lineage [20,33].

The growth and penetration of these cells were assessed by using the light microscopic examination [hematoxylin and eosin (H&E) staining], scanning electron microscopic examination and by examining under inverted phase contrast microscopic examination. Light microscopic examination revealed the growth of mesenchymal stem cells over the acellular matrices. The scanning electron microscopic examination also showed the clear collagen fibers of the acellular matrices and the clumps of mesenchymal stem cells present around the collagen fibers and also over the matrices. Examination of cultured matrix under inverted phase contrast microscope twice daily also revealed proliferation and viability. These findings were similar to the findings observed by Xiaojun et al. [35]. Similar finding was also observed by Chun et al. [36]. In case of G-BMSC, light microscopic and scanning electron microscopic examination revealed the attachment and proliferation of mesenchymal stem cells over the acellular matrices at an early stage as compared to mesenchymal stem cells of rat origin. The count of cells (2.7×10^5 cells/cm^2 on the 2×2 cm^2 size acellular FSB matrices) used for seeding over unit area was kept constant for both the type of cells. G-BMSC takes around 3–5 days for the cells to attach and proliferate over the acellular matrix but in case of R-BMSC even after 6–7 days of seeding the attachment and proliferation was not good as compared to G-BMSC. This tendency is comparable with the results obtained in case of primary culture and subculturing of R-BMSC, where it also takes more time to reach confluence as compared to G-BMSC. The protocols for decellularization of fish swim bladder was found to be less toxic for the mesenchymal stem like cells to grow and proliferate [20].

11.6 Evaluation of bioengineered fish swim bladder matrices for skin wound healing in a rat model

The bioengineered fish swim bladder scaffolds were evaluated for the repair of full-thickness skin defects in rats. The study was conducted on 32 clinically healthy adult Wistar rats of either sex. The animals were provided standard diet and water ad libitum and were maintained under uniform managerial conditions. The animals were acclimatized for approaching and handling by 10–15 days, prior to the commencement of study. The animals were randomly divided into four groups of eight animals each. Animals were anaesthetized using xylazine (4 mg/kg body weight) and ketamine (40 mg/kg body weight) combination and were restrained in sternal recumbency. Dorsal thoracic area was prepared for aseptic surgery. One 2×2 cm^2 size full-thickness skin wound was created on the dorsum of each animal for assessing the healing potential of bioengineered scaffolds. The defect in different wounds was repaired immediately as per the detailed protocol given below [20,33].

11.6.1 Evaluation of wound healing

The wound healing was evaluated on the basis of gross, immunological, and histopathological observations at different intervals. Graph pad prism software version 5.0 was used for data analysis. Two-way ANOVA was used to compare the means at different time intervals among different groups [20,33].

11.6.2 Gross observations

Gross observations included the measurement of wound area, percentage contraction of wound area with respect to original wound size.

11.6.2.1 Wound area

Mean ± SE of the total wound area (mm^2) of the skin wounds at different time intervals are presented in Fig. 11.9. With the passage of time, a gradual decrease in wound area (mm^2) was observed in all the groups during the entire observation period. A significant decrease ($P < .001$) in wound area was observed between control group and treatment groups on days 7, 14, 21, and 28. There is no significant difference ($P > .05$) between groups III and IV in later stages (day 28) [20,33].

11.6.2.2 Wound contracture

Percent contraction (mean ± SE) of wound area at different time intervals in different groups are presented in Fig. 11.10. With the passage of time, a gradual increase in percent wound contraction was observed in all the groups during the observation period.

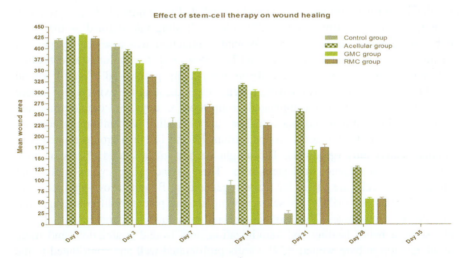

Figure 11.9 Mean ± SE of the wound area (mm^2) at different time intervals in various groups.

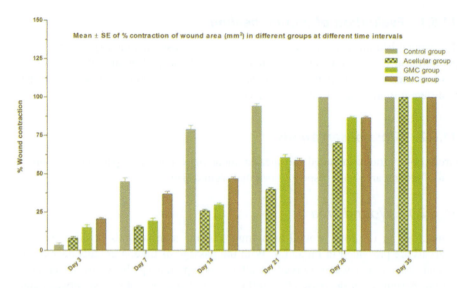

Figure 11.10 Mean ± SE of percent contraction of wound (%) at different time intervals in various groups.

A significant increase ($P < .001$) in percent contraction was observed between group I and II, group I and III and group I and IV on days 7, 14, 21, and 28. There is no significant difference ($P > .05$) in percent contraction between group III and IV in later stages (day 28) [20,33].

Wound contraction is the centripetal displacement of the wound edges that facilitates its closure after trauma. This process is carried out by myofibroblasts that contain alpha-actin from smooth muscle and is mediated by contractile forces produced by granulation tissue from wound [37]. Wound healing rate defined as the gross epithelialization of the wound bed. Wound contraction was assessed by percent retention of the original wound area and has been used to monitor wound healing [38]. Wound area decreased gradually as the healing progressed. Control wound healed completely by 28 days leaving a large scar indicating the existence of severe contraction. Acellular matrix-implanted group took 35 days for complete healing, but it was with minimum contraction. The rate of healing with minimum contraction was more in bioengineered groups (groups III and IV) than groups treated with acellular matrix only (group II). This might be due to the added effects of bone marrow−derived mesenchymal stem cells. There is increasing evidence showing that adult stem cells are useful for tissue regeneration. Bone marrow−derived mesenchymal stem cells (MSCs) are self-renewing and are potent in differentiating into multiple cells and tissues. The tissue-engineered skin containing mesenchymal stem cells showed better healing and keratinization, less wound contraction, and more vascularization in burn wounds [39]. Grafts proliferated well and contributed to the neotissues. The data suggest that tissue-engineered skin containing MSCs in a burn defect can accelerate wound healing and receive satisfactory effects. In the present study also, MSC-implanted groups showed better healing potential as compared to non-MSC-implanted group [39].

11.6.3 Computerized planimetry (color digital image processing)

Color digital image processing was done by taking color photographs of repair site with the help of digital camera at a fixed distance. Analysis of shape, size, irregularity, and color of the lesion was determined. The parameters were recorded on days 0, 3, 7, 14, 21, and 28. The observations recorded in different groups are presented in Fig. 11.11 [20,33].

In group I, on day 3, the wounds were covered with soft and fragile pinkish mass with serum like exudates oozing out. On day 7, the surface became more desiccated and necrosed with some exudate. By day 14, the wound size decreased markedly and a clear-cut thick crust developed and it was detaching leaving a raw granular pink tissue. By day 21, wound healed up completely leaving a large scar.

In group II, on day 3 the top layer of acellular graft appeared yellow in color with necrosed margin. By day 7, the top layer of graft got dried and turned up brown. By day 14, the upper layer became more desiccated and shriveled. On day 21, the top layer was seen in a stage of detachment from the underlying tissue. On day 28, the dried-up top layer was completely sloughed off exposing the healing tissue. On day 35, the wound healed up completely and no scar was observed.

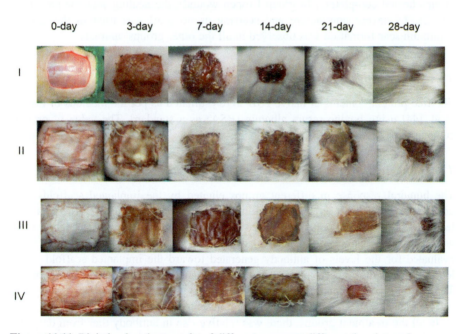

Figure 11.11 Digital color photographs of different groups at different time intervals.

In group III, on day 3, the top layer of acellular grafts appeared yellow in color. By day 7, the top layer of graft got dried and turned up brown. On day 14, the top layer was seen in a stage of detachment from the underlying tissue as only the sutured end holding it in place. On day 21, the dried-up top layer was completely sloughed off and newly formed granulation tissue within the underlying acellular graft covered the entire surface of the wound. By day 28, size of the wound decreased markedly. The wound edges healed completely and the remaining granulation tissue in the center became dried up indicating the healing beneath it. By day 35, the implanted area appeared similar to the normal skin.

In group IV, on day 3 some yellowish patchy areas were observed over the top acellular layer. On day 7, the top layer became dried and brown in color. On day 14, the top layer appeared greenish yellow in color indicating its slough off. By day 21, upper layer sloughed off completely exposing the healing tissue. By day 28, wound margins healed leaving a small central area. By day 35, the wound healed up completely with no scar and the implanted area appeared similar to the normal skin.

The color of the implanted samples changed from white to dark brown and finally dark revealed that on subsequent time intervals, decrease in vascularization and continuous loss of moisture contents of the graft lead to change its color in to black. Similar findings have been reported after the repair of full-thickness skin defects in rabbits [40,41] and in rats [42,43]. Upper graft layer detached from the body in form of the scar and the underlying layer of graft was completely absorbed and newly formed granulation tissue within the graft covered the whole surface of the wound in implanted group animals. On day 35, all the wounds of the implanted groups healed completely. In group I (open wound), the healing was completed on day 28, by severe contracture and scarring as no graft was used in this group. Minimum scar formation was observed in all the other groups' animals.

11.6.4 Immunological observations

The free protein contents of native FSB were estimated as per the methods of Lowry et al. [44] using bovine serum albumin (BSA) as a standard. The value of protein contents was 20.34 mg/mL.

11.6.4.1 Indirect enzyme-linked immunosorbent assay

The humoral response in different groups elicited by the implanted scaffolds was determined by indirect enzyme-linked immunosorbent assay (ELISA). The sera collected from different groups were tested in ELISA using native FSB as coating antigen. The serum samples collected on days 0, 21, and 28 post implantations were evaluated for the levels of antibody generated toward the implanted scaffolds. The levels of antibodies present in serum prior to implantation were taken as basal values. The scaffolds specific antibodies were expressed as mean ± SE absorbance at 492 nm wavelengths (OD492). ELISA reactions are presented in Fig. 11.12 [21,33].

In all the treatment groups, there was relative rise in antibody titer from day 0 to day 28 post implantation. There exists significantly higher ($P < .001$) B-cell

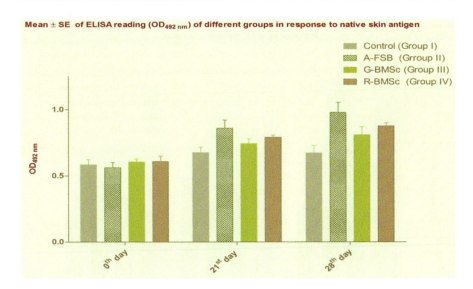

Figure 11.12 Mean ± SE absorbance at 492 nm wavelengths (OD492).

response in group II when compared with control group on day 21 and day 28 post implantation. Similarly, group IV also exhibited a significantly higher ($P < .001$) B-cell response when compared to control group on postimplantation day 28. No significant difference ($P > .05$) exists between remaining groups. The B-cell responses were highest in the group II (implanted with acellular FSB) on day 28 as evidenced by higher absorbance values when compared with other groups [21,33].

11.6.4.2 Lymphocyte proliferation assay

The cell-mediated immune response toward implanted scaffolds in all the experimental animals was assessed by MTT colorimetric assay. The stimulation of rat splenocytes with concanavalin A (Con A) and phytohemagglutinin (PHA) was considered as positive control, whereas unstimulated culture cells were taken as negative control. At day 45 post implantation, the rats from all the four groups were sacrificed and spleens were separated. The splenocyte culture was performed to grow cells in vitro. These spleenocytes were stimulated with native FSB antigen, Con A, and PHA. Mean ± SE stimulation index (SI) values of the control group, rats implanted with acellular FSB, G-BMSC-seeded acellular FSB, and R-BMSC-seeded acellular FSB at 45 days post implantation are presented in Fig. 11.13 [21,33].

Against the native FSB antigen the group II showed significant amount of stimulation ($P < .001$) when compared to groups I, III and IV. As compared to the SI values of Con A and PHA, all the groups exhibited considerable rise in SI values except the control group. The T-cell responses were highest in the group II (implanted with acellular FSB) as evidenced by higher SI values in comparison to control animals [21,33].

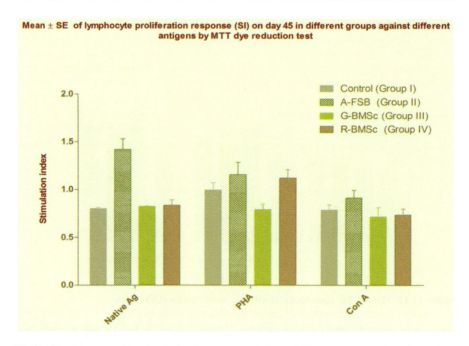

Figure 11.13 Mean ± SE stimulation index (SI) values of different groups at 45 days post implantation.

Biological scaffolds are mainly composed of mammalian ECM and can be used for the reconstruction of various tissues and organs. These scaffolds are typically allogenic or xenogeneic in origin and derived from tissues such as fish swim bladder, dermis, pericardium, diaphragm, small intestine submucosa, etc. Cells in the ECM have classes I and II histocompatibility antigens capable of eliciting rejection reactions. Also, the cells have glycoproteins recognized by the immune system of hosts, which elicit rejection reactions. Therefore, if these substances are eliminated from ECM, rejection reactions can be prevented. Collagens are weakly immunogenic as compared to other proteins. However, complete elimination of alloantigens is considerably difficult to perform and verify [27]. Acellular grafts were less immunogenic having better tolerance by allogenic hosts and equally effective as isograft [45]. The tissues are processed by decellularization and/or crosslinking to remove or mask antigenic epitopes and DNA [9]. Biologic scaffold processing methods plays an important role in determining the host response.

An immune response against nonself and self-antigens is initiated by presentation of the antigen in a suitable form to T cells. Antigen can only be presented to T cells in the context of molecules of MHC [46]. Hence, practically each nucleated cell of the body is able to present antigen, first by virtue of a constitutive MHC class I expression, and second by a de novo expression of MHC class II molecules on the surface of the cell. It is well known that when an antigen enters the body, two different types of immune response may occur. The first type, known as the "humoral immune response," involves the synthesis and release of free antibody into the blood and other body fluids. The second type of immune response, known as the "cell-mediated immune response," involves the production of "sensitized" lymphocytes which are the effectors of this type of immunity.

In the present study, the humoral responses in rats elicited by the cutaneous implanted grafts were determined by using indirect ELISA. The cell-mediated immune response toward cutaneous implanted scaffolds in all the experimental animals was assessed by lymphocyte proliferation assay/MTT colorimetric assay. The ELISA was performed to check the extent of antibody generated toward the graft components. The absorbance values were taken as a measure to compare the magnitude of immune response. Lymphocyte proliferation assay (LPA) measures the ability of lymphocytes placed in short-term tissue culture to undergo a clonal proliferation when stimulated in vitro by a foreign molecule (antigen/mitogen). Th1 lymphocytes produce cytokines such as IL-2, IFN-alpha, and TNF-beta leading to macrophage activation, stimulation of complement fixing Ab isotypes (IgG2a and IgG2b in mice) and differentiation of $CD8^+$ cells to a cytotoxic T cells [47]. Activation of this pathway is associated with both allogeneic and xenogeneic transplant rejection [48]. Th2 lymphocytes produce IL-4, IL-5, IL-6, and IL-10, cytokines that do not activate macrophages and that lead to production of noncomplement fixing Ab isotypes (IgG1 in mice). Activation of the Th2 pathway is associated with transplant acceptance [49]. $CD4^+$ lymphocytes proliferate in response to antigenic peptides in association with class II major histocompatibility complex (MHC) molecules on antigen presenting cells (APCs).

Animals of group II (acellular matrix-implanted group) showed higher immune response as compared to groups I, III, and IV. The least immune response was in animals of group I, where no graft was used. Whatever, immune response present in that group might be due to bacterial and viral contamination in the wound as it was uncovered. More immune response in group II animals might be due to the immunogenic nature of decellularized ECM. Even though the tissue-engineered groups (groups III and IV) also have the same biomaterial as scaffold, they elicit a lesser immune response than group II animals. This might be due to the capacity of mesenchymal stem cells (MSCs) to regulate the immune response.

Three broad mechanisms contribute to this effect. First, mesenchymal stem cells are hypoimmunogenic, often lacking MHCII and costimulatory molecule expression. Second, these stem cells prevent T-cell responses indirectly through modulation of dendritic cells and directly by disrupting NK as well as $CD8^+$ and $CD4^+$ T-cell function. Third, mesenchymal stem cells induce a suppressive local microenvironment through the production of prostaglandins and interleukin-10 as well as by the expression of indoleamine 2,3,-dioxygenase, which depletes the local milieu of tryptophan. In contrary to expectation the group IV animals, treated with allogeneic stem cell (R-BMSC)-seeded graft exhibits a bit more immune response than group III animals, where the stem cell source was xenogeneic (G-BMSC). This might be due to the slow rate of attachment and proliferation of R-BMSC.

11.6.5 Histopathological observations

The biopsy specimen from the implantation site were collected on 3, 7, 14, 21, and 28 days postoperatively. The biopsies were then processed for hematoxylin and eosin staining. The results are presented in Fig. 11.14. The biopsies were also subjected to Masson's trichrome staining for the appreciation of the collagen formation and results are presented in Fig. 11.15.

In group I, on day 3, high degree inflammatory cell infiltration, edema, and congestion were observed. Fibroblast proliferation started and some neovascularization was also observed. On day 7, proliferation of fibroblasts and angioblasts became more and there was partial epithelialization at the margin of wound area covering the granulation tissue. On day 14, severe proliferation of fibroblasts and neovascularization was observed along with inflammatory cell infiltration. On day 21, collagen formation was evident in some areas (immature collagen) and inflammation was greatly reduced. On day 28, a high degree of collagen deposition was observed [20,33].

In group II, inflammatory changes were observed on day 3 postoperatively. Fibroblast proliferation was found in graft tissue and some neovascularization was also observed. There was necrosis and sloughing of the superficial graft. On day 7, the grafted tissue nearer to host tissue was completely infiltrated by proliferating fibroblasts, which were intervened in the matrix. The fibrous tissue also revealed numerous

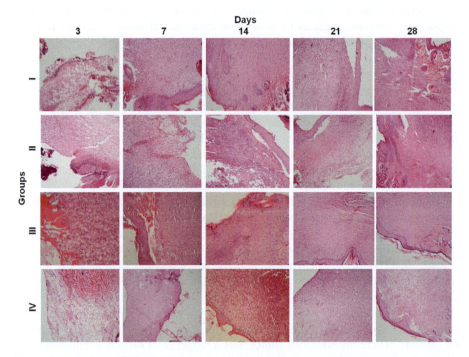

Figure 11.14 Histopathological observations of wound healing of different groups at various time intervals (H&E staining, × 100).

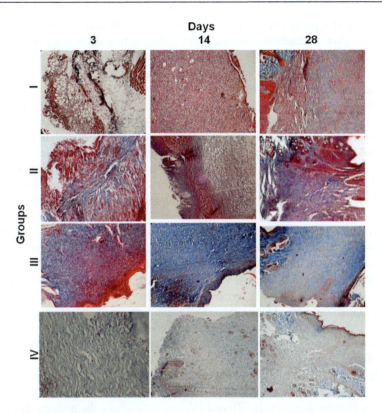

Figure 11.15 Histopathological observations of wound healing of different groups at various time intervals (Masson's trichrome staining, ×100).

blood vessels. Sloughing changes of the superficial graft has become more prominent. On day 14, the deep layer of dermis revealed deposition of new collagen fibers in which fibroblasts were dispersed. By day 21, epithelialization started. Inflammatory changes and neovascularization reduced. At this stage, the grafted tissue was replaced by collagenous connective tissue. On day 28, superficial epithelialization was observed under superficial graft tissue. In the dermis, there was massive deposition of collagen fibers. Hair follicles and skin gland could also be observed [20,33].

In group III, on day 3, severe inflammatory changes were observed. Fibroblast proliferation and neovascularization were found in the graft tissue. Necrosis and sloughing of the superficial graft were observed. By day 7, fibroblast proliferation became more prominent. By day 14, epithelialization started. By day 21, epithelialization was observed in margin along with new collagen deposition in matrix. By day 28, epithelialization was almost completed covering the granulation tissue. Hair follicles and skin glands were also observed [20,33].

In group IV, on day 3, inflammatory changes were seen. Fibroblast proliferation and neovascularization was found in graft tissue. Necrosis and sloughing changes of the superficial graft were observed. By day 7, more cells had started migrating inside the grafts and sloughing changes of the superficial layer became more prominent. On day 14, marginal epithelialization and mild collagen formation were observed. On day 21, the collagen formation was appreciable. Hair follicles and skin glands were seen. By day 28, superficial epithelialization was almost completed and massive deposition of collagen fibers was found [20,33].

Full-thickness skin wound healing occurs by granulation tissue formation, contraction, and epithelialization [50]. Epithelialization occurs by migration of undamaged epidermal cells from the wound margins across the granulation bed [51]. Exogenous collagen supplementation enabled faster migration of cells that are involved in cutaneous wound healing. Since the exogenous collagen is molecular in nature [52] and supplies endogenous collagen in vivo, it readily integrates with the wound tissue and facilitates the attachment, migration, and proliferation of cells on the wound site [53].

Grafts applied in groups II, III, and IV showed necrosis of the superficial graft and underlying graft showed severe fibroblastic proliferation with inflammatory cells. The tissues of groups I exhibited severe proliferation of fibroblast with inflammatory cells in the wound area. Moderate degree of neovascularization was also detected in all the groups. No new collagen formation was detected in any of the groups. But the graft-implanted samples showed the existence of mature collagen of FSB origin. Although inflammation is necessary for healing by fighting infection and inducing the proliferation phase, healing proceeds only after inflammation is controlled [54]. On day 3, minimum score was observed in groups III and IV (18).

On day 7, post implantation, moderate-to-severe inflammation was present in all the groups; it was minimum in groups II, III, and IV. The early control of inflammation, as in case of groups II, III, and IV, might facilitate the progress to the next phase of wound healing. Proliferation of fibroblasts was more in the bioengineered acellular matrices. Similar findings were observed by Perme et al. [55]. On day 14, least scores were observed in groups III (13) followed by group IV (15). Sloughing of the upper layer of graft was observed in all the groups except control group where wound remained open. It may be either due to the desiccation of the graft in high environmental temperature or due to an impaired formation of new blood vessels. Meanwhile, the underlying granulation tissue increased in mass that pushed up the graft upward. The animals of groups II, III, and IV showed well-formed collagen and neovascularization with superficial epithelialization. Epithelialization and neovascularization were faster in group III and IV as compared to other groups. Group I also showed a significant epithelialization. The enhanced rate of wound contraction and significant reduction in healing time might be due to enhanced epithelialization.

On day 21, the least score was observed in group III (10) followed by group IV (12). Epithelialization was more similar to the normal skin in the wounds of these groups. At 21 days of post implantation, the bioengineered groups showed higher collagen synthesis as compared to the control groups and can be appreciated by the Masson's trichrome staining. Purohit et al. [41] on day 21 found that the acellular dermal matrix throughout the width and length was replaced by mature collagenous connective tissue in experimentally created wounds in rabbits.

On day 28 post implantation, the control group healed completely leaving abundant scar tissue but no complete healing was observed for other groups. The collagen fiber arrangement was almost similar to normal skin for acellular group and tissue-engineered group. Marked fibroblastic response associated with an abundant new collagenous fibrous tissue was observed in these groups. The deposition of new collagen was oriented parallel to the skin surface. In acellular group and tissue-engineered groups, hair follicle and skin glands could be detected as in case of normal skin indicating the culmination of repair process. Here the healing was with minimum contraction and scar formation [20,33].

11.7 Evaluation of acellular and crosslinked acellular swim bladder matrix for skin wound healing in rabbit model

11.7.1 Introduction

Collagen-rich matrices prepared after removal of cellular components from native tissues are currently being used to facilitate wound healing and tissue regeneration. Those matrices provide a native framework for cell adhesion at the site of tissue defect and allow local cells to migrate into the matrix and adhere before undergoing differentiation [13]. Collagen-rich matrices are slowly degraded by cellular proteases at the implantation site and are replaced by new endogenous ECM proteins secreted by ingrowing fibroblasts. Furthermore, the matrices stimulate rapid neovascularization during tissue regeneration [56] and are relatively inert immunologically [14]. They are resistant to infections [57–60], a feature that may be attributable to their inherent antimicrobial activity [61] and their ability to rapidly vascularize, therefore clear bacteria. They have been traditionally prepared from native tissues of land-based mammalian species [58–60,62–65]. However, its biomedical uses are associated with a risk of disease transfer [66] and may also carry an ethnocultural stigma. Fish processing wastes such as swim bladder may serve as economical, viable, and safer alternative for collagen-rich matrices to mammalian sources. Moreover, the environmental issue related to pollution from fish wastes can also be addressed [67].

The swim bladder tissues are mainly composed of fibrillar or type I collagen (80%) and multilayered transitional epithelial cells [68]. Previously, antigenic epitopes associated with the cellular elements have been shown to elicit proinflammatory response and overt immune mediated rejection of the tissues [69]. Hence, the removal of antigenic epitopes is necessary to minimize or avoid an adverse immunologic response by xenogeneic recipients [9]. However, even after a complete extraction of cellular proteins, a cross-species response was appreciated after use of xenogeneic acellular tissues [69]. Crosslinking is another technique that is used to reduce the cross-species immune response toward the structural proteins [8,14]. Despite having high collagen content, the swim bladder tissue has not been investigated till date to the best of our knowledge. The wound healing effects of biologic scaffolds are generally evaluated in full-thickness skin animal models including rabbits [64,70,71]. Rabbits possess a subcutaneous panniculus carnosus muscle that contributes to skin wound healing by both contraction and reepithelialization, whereas reepithelialization is the only mechanism of skin wound healing in humans. Despite differences in wound healing pattern with humans, rabbits are often used as full-thickness skin wound model for its ready availability, low costs, ease of handling, and postoperative care [72]. To mimic the skin wound healing pattern in rabbits, panniculus carnosus muscle is generally being excised [64,70,71].

Healing potential of collagen-rich acellular swim bladder matrix (ASBM) in tissue repair and regeneration in rabbit was investigated. We also crosslinked ASBM with epoxy compounds to examine whether it reduces immunogenicity further. Besides, we compared the healing potential between ASBM and 1,4-butanediol diglycidyl ether (BDDGE) crosslinked ASBM-implanted full-thickness skin wounds in a rabbit model.

11.7.2 Materials and methods

The protocols used in this study were approved by the Institute Animal Ethics Committee of the ICAR-Indian Veterinary Research Institute, Izatnagar, Uttar Pradesh, India. The National Institutes of Health "Guide for the Care and use of Laboratory Animals" was followed in all animal experiments. Eighteen New Zealand white rabbits *(Oryctolagus cuniculus)* of either sex (aged 5–6 months, 1.3–2.2 kg body weight) were procured from the Laboratory Animals Resource section of the ICAR-Indian Veterinary Research Institute. Animals were housed individually in rabbit cages maintained at 24°C and 55%–65% humidity with 12–12 hours light-dark cycle. The rabbits had free access to food and water. The animals were acclimatized to laboratory conditions for 10 days before start of experiment.

11.7.3 Preparation of acellular swim bladder matrix and crosslinking with epoxy compounds

Preparation and crosslinking of ASBM from Rohu fish (*Labeo rohita*) were based on previous reports [57,73]. The representative images of Rohu fish and native swim bladder are presented in Fig. 11.16.

Briefly, swim bladder of Rohu fish (a fresh water fish) was collected under clean conditions, transferred to the laboratory in chilled (4°C) sterile phosphate-buffered saline (PBS, pH 7.4) supplemented with 0.1% amikacin and 0.02% ethylene diamine tetra acetic acid, and immediately processed. Swim bladders were cut into 20×20 mm^2 pieces and washed thoroughly in sterile PBS. They were placed in 0.5% sodium deoxycholate detergent and were subsequently agitated in a shaker for 24 hours at 37°C. Prepared ASBM was crosslinked with 1% BDDGE for 48 hours at 37°C, followed by washing with sterile PBS. Crosslinked ASBM was stored in PBS supplemented with 0.1% amikacin at -20°C until further use.

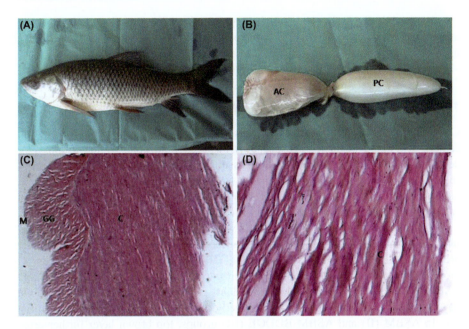

Figure 11.16 (A) Image of Rohu fish (*Labeo rohita*), (B) Native swim bladder, (C) Microphotograph of the acellular fish swim bladder (H&E stain, ×40), (D) Microphotograph of the acellular fish swim bladder (H&E stain, ×200).

11.7.4 Wound creation and implantation

Perioperative antibiotic prophylaxis was provided with intramuscular enrofloxacin (10 mg/kg), and perioperative analgesia was provided with intramuscular meloxicam (0.3 mg/kg) for each rabbit. Wound creation and implantation were performed as described earlier [64]. Briefly, rabbits were anesthetized with an intramuscular injection of xylazine hydrochloride (Indian Immunologicals Ltd, Hyderabad, India) and ketamine hydrochloride (Neon Laboratories Ltd, Thane, India) (10 and 50 mg/kg, respectively). Furthermore, animals were restrained in sternal recumbency and the dorsum (thoracic lumbar region) was prepared for aseptic surgery. Four full-thickness skin wounds (20×20 mm^2 each), two on the right and two on the left side, were created on the dorsum of each rabbit by excising down to the panniculus carnosus. A gap of 20 mm was kept between the two wounds and the wounds were created 20 mm away from the midline on either side of the dorsum. Subsequently, rabbits were randomly divided into three equal groups according to the treatment of the wounds: untreated sham control (I), implanted with double layers of ASBM (II), and ASBM-BDDGE (III). The scaffolds were secured to the edge of the skin wounds with eight simple interrupted 4–0 nylon sutures. After the procedure, the wounds were bandaged with surgical dressing containing 1% chlorhexidine (Gujarat Healthcare, Ahmedabad, India). After recovery from anesthesia, rabbits were housed individually in properly disinfected cages. Enrofloxacin (10 mg/kg intramuscularly once a day) was continued for 5 days and meloxicam (0.3 mg/kg intramuscularly once a day) for 3 days.

11.7.5 Gross observations and planimetry

Color photographs were taken on days 7, 14, 21, and 28 with the help of digital camera at a fixed distance. The shape, irregularity, and color of the lesion were determined. Gross observation of wounds during the entire period of experiment showed no visible suppurative inflammation. None of the animals became sick or died during this experiment. The representative images of wound of one animal in each group at baseline (day 0) and end of healing are presented in Fig. 11.17.

On day 7, in sham (I) group, wounds were covered with soft and fragile pale-pink mass having mildly desiccated top surface. The color of the implant became brown in ASBM (II) and ASBM-BDDGE (III) groups. On day 14, in sham (I) group, a clear-cut thick crust developed on the wound surface. In ASBM (II) and ASBM-BDDGE (III) groups, color of implants remained same as that of day 7. On day 21, in sham (I) group, the crust detached off leaving a raw granular pinkish tissue. In ASBM (II) and ASBM-BDDGE (III) groups, top brown layer further dried and got detached, whereas inner layer of the scaffolds appeared to be integrated well with the wounds, and newly formed epidermis covered the whole surface of the wounds. On day 28, in sham (I) group, the wound healed up completely by severe contraction leaving a large scar, whereas in ASBM (II) and ASBM-BDDGE (III) groups, complete wound healing was observed without contraction.

Visible appearances of wound are reliable parameters in macroscopic evaluation for wound healing. In this study, the color of the implanted matrices changed from white (on day 0) to brown in the course of wound healing and finally dried-up top layer got detached (on day 21) in ASBM (II) and ASBM-BDDGE (III) groups. Detachment of top layer may be due to lack of vascularization and evaporative water loss from the top layer of the grafts. But at the same time, it fulfilled our intention of protecting the underneath layer from desiccation. By the time of detachment of the top layer, underneath layer gets disintegrated and merged with neoformed collagen of wound bed by blocking the contraction, thus serving our purpose. This is consistent with the result of a recent study, in which 1-ethyl-3-(3-dimethylaminopropyl) carbodiimide hydrochloride (EDC) crosslinked acellular small intestinal matrix was used to repair full-thickness skin wounds in rabbits [64]. The bilayer concept of wound coverage in which both epidermal and dermal analogs are used is widely accepted [64,74]. Acellular matrices rapidly stimulate neoangiogenesis during tissue regeneration [56]; immunologically inert [14] and resistant to infections [57–60] are the features that may be attributable to their inherent antimicrobial activity [61].

Fish swim bladder−derived tissue scaffolds 331

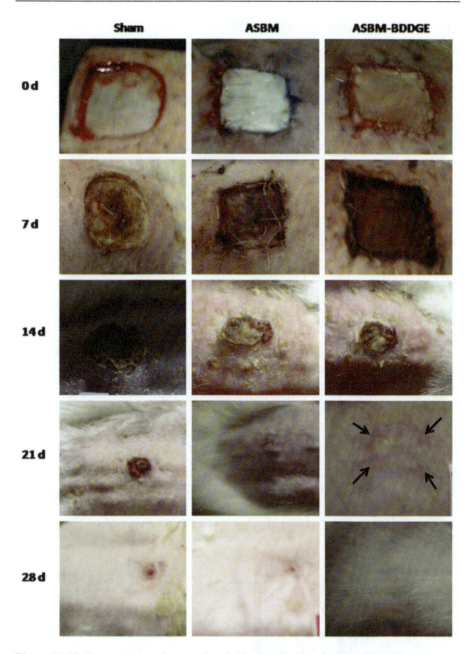

Figure 11.17 Representative photographs of skin wounds of sham (I), ASBM (II), and ASBM-BDDGE (III) groups on days 0, 7, 14, 21, and 28. Arrows indicate repaired skin.

11.7.6 Wound area and contraction

Wound area of all groups was measured by tracing its contour using transparent sheets with graph paper on days 0, 7, 14, 21, and 28 postoperatively. The area (millimeter square) within the boundaries of each tracing was determined planimetrically, and percent contraction was measured by a rate of wound reduction as well as percent reduction in wound area. During the course of wound healing, the mean wound area decreased significantly ($P < .05$) in sham (I) group compared with ASBM (II) and ASBM-BDDGE (III) groups on postoperative days 7 and 14 (Fig. 11.18).

On postoperative day 21, wounds healed completely in ASBM (II) and ASBM-BDDGE (III) groups compared with sham (I) group. Significantly decreased ($P < .05$) wound area was recorded in the sham (I) group compared with ASBM (II) and ASBM-BDDGE (III) groups on postoperative days 7 and 14 (Fig. 11.18). On postoperative days 21, wounds healed completely in ASBM (II) and ASBM-BDDGE (III) groups compared with sham (I) group. Percent wound contraction was significantly ($P < .05$) lesser in ASBM (II) and ASBM-BDDGE (III) groups compared with sham (I) group at all time intervals (Fig. 11.19).

Wound area measurement is frequently used in the clinical and research setting to monitor progress and determine efficacy of treatment. In our study, both the matrices were found to give approximately similar results with regard to wound contraction. As the healing progressed, significantly ($P < .05$) lesser contraction was observed in wounds with implants compared with open wounds. This is in agreement with the findings of previous report [64]. This result shows healing of open wounds through contraction, whereas normal healing of wounds with implants.

11.7.7 Immunological observations

11.7.7.1 Enzyme-linked immunosorbent assay

Rabbit sera, harvested on days 0, 20, 40, and 60, were analyzed by a standard ELISA for antibodies against the implants as per the previous report [64]. Briefly,

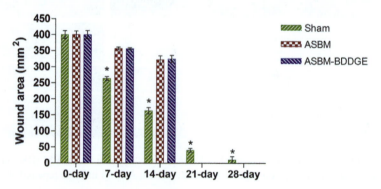

Figure 11.18 Wound area (millimeter square) in sham (I), ASBM (II), and ASBM-BDDGE (III) groups on days 0, 7, 14, 21, and 28. Data are expressed as the mean ± standard deviation of the mean. *$P < .05$ versus other groups on the same day.

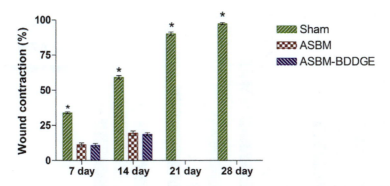

Figure 11.19 Wound contraction (%) in sham (I), ASBM (II), and ASBM-BDDGE (III) groups on days 0, 7, 14, 21, and 28. Data are expressed as the mean ± standard deviation of the mean. *$P < .05$ versus other groups on the same day.

for all sera, a 1:100 dilution was examined against the protein (derived from implant material). The secondary antibody used was a 1:30,000 dilution of an affinity-purified goat antirabbit immunoglobulin G (IgG) coupled to horseradish peroxidase, using o-phenylene diamine as substrate. The absorbance at 492 nm was determined with an automated ELISA plate reader (ECIL, Hyderabad, India). The antibodies titer present in serum samples harvested on day 0 was taken as basal values.

In sham (I) group, total IgG response was unchanged ($P > .05$) on days 20, 40, and 60 compared with day 0 value. In ASBM (II) and ASBM-BDDGE (III) groups, total IgG response was increased significantly ($P < .05$) on postimplantation days 20 and 40 compared with day 0 value. Thereafter, the values reached to normal on postimplantation day 60 compared with day 0 values (Fig. 11.20).

On days 20 and 40, total IgG response was increased significantly ($P < .05$) in ASBM (II) and ASBM-BDDGE (III) groups compared with sham (I) group. However, total IgG response was decreased significantly ($P < .05$) in the ASBM-BDDGE (III) group compared with ASBM (II) group. On days 60, total IgG response remained unchanged ($P > .05$) in the ASBM (II) group compared with sham (I) and ASBM-BDDGE (III) groups.

Antibody responses against both collagen-rich matrix of fish origin were assessed in New Zealand white rabbits. In this study, the total IgG response in ASBM-BDDGE (III) rabbit sera was decreased significantly ($P < .05$) compared with ASBM (II) on postimplantation days 20 and 40 as evidenced in ELISA assay. Reduced total IgG response in sera was also observed in rabbits with EDC crosslinked acellular small intestinal matrix implant [64]. Similarly, a reduced total IgG response was also observed in sera of rabbits with subcutaneously implanted glutaraldehyde crosslinked bovine pericardium [63]. Reduced inflammatory response toward EDC crosslinked acellular aortic matrix implant was also reported in equine study [75]. Additionally, no adverse immunologic reactions were observed in rabbits during entire study periods. This study further confirmed reduced humoral immune response toward crosslinked collagen-rich matrix. Crosslinking has been

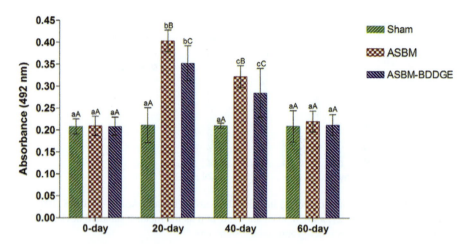

Figure 11.20 Enzyme-linked immunosorbent assay reading (absorbance at 492 nm) of sera from rabbits of sham (I), ASBM (II), and ASBM-BDDGE (III) groups on days 0, 20, 40, and 60. Data are expressed as the mean ± standard deviation of the mean. Capital superscripts (alphabets) indicate level of significance ($P < .05$) among the groups on the same day. Small superscripts (alphabets) indicate level of significance ($P < .05$) within the group on different day.

suggested as a technique to reduce antigenicity of collagen-rich matrices [8]. The Rohu fish collagen is regarded as nonimmunogenic [67]. However, its denaturation or degeneration may lead to exposure of antigenic determinants, which can elicit a severe host immune response [76].

11.7.7.2 Lymphocyte proliferation evaluation

Lymphocyte proliferation evaluation was done using 3-(4,5-dimethylthiazol-2-yl)-2,5-diphenyltetrazolium bromide assay. Blood collected from rabbits with implants (ASBM and ASBM-BDDGE) on days 0, 21, and 42 was used for lymphocyte proliferation. Furthermore, proliferated lymphocytes were measured using 3-(4,5-dimethylthiazol-2-yl)-2,5-diphenyltetrazolium bromide (MTT) colorimetric assay as per the previous report. Briefly, 2 mL of blood was aseptically collected in a heparinized tube from ear vein of each rabbit and mixed with equal volume of $1 \times$ PBS. It was carefully layered over 4 mL of lymphocyte separation medium (Histopaque 1077) and centrifuged at 2000 rpm for 30 minutes. The buffy coat (lymphocyte) was harvested and washed twice with $1 \times$ PBS at 1000 rpm for 5 minutes. Supernatant was removed and cell pellet was resuspended in Rosewell Park Memorial Institute (RPMI) 1640 medium supplemented with 10% fetal bovine serum. The cells were adjusted to a concentration of 2×10^6 viable cells per milliliter in RPMI 1640 growth medium supplemented with 10% fetal bovine serum and seeded in a 96-well tissue culture plate (100 μg/well). The cells were incubated at 37°C in 5% CO_2 environment. Lymphocytes from each rabbit were stimulated with antigens (native swim bladder and ASBM) (10–20 μg/mL) and phytohemagglutinin (10 μg g/mL), a T-cell mitogen in triplicates, and three wells were left unstimulated for each sample. Forty microliters of MTT solution (5 mg/mL) were added to each well, and cultures were incubated for an additional 4 hours at 37°C. For the viability assay, the formazan product was dissolved in 1.5 mL dimethyl sulfoxide and the absorbance at 570 nm was determined in each well with a 96-well plate reader. The growth of the stimulated lymphocyte was compared with that of unstimulated cells. The stimulation index (SI) was calculated as absorbance of stimulated cultures divided by the absorbance of unstimulated cultures. The SI values on postimplantation days 21 and 42 were found significantly ($P < .05$) lesser in ASBM-BDDGE (III) group compared with ASBM (II) group when stimulated with both native and acellular antigens as detected by MTT colorimetric assay (Fig. 11.21).

The lymphocyte proliferation assay is a technique to determine T-lymphocyte functions in vitro. CD4þ lymphocytes proliferate in response to antigenic peptides in association with class II major histocompatibility complex (MHC) molecules on antigen presenting cells. The T-cell subsets (Th1 and Th2) generated by naive T cell on MHC antigen stimulation play a major role in graft rejection through activity of different sets of cytokines that activate macrophages and B cells. Cells in ECM have classes I and II histocompatibility antigens capable of eliciting rejection reactions. Also, the cells have glycoproteins recognized by the immune system of hosts, which elicit rejection reactions. Therefore, if these substances are eliminated from ECM, rejection reactions can be prevented. However, complete elimination of all antigens is considerably difficult to perform and verify [27]. In this study, SI values in ASBM-BDDGE (III) rabbit blood on postimplantation days 21 and 42 were found significantly ($P < .05$) lesser compared with ASBM (II) when stimulated with both native and acellular antigens as detected by MTT colorimetric assay. Additionally, on postimplantation days 21 and 42, SI values were maximal

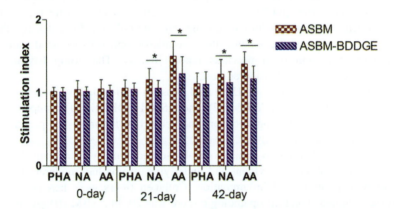

Figure 11.21 Stimulation index values (mean ± standard deviation of the mean) of rabbits peripheral blood lymphocytes in ASBM (II) and ASBM-BDDGE (III) groups on days 0, 21, and 42. *$P < .05$ versus other groups on the same day. AA, acellular antigens; ASBM, acellular swim bladder matrix; BDDGE, 1,4-butanediol diglycidyl ether; NA, native antigens; PHA, phytohemagglutinin.

in ASBM (II) rabbit blood when it was in vitro cultured and stimulated with acellular antigen. The greater ability of acellular antigen to stimulate the lymphocytes in vitro may be attributed to the biological detergent (sodium deoxycholate) treatment. After detergent treatment, the bonds between protein molecules were broken and resulted into a change from quaternary and tertiary structure to primary and secondary structures.

Therefore, the acellular antigen had greater ability to trigger cell-mediated immune (CMI) response in host because of the presence of shorter peptide fragments, which can be presented to the immune system by MHC class II pathway and stimulate the CD4þ lymphocytes, whereas the crosslinked tissue was not processed in the body to form shorter immunogenic fragments, which can elicit the CMI in host. This may also be because of the fact that on crosslinking tissues with different chemicals, the site where biological enzymes act in vivo are masked and the crosslinked tissue is no longer broken down into smaller peptide fragments to elicit host immune response. Crosslinking of the collagen-rich ASBM with BDDGE might have masked immunogenic epitopes and causing reduced CMI in host body. Crosslinking may prove effective for lowering immunogenicity by altering the display of antigenic determinants [77].

11.7.8 Histological observations

Tissue biopsies from healing wounds in each treatment group were collected on days 7, 14, 21, and 28. The tissue specimens were then fixed in 10% neutral-buffered formalin solution and processed for paraffin embedding. Sections were cut to a thickness of 5 micron, deparaffinized, and stained with H&E following established procedures. Epithelialization, inflammation, fibroplasia, and neovascularization in the H&E sections were microscopically evaluated using a histologic scoring system described earlier [64,74]. The duplicate sections from each treatment group also underwent Masson's trichrome staining, and the collagen fiber density, thickness, and arrangement at the healing site were microscopically evaluated using histologic scoring system of a previous report [64,74]. The group having least histopathologic score was considered best.

The representative images of H&E and Masson's trichrome-stained histologic wound sections of sham (I), ASBM (II), and ASBM-BDDGE (III) groups on days 7, 14, 21, and 28 ($\times 40$ magnification and scale bar 100 mm) are presented in Figs. 11.22 and 11.23, respectively.

Healing of wounded tissues is histologically scored. On day 7, in sham (I) group, partial epithelialization was evident and underlying stroma had moderate fibroplasia, neovascularization, and inflammation. Collagen fibers were less dense, thin, and worse arranged. Total histopathologic score was 20. In ASBM (II) group, histologic changes are similar to the sham (I) group except mild neovascularization and fibroplasia were observed and the total histopathologic score was 18. In ASBM-BDDGE (III) group, partial epithelialization was evident and underlying stroma had mild fibroplasia, neovascularization, and mild inflammation. Collagen fibers were dense, thin, and better arranged. Total histopathologic score was 15. On day 14, in

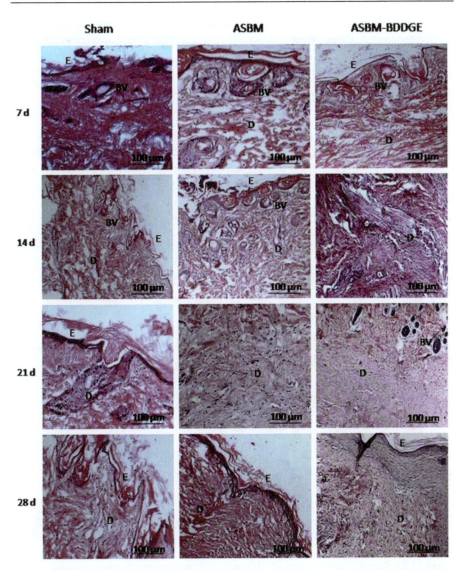

Figure 11.22 Histologic images of hematoxylin and eosin-stained histologic wound sections of sham (I), ASBM (II), and ASBM-BDDGE (III) groups on days 7, 14, 21, and 28 (340 magnification and scale bar 100 mm). BV, blood vessels; D, dermis; E, epithelial layer; I, inflammatory cells.

sham (I) group, partial epithelialization was observed and dermis had mild neovascularization and inflammation. Moderate fibroplasia was appreciated and collagen fibers were less dense, thin, and better arranged. Total histologic score was 17.

In ASBM (II) group, partial epithelialization was present. Dermis had mild fibroplasia, neovascularization, and inflammation. Collagen fibers were denser, thick,

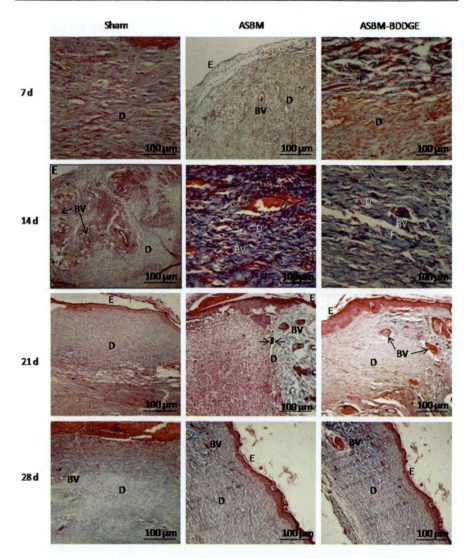

Figure 11.23 Representative images of Masson's trichrome-stained histologic wound sections of sham (I), ASBM (II), and ASBM-BDDGE (III) groups on days 7, 14, 21, and 28 (340 magnification and scale bar 100 mm). BV, blood vessels; D, dermis; E, epithelial layer; F, fibroblasts.

and better arranged. Histopathologic score was 14 at this stage. In ASBM-BDDGE (III) group, histologic changes were similar to the ASBM (II) group except increased collagen fibers density, and the histopathologic score was 13 at this stage. On days 21, epithelialization was complete and inflammation was absent in all the groups. In sham (I) group, underlying epidermis, mild fibroplasias and neovascularization were observed. Collagen fibers were more dense, thick, and better arranged.

Histopathologic score was 13. In ASBM (II) and ASBM (III) groups, fibroplasia and neovascularization resembled normal skin. Collagen fibers were denser, thick, and best arranged. Wound healing was complete without regeneration of adnexal elements. The total histologic scores were further reduced to 8 in both groups. On days 28, in sham (I) group, epithelialization was complete and inflammation was absent. Fibroplasia resembled normal skin with mild neovascularization. Stroma had dense, thick, and better arranged collagen fibers. The dermis was with normal adnexa and the histologic score was further reduced to 11. In ASBM (II) and ASBM (III) groups, healing was complete and histologic scores were 8. Neovascularization that resembled normal skin and collagen fibers was best arranged. The histopathologic score was 8 at this stage. ASBM (III) group showed lowest histopathologic scores on postimplantation days 7 and 14. Thereafter, in ASBM (II) and ASBM (III) groups, the total histopathologic scores remained same (8) on postimplantation days 21 and 28.

Implantation of ASBM and ASBM-BDDGE on the full-thickness skin wounds in rabbits provided better epithelialization, improved blood vessels supply, and collagenization on all postimplantation days compared with open wounds. The advantages of the use of ECM on full-thickness skin wounds are well established in many experimental models [64,74]. In our study, BDDGE crosslinked ASBM from Rohu fish promoted the biological activities that resulted in faster healing of full-thickness skin wounds in rabbits. In this study, BDDGE crosslinked ASBM provided a significant decrease in the severity of inflammatory response. This activity might be related to the antigen masking activity of the BDDGE on lymphocyte proliferation. The BDDGE crosslinked scaffolds have been shown in experimental trial to be well tolerated [78]. In ASBM (II) and ASBM-BDDGE (III) groups, the epithelialization, neovascularization, and collagen fiber arrangement and its density were resembled to that of normal skin by day 21, whereas the sham (I) group performed suboptimally in both gross and histologic grades of healing in comparison with both the treatment groups.

The healing of the injured mammalian skin is an extremely complex and dynamic process involving a series of sequential and overlapping phases, including inflammation, proliferation, and remodeling. The process requires interactions between a variety of cell types, multiple cytokines, growth factors, and ECM. The spontaneous healing of full-thickness skin wounds takes a long time and frequently results in scar formation, which becomes clinically apparent as hypertrophy, poor skin elasticity, and undesirable cosmetic appearance [79]. Hence, dermal regeneration is essential to improve the functional and cosmetic outcomes in the healing of full-thickness skin defects. Allogenic or xenogeneic collagen-rich ECM is currently used for wound healing and tissue regeneration. Their role in tissue regeneration is to provide the microarchitecture required for cellular infill and mechanical support to maintain tissue integrity during regeneration process [80]. In this study, we have prepared a collagen-rich ECM from swim bladder of Rohu fish using an ionic biological detergent (sodium deoxycholate) and chemically crosslinked with BDDGE [57,73]. A cell-free intact structural framework of collagen remained after chemical decellularization of fish swim bladder. Chemical processing can lead to matrix denaturation, which further leads to exposure of antigenic determinants [76]. Immunogenicity of the matrix that resulted from exposed antigenic determinants can be overcome by chemical crosslinking [8]. Both the matrices (ASBM and ASBM-BDDGE) were used for the repair of full-thickness skin wounds in rabbits.

11.8 Conclusion

The protocol for decellularization of fish swim bladder was found to be less toxic for the mesenchymal stem cells to grow and proliferate. Protocols for the collection and in vitro culture of bone marrow—derived mesenchymal stem like cells of rat and goat origin have been standardized in the laboratory. Acellular fish swim bladder itself has wound healing potency; however, seeding with bone marrow—derived mesenchymal stem like cells increases the wound healing potency. Acellular fish swim bladder seeded with R-BMSC and G-BMSC found to be novel bioengineered biomaterials in full-thickness skin repair. Acellular fish swim bladder matrix seeded with G-BMSC has an edge over R-BMSC in full-thickness skin wound repair. Preservative technique should be determined or evaluated for both stem cells and biomaterial construct so that it can be commercialized. Clinical trials should be attempted in future. The total IgG response in sera of rabbits with ASBM-BDDGE implant was minimal compared with ASBM as evidenced in ELISA. The SI value of blood lymphocytes of rabbits with ASBM-BDDGE implant was minimal for more than 42 days as detected by MTT assay. Implantation of ASBM and ASBM-BDDGE provided better epithelialization, improved neovascularization, and collagenization of full-thickness skin wounds in rabbits. Altogether, present findings indicate that BDDGE crosslinked ASBM derived from Rohu fish has potential for the clinical applications. Being of fish origin, it is expected that their clinical applications will not be limited by ethnocultural stigma. Further studies are warranted to identify various growth factors present within the material and dissect their bioactivity

References

[1] Lanza R, Klimanskaya I. Essential stem cell methods: tissue engineering. 1st edn Burlington: Academic Press; 2009. p. 282—313.
[2] Dahms SE, Piechota HJ, Dahiya R, Lue TF, Tanagho EA. Composition and biomechanical properties of the bladder acellular matrix graft: comparative analysis in rat, pig and human. Br J Urol, 82. 1998. p. 411—9.
[3] Piechota HJ, Dahms SE, Nunes LS, Dahiya R, Lue TF, Tanagho EA. In vitro functional properties of the rat bladder regenerated by the bladder acellular matrix graft. J Urol 1998;159:1717—24.
[4] Yoo JJ, Meng J, Oberpenning F, Atala A. Bladder augmentation using allogenic bladder submucosa seeded with cells. Urology 1998;51:221—5.

[5] Chen F, Yoo JJ, Atala A. Acellular collagen matrix as a possible "off the shelf" biomaterial for urethral repair. Urology 1999;54:407−10.
[6] Van der Laan JS, Lopez GP, Van PB, Nieuwenhuis P, Ratner BD, Bleichrodt RP, et al. TFE plasma polymerized dermal sheep collagen for the repair of abdominal wall defects. Int J Artif Org 1991;14:661−6.
[7] Maiti SK, Hoque M, Kalicharan KN, Singh GR. Evaluation of a new absorbable suture fish-gut. Indian J Anim Sci 2001;71:352−4.
[8] Schmidt CE, Baier JM. Acellular vascular tissue: natural biomaterials for tissue repair and tissue engineering. Biomaterials 2000;21:2215−31.
[9] Gilbert TW, Sellaroa TL, Badylak SF. Decellularization of tissues and organs. Biomaterials 2006;27:3675−83.
[10] Bianchi ME. DAMPs, PAMPs and alarmins: all we need to know about danger. J Leukoc Biol 2007;81:1−5.
[11] Lotze MT. Damage-associated molecular pattern molecules. Clin Immunol 2007;124:1−4.
[12] Brown-Etris MRN, Cutshall WD, Hiles MC. A new biomaterial derived from small intestine submucosa and developed into a wound matrix device. Wounds 2002;14:150−66.
[13] Badylak SF. The extracellular matrix as a biologic scaffold material. Biomaterials 2007;28:3587−93.
[14] Badylak SF, Gilbert TW. Immune response to biologic scaffold materials. Semin Immunol 2008;20:109−16.
[15] Badylak SF, Freytes DO, Gilbert TW. Extracellular matrix as a biological scaffold material: Structure and function. Acta Biomat 2009;5:1−13.
[16] Oberpenning F, Meng J, Yoo JJ, Atala A. De novo reconstitution of a functional mammalian urinary bladder by tissue engineering. Nat Biotechnol 1999;17:149−55.
[17] Atala A, Lanza RP. Methods of tissue engineering. San Diego: Academic Press; 2001. p. 123.
[18] Atala A. Bladder regeneration by tissue engineering. Br J Urol 2001;88:765−70.
[19] Schultz SS, Abraham S, Lucas PA. Stem cells isolated from adult rat muscle differentiate across all three dermal lineages. Wound Repair Regen 2006;4:224−31.
[20] Remya V. Tissue engineered fish swim bladder scaffold seeded with bone marrow derived mesenchymal stem cells of rat and goat for dermal reconstruction [MVSc thesis]. Izatnagar, Uttar Pradesh, India: Deemed University, Indian Veterinary Research Institute; 2012.
[21] Remya V, Kumar N. Immunomodulatory potential of bone marrow derived mesenchymal stem cells against a biological graft. Int J Sci Res 2014;3(7):38−40.
[22] Courtman DW, Pereira CA, Kashef V, McComb D, Lee JM, Wilson GJ. Development of a pericardial acellular matrix biomaterial-biochemical and mechanical effects of cell extraction. J Biomed Mater Res 1994;28:655−66.
[23] Jorge-Herrero E, Fernandez P, Turnay J, Olmo N, Calero P, Garcia R, et al. Influence of different chemical crosslinking treatments on the properties of bovine pericardium and collagen. Biomaterials 1991;20:539−45.
[24] Sacks MS, Gloeckner DC. Quantification of the fiber architecture and biaxial mechanical behaviour of porcine intestinal submucosa. J Biomed Mater Res 1999;46:1−10.
[25] Voytik-Harbin SL, Brightman AO, Waisner BZ, Robinson JP, Lamar CH. A tissue-derived extracellular matrix that promote tissue growth and differentiation of cells invitro. Tissue Eng 1998;4:157−74.
[26] Wilson GJ, Yeger H, Klemet P, Lee JM, Courtman DW. Acellular matrix: a biomaterials approach for coronary artery bypass and replacement. Ann Thorac Surg 1990;60:353−9.

[27] Malone JM, Brendel K, Duhamil RC, Reinert RL. Detergent-extracted small diameter vascular prostheses. J Vasc Surg 1984;1:181−91.
[28] Parien JL, Kim BS, Atala A. In-vitro and in-vivo biocompatibility assessment of natural derived and polymeric biomaterials using normal human urothelial cells. J Biomed Mat Res 2001;55:33−9.
[29] Coito AJ, Kupiec-Weglinsky JW. Extracellular matrix protein by standers or active participants in the allograft rejection cascade? Ann Transpl 1996;1:14−18.
[30] Seddon AM, Curnow P, Booth PJ. Membrane proteins, lipids and detergents: not just a soap opera. Biochem Biophys Acta2004 1666;105−17.
[31] Kumar V. Acellular buffalo small intestinal submucosa and fish swim bladder for the repair of full-thickness skin wounds in rabbits [MVSc thesis]. Izatnagar, Uttar Pradesh, India: Deemed University, Indian Veterinary Research Institute; 2010.
[32] Remya V, Kumar N, Kutty MVH. A method for cell culture and RNA extraction of rabbit bone marrow derived mesenchymal stem cells. Int J Sci Res 2014;3(7):31−3.
[33] Remya V, Kumar N, Sharma AK, Sonal, Negi M, Maiti SK, et al. Acellular fish swim bladder biomaterial construct seeded with mesenchymal stem cells for full thickness skin wound healing in rats. Trends Biomat Artific Organs 2014;28(4):127−35.
[34] Lovitt CJ, Shelper TB, Avery VM. Advanced cell culture techniques for cancer drug discovery. Biology 2014;3(2):345−67.
[35] Xiaojun Z, Zhihong D, Hailun Y, Weihua G, Yuan L, Dandan MM, et al. Expansion of delivery of human fibroblasts on micronized acellular dermal matrix for skin regeneration. Biomaterials 2009;30:2666−74.
[36] Chun MD, Yan Z, Yan NH, Wen ST. Perfusion seeding of collagen-chitosan sponges for dermal tissue engineering. Process Biochem 2008;43:287−96.
[37] Neagos D, Mitran V, Chiracu G, Ciubar R, Iancu C, Stan C, et al. Skin wound healing in a free-floating fibroblast populated collagen latice model. Romanian J Biophys 2006;16:157−68.
[38] Schallberger SP, Stanley BJ, Hauptman JG, Steficek BA. Effect of porcine small intestinal submucosa on acute full-thickness wounds in dogs. Vet Surg 2008;37:515−24.
[39] Li B, Wang JHC. Fibroblasts and myofibroblasts in wound healing: Force generation and measurement. J Tissue Viability 2011;20:108−20.
[40] Purohit S. Biocompatibility testing of acellular dermal grafts in a rabbit model: an invitro and In-vivo study [PhD thesis]. Izatnagar, Uttar Pradesh, India: Deemed University Indian Veterinary Research Institute; 2008.
[41] Purohit S, Kumar N, Sharma AK, Maiti SK, Shrivastava S, Saxena S, et al. Fibroblast therapy for early healing of full thickness skin defects in mice: a new approach. Trends Biomat. Artif Organs 2019;33(3):69−76.
[42] Kaarthick D.T. Repair of cutaneous wound using acellular pericardium and diaphragm of buffalo origin seeded with mouse embryo fibroblast cells in a rat model [MVSc thesis]. Izatnagar, Bareilly, Uttar Pradesh, India: Deemed University, Indian Veterinary Research Institute; 2012.
[43] Kaarthick DT, Sharma AK, Kumar N, Kumar V, Gangwar AK, Maiti SK, et al. Accelerating full-thickness dermal wound healing using primary mouse embryonic fibroblasts seeded bubaline acellular diaphragm matrix. Trends Biomat. Artif Organs 2017;33(1):16−23.
[44] Lowry OH, Rosebrough NJ, Farr AL, Randall RJ. Protein measurement with the folin phenol reagent. J Biol Chem 1951;193:265−75.
[45] Gulati AK, Cole GP. Nerve graft immunogenicity as a factor determining axonal regeneration in the rat. J Neurosurg 1990;72:114−22.

[46] Townsend A, Bodmer H. Antigen recognition by class I restricted T lymphocytes. Annu Rev Immunol 1989;7:601−24.
[47] Abbas AK, Murphy KM, Sher A. Functional diversity of helper T-lymphocytes. Nature 1996;383:787−93.
[48] Chen N, Gao Q, Field EH. Prevention of Th1 response is critical for tolerance. Transplantation 1996;61:1076−83.
[49] Piccotti JR, Chan SY, Van Buskirk AM, Eichwald EJ, Bishop DK. Are Th2 helper T-lymphocytes beneficial, deleterious, or irrelevant in promoting allograft survival? Transplantation 1997;63:619−24.
[50] Fossum TW, Hedlund CS, Johnson AL, Schulz KS, Seim HB, Willard MD, et al. Surgery of the integumentary system. Manual of small animal surgery. Mosby; 2007. p. 159−75.
[51] Swaim SF, Henderson RA. Wound management. Small animal wound management. Lea and Febiger; 1990. p. 9−33.
[52] Nithya M, Suguna L, Rose C. The effect of nerve growth factor on the early responses during the process of wound healing. Biochemica et Biophysica Acta 2003;1620:25−31.
[53] Judith R, Nithya M, Rose C, Mandal AB. Application of a PDGF containing novel gel for cutaneous wound healing. Life Sci 2010;87:1−8.
[54] Midwood KS, Williams LV, Schwarzbauer JE. Tissue repair and the dynamics of the extracellular matrix. Int J Biochem Cell Biol 2004;36:1031−7.
[55] Perme H, Sharma AK, Kumar N, Singh H, Dewangan R, Maiti SK. In-vitro biocompatibility evaluation of crosslinked cellular and acellular bovine pericardium. Trends Biomater Artif Organs 2009;23:66−75.
[56] Irvine SM, Cayzer J, Todd EM, Lun S, Floden EW, Negron L, et al. Quantification of in vitro and in vivo angiogenesis stimulated by ovine forestomach matrix biomaterial. Biomaterial 2011;32(27):6351−61.
[57] Kumar V, Devarathnam J, Gangwar AK, Kumar N, Sharma AK, Pawde AM, et al. Use of acellular aortic matrix for reconstruction of abdominal hernias in buffaloes. Vet Rec 2012;170:392−6.
[58] Kumar V, Kumar N, Mathew DD, Gangwar AK, Saxena AC, Remiya V. Repair of abdominal wall hernias using acellular dermal matrix in goats. J Appl Anim Res 2013;41:117−21.
[59] Kumar V, Gangwar AK, Mathew DD, Ahmad RA, Saxena AC, Kumar N. Acellular dermal matrix for surgical repair of ventral hernia in horses. J Equine Vet Sci 2013;33:238−41.
[60] Kumar V, Kumar N, Gangwar AK, Saxena AC. Using acellular aortic matrix to repair umbilical hernias of calves. Aust Vet J 2013;91:251−4.
[61] Sarikaya A, Record R, Wu CC, Tullius B, Badylak S, Ladisch M. Antimicrobial activity associated with extracellular matrices. Tissue Eng 2002;8:63−9.
[62] Singh J, Kumar N, Sharma AK, Maiti SK, Goswami TK, Sharma AK. Acellular biomaterials of porcine origin for the reconstruction of abdominal wall defects in rabbits. Trends Biomat. Artif Organs 2008;22:33−8.
[63] Singh H, Kumar N, Sharma AK, Kataria M, Munjal A, Kumar A, et al. Activity of MMP-9 after repair of abdominalwall defectswith acellular and crosslinked bovine pericardiumin rabbit. Int Wound J 2012;11:5. Available from: https://doi.org/10.1111/j.1742-481X.2012.01031.x.
[64] Kumar V, Kumar N, Gangwar AK, Singh H. Comparison of acellular small intestinal matrix (ASIM) and 1-ethyl-3-(3-dimethylaminopropyl) carbodiimide cross linked

ASIM (ASIM-EDC) for repair of full-thickness skin wounds in rabbits. Wound Med 2015;2015(7):24–33. Available from: https://doi.org/10.1016/j.wndm.2015.01.001.
[65] Kumar V, Gangwar AK, Kumar N, Singh H. Use of the bubaline acellular diaphragm matrix for umbilical hernioplasty in pigs. Vet Arh 2015;85:49–54.
[66] Trevitt CR, Singh PN. Variant Creutzfeldt-Jakob disease: pathology, epidemiology, and public health implications. Am J Clin Nutr 2003;78:651S.
[67] Pati F, Datta P, Adhikari B, Dhara S, Ghosh K, Das Mohapatra PK. Collagen scaffolds derived from fresh water fish origin and their biocompatibility. J Biomed Mater Res A 2012;100:1068–75.
[68] Rose C, Mandal AB, Joseph KT. Characterization of collagen from the swim bladder of catfish (*Tachysurus maculates*). Asian. Fish Sci 1998;1998(11):1–17.
[69] Gock H, Murray-Segal L, Salvaris E, Cowan P, D'Apice AJ. Allogeneic sensitization is more effective than xenogeneic sensitization in eliciting Gal-mediated skin graft rejection. Transplantation 2004;77:751–8.
[70] Revi D, Vineetha VP, Muhamed J, Surendran GC, Rajan A, Anilkumar TV. Porcine cholecyst derived scaffold promotes full-thickness wound healing in rabbit. J Tissue Eng 2013;. Available from: https://doi.org/10.1177/2041731413518060.
[71] Revi D, Vineetha VP, Muhamed J, Surendran GC, Rajan A, Kumary TV, et al. Wound healing potential of scaffolds prepared from porcine jejunum and urinary bladder by a non-detergent/enzymatic method. J Biomater Appl 2014;29:1218. Available from: http://doi.org/10.1177/0885328214560218.
[72] Lemo N, Marignac G, Reyes-Gomez E, Lilin T, Crosaz O, Dohan Ehrenfest DM. Cutaneous epithelialization and wound contraction after skin biopsies in rabbits: a mathematical model for healing and remodeling index. Vet Arh 2010;80:637.
[73] Kumar V, Kumar N, Singh H, Gangwar AK, Dewangan R, Kumar A, et al. Effects of crosslinking treatments on the physical properties of acellular fish swim bladder. Trends Biomater. Artif Organs 2013;27:93–101.
[74] Gangwar AK, Kumar N, Sharma AK, Devi SKH, Negi M, Shrivastava S, et al. Bioengineered acellular dermal matrix for the repair of full thickness skin wounds in rats. Trends Biomat Artif Organs 2013;27(2):67–80.
[75] Kumar V, Kumar N, Singh H, Mathew DD, Singh K, Ahmad RA. An acellular aortic matrix of buffalo origin crosslinked with 1-ethyl-3-3-Dimethyl amino propyl carbodiimide hydrochloride for the repair of inguinal hernia in horses. Equine Vet Edu 2013;25:398–402.
[76] Chevallay B, Herbage D. Collagen-based biomaterials as 3-D scaffold for cell cultures: Applications for tissue engineering and gene therapy. Med Biol Eng Comput 2000;38:211–17.
[77] Yannas IV. Natural materials. In: Ratner BD, Hoffman AS, Schoen FJ, Lemons JE, editors. Biomaterial science. San Diego: Academic Press; 1996. p. 84–9.
[78] Devarathnam J, Sharma AK, Kumar N, Shrivastava S, Sonal B, Rai RB. In vivo biocompatibility determination of acellular aortic matrix of buffalo origin. Prog Biomater 2014;3:115. Available from: http://doi.org/10.1007/s40204-014-0027-6.
[79] Gurtner GC, Werner S, Barrandon Y, Longaker MT. Wound repair and regeneration. Nature 2008;453:314–19.
[80] Lun S, Irvine SM, Johnson KD, Fisher NJ, Floden EW, Negron L, et al. A functional extracellular matrix biomaterial derived from ovine forestomach. Biomaterials 2010;31 (16):4517–29.

Stem cell loading multiwalled carbon nanotubes-based bioactive scaffold for peripheral nerve regeneration in a rat model

Mamta Mishra[1], Merlin Mamachan[2], Manish Arya[1], Swapan Kumar Maiti[2] and Naveen Kumar[2,*]

[1]Department of Veterinary Surgery and Radiology, College of Veterinary and Animal Sciences, Bihar Animal Sciences University, Kishanganj, Bihar, India, [2]Division of Surgery, ICAR-Indian Veterinary Research Institute, Izatnagar, Uttar Pradesh, India

12.1 Introduction

All body functions are activated, synchronized, and controlled by a substantial, complex network, the nervous system. The brain, spinal cord, and cranial and spinal nerves, respectively, comprise the central nervous system (CNS) and peripheral nervous system (PNS). Unlike CNS, the PNS is neither protected by bone covering nor lined by the blood-brain barrier (BBB), being much more susceptible to traumatisms and to destructive substances [1]. Peripheral nerve injury (PNI) is one of the paramount pathologies encountered in animals leading to partial or total loss of motor, sensory, and autonomic functions transmitted by the injured nerves to the denervated segments of the body, due to interference with the axonal continuity, neuronal degeneration distal to the site of injury, and ultimate apoptosis of axotomized neurons [2].

PNI cataloging is actually based upon the severity of damage in nerve layers. Seddon first classified peripheral nerve injury (PNI) into three types, mild degree neurapraxia, moderate degree axonotmesis, and severe degree neurotmesis, based upon the severity of tissue injury, prognosis, and time to recovery [3]. Further, expansion of this classification was done by Sunderland who further stratified Seddon's three categories into five categories according to severity of the injury. First-degree injury is equivalent to Seddon's neurapraxia. The distinction between Seddon's axonotmesis and second to fourth degree traumas is the degree of mesenchymal nerve injury. Seddon's neurotmesis is the same as fifth-degree damage [4,5].

The formation of new axons, neurons, glia, myelin, and synapses is related with neuroregeneration, which is the regrowth, restoration, or repair of deteriorated nerves and nervous tissues. PNS has the innate capacity for auto repair and regeneration but this capacity is not

---

*Present affiliation: Veterinary Clinical Complex, Apollo College of Veterinary Medicine, Jaipur, Rajasthan, India.

efficient as desirable, while CNS is unable to auto regulate self-repairment and regeneration [6]. The extent of regenerative capacity of nervous tissue proportionates the number of glial cells present in them. Oligodendrocytes, astrocytes, and microglia make up the CNS glial cell population, while Schwann cells make up the PNS glial cell population [7].

Mesenchymal stem cells have a high degree of flexibility, which makes them useful in tissue engineering. Their exceptional immunomodulatory characteristic, as well as their capacity to attract themselves to the site of injury, makes them a "natural in vivo system for tissue healing" [8].

Polycaprolactone scaffolds have a higher mechanical strength than natural polymers, are highly biocompatible, have a flexible shape, and can be processed in a controlled manner, allowing for optimal anatomical fit [9]. As it has been proven in comparable biomaterials, carbon nanotubes (CNT) + collagen polymer improves electrical conductivity, resulting in good survival of neuronal cells [10]. The CNTs exhibit morphological similarities to neurites, and tiny CNT bundles have dendritic-like diameters, boosting possibilities for not just exploring, mending, activating, or reconfiguring neural networks, but also learning about basic neuronal processes [11,12].

By modulating the inflammatory and early proliferative stages of nerve repair, local augmentation of insulin-like growth factor (IGF) levels at the crush injury site may promote axonal sprouting. IGF-I administration to the crush-injured location may hasten the functionalization of paralyzed muscle by enhancing the pace of recovery [13].

12.2 Isolation culture and expansion of bone marrow-derived mesenchymal stem cell

12.2.1 Stem cell collection

There are numerous techniques for harvesting and cultivating stem cells, and they vary depending on the species. The species from which bone marrow is obtained influences how long it takes for a primary cell culture to attain confluence [14]. Aseptic preparation is made at the collection site (Fig. 12.1A). The epiphysis and metaphysis of the long bones were cut with an electric/manual bone saw after the femur and tibia were gathered and meticulously cleaned (Fig. 12.1B).

After that, open bones were put in falcon tubes with DMEM/cell culture media but only with 1% PS instead of FBS (penicillin-streptomycin). Then, 5 mL of medium without serum (FBS) and an antibiotic were added to a Petri dish with the bones (40–50 mm Petri dish). By flushing via the diaphyseal canal into a falcon tube with a 15 mL capacity and complete culture media, bone marrow was obtained (10% FBS, 15% PS, 90% DMEM-LG).

12.2.2 Stem cell isolation

Following the collection of marrow aspirates, the cell suspension was centrifuged at 980 rpm for 5 minutes to concentrate the cells. The cell pellets were resuspended

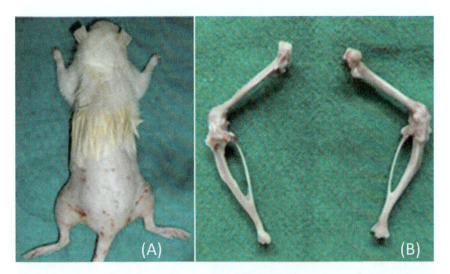

Figure 12.1 Collection of bone marrow from rat: (A) aseptic site preparation, (B) collection of femur and tibia for isolation of bone marrow.

with 5 mL of complete DMEM medium, then layered over, histopaque-1077 (Sigma), and centrifuged at 2500 rpm for 30 minutes. Mononuclear cells were collected from the interface of ficoll by gradient centrifugation (Fig. 12.2A) and washed with Ca^+ and Mg^+ free Dulbecco's phosphate buffer saline (DPBS). The cell count and viability was done after addition of trypan blue dye in the automatic cell counter and MTT analysis in spectrophotometer [15].

12.2.3 Stem cell culture

The supernatant was removed after two to three washings, and the pellet was then resuspended in full culture medium (90 mL DMEM-LG; 10 mL FBS, 1 mL PS) (Fig. 12.2B).

However, the number of nucleated cells was counted under the Neubauer chamber before cells were measured using a prevolume fixed pipette and loaded into T-25 flasks. After being gently mixed with fresh complete culture media (P0) using a pipette in and out technique, the cells were then put in a T-25 flask and kept in a CO_2 incubator at 5% CO_2, 21% O_2 at 37°C. In 4–5 days, the MSC was affixed to the flask's plastic wall.

12.2.4 Stem cell expansion

Every third day starting from day 0 of loading for the first 7 days, cell adherence was assessed, and any unabsorbed cells were rinsed out. Adherent cells present in the T-25 flask were supplied with growth medium. After 7 days, the flasks were kept at a confluence level of up to 80%, and full media changes were made every 3–4 days. When the BMSC cell population reached 80%, the flasks were trypsinized to raise the cells for subsequent passaging using trypsin (0.25% EDTA) (Fig. 12.3A and B).

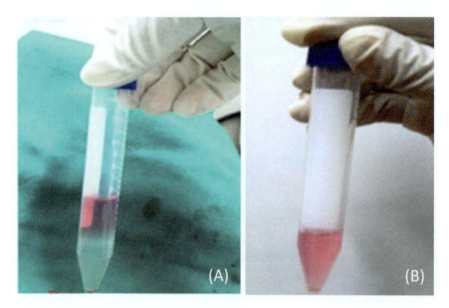

Figure 12.2 (A) Stratified layer of bone marrow after gradient centrifugation, (B) mono- and polymorphonuclear cell pellet.

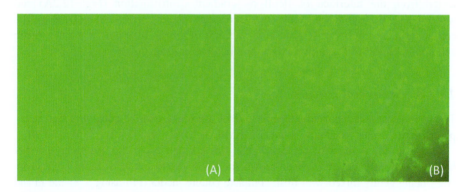

Figure 12.3 Morphological features of rBM-MSC at different day intervals. (A) Day 0 showing round bone marrow mononuclear cells, (B) day 20 showing spindle shaped isomorphic cells.

After cells were removed from the parent flask, trypsin neutralization was carried out using two times as much trypsin as entire medium. After 5 minutes of centrifugation at 980 rpm (G force 112), the supernatant was removed, and the cell pellet was once again centrifuged with PBS for 5 minutes. To increase the population in a filled complete growth media, we loaded a T-75 flask according to the research needs (8 mL). Similar to the previous steps, the second and third passages' cells have been expanded. For the in vivo investigation, the enlarged cells of the "P3" population were utilized.

12.3 Stem cell characterization

12.3.1 CD marker-based expression via reverse transcription polymerase chain reaction

Materials required: RNA isolation kit (Qiagen RNeasy plus mini kit), cDNA synthesis kit (Fermentas Kit), PCR machine, agar gel electrophoresis machine, Beta actin and CD primers (chromus, United States): CD105, CD44, CD45 (negative), CD34 (negative), and CD73.

Method for characterizing stem cells using CD markers.

The following primer listed in the Table 12.1 was used in the PCR for the 4 million "P3" rat MSC. Amplicons were subjected to electrophoresis in 0.1% agarose gel at 90 volts after amplification. The final visualization of the products in gel doc was as bands.

12.4 Cell viability assay

This was performed by two methods: MTT Assay (Qualitative) and Trypan blue exclusion method (Quantitative).

12.4.1 Trypan blue exclusion method

The vitality and concentration of the cells were estimated using this assay. Starting at the P3 stage, 30 μL of cell suspension were mixed with 30 μL of trypan blue solution (50 μL of sodium chloride at 4.25% in 200 μL of trypan blue). Following thorough mixing, a 10 μL aliquot was placed into a Neubauer chamber, and using an optical microscope and a low power (10 ×) objective lens, cell counting was carried out in the WBC chamber.

Table 12.1 PCR primer for amplification.

CD44	F	5'-GCCTGCCCACCATGGCTCAG-3'	377 bp
	R	5'-CCCGGGAGACCCACTGCTCA-3'	
CD73	F	5'-CACTCAGTCATGCCGCTTTA-3'	533 bp
	R	5'-CGCTGATATCTTGCTCACCA-3'	
CD105	F	5'-CCGGCGAATACTCTCTCAAG-3'	342 bp
	R	5'-AGGTCAGGTTCAGGATGGTG-3'	
CD45	F	5'-TCCCGCCGACACAGCTCTCA-3'	531 bp
	R	5'-GGCTCGGCATTCACGTCCCA-3'	
CD34	F	5'-TAGGGCTCAGTGCCTGCTGCT-3'	400 bp
	R	5'-GCCGTTTCTGGAGGTGGCCT-3'	
Beta actin	F	5'-AAGGACCTGTAGGCCAACAC-3'	432 bp
	R	5'-CACCTTCACCGTTCCAGTTT-3'	

12.4.2 MTT (3-(4,5-dimethylthiazolyl-2)-2,5-diphenyltetrazolium bromide) assay

3-(4,5-dimethylthiazolyl-2)-2,5-diphenyltetrazolium bromide was used to count the number of proliferating cells (MTT). Mitochondrial dehydrogenases, which is found in living cells, decreases MTT. Cells from the third passage were plated into 96-well ELISA plates with complete culture media at a cell density of 1×10^4 cells/well and kept in a CO_2 incubator at 37°C with 5% CO_2 for 0, 3, 7, and 14 days. Following incubation, 20 μL of MTT (5 mg/mL) were added to each well, which underwent an additional 4 hours of incubation in a CO_2 incubator. After incubation, the monolayer was dried for a little while before the media was gently removed without disturbing it. 150 μL of dimethyl sulfoxide were added to each well and agitated without the formation of air bubbles in order to dissolve the formazan crystals. The absorbance was measured at 570 nm using an ELISA reader. The experiment was carried out using DMSO as a control.

$$\text{Viable cells}(\%) = 1 - (\text{number of blue cells/total cells}) \times 100.$$

12.5 Colony forming assays

12.5.1 Colony forming units-fibroblasts assay

This was done to count the MSC present in the P0 bone marrow aspirates. 1×10^6 cells from primary culture (P0) were utilized for the cell seeding population in DMEM with low glucose, 100 U/mL penicillin, 100 g/mL streptomycin, and 15% FBS at 37°C in an incubator with 5% CO_2. After being fed every 3 days, colonies were stained with Giemsa on day 14 and counted using a $10 \times$ optical microscope [16].

12.5.2 Colony forming units-osteoblasts assay

In plates containing DMEM, 10% fetal bovine serum (FBS), 50 g/mL l-ascorbic acid, and 2.0 mM-glycerophosphate osteoblast development media, the "P0" bone marrow MSC cells (1×10^5 cells/Petri dish) were planted. The media was changed every 3 days. On day 14, alizarin red staining was used to count the mineralized bone matrix colonies [17]

12.6 Tri-lineage staining characterization

Optical microscope, distilled water, ice-cold ethanol 70%, Alizarin red S (Sigma Aldrich) for osteogenic lineage, alcian blue for chondrogenic lineage, and Oil O red for adipogenic lineage.

12.6.1 Alizarin red staining

For osteogenic differentiation, cells from the third passage were counted and planted in a 24 well culture plate at a rate of 2×10^6 cells per well. Osteogenic differentiation medium (Stempro) was added to four wells once the cells had reached 70%–80% confluency, whereas the negative controls only received complete media. Weekly media replacements were made, and the culture might last up to 21 days. Alizarin red staining was used to measure calcium deposition in order to evaluate the ability of cells to differentiate [14].

12.6.2 Alcian blue staining

Similar cell seeding procedures to those employed previously for osteogenic differentiation were used. Four wells were supplied with chondrogenic development medium (Stempro) once it reached 70%–80% confluency, while the control wells only received full media. Media was added twice weekly, and culture was sustained for 18 days. Chondrogenic differentiation was determined using alcian blue staining.

12.6.3 Oil red O staining

Cell plating was employed to promote adipogenic differentiation in a manner similar to that which promotes osteogenic differentiation. After reaching 70%–80% confluency, four wells got an addition of adipogenic differentiation medium (Stempro), while the control wells only received full media. The culture was retained for a month while the media were changed twice a week. Oil red O staining revealed lipid droplets inside the cell, confirming that the adipogenic differentiation was positive.

12.7 Scaffold preparation and characterization

The composite nanoneural scaffold with the dimension of 20 mm × 5 mm × 0.05 mm was made of collagen + polycaprolactone (PCL) + multiwalled carbon nanotubes (MWCNT) in 7.5:2.5 percentage by weight (wt.%) blend of collagen (collagen from natural source) and polycaprolactone along with 0.5 wt.% concentration of randomly arranged carbon nanotubes (Fig. 12.4).

Crosslinking strategies of pure collagen scaffold enhance the mechanical and structural properties, but may introduce negative effects too on cellular response in vivo. Thus in this study, a mixture of natural and synthetic polymer (PCL) reinforced with MWCNT has been used. It was prepared using electrospinning process at Indian Institute of Technology (IIT) at Roorkee. The scaffolds were supplied to IVRI for this research work.

Figure 12.4 Composite collagen + polycaprolactone + MWCNT nanoneural scaffold. *MWCNT*, Multiwalled carbon nanotubes.

12.7.1 Electrical property of the neural scaffold

Scaffold's electrical property was measured in form of I-V characteristics in which specific voltage was applied to 1×1 cm scaffold and current was measured. The experiment was performed at room temperature in triplicates for each scaffold. The following equation was used to calculate the resistivity of the scaffolds using I–V characteristics:

$$\rho = V \times A / I \times L,$$

where ρ represent desired resistivity, V represents voltage applied, A represents the electrode area, I represent the obtained current, and L represents the distance between electrodes. The obtained resistivity was used to calculate the conductivity. The following equation was used to calculate the conductivity:

$$\sigma = 1/\rho.$$

In MWCNT-based scaffolds, the electrical conductivity in the direction of alignment was 0.000035 ($\pm$ 0.0000049) S/m. Overall, these findings show that the scaffolds can be given greater conductivity, which stimulates the differentiation of neurons, by strengthening the MWCNTs within the polymeric matrix.

12.7.2 Surface roughness of the neural scaffold

Surface roughness of the scaffolds was measured at a relative height along a random line on scaffold's surface. Three random lines were selected for each surface.

The surface roughness may be a contributing factor in the deterioration of the scaffolds. In comparison to scaffolds with less rough surfaces, those with enhanced surface roughness may have more exposed polymeric chain ends for the enzymes to act on, leading to increased breakdown. Surface roughness was reported to be 83% higher for MWCNT scaffolds.

12.7.3 Neural scaffold degradation test

To check the biodegradability of the scaffolds, they were subjected to degradation test in a modified buffer system, simulating the biochemical composition of the tissue fluid. These scaffolds were expected to encounter in vivo conditions. Degradation of the scaffolds was checked in single modified buffer system, namely, mPBS, which contains collagenase and lipase. Each type of scaffold (L: 20 mm × W: 5 mm × H: 1 mm) was immersed in the mPBS individually and incubated for 60 days at 37°C in a sterilized environment. Every 5 days of interval, old mPBS were replaced with fresh mPBS and scaffolds were washed thoroughly with deionized water, air-dried, and weighed. The percent weight loss of the scaffolds was plotted against time (in days).

Another crucial element for the appropriateness of the scaffold is the biodegradation of the tissue-engineered scaffolds. MWCNT containing scaffolds deteriorated more quickly. Briefly, the degradation of collagen + PCL + MWCNT scaffold was found to be $41.1 \pm 2.2\%$ after 60 days.

12.8 In vitro cytotoxicity testing

Collagen + PCL + Br + MWCNT scaffolds were maintained in BM-MSC cell culture in DMEM with 15% FBS and 100 IU/ mL penicillin-streptomycin to test for cytotoxicity and biocompatibility. The sample scaffold was sterilized for 4 hours in ethanol before cell seeding, followed by 20 minutes in a type C UV laminar chamber. The scaffold and cell-containing flask were incubated in an incubator with 5% CO_2 and 85% humidity for 4 hours following cell loading. It is examined for attachment after 4 hours, and proliferating cells are seen by 4', 6-diamidino-2-phenylindole (DAPI) staining after 14 days [18].

After 4 hours and 14 days of nuclei staining with DAPI, the cell attachment revealed proliferating cells on and into the porous area of the scaffold. Additional in vivo effects of stem cells would be positive evidence of increased healing and cell proliferation. The proliferating cells with expanded cytoplasm displayed a blue cytoskeleton.

12.9 Preparation of stem cell loaded MWCNT-based bioactive nanoneural construct and assessment

12.9.1 Stem cell loading on nanobiomaterial construct

12.9.1.1 Creation of fibronectin attachment sites

Prior to incubation, the nanocomposite scaffold was exposed to type C UV radiation. In a phosphate buffered solution containing 50 μM^{-1} fibronectin, the scaffold was then incubated for 24 hours at 37°C [19] (Fig. 12.5A).

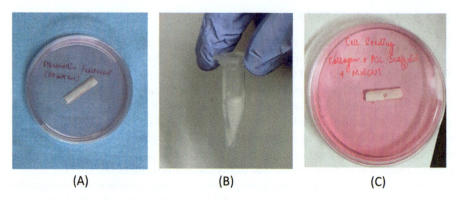

Figure 12.5 (A) Scaffold incubation in fibronectin solution. (B) Stem cell loading into scaffold. (C) Different stem cell loaded scaffold in complete media receptacle.

12.9.1.2 Centrifugal seeding of mesenchymal stem cells in the scaffold

In a 2.5 mL centrifuge tube, a scaffold that had been prepared with fibronectin was placed. Then, 2 mL of complete media containing cytosuspension (5×10^5) were introduced (Fig. 12.5B).

Three cycles of centrifugation were carried out, with each cycle showing a tube being spun at 500 rpm for 2 minutes and pausing for one minute in between. The scaffold was transferred to a Petri dish plate with complete culture conditions for 14 days in preparation for DAPI staining. The cell-seeded scaffold was carried in full media for in vivo administration following the centrifugation cycle [15,20] (Fig. 12.5C).

12.9.1.3 Percentage of cell seeding assessment

A 2.5 mL centrifuge tube was filled with the 5 million cells/2 mL complete media that the hemocytometer had counted. After three cycles of seeding centrifugation, the remaining complete media was utilized to measure the number of cells per microliter that were not attached to the scaffold using a hemacytometer. The seeding percentage of cells was estimated using the formula below:

% of cell seeding = (No. of cells/mL in centrifugation tube present before seeding on to the scaffold − No. of cells/mL in centrifugation tube present after seeding on to the scaffold)/No. of cells/mL in centrifugation tube present before seeding on to the scaffold × 100 [20,21].

12.9.2 Diamidino-2-phenylindole staining

A stem cell-seeded composite random polymeric neural scaffold made of PCL, collagen and MWCNT was preserved with 4% paraformaldehyde at 4°C for 30 minutes. Following two phosphate buffer saline (PBS) cleanings, the scaffolds were treated with 4, 6-diamidino-2-phenylindole for 2 minutes (DAPI). The scaffold was then cleaned twice with PBS after that. Three times, the experiment was run with fluorescent microscope visualization. The scaffold's blue immunofluorescence was caused by the stem cells' DAPI-labeled nuclei (Fig. 12.6A and B).

12.9.3 Preparation of IGF-I

Insulin-like growth factor I was used as a biostimulator for this study. Recombinant human IGF-I (Sigma) was used for this purpose. As the $t1/2$ (half-life) of the free IGF-I is very less, so it was reconstituted in 0.1% bovine serum albumin (BSA) and phosphate buffer saline for its longer duration of action. The concentration used in this study was 234.74 ng/0.5 mL of IGF-I.

12.10 Preparation of sciatic nerve injury model

These scaffolds were left so as cells can proliferate. After 14—18 days, these can be utilized for peripheral nerve injury studies. Healthy adult male Wister rats weighing around 200—300 g between 2 and 3 months of age were used throughout the study. Animals were procured from the laboratory Animal Resources (LAR) section of the institute. Rats were maintained in polyethylene cages with food and water ad libitum in a laboratory with controlled ambient temperature. The experimental animal models of sciatic nerve crush injury were prepared in 75 animals and randomly divided into 5 groups, namely, groups A, B, C, D, and E, having 15 animals in each group. Animals of each group were given treatment according to Table 12.2.

Using standard aseptic surgical procedures, intraperitoneal xylazine and ketamine (@ 3 mg/kg and 30 mg/kg, respectively) were administered. Anesthetized animals were approached caudolaterally over left hindlimb in order to gain access to peripheral sciatic nerve [22]. At the back of the thigh, one longitudinal cutaneous incision was made followed by sciatic nerve exposure through a crush window (Fig. 12.7A), which was

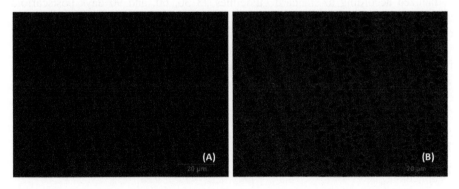

Figure 12.6 DAPI staining of rBM-MSC in cytoskeleton: (A) Control: scaffold without rBM-MSC, (B) PCL + collagen + MWCNT scaffold. *DAPI*, 6-Diamidino-2-phenylindole; *MWCNT*, multiwalled carbon nanotubes; *PCL*, polycaprolactone.

Table 12.2 Treatment protocol for animals in different groups.

\S. No.	Groups	Number of animals (rat)	Treatment
1	A	15	Phosphate buffered saline (control)
2	B	15	Implanted with randomly aligned collagen + PCL + MWCNT at the site of injury
3	C	15	Implanted with randomly aligned collagen + PCL + MWCNT and instilled with IGF-I on 0, 3, 7, 14, 21, and 28 days at the site of injury
4	D	15	Implanted with stem cell-laden randomly aligned collagen + PCL + MWCNT at the site of injury
5	E	15	Implanted with stem cell-laden randomly aligned collagen + PCL + MWCNT and instilled with IGF-I on 0, 3, 7, 14, 21, and 28 days at the site of injury

MWCNT, Multiwalled carbon nanotubes; *PCL*, polycaprolactone.

created by dissecting along a plane between the muscle bellies of biceps femoris and semitendinosus muscles. The left sciatic nerve was crushed for ninety seconds using the tip (3 mm) of curved hemostatic forceps, and the compression strength was calibrated at the second locking position (Fig. 12.7B). Scaffold of size 1 mm $\times$ 0.5 mm was wrapped around the crushed nerve site (Fig. 12.7C and D). A 3−0 Vicryl suture was utilized to suture muscle followed by 2−0 polyamide for sealing the skin (Fig. 12.7E and F).

Postoperatively, at the site of injury, local infusion of cells @ one million cells at 0-, 15-, 30-, 45-, 60-, 75-, and 90-day intervals and IGF-I at the rate of 140 ng/mL at 0-, 4-, 7-, 10-, 14-, 21-, and 28-day intervals was given in respective subgroups.

The animals were housed individually after recovering from anesthesia, with antibiotics, analgesics, and antiseptic ointment applied topically. The animals were euthanized according to standard guidelines 23 30, 60, and 90 days postscaffold treatment, and the tissue of interest for study sciatic nerve was collected and processed for scanning electron microscopic studies and molecular testing.

Throughout the study, animals were kept in separate cages with free access to food and water in accordance with lab animal care rules. The mixed smash chows with the recommended composition were obtained from the ICAR-Division IVRI's of Animal Nutrition.

12.10.1 Neurological evaluation

12.10.1.1 Sciatic function index

Preoperatively, 30 days, 60 days, and 90 days postsurgery, a walking track analysis was undertaken. The rat's hind feet were painted with nankin ink, and the animals were placed in a walking route to leave their imprints (Fig. 12.8) [24].

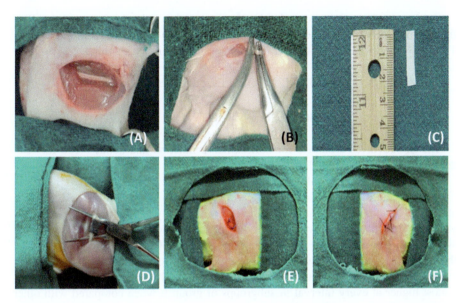

Figure 12.7 Sciatic nerve injury operative procedures: (A) Longitudinal cutaneous incision was made to expose sciatic nerve through a crush window, (B) creation of sciatic nerve crush injury (90 s) using the tip of 3 mm curved hemostatic forceps, (C, D) crushed site was exposed and scaffold of size 1 mm × 0.5 mm was wrapped around the crushed nerve site, (E, F) muscle and skin suturing postscaffold treatment.

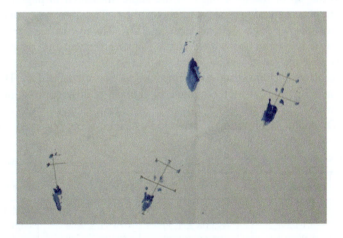

Figure 12.8 Walking track analysis (Sciatic Function Index).

On the test (E) and the control side (N), the lengths of the third toe to heel (PL), first to fifth toe (TS), and second toe to fourth toe (IT) were measured in each rat. The following formula was utilized to compute the SFI in each animal.

$$SFI = -38{:}33(EPL\text{-}NPL)/NPL + 109{:}5 \times (ETS\text{-}NTS)/NTS$$

$$+ 13{:}33(EIT\text{-}NIT)/NIT\text{-}8{:}8.$$

The SFI revolves around 0 for normal nerve function and about -100 for total nerve impairment in general.

Prior to surgery, SFI values in all the groups were near zero. After the nerve transection, the mean SFI decreased near to -100 due to the complete loss of sciatic nerve function in all animals. At the end of the study period, the statistical analyses revealed that the recovery of nerve function in groups A, B, and C remained significantly lower than the base values score up to 60 days. Comparison among the different groups and subgroups showed a significant functional recovery in groups D and E from group A 60th day onward. The functional recovery was almost higher at every interval in subgroups of group E when compared with other subgroups. The mean ± SE values of toe out angle of animals of different treatment groups are shown in Table 12.3.

12.10.1.2 Toe out angle

In rats, functional improvement of the sciatic nerve is thought to be associated with external rotation of the leg. The TOA, which is defined physiologically from the calcaneus to the tip of the third digit, is the angle, measured in degrees, between the direction of progression and a reference line. The rats were placed on a $15 \times 15 \times 3$ cm acrylic glass sheet to measure TOA. The plantar surface of the animals' paws was photographed with a camera placed beneath the clear base plate [25]. Angles were measured and recorded (Fig. 12.9).

The angle between normal and experimental feet was calculated by TOA analysis. The mean ± SE values of toe out angle of animals of different treatment groups are shown in Table 12.4. On day 90, highly significant changes were observed in groups C, D, and E. Considerable temporal changes were observed in case of

Table 12.3 Sciatic Function Index (SFI) of different treatment groups at various time intervals.

Groups	Preoperative	Day 30	Day 60	Day 90
A	−2.13 ± 1.1	−93.74 ± 2.64	−89.18 ± 1.28	−84.48 ± 1.31
B	−5.15 ± 1.2	−93.08 ± 1.50	−79.71 ± 2.48	−66.9 ± 1.94
C	−5.09 ± 1.5	−91.68 ± 1.79	−78.79 ± 1.48	−64.55 ± 2.33
D	−4.25 ± 2.9	−89.06 ± 2.10	−77.33 ± 2.42	−60.48 ± 2.37[b]
E	−8.02 ± 1.1	−92.15 ± 1.6	−76.52 ± 1.53[a]	−41.79 ± 2.34[b]

[a]Mean value differs significantly at $P < .05$ within group.
[b]Mean value differs significantly at $P < .01$ within the group.

Figure 12.9 Performing toe out angle analysis.

Table 12.4 Toe Out Angle (TOA) of different treatment groups at various time intervals.

Groups	Pre	Day 30	Day 60	Day 90
A	14.18 ± 2.1	58.44 ± 2.09	48.66 ± 1.05	42.99 ± 0.90
B	7.99 ± 1.3	52.84 ± 1.66	38.26 ± 1.28	32.00 ± 1.98
C	12.9 ± 0.6	51.21 ± 1.04	35.27 ± 1.50	30.69 ± 1.35
D	7.6 ± 1.6	50.95 ± 1.48	29.65 ± 2.11	20.04 ± 1.66
E	8.2 ± 2.0	49.05 ± 0.70[a]	25.90 ± 2.43[b]	18.27 ± 1.28[b]

[a]Mean value differs significantly at $P < .05$ within group.
[b]Mean value differs significantly at $P < .01$ within the group.

groups C, D, and E showing good rate of recovery in toe angle. The results showed significantly better angle for the contralateral foot than the experimental foot in animals implanted with scaffold having collagen + PCL + Br + MWCNT with BMSCs and IGF-I and it showed a better recovery compared to the control group. In a similar study, analysis of toe out angle in autograft and the nanofibrous conduit with SCs group was done and better recovery was obtained experimental group compared to the control group [26].

12.10.2 Biochemical parameters

Blood was collected from the orbital plexus of rats using capillary tubes and two mL of blood was drawn at 30th, 60th, and 90th day postoperatively. For C-reactive protein, estimation blood was collected at 7-, 14-, and 21-day time intervals. Serum separation was done in order to quantitative inflammatory mediators like C-reactive protein (CRP) using rat-CRP-ELISA kit (Sigma)

The mean ± SE values of the C-reactive protein values of animals of different groups are shown in Table 12.5. In serum C-reactive protein estimation, CRP levels were increased in all groups postcrush injury and scaffold implantation and the high levels remained up to 7 days in all groups except in group A where it further increased till day 14. Highly significant ($P < .01$) changes were observed in groups B and E while significant changes ($P < .05$) in terms of CRP level were noticed in

Table 12.5 Mean ± SE values of the C-reactive protein values of animals of different groups.

Groups	Pre	Day 7	Day 14	Day 21
A	335 ± 61	721 ± 62	797 ± 78	602 ± 53
B	287 ± 45	798 ± 51[b]	338 ± 54	401 ± 51
C	322 ± 54	802 ± 61[a]	427 ± 63	414 ± 62
D	329 ± 71	795 ± 68[a]	346 ± 63	474 ± 59
E	292 ± 65	834 ± 74[b]	389 ± 71	375 ± 58

[a] Mean value differs significantly at $P < .05$ within group.
[b] Mean value differs significantly at $P < .01$ within the group.

Table 12.6 Primers used for quantitative real time PCR.

Gene	Primer pairs (5'–3')	Annealing temperature (°C)	Amplicon size (bp)
NRP-1	F-GGAGCTACTGGGCTGTGAAG	58	135 bp
NRP-1	R-CCTCCTGTGAGCTGGAAGTC		
NRP-2	F-GCGCAAGTTCAAAGTCTCCT	60	216 bp
NRP-2	R-TCACAGCCCAGCACTTC		
GAP-43	F- CAGGAAAGATCCCAAGTCCA	58	207 bp
GAP-43	R-GAACGGAACATTGCACACAC		
β-actin	F-TATTGGCAACGAGCGG	60	54 bp
β-actin	R-CGGATGTCAACGTCAC		

groups C and D at day 30. The CRP levels raised up to twofolds comparing to its prevalue of the trial in almost all groups at day 7. No significant changes were seen at day 14 and 21 in groups B, C, D and E.

12.10.3 Relative expression of different genes by Real Time PCR

Sciatic nerve samples were harvested and preserved in RNA Later (Sigma) at −80°C posteuthanization. The relative mRNA expression profile of three different genes, namely, neuropillin-1 (NRP-1), neuropillin-2 (NRP-2), and GAP-43 (Growth Associated Protein-43) was performed using DyNAmo SYBR green (Thermo Scientific, United States) and Real Time qPCR machine (Bio-Rad, United States) at 30th, 60th, and 90th days postoperatively. β-actin was kept as housekeeping gene. Primers used in the study are mentioned in Table 12.6.

Relative expression of NRP-1, NRP-2, and GAP-43 genes was quantified by real time PCR and results are shown in Fig. 12.10.

At day 90, expression profiles of groups B and C were lower in comparison to control group A, while a higher value of expression was seen in groups D and E. Relative mRNA expression of NRP-2 showed a trend similar to NRP-1 where a lower value of expression was seen in groups B and C while a higher value of expression was seen in groups D and E. The profile of GAP-43 gene transcripts in this study on

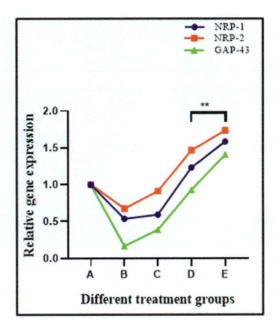

Figure 12.10 Showing relative gene expression profile of NRP-1, NRP-2, and GAP-43 at day 90 postscaffold treatment.

comparison with control group A showed a higher expression. Overall, the relative mRNA expression of subgroups of group E dominated all the other groups and had a higher mean score. Similar findings were reported while treating sciatic nerve injuries of rabbits with stem cells and stem cell conditioned media [27].

12.10.4 Scanning electron microscopic evaluation of nerve

SEM was used to examine the surface morphology of the left sciatic nerve crush site in one randomly chosen rat from each group on days 30, 60, and 90. Following animal sacrifice, the specimens were kept in 2.5% glutaraldehyde buffer. At the injury site's center, the preserved samples were divided into two sections for longitudinal and transverse scanning. The samples were examined using a scanning electron microscope (Jeol JSM 6610 LV type) with the proper acceleration voltage and magnification range [28].

The scanning electron microscopy of the sciatic nerve sample collected at day 90 was done to assess the orientation of the fibrillar network and surface morphology of the injured nerve. None of the nerve samples showed a completely normal electron microscopic appearance of the sciatic nerve. The longitudinal and transverse sections of the nerve samples were compared with the normal sciatic nerve collected from the contra lateral limb. In group A, smooth fibers with randomly arranged fibrils were noticed. In almost groups B and C fibers could be visualized but had a rough surface morphology without any clear demarcation of the fiber

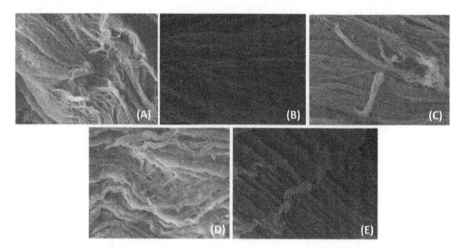

Figure 12.11 SEM images of rat sciatic nerve of various treatment groups: (A) Gp A, (B) Gp B, (C) Gp C, (D) Gp D, (E) Gp E at day 90 (small arrow—site of injury, block arrow—cell entrapped within fibers, curly bracket—radially arranged fibers in normal orientation and architecture).

when compared to the normal sciatic nerve. In group D, randomly arranged fibrillar network with rough surface morphology similar to groups B and C was noticed, whereas in group E radially arranged fiber could be visualized with rough connective tissue network and a surface morphology resembling normal nerve architecture could be visualized but complete healing could not be recorded (Fig. 12.11A–E).

12.11 Conclusion

In the present chapter, in vitro and in vivo systems to induce nerve regeneration in sciatic nerve crush injury model were developed using bone marrow-derived mesenchymal stem cell-laden nanoneural construct. Nano neural scaffold characterization was done. The third passage BMSCs were seeded onto collagen + polycaprolactone + multi-walled carbon nanotubes scaffold and then implanted in animals at the level of induced crush injury. The animal models were equally distributed into five different groups, namely, A, B, C, D, and E and treated with phosphate buffer saline (PBS), carbon nanotube-based neural scaffold only, scaffold with IGF-I, stem cell-laden scaffold, and stem cell-laden scaffold with IGF-I, respectively. The nerve regeneration was assessed based on physiconeuronal, biochemical, relative expression of NRP-1, NRP-2, and GAP-43 and scanning electron microscopy. Sciatic nerve injury model with crush injury produced for 90 seconds was standardized and successfully used in this study. All the biochemical parameters were in normal range in all the groups indicating no scaffold related changes. Physiconeuronal, histopathological, relative gene expression and scanning electron microscopy observations revealed appreciable nerve regeneration in group

E, followed by groups D, C, and B. Restricted to no regeneration was observed in group A. Carbon nanotube-based scaffold provided electroconductivity for proper neuronal regeneration while rat bone marrow-derived mesenchymal stem cells were found to induce axonal sprouting, cellular transformation. IGF-I in this study induced stem cell differentiation, myelin synthesis, angiogenesis, and muscle differentiation.

References

[1] Tamilmahan P, Maiti SK, Sangeetha P, Rashmi, Singh P, Kumar N, et al. Culture, characterization and differentiation potential of rat bone marrow derived mesenchymal stem cells. J Stem Cell Res Therapeut. 2016;(5).
[2] Maiti SK. Mesenchymal stem cells derived from rat bone marrow (rBM MSC): techniques for isolation, expansion and differentiation. J Stem Cell Res Therap 2017;3(3). Available from: https://doi.org/10.15406/jsrt.2017.03.00101.
[3] Seddon HJ. Three types of nerve injury. Brain 1943;66(4):237–88. Available from: https://doi.org/10.1093/brain/66.4.237.
[4] Tiwary R. Evaluation of nucleated marrow cells along with TGF-β1/IGF-1 for cartilage and nerve repair in rabbits. Thesis PhD. Izatnagar, India: Deemed University, Indian Veterinary Research Institute; 2011.
[5] Varejão ASP, Cabrita AM, Geuna S, Melo-Pinto P, Filipe VM, Gramsbergen A, et al. Toe out angle: a functional index for the evaluation of sciatic nerve recovery in the rat model. Exp Neurol 2003;183(2):695–9. Available from: https://doi.org/10.1016/S0014-4886(03)00208-5, http://www.elsevier.com/inca/publications/store/6/2/2/8/2/8/index.htt.
[6] Biazar E, Keshel SH, Pouya M. Behavioral evaluation of regenerated rat sciatic nerve by a nanofibrous PHBV conduit filled with Schwann cells as artificial nerve graft. Cell Commun Adhes 2013;20(5):93–103. Available from: https://doi.org/10.3109/15419061.2013.833191.
[7] Committee for the Purpose of Control and Supervision on Experiments on Animals. CPCSEA guidelines for laboratory animal facility. Indian J Pharmacol 2003;35(4):257–74.
[8] Cameron RE, Kamvari-Moghaddam A. Synthetic Bioresorbable Polymers. Elsevier BV; 2008. p. 43–66. Available from: 10.1533/9781845695033.2.43.
[9] Mohan D. Evaluation of mesenchymal stem cells with conditioned media and m-EGF for regeneration of liver tissue after partial hepatectomy in rats. PhD Thesis. Indian Veterinary Research Institute; 2018.
[10] Ninu AR, Maiti SK, Shiva Kumar MU, Kumar S, Sangeetha P, Kritaniya D, et al. Isolation, proliferation, characterization and in vivo osteogenic potential of bone-marrow derived mesenchymal stem cells (rBMSC) in rabbit model. Indian J Exp Biol 2017;55(2):79–87. Available from: http://nopr.niscair.res.in/bitstream/123456789/40227/2/IJEB%2055(2)%2079-87.pdf.
[11] Stringer SK, Seligmann BE. Effects of two injectable anaesthetic agents on coagulation assays in the rat. Lab Ani Sci. 1966;46:430–3.
[12] Ninu A.R. Mesenchymal stem cell construct with or without growth factors for osteogenesis in critical sized bone defect in rabbit model. Izatnagar, Izatnagar; 2014.
[13] Serpell CJ, Kostarelos K, Davis BG. Can carbon nanotubes deliver on their promise in biology? Harnessing unique properties for unparalleled applications. ACS Central Science 2016;2(4):190–200. Available from: https://doi.org/10.1021/acscentsci.6b00005, http://pubs.acs.org/journal/acscii.

[14] Emel E, Ergün SS, Kotan D, Gürsoy EB, Parman Y, Zengin A, et al. Effects of insulin-like growth factor−I and platelet-rich plasma on sciatic nerve crush injury in a rat model. J Neurosurg 2011;114(2):522−8. Available from: https://doi.org/10.3171/2010.9.jns091928.

[15] Redondo-Gómez C, Orozco F, Noeske PLM, Soto-Tellini V, Corrales-Ureña YR, Vega-Baudrit J. Cholic acid covalently bound to multi-walled carbon nanotubes: improvements on dispersion stability. Mater Chem Phys 2017;200:331−41. Available from: https://doi.org/10.1016/j.matchemphys.2017.07.089, http://www.journals.elsevier.com/materials-chemistry-and-physics.

[16] MacDonald RA, Voge CM, Kariolis M, Stegemann JP. Carbon nanotubes increase the electrical conductivity of fibroblast-seeded collagen hydrogels. Acta Biomater 2008;4 (6):1583−92. Available from: https://doi.org/10.1016/j.actbio.2008.07.005, http://www.journals.elsevier.com/acta-biomaterialia.

[17] Rosso G, Liashkovich I, Gess B, Young P, Kun A, Shahin V. Unravelling crucial biomechanical resilience of myelinated peripheral nerve fibres provided by the Schwann cell basal lamina and PMP22. Sci Reports. 2015;4:7286−98.

[18] Kronenberg H. Adult mesenchymal stem cells. StemBook 2009;. Available from: https://doi.org/10.3824/stembook.1.38.1.

[19] Navarro X, Vivó M, Valero-Cabré A. Neural plasticity after peripheral nerve injury and regeneration. Prog Neurobiol 2007;82(4):163−201. Available from: https://doi.org/10.1016/j.pneurobio.2007.06.005.

[20] Lenze U, Pohlig F, Seitz S, Ern C, Milz S, Docheva D, et al. Influence of osteogenic stimulation and VEGF treatment on in vivo bone formation in hMSC-seeded cancellous bone scaffolds. BMC Musculoskelet Disord 2014;15(1). Available from: https://doi.org/10.1186/1471-2474-15-350.

[21] Zhang E. Endoplasmic reticulum stress impairment in spinal dorsal horn of a neuropathic pain model. Science Reports. 2015;5.

[22] Bain JR, Mackinnon SE, Hunter DA. Functional evaluation of complete sciatic, peroneal, and posterior tibial nerve lesions in the rat. Plast Reconstr Surg 1989;83 (1):129−36. Available from: https://doi.org/10.1097/00006534-198901000-00024, http://journals.lww.com/plasreconsurg/pages/issuelist.aspx.

[23] Schmidt CE, Leach JB. Neural tissue engineering: strategies for repair and regeneration. Annu Rev Biomed Eng 2003;5:293−347. Available from: https://doi.org/10.1146/annurev.bioeng.5.011303.120731.

[24] Sunderland S. A classification of peripheral nerve injuries producing loss of function. Brain 1951;74(4):491−516. Available from: https://doi.org/10.1093/brain/74.4.491.

[25] Sivanarayanan T.B. Evaluation of mesenchymal bone marrow derived mesenchymal stem cells with stem cell conditioned media for the repair of acute and subacute nerve injuries. Thesis PhD. Izatnagar, Izatnagar; 2015.

[26] Cellot G, Cilia E, Cipollone S, Rancic V, Sucapane A, Giordani S, et al. Carbon nanotubes might improve neuronal performance by favouring electrical shortcuts. Nature Nanotechnology 2009;4(2):126−33. Available from: https://doi.org/10.1038/nnano.2008.374.

[27] L.R. Robinson, Spencer Stelfa, E., Diagnosis and rehabilitation of peripheral nerve injuries Robinson LR (ed) Trauma rehabilitation. Philadelphia, Philadelphia, (2006), 160−216.

[28] Peter M, Binulal NS, Nair SV, Selvamurugan N, Tamura H, Jayakumar R. Novel biodegradable chitosan−gelatin/nano-bioactive glass ceramic composite scaffolds for alveolar bone tissue engineering. Chemical Engineering Journal 2010;158(2):353−61. Available from: https://doi.org/10.1016/j.cej.2010.02.003.

Cellular architects: mesenchymal stem cells crafting the future of regenerative medicine

Rahul Kumar Udehiya[1] and Sarita Kankoriya[2]
[1]Department of Veterinary Surgery and Radiology, Faculty of Veterinary and Animal Sciences, Institute of Agricultural Sciences, Rajiv Gandhi South Campus, Banaras Hindu University, Barkachha, Uttar Pradesh, India, [2]Veterinary Hospital, Shivpur, Department of Animal Husbandry, Mirzapur, Uttar Pradesh, India

13.1 Introduction

The field of stem cell research has experienced rapid growth in the 21st century, marked by significant discoveries and advancements. Stem cell research is expanding more than twice as fast as the world average growth in research. Notably, the annual growth rate of studies on induced pluripotent stem cells, a rapidly growing type, is an astonishing 77% [1]. Major breakthroughs include the generation of the first functioning whole organ, the thymus, and the birth of the first documented human baby girl through in vitro fertilization, who now has children of her own [2]. Stem cell research aims to create other functioning whole organs like the kidney and intestine [3].

This progress has given rise to the field of regenerative medicine, a multidisciplinary branch dealing with the replacement, engineering, or regeneration of human and animal cells, tissues, or organs to restore normal function. Regenerative medicine holds the promise of repairing injured tissues and restoring normal cellular function. Mesenchymal stem cells (MSCs), a type of self-renewing multipotent cell, are widely studied and used in clinical trials due to their regenerative effects [4]. MSCs show promise in treating conditions such as diabetes, with their ability to differentiate into multiple cell types, low immunogenicity, and secretion of biologic factors to restore and repair tissues [5].

The MSCs, found in various parts of the body, can be isolated from multiple sources, including bone marrow, bodily fluids, and perinatal tissues. They respond to microenvironmental changes by releasing immune modulatory and trophic factors, supporting the regeneration of injured cells and tissues. The understanding of MSC biology and their mechanisms of development and function holds potential for innovative solutions in treating stem-based diseases and disorders [6].

Despite the hope and progress, challenges remain in applying MSC biology and regenerative medicine to improve human and animal diseases. Various states and countries have invested significantly in stem cell and regenerative medicine research, with notable initiatives [1] This chapter presents a global collection of essays from research scientists worldwide, discussing progress in MSC biology and regenerative medicine, including applications in treating diseases. It explores topics ranging from MSC biology and development to their applications in tissue repair, regeneration, and addressing important diseases, providing insights into the promise of MSC therapeutics and regenerative medicine in the real world.

13.2 History of mesenchymal stem cells

In 1968, Friedenstein conducted bone fragment transplants and observed the formation of nonhematopoietic mesenchymal tissue in heterotopic areas, terming the cells osteoblasts [7]. Subsequent studies in 1970 and 1976 confirmed the colony formation and plastic adherent ability of these cells [8,9]. Human bone marrow was later found to contain similar cells capable of self-renewal and in vitro multilineage differentiation, named "mesenchymal stem cells" by Caplan in 1991 [10,11]. The verification of the in vivo bone formation capacity of human bone marrow MSCs occurred in 1997 [12].

Concerns about the correctness of the name "mesenchymal stem cells" arose, leading to the introduction of a new terminology, "mesenchymal stromal cells" by the International Society for Cellular Therapy (ISCT) in 2005. The ISCT stipulated that only cells demonstrating stemness by specific criteria could be termed mesenchymal stem cells, despite the continued use of the acronym MSC [13]. In 2006, the ISCT outlined minimal criteria for defining human MSCs (referred to as mesenchymal stromal cells), encompassing adherence to plastic, specific surface antigen expression, and multipotent differentiation potential [14]. Despite attempts to distinguish between mesenchymal stromal cells and MSCs, confusion persisted in their usage, with researchers employing minimal criteria without evidence of stem cell activity [15,16].

The subsequent decade saw increased emphasis on the functions of MSCs, with studies revealing their ability to enhance hematopoietic stem cell engraftment and contribute to tissue repair due to their multipotential capacity [15,17–19]. The discovery of MSCs' highly active cytokine-secreting property prompted investigations into their role in immunomodulation and homeostasis maintenance [20,21]. Recognizing the potential for MSCs as therapeutic cells, discussions arose about large-scale proliferation for commercialization. In 2016, the ISCT issued a statement emphasizing that culture conditions can impact MSC functions, urging guidance from clinical applications for large-scale proliferation and optimization of the culture system [22].

13.3 Isolation and culture of mesenchymal stem cells

This initiates with the isolation of mesenchymal stem cells from diverse sources, including bone marrow and adipose tissue. Standardized protocols for cell isolation and culture are rigorously adhered to, ensuring the maintenance of MSC characteristics. Parameters such as cell viability, morphology, and surface marker expression are continuously monitored to validate the purity and identity of the cultured cells Table 13.1

13.3.1 Isolation methods
13.3.1.1 Tissue source
MSCs can be isolated from various tissues, including bone marrow, adipose tissue, umbilical cord blood, and dental pulp.

13.3.1.2 Bone marrow–derived mesenchymal stem cells
Aspirate bone marrow from iliac crest (Fig. 13.1A) or other bones. Isolate mononuclear cells by density gradient centrifugation. Culture cells in plastic dishes, and MSCs adhere while non-MSCs are removed [33,34].

13.3.1.3 Adipose tissue–derived mesenchymal stem cells
Obtain adipose tissue through liposuction or biopsy (Fig. 13.1B). Digest tissue to release stromal vascular fraction (SVF). MSCs adhere during culture, allowing for isolation [35].

13.3.1.4 Umbilical cord blood–derived mesenchymal stem cells
Isolate mononuclear cells from umbilical cord blood (Fig. 13.1C). Culture cells in adherent conditions to select MSCs [36].

13.3.2 Culture conditions
13.3.2.1 Basal medium
Typically use α-MEM, DMEM, or similar basal medium [33,34].

13.3.2.2 Serum supplementation
Fetal bovine serum (FBS) is commonly used for its rich nutrient content. Serum-free media are also employed for defined conditions.

13.3.2.3 Growth factors and supplements
Add basic fibroblast growth factor (bFGF), epidermal growth factor (EGF), and platelet-derived growth factor (PDGF) to enhance proliferation. Include dexamethasone, ascorbic acid, and β-glycerophosphate for osteogenic differentiation. For adipogenic differentiation, use indomethacin, insulin, and isobutyl methylxanthine. Chondrogenic differentiation involves TGF-β1, insulin, transferrin, and selenous acid [37].

Table 13.1 The mesenchymal stem cell (MSCs) isolated from different sources using different isolation techniques.

S. no.	Source	Isolation technique	Media and serum	Cell-surface markers	Lineage differentiation
1.	Bone marrow [23,24].	Ficoll density gradient centrifugation	Knockout DMEM, DMEM 1 10% FBS	Positive: SH2, SH3, CD29, CD44, CD49e, CD71, CD73, CD90, CD105, CD106, CD166, CD120a, CD124 Negative: CD34, CD45, CD19, CD3, CD31, CD11b, HLA-DR	Adipogenic, chondrogenic, osteogenic
2.	Umbilical cord, umbilical cord blood [25,26].	Ficoll-Hypaque density gradient centrifugation, enzymatic digestion (0.25% trypsin-EDTA), explant culture	DMEM 1 10% FBS, 10% FCS, MSCGM 1 10% FCS	Positive: CK8, CK18, CK19, CD10, CD13, CD29, CD44, CD73, CD90, CD105, CD106, HLA-I, HLA-II Negative: CD14, CD31, CD33, CD34, CD45, CD38, CD79, CD133, vWF, HLA-DR	Adipogenic, chondrogenic, osteogenic, endothelial-like cells, neuron-like cells
3.	Wharton's jelly [27,28].	Explant culture, enzymatic digestion (0.1% collagenase II, 1 mg/mL collagenase B1 trypsin, collagenase 1 hyaluronidase 1 trypsin, trypsin-EDTA, 0.26% collagenase I 1 0.07% hyaluronidase 1 0.125% trypsin)	DMEM/DMEM-F12 1 10% FBS	Positive: CD13, CD29, CD44, CD73, CD90, CD105, HLA-I Negative: CD14, CD34, CD45, CD31, HLA-II	Adipogenic, oseogenic
4.	Adipose tissue [29,30].	Enzymatic digestion (1.5 mg/mL collagenase I, 1 mg/mL collagenase I in 0.1% BSA)	DMEM-low glucose 1 MCDB201 1 2% FCS, DMEM 1 20% FBS, Mesenpro RS	Positive: CD13, CD29, CD44, CD73, CD90, CD105, CD166, HLA-I, HLA-ABC Negative: CD10, CD14, CD24, CD31, CD34, CD36, CD38, CD45, CD49d, CD117, CD133, SSEA4, CD106, HLA-II, HLA-DR	Adipogenic, chondrogenic, osteogenic, neurogenic, muscular

(Continued)

Table 13.1 (Continued)

S. no.	Source	Isolation technique	Media and serum	Cell-surface markers	Lineage differentiation
5.	Amniotic fluid [31,32].	Density gradient centrifugation, enzymatic digestion (0.25% trypsin 1 1.2 units/mL of dispase 1 2 mg/mL collagenase 1)	α-MEM 1 20% FBS, DMEM-F12 1 10% FBS, high glucose DMEM 1 20% hESC-defined FBS, KSR-based media	Positive: SH2, SH3, SH4, CD29, CD44, CD49, CD54, CD58, CD71, CD73, CD90, CD105, CD123, CD166, HLA-ABC Negative: CD10, CD11, CD14, CD31, CD34, CD49, CD50, CD117, HLA-Dr, DP, DQ, EMA	Adipogenic, osteogenic, neurogenic

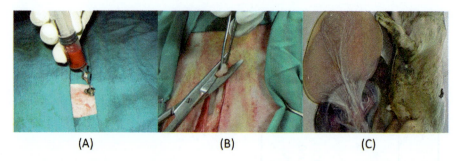

Figure 13.1 Collection of tissues from different sources from animals: (A) bone marrow, (B) adipose tissue, and (C) umbilical cord.

13.3.2.4 Incubation conditions

Maintain a humidified atmosphere with 5% CO_2. Control temperature (37°C) for optimal cell growth [33,34].

13.3.2.5 Passaging

Detach adherent MSCs using trypsin-EDTA for subsequent passages. Control passage number to avoid senescence.

13.4 Characterization of mesenchymal stem cells: key aspects

13.4.1 Cell-surface marker expression

The identification of mesenchymal stem cells (MSCs) relies on specific cell-surface marker expression criteria, as outlined by Dominici et al. [38] in 2006. According to this guideline, MSCs should express CD105, CD73, and CD90 ($\geq$95%), while CD45, CD34, CD14 or CD11b, CD79a or CD19, and HLA class II should not be expressed ($\leq$2%) [14]. However, the practicality and reliability of these markers warrant closer examination.

13.4.1.1 CD73 marker

CD73, an ecto-5'-nucleotidase, has been suggested as a potential MSC marker. Despite initial indications of specificity for MSCs, there is a lack of verifiable evidence supporting the use of anti-CD73 antibodies for in vivo MSC detection [39,40].

13.4.1.2 CD90 marker

CD90, also known as Thy1, has been recommended as an MSC marker, but its lack of cell type-specificity and limited evolutionary conservation raise concerns.

Commonly used anti-CD90 antibodies may not reliably react with MSCs across different species [41].

13.4.1.3 CD105 marker

CD105, or endoglin, is a glycoprotein highly expressed in vascular endothelial cells. While initially recommended as a positive marker, its variability in expression levels among MSCs from different tissues and its increase during culture passages question its utility as an in vivo MSC marker [42,43].

13.4.1.4 CD34 marker

CD34, recommended as a negative MSC marker, is contentious. The guideline lacks explicit reasons or references, and evidence suggests that CD34 negativity in MSCs may be a cell culture-induced phenomenon rather than an accurate reflection of their in vivo status [44]. Notably, adipose tissue—derived MSCs (ADSCs) are generally classified as CD34 + , challenging the generalization of CD34 as a negative marker [45]. Detailed histological studies propose the existence of CD34 + adventitial progenitor cells (APCs) in larger blood vessels and CD34 + vascular stem cells (VSCs) in capillaries and larger vessels, potentially challenging the notion of CD34 as a negative MSC marker [46,47].

13.4.1.5 Stro-1 marker

Stro-1, widely recognized as an MSC marker, has been a cornerstone in MSC studies since 1991. However, concerns about its specificity arise from its origin as a monoclonal antibody generated using human CD34 + bone marrow cells as immunogen. Evidence indicates significant overlaps between Stro-1 and endothelial markers, compromising its reliability as an in vivo MSC marker [48]. In the current understanding of MSC markers requires careful reconsideration, emphasizing the need for comprehensive and context-specific evaluations to ensure accurate identification and characterization of MSCs in various tissues and species.

13.4.2 Differentiation potential—beyond traditional lineages

Mesenchymal stem cells (MSCs) exhibit remarkable versatility by differentiating into osteocytes, chondrocytes, and adipocytes, a hallmark of their identity. In vitro, this differentiation is orchestrated by specific supplements tailored for each lineage (Fig. 13.2).

13.4.2.1 Adipogenesis:adipogenesis

Adipogenesis:adipogenesis is the process by which mesenchymal stem cells (MSCs) differentiate into mature fat cells, also known as adipocytes. This process is essential for maintaining energy balance and regulating metabolism in the body. The process of adipogenesis by MSCs can be divided into three main stages:

1. *Commitment:* In this stage, MSCs receive signals that commit them to the adipocyte lineage. These signals can come from various factors, such as growth factors, hormones, and cytokines.

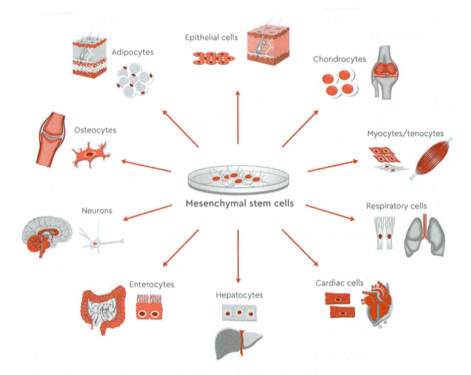

Figure 13.2 Differentiation potential of mesenchymal stem cells.
Source: From https://promocell.com/blog/using-mesenchymal-stem-cells-in-regenerative-medicine/.

2. *Proliferation:* Once committed, MSCs undergo several rounds of cell division to increase their number. This expansion of the preadipocyte population is crucial for generating sufficient fat cells to meet the body's needs.
3. *Terminal differentiation:* In the final stage, preadipocytes mature into functional adipocytes. This involves the accumulation of triglycerides (fats) within the cells, the expression of adipocyte-specific genes, and the development of characteristic features such as a large lipid droplet and insulin sensitivity.

Adipogenesis can be induced by dexamethasone, indomethacin, insulin, and isobutyl methylxanthine. Differentiation confirmed by fat droplet production [49–51].

13.4.2.2 Chondrogenesis

Chondrogenesis, the process of mesenchymal stem cells (MSCs) differentiating into mature cartilage cells (chondrocytes), is a fascinating and valuable area of research with potential applications in cartilage repair and regeneration. The process of chondrogenesis by MSCs:

1. *Induction:* MSCs require specific signals to "commit" to the chondrogenic lineage. These signals can come from growth factors like TGF-β, specific culture conditions (3D aggregates, biomaterials), or mechanical stimuli.

2. *Proliferation and condensation:* Under these influences, MSCs proliferate and condense into aggregates, enhancing cell-cell interactions crucial for further differentiation.
3. *Matrix production and maturation:* Chondrocytes begin producing cartilage-specific extracellular matrix (ECM) components like collagen type II and aggrecan, gradually replacing the initial cell mass.
4. *Hypertrophy (optional):* In some cases, chondrocytes undergo hypertrophy, enlarging and transitioning toward bone formation. This is generally undesirable for cartilage repair as it leads to weaker cartilage-like tissue.

The factors influencing chondrogenesis are:

- Growth factors: TGF-β family (especially TGF-β3), BMPs, IGF-1, FGF-18
- Biomolecules: Extracellular matrix proteins (collagen, fibronectin), hyaluronic acid
- Culture conditions: 3D culture systems (pellets, hydrogels), hypoxia, mechanical loading
- Genetic factors: Variations in individual MSCs can affect their chondrogenic potential

Cell culture in medium is supplemented with insulin, transferrin, selenium, linoleic acid, selenium acid, pyruvate, ascorbic phosphate, dexamethasone, and TGF-β III. Additional aid is from IGF-1 and bone morphogenetic proteins (BMP)-2. Differentiation is validated by proteoglycan and type II collagen synthesis [38,52,53].

13.4.2.3 Osteogenesis

Osteogenesis, the formation of new bone, is a complex process that involves the differentiation of mesenchymal stem cells (MSCs) into mature bone-forming cells called osteoblasts. The process of osteogenesis by MSCs can be broadly divided into four stages:

1. *Induction:* MSCs are exposed to various signals, such as growth factors and cytokines, that induce them to commit to the osteogenic lineage. This stage is primarily regulated by the bone morphogenetic protein (BMP) signaling pathway.
2. *Proliferation:* The committed MSCs, now called preosteoblasts, begin to rapidly divide and increase in number. This stage is essential for generating a sufficient number of osteoblasts for bone formation.
3. *Matrix maturation:* Preosteoblasts differentiate into mature osteoblasts, which synthesize and secrete the extracellular matrix (ECM) of bone. The ECM is primarily composed of collagen type I, which provides a scaffold for bone mineralization.
4. *Mineralization:* Osteoblasts deposit minerals, such as calcium and phosphate, onto the collagen fibers in the ECM, leading to the formation of hard, mineralized bone. This stage is regulated by several factors, including alkaline phosphatase activity and the expression of osteocalcin, a bone-specific protein.

Triggered by ascorbic acid, β-glycerophosphate, and dexamethasone the differentiation markers include mineralization of calcium deposits and increased alkaline phosphatase expression [54–57]. However, beyond traditional lineages, MSCs, under specific conditions, demonstrate potential to differentiate into diverse cell types.

13.4.2.4 Myogenesis

Myogenesis, the formation of new muscle tissue, can also occur through the differentiation of mesenchymal stem cells (MSCs). While not as extensively researched as osteogenesis, it holds significant potential for muscle repair and regeneration. MSCs receive signals, primarily from growth factors like hepatocyte growth factor (HGF) and insulin-like growth factor-1 (IGF-1), pushing them toward the myogenic lineage. This stage involves the activation of specific genes, like MyoD and Myogenin, which act as master regulators of muscle development. Committed MSCs, now known as myoblasts, begin to divide rapidly, increasing the population of muscle precursor cells. Myoblasts fuse with each other to form elongated, multinucleated structures called myotubes, the early stages of muscle fibers. This process is mediated by cell adhesion molecules and requires close contact between myoblasts. Within myotubes, muscle proteins like myosin and actin are expressed and organized, leading to the formation of mature, contractile muscle fibers. This stage involves complex interactions between various signaling pathways and gene expression programs. Addition of 5-azacytidine allows MSCs to yield muscle cells, including cardiomyocytes and myoblasts expressing muscle-specific markers [58,59].

13.4.2.5 Endodermal derivatives

Trans differentiation into hepatocytes involves multistage culture with specific factors, yielding cells expressing hepatic markers. β cells of pancreatic islets are obtained with the use of growth factors and specific chemicals, capable of producing insulin [60].

13.4.2.6 Ectodermal derivatives

MSCs can differentiate into neurons in the presence of β-mercaptoethanol (BME) stimulation followed by NGF. Other factors like insulin, retinoic acid, bFGF, EGF, valproic acid, BME, and hydrocortisone also induce neural differentiation [61].

13.4.2.7 Mesengenesis

Described by Caplan and Dennis as a process where MSCs give rise to myoblasts, bone marrow stromal cells, fibroblasts, and cells contribute to connective tissues, ligaments, and tendons. Acid, BME, and hydrocortisone also induce neural differentiation [62].

13.4.2.8 Mature neurons

The differentiation of mesenchymal stem cells (MSCs) into mature neurons is a promising area of research with potential applications in treating neurodegenerative diseases and injuries. The process of differentiating MSCs into neurons involves exposing them to specific growth factors and other molecules that stimulate the expression of neuronal genes and proteins. This process can be challenging, as it requires carefully controlled conditions to ensure that the MSCs differentiate into the desired type of neuron. However, if successful, this process could lead to the development of new therapies for a variety of neurological conditions. For example, MSCs could be differentiated into dopaminergic neurons to treat Parkinson's disease, or into oligodendrocytes to treat multiple sclerosis.

Here are some of the key challenges and potential applications of differentiating MSCs into mature neurons. First is identifying the optimal conditions for differentiation: different types of MSCs may require different conditions for differentiation into neurons. Second is ensuring the purity of the differentiated cells. It is important to ensure that the differentiated cells are mature neurons and not simply precursors and third is preventing tumor formation. There is a risk that MSCs could form tumors if they are not properly differentiated. Stimulation with glial cell−derived neurotrophic factor, BDNF, retinoic acid, 5-azacytidine, isobutylmethylxanthine, and indomethacin transforms MSCs into mature neurons expressing neural markers [61,63]. Under controlled in vitro conditions and specific stimuli, MSCs exhibit the potential to differentiate into cell types representing all three embryonic germ layers, showcasing their remarkable plasticity.

13.4.3 Immunomodulatory properties of mesenchymal stem cells

Mesenchymal stem cells (MSCs) exhibit diverse immunomodulatory properties, playing a crucial role in regulating immune responses. Here is a brief overview of the key immunomodulatory characteristics of MSCs:

13.4.3.1 Innate immune response interaction

MSCs interact with monocytes, influencing their differentiation and function. They can modulate macrophages toward an antiinflammatory (M2) phenotype, dampening proinflammatory responses [64].

13.4.3.2 Adaptive immune response modulation

MSCs suppress T-cell proliferation and activation, contributing to immune tolerance. They promote the generation of regulatory T cells (Tregs), which further suppress immune responses. MSCs help maintain a balance between proinflammatory Th17 cells and antiinflammatory Tregs [65].

13.4.3.3 Cytokine and chemokine regulation

MSCs secrete antiinflammatory cytokines, such as IL-10 and TGF-β, reducing immune responses. They influence chemokine production, impacting immune cell recruitment and migration [66].

13.4.3.4 Inhibition of immune cell functions

MSCs suppress natural killer (NK) cell activity, reducing cytotoxic effects. They regulate B-cell proliferation and antibody production [66].

13.4.3.5 Modulation of dendritic cells

MSCs inhibit the maturation and antigen-presenting capacity of dendritic cells. They promote the generation of tolerogenic dendritic cells with immunosuppressive properties [67].

13.4.3.6 Induction of immune tolerance

MSCs contribute to the acceptance of allografts, suppressing alloreactive immune responses. They are used to prevent or treat graft-versus-host disease (GvHD) after hematopoietic stem cell transplantation [68].

13.4.3.7 Modulation of inflammatory environment

MSCs contribute to the resolution of inflammation by interacting with inflammatory mediators. They may mitigate cytokine storms associated with severe immune reactions [67].

13.4.3.8 Immunosuppressive factors

MSCs can directly interact with key innate immune cells like macrophages, dendritic cells, and natural killer (NK) cells, suppressing their proinflammatory activity. This is achieved through mechanisms like:

- *Blocking inflammatory signaling pathways:* They inhibit the activation of NF-κB, a key transcription factor for inflammatory genes.
- *Inducing Treg differentiation:* They promote the development of regulatory T cells (Tregs), which dampen immune responses.
- *Modulating cytokine production:* They decrease the production of proinflammatory cytokines like IL-1, IL-6, and TNF-α while increasing antiinflammatory cytokines like IL-10 and TGF-β.

MSCs produce immunosuppressive factors such as prostaglandin E2 (PGE2) and indoleamine 2,3-dioxygenase (IDO). They secrete tumor necrosis factor-inducible gene 6 protein (TSG-6), exhibiting antiinflammatory effects [69].

13.4.3.9 Microenvironment modulation

MSC-derived exosomes carry immunomodulatory cargo, influencing nearby immune cells. MSCs create a regenerative microenvironment, supporting tissue repair [70].

13.4.3.10 Role in autoimmune diseases

MSCs show promise in modulating aberrant immune responses in autoimmune diseases like rheumatoid arthritis and multiple sclerosis [71].

Understanding these immunomodulatory properties positions MSCs as potential therapeutic tools for immune-related disorders, transplantation, and tissue regeneration. Ongoing research aims to uncover the specific mechanisms underlying these effects, enhancing their clinical application.

13.5 Mesenchymal stem cells in disease treatment: a multifaceted approach

Mesenchymal stem cells (MSCs) have demonstrated promising therapeutic potential across various medical conditions. The ability of MSCs to differentiate into diverse cell lineages, coupled with their ethical advantages over embryonic and induced pluripotent stem cells (iPSCs), positions them as excellent candidates for cell therapy. The versatility of MSCs is highlighted in their efficacy for dermatological, musculoskeletal, neurological, cardiovascular, respiratory, renal, gastroenterological, urological conditions, and more. The accumulated evidence from these trials underscores the considerable potential of MSCs in advancing the treatment landscape across various medical disciplines.

Research on bone regeneration, crucial in cases like fractures, defects, osteoarthritis, and osteoporosis, has gained significant attention. While autogenous bone grafts are a standard approach, they come with drawbacks such as unpredictable absorption, prolonged recovery, and potential pain and nerve injury for the patient. Mesenchymal stem cells (MSCs) have become an appealing solution for bone tissue formation due to advancements in understanding bone tissue biology and tissue regeneration [72].

In a pilot study using allogeneic bone marrow−derived MSCs (BMSCs), improvements were observed in bone fracture participants, showing enhanced Tomographic Union Score and Global Disease Evaluation scores, along with reduced pain [33]. Another study combining Wharton's Jelly-derived MSCs (WJ-MSCs) with teriparatide demonstrated positive outcomes for osteoporotic vertebral compression fractures, indicating improved scores in pain and disability indices [73].

For osteoarthritis (OA), studies with BMSCs and umbilical cord−derived MSCs (UC-MSCs) showed promising results, with improved Knee Injury and Osteoarthritis Outcome Scores (KOOS), reduced cartilage catabolic biomarkers, and lower levels of proinflammatory factors. Subchondral infusion of BMSCs postponed or avoided total knee arthroplasty in OA patients, and intraarticular administration of BMSCs with platelet-rich plasma proved to be a safe and feasible alternative treatment [74].

The skin, consisting of the epidermis, dermis, and subcutaneous layer, undergoes a defined wound healing process. Mesenchymal stem cells (MSCs) play a crucial role in each stage, aiding in skin regeneration and reducing scarring by migrating to the damaged site, suppressing inflammation, and enhancing cell growth [75,76]. Clinical studies affirm the efficacy and safety of MSCs in wound healing, demonstrating their positive impact on ulcer size reduction and closure [77]. Studies on vocal fold scarring, diabetic foot ulcers, and intrauterine adhesion show significant improvements with MSC administration, emphasizing their therapeutic potential in skin-related conditions [6,78].

In the context of myocardial injury, hypoxia-treated MSCs preconditioned for 24 hours significantly improved infarct size, left ventricular function, cardiomyocyte proliferation, vascular density, and myocardial glucose uptake in a monkey's infarcted heart after 90 days [79]. Similarly, intravenously administered human MSCs under 5% O_2 conditions improved left ventricular ejection fraction and reduced infarct size in mice with left ventricular dysfunction after 21 days [80]. In a double-blind, placebo-controlled trial involving 53 patients, human MSC treatment resulted in improved forced expiratory volume in 1 second and left ventricular ejection fraction [81].

Pulmonary diseases, including chronic obstructive pulmonary disease and bronchopulmonary dysplasia, have been targeted with MSC therapies. A Phase I study demonstrated that combined one-way endobronchial valve insertion and MSC treatment reduced circulating C-reactive protein levels and improved the BODE index in severe chronic obstructive pulmonary disease patients [82]. In a Phase II study, bronchopulmonary dysplasia patients receiving intratracheal transplantation of allogeneic human umbilical cord blood−derived MSCs exhibited lower severity than the control group [83]. MSC-based therapies for acute respiratory distress syndrome are currently under investigation in a Phase II study [84].

For liver diseases, adipose-derived MSCs and bone marrow MSCs seeded in regenerated silk fibroin scaffolds showed promise in a CCl4-induced liver injury mouse model, demonstrating neovascularization and partial improvement of liver function [85]. Alcoholic cirrhosis patients treated with MSCs exhibited histological improvements and decreased levels of transforming growth factor β1, type I collagen, and α-smooth muscle actin [86].

In gastrointestinal diseases, MSC-loaded hydrogels improved colonic epithelial structure and hyperpermeability in a rat model of radiation-induced severe colonic damage [87]. Crohn's disease patients receiving intravenous infusions of allogeneic MSCs exhibited clinical remission and endoscopic improvement. In a phase III trial for complex perianal fistulas in Crohn's disease, MSC therapy resulted in a significant remission compared to the placebo group after 24 weeks [88].

13.6 Summary and conclusions

In summary, this chapter provides a thorough examination of mesenchymal stem cells (MSCs) and their significant transformative potential in the realm of regenerative medicine. Employing a multidisciplinary methodology encompassing cell isolation, characterization, immunomodulatory studies, tissue regeneration assays, and clinical investigations, the research unveiled the distinctive attributes of MSCs, underscoring their adaptability in differentiation, immunomodulation, and tissue repair.

Through meticulous experimentation and an exhaustive literature review, intricate mechanisms orchestrating cellular regeneration by MSCs were elucidated, establishing them as central entities in therapeutic interventions. This section contextualized the findings within the broader landscape of regenerative medicine, drawing comparisons with existing literature and accentuating the clinical and translational relevance of the study. Transparently addressing limitations and challenges offered a reflective perspective on the research process.

The forward-looking perspective delineated potential future directions and research recommendations, thereby facilitating continued innovation in the field. In conclusion, this chapter underscores the pivotal role of mesenchymal stem cells as architectural entities shaping the future of regenerative medicine. The elucidated mechanisms, immunomodulatory properties, and regenerative capabilities of MSCs furnish valuable insights promising diverse clinical applications.

As the field advances, the study advocates for a judicious yet optimistic approach, acknowledging the transformative potential of MSCs while recognizing the complexities necessitating further exploration. Serving as a foundational resource, this chapter caters to researchers, clinicians, and scholars actively involved in the dynamic landscape of regenerative medicine. By unraveling the intricacies of cellular therapy and MSC biology, the study significantly contributes to the ongoing discourse on the potential of MSCs to revolutionize approaches to healing, tissue repair, and the future of medical interventions.

References

[1] Zakrzewski W, Dobrzyński M, Szymonowicz M, Rybak Z. Stem cells: past, present, and future. Stem Cell Res Ther 2019;10:68. Available from: https://doi.org/10.1186/s13287-019-1165-5.
[2] Eskew AM, Jungheim ES. A history of developments to improve in vitro fertilization. Mo Med 2017;114(3):156−9.
[3] Liu Y, Yang R, He Z, Gao WQ. Generation of functional organs from stem cells. Cell Reg 2013;2(1):1. Available from: https://doi.org/10.1186/2045-9769-2-1.
[4] Margiana R, Markov A, Zekiy AO, Siahmansouri H. Clinical application of mesenchymal stem cell in regenerative medicine: a narrative review. Stem Cell Res Ther 2022;13:366. Available from: https://doi.org/10.1186/s13287-022-03054-0.
[5] Moreira A, Kahlenberg S, Hornsby P. Therapeutic potential of mesenchymal stem cells for diabetes. J Mol Endocrinol 2017;59(3):R109−20. Available from: https://doi.org/10.1530/JME-17-0117.
[6] Huang J, Li Q, Yuan X, Liu Q, Zhang W, Li P. Intrauterine infusion of clinically graded human umbilical cord-derived mesenchymal stem cells for the treatment of poor healing after uterine injury: a phase I clinical trial. Stem Cell Res Ther 2022;13:85. Available from: https://doi.org/10.1186/s13287-022-02756-9.

[7] Friedenstein AJ, Petrakova KV, Kurolesova AI, Frolova GP. Heterotopic of bone marrow. Analysis of precursor cells for osteogenic and hematopoietic tissues. Transplanttion 1968;6:230–47.
[8] Friedenstein AJ, Chailakhjan RK, Lalykina KS. The development of fibroblast colonies in monolayer cultures of guinea-pig bone marrow and spleen cells. Cell Tissue Kinetics 1970;3:393–403.
[9] Friedenstein AJ, Gorskaja JF, Kulagina NN. Fibroblast precursors in normal and irradiated mouse hematopoietic organs. Exptl Hematol 1976;4:267–74.
[10] Castro-Malaspina H, Gay RE, Resnick G, Kapoor N, Meyers P, Chiarieri D, et al. Characterization of human bone marrow fibroblast colony-forming cells (CFU-F) and their progeny. Blood 1980;56:289–301.
[11] Caplan AI. Mesenchymal stem cells. J Ortho Res 1991;9:641–50.
[12] Kuznetsov SA, Krebsbach PH, Satomura K, Kerr J, Riminucci M, Benayahu D, et al. Single-colony derived strains of human marrow stromal fibroblasts form bone after transplantation in vivo. J Bone Miner Res 1997;12:1335–47.
[13] Horwitz EM, Le BK, Dominici M, Mueller I, Slaper-Cortenbach I, Keating A. Clarification of the nomenclature for MSC: the international society for cellular therapy position statement. Cytotherapy 2005;7:393–5.
[14] Dominici M, Le BK, Mueller I, Slaper-Cortenbach I, Marini F, Krause D, et al. Minimal criteria for defining multipotent mesenchymal stromal cells, International Society for Cellular Therapy Position statement. Cytotherapy 2006;8:315–17.
[15] Le BK, Samuelsson H, Gustafsson B, Remberger M, Sundberg B, Arvidson J, et al. Transplantation of mesenchymal stem cells to enhance engraftment of hematopoietic stem cells. Leukaemia 2007;21:1733–8.
[16] Soleimani M, Nadri S. A protocol for isolation and culture of mesenchymal stem cells from mouse bone marrow. Nat Protoc 2009;4:102–6.
[17] Chao YH, Wu HP, Chan CK, Tsai C, Peng CT, Peng CT, et al. Umbilical cord-derived mesenchymal stem cells for hematopoietic stem cell transplantation. J Biomed Biotechnol 2012;759503.
[18] Han J, Park J, Kim BS. Integration of mesenchymal stem cells with nano-biomaterials for the repair of myocardial infarction. Adv Drug Deliv Rev 2015;95:15–28.
[19] Martino MM, Maruyama K, Kuhn GA, Satoh T, Takeuchi O, Müller R, et al. Inhibition of IL-1R1/MyD88 signalling promotes mesenchymal stem cell-driven tissue regeneration. Nat Commun 2016;7:11051.
[20] Davies LC, Heldring N, Kadri N, Le BK. Mesenchymal stromal cell secretion of programmed death-1 ligands regulates T cell mediated immunosuppression. Stem Cell 2017;35:766–76.
[21] Zhao K, Lou R, Huang F, Peng Y, Jiang Z, Huang K, et al. Immunomodulation effects of mesenchymal stromal cells on acute graft-versus-host disease after hematopoietic stem cell transplantation. Bio Blood Marrow Transpl 2015;21:97–104.
[22] Martin I, De Boer J, Sensebe L. A relativity concept in mesenchymal stromal cell manufacturing. Cytotherapy 2016;18:613–20.
[23] Pittenger MF, Mackay AM, Beck SC, Jaiswal RK, Douglas R, Mosca JD, et al. Multilineage potential of adult human mesenchymal stem cells. Science 1999;284:143–7.
[24] Mamidi MK, Nathan KG, Singh G, Thrichelvam ST, Yusof M, Nasim NA, et al. Comparative cellular and molecular analyses of pooled bone marrow multipotent mesenchymal stromal cells during continuous passaging and after successive cryopreservation. J Cell Biochem 2012;113:3153–64.

[25] Miao Z, Jin J, Chen L, Zhu J, Huang W, Zhao J, et al. Isolation of mesenchymal stem cells from human placenta: comparison with human bone marrow mesenchymal stem cells. Cell Biol Intern 2006;30:681−7.

[26] Bieback K, Kern S, Klüter H, Eichler H. Critical parameters for the isolation of mesenchymal stem cells from umbilical cord blood. Stem Cell 2004;22:625−34.

[27] Salehinejad P, Alitheen NB, Ali AM, Omar AR, Mohit M, Janzamin E, et al. Comparison of different methods for the isolation of mesenchymal stem cells from human umbilical cord Wharton's jelly. Vitro Cell Dev Biol-Animal 2012;48:75−83.

[28] Yoon JH, Roh EY, Shin S, Jung NH, Song EY, Chang JY, et al. Comparison of explant-derived and enzymatic digestion-derived MSCs and the growth factors from Wharton's jelly. BioMed Res Inter 2013;428726.

[29] Baglioni S, Francalanci M, Squecco R, Lombardi A, Cantini G, Angeli R, et al. Characterization of human adult stem-cell populations isolated from visceral and subcutaneous adipose tissue. FASEB J 2009;23:3494−505.

[30] Wagner W, Wein F, Seckinger A, Frankhauser M, Wirkner U, Krause U, et al. Comparative characteristics of mesenchymal stem cells from human bone marrow, adipose tissue, and umbilical cord blood. Exptl Hematol 2005;33:1402−16.

[31] In't Anker PS, Scherjon SA, Kleijburg-van der Keur C, Noort WA, Claas FH, Kanhai HH. Amniotic fluid as a novel source of mesenchymal stem cells for therapeutic transplantation. Blood 2003;102:1548−9.

[32] Cai J, Li W, Su H, Qin D, Yang J, Zhu F, et al. Generation of human induced pluripotent stem cells from umbilical cord matrix and amniotic membrane mesenchymal cells. J Biol Chem 2010;285:11227−34.

[33] Udehiya RK, Amarpal, Kinjavdekar P, Aithal HP, Nath A, Pawde AM, et al. Comparison of autogenic and allogenic bone marrow-derived mesenchymal stem cells for the repair of segmental bone defects in rabbits. Res Vet Sci 2013;94(3):743−52. Available from: https://doi.org/10.1016/j.rvsc.2013.01.011.

[34] Udehiya RK, Amarpal, Kinjavdekar P, Aithal HP, Nath A, Pawde AM, et al. Isolation, ex vivo expansion and characterization of rabbit bone marrow-derived mesenchymal stem cells (rBM-MSCs). Indian J Vet Surg 2013;34(1):34−6.

[35] Varghese V.A. Clinical application of adipose-derived mesenchymal stem cells for the management of osteoarthritis in dogs [MVSc thesis]. Ludhiana, India: Guru Angad Dev Veterinary and Animal Sciences University; 2019.

[36] Bieback K, Netsch P. Isolation, culture, and characterization of human umbilical cord blood-derived mesenchymal stromal cells. Methods Mol Biol 2016;1416:245−58. Available from: https://doi.org/10.1007/978-1-4939-3584-0_14.

[37] Gharibi B, Hughes FJ. Effects of medium supplements on proliferation, differentiation potential, and in vitro expansion of mesenchymal stem cells. Stem Cell Trans Med 2012;1(11):771−82. Available from: https://doi.org/10.5966/sctm.2010-0031.

[38] Danisovic L, Varga I, Polak S, Ulicna M, Hlavackova L, Bohmer D, et al. Comparison of in vitro chondrogenic potential of human mesenchymal stem cells derived from bone marrow and adipose tissue. Gen Physiol Biophy 2009;28:56−62.

[39] Barry F, Boynton R, Murphy M, Haynesworth S, Zaia J. The sh-3 and sh-4 antibodies recognize distinct epitopes on cd73 from human mesenchymal stem cells. Biochem Biophy Res Commun 2001;289:519−24.

[40] Singh H, Lonare MK, Sharma M, Udehiya R, Singla S, Saini SP, et al. Interactive effect of carbendazim and imidacloprid on buffalo bone marrow-derived mesenchymal stem cells: Oxidative stress, cytotoxicity, and genotoxicity. Drug Chem Toxicol 2012;46(1):35−49. Available from: https://doi.org/10.1080/01480545.2021.2007023.

[41] Boxall SA, Jones E. Markers for characterization of bone marrow multipotential stromal cells. Stem Cell Intern 2012;975871.
[42] Varma MJ, Breuls RG, Schouten TE, Jurgens WJ, Bontkes HJ, Schuurhuis GJ, et al. Phenotypical and functional characterization of freshly isolated adipose tissue-derived stem cells. Stem Cell Dev 2007;16:91−104.
[43] Nassiri F, Cusimano MD, Scheithauer BW, Rotondo F, Fazio A, Yousef GM, et al. Endoglin (CD105): A review of its role in angiogenesis and tumor diagnosis, progression, and therapy. Anticancer Res 2011;31:2283−90.
[44] Lin CS, Ning H, Lin G, Lue TF. Is CD34 truly a negative marker for mesenchymal stromal cells? Cytotherapy 2012;14:1159−63.
[45] Gimble JM, Katz AJ, Bunnell BA. Adipose-derived stem cells for regenerative medicine. Circulation Res 2007;100:1249−60.
[46] Lin G, Garcia M, Ning H, Banie L, Guo YL, Lue TF, et al. Defining stem and progenitor cells within adipose tissue. Stem Cell Dev 2008;17:1053−63.
[47] Lin CS, Lue TF. Defining vascular stem cells. Stem Cell Dev 2013;22(7):1018−26.
[48] Ning H, Lin G, Lue TF, Lin CS. Mesenchymal stem cell marker STRO-1 is a 75 kD endothelial antigen. Biochem Biophy Res Commun 2011;413:353−7.
[49] Schipper BM, Marra KG, Zhang W, Donnenberg AD, Rubin JP. Regional anatomic and age effects on cell function of human adipose-derived stem cells. Ann Plastic Surg 2008;60:538−44.
[50] Shahparaki A, Grunder L, Sorisky A. Comparison of human abdominal subcutaneous versus omental preadipocyte differentiation in primary culture. Metabolism 2002;51:1211−15.
[51] Ferguson RE, Cui X, Fink BF, Vasconez HC, Pu LL. The viability of autologous fat grafts harvested with the LipiVage system: a comparative study. Ann Plastic Surg 2008;60:594−7.
[52] Huang JI, Kazmi N, Durbhakula MM, Hering TM, Yoo JU, Johnstone B. Chondrogenic potential of progenitor cells derived from human bone marrow and adipose tissue: a patient-matched comparison. J Ortho Res 2005;23:1383−9.
[53] Afizah H, Yang Z, Hui JH, Ouyang HW, Lee EH. A comparison between the chondrogenic potential of human bone marrow stem cells (BMSCs) and adipose-derived stem cells (ADSCs) taken from the same donors. Tissue Engin 2007;13:659−66.
[54] Shi YY, Nacamuli RP, Salim A, Longaker MT. The osteogenic potential of adipose-derived mesenchymal cells is maintained with aging. Plastic Recon Surg 2005;116:1686−96.
[55] Khan WS, Adesida AB, Tew SR, Andrew JG, Hardingham TE. The epitope characterization and the osteogenic differentiation potential of human fat pad-derived stem cells is maintained with ageing in later life. Injury 2009;40:150−7.
[56] Weinzierl K, Hemprich A, Frerich B. Bone engineering with adipose tissue-derived stromal cells. J Cranio-Maxillo-Facial Surg 2006;34:466−71.
[57] Jurgens WJ, Oeday RS, Varma MJ, Helder MN, Zandiehdoulabi B, Schouten TE, et al. Effect of tissue-harvesting site on yield of stem cells derived from adipose tissue: Implications for cell-based therapies. Cell Tissue Res 2008;332:415−26.
[58] Dumont NA, Bentzinger CF, Sincennes MC, Rudnicki MA. Satellite cells and skeletal muscle regeneration. Comp Physiol 2015;5(3):1027−59. Available from: https://doi.org/10.1002/cphy.c140068.
[59] Yin H, Price F, Rudnicki MA. Satellite cells and the muscle stem cell niche. Physiol Rev 2013;93(1):23−67. Available from: https://doi.org/10.1152/physrev.00043.2011.

[60] Cheng X, Tiyaboonchai A, Gadue P. Endodermal stem cell populations derived from pluripotent stem cells. Curr Opin Cell Biol 2013;25(2):265−71. Available from: https://doi.org/10.1016/j.ceb.2013.01.006.
[61] Hernández R, Jiménez-Luna C, Perales-Adán J, Perazzoli G, Melguizo C, Prados J. Differentiation of human mesenchymal stem cells towards neuronal lineage: clinical trials in nervous system disorders. Biomol Ther 2020;28:34−44.
[62] Scuteri A, Donzelli E, Foudah D, Caldara C, Redondo J, D'Amico G, et al. Mesengenic differentiation: comparison of human and rat bone marrow mesenchymal stem cells. Inter J Stem Cells 2014;7(2):127−34. Available from: https://doi.org/10.15283/ijsc.2014.7.2.127.
[63] Bueno C, Martínez-Morga M, García-Bernal D, Moraleda JM, Martínez S. Differentiation of human adult-derived stem cells towards a neural lineage involves a dedifferentiation event prior to differentiation to neural phenotypes. Sci Rep 2012;11:12034. Available from: https://doi.org/10.1038/s41598-021-91566-9.
[64] Hu C, Li L. The immunoregulation of mesenchymal stem cells plays a critical role in improving the prognosis of liver transplantation. J Transl Med 2019;17:412. Available from: https://doi.org/10.1186/s12967-019-02167-0.
[65] de Witte SFH, Luk F, Sierra Parraga JM, Gargesha M, Merino A, Korevaar SS, et al. Immunomodulation by therapeutic mesenchymal stromal cells (MSC) is triggered through phagocytosis of MSC by monocytic cells. Stem Cells 2018;36:602−15. Available from: https://doi.org/10.1002/stem.2779.
[66] Kyurkchiev D, Bochev I, Ivanova-Todorova E, Mourdjeva M, Oreshkova T, Belemezova K, et al. Secretion of immunoregulatory cytokines by mesenchymal stem cells. World J Stem Cell 2014;6(5):552−70. Available from: https://doi.org/10.4252/wjsc.v6.i5.552.
[67] Jiang W, Xu J. Immune modulation by mesenchymal stem cells. Cell Prolif 2020;53 (1):e12712. Available from: https://doi.org/10.1111/cpr.12712.
[68] Iwai S, Okada A, Sasano K, Endo M, Yamazaki S, Wang X, et al. Controlled induction of immune tolerance by mesenchymal stem cells transferred by maternal microchimerism. Biochem Biophy Res Commun 2012;539:83−8. Available from: https://doi.org/10.1016/j.bbrc.2020.12.032.
[69] Ghannam S, Bouffi C, Djouad F, Jorgensen C, Noël D. Immunosuppression by mesenchymal stem cells: mechanisms and clinical applications. Stem Cell Res Ther 2010;1 (1):2. Available from: https://doi.org/10.1186/scrt2.
[70] Zhang L. The role of mesenchymal stem cells in modulating the breast cancer microenvironment. Cell Transpl 2023;32. Available from: https://doi.org/10.1177/09636897231220073.
[71] Rad F, Ghorbani M, Mohammadi Roushandeh A, Habibi Roudkenar M. Mesenchymal stem cell-based therapy for autoimmune diseases: emerging roles of extracellular vesicles. Mol Biol Rep 2019;46(1):1533−49. Available from: https://doi.org/10.1007/s11033-019-04588-y.
[72] Čamernik K, Mihelič A, Mihalič R, Haring G, Herman S, Marolt Presen D, et al. Comprehensive analysis of skeletal muscle- and bone-derived mesenchymal stem/stromal cells in patients with osteoarthritis and femoral neck fracture. Stem Cell Res Ther 2020;11:146. Available from: https://doi.org/10.1186/s13287-020-01657-z.
[73] Shim J, Kim KT, Kim KG, Choi UY, Kyung JW, Sohn S, et al. Safety and efficacy of Wharton's jelly-derived mesenchymal stem cells with teriparatide for osteoporotic vertebral fractures: a phase I/IIa study. Stem Cell Transl Med 2021;10(4):554−67. Available from: https://doi.org/10.1002/sctm.20-0308.

[74] Lamo-Espinosa JM, Blanco JF, Sánchez M, Moreno V, Granero-Moltó F, Sánchez-Guijo F, et al. Phase II multicenter randomized controlled clinical trial on the efficacy of intra-articular injection of autologous bone marrow mesenchymal stem cells with platelet-rich plasma for the treatment of knee osteoarthritis. J Transl Med 2020;18 (1):356. Available from: https://doi.org/10.1186/s12967-020-02530-6.

[75] Hu MS, Borrelli MR, Lorenz HP, Longaker MT, Wan DC. Mesenchymal stromal cells and cutaneous wound healing: a comprehensive review of the background, role, and therapeutic potential. Stem Cell Inter 2018;6901983. Available from: https://doi.org/10.1155/2018/6901983.

[76] Marfia G, Navone SE, Di Vito C, Ughi N, Tabano S, Miozzo M, et al. Mesenchymal stem cells: potential for therapy and treatment of chronic non-healing skin wounds. Organogenesis 2015;11(4):183−206. Available from: https://doi.org/10.1080/15476278.2015.1126018.

[77] Falanga V, Iwamoto S, Chartier M, Yufit T, Butmarc J, Kouttab N, et al. Autologous bone marrow-derived cultured mesenchymal stem cells delivered in a fibrin spray accelerate healing in murine and human cutaneous wounds. Tissue Engin 2007;13 (6):1299−312. Available from: https://doi.org/10.1089/ten.2006.0278.

[78] Shi R, Jin Y, Cao C, Han S, Shao X, Meng L, et al. Localization of human adipose-derived stem cells and their effect in repair of diabetic foot ulcers in rats. Stem Cell Res Ther 2016;7(1):155. Available from: https://doi.org/10.1186/s13287-016-0412-2.

[79] Hu X, Xu Y, Zhong Z, Wu Y, Zhao J, Wang Y, et al. A large-scale investigation of hypoxia-preconditioned allogeneic mesenchymal stem cells for myocardial repair in nonhuman primates: paracrine activity without remuscularization. Circulation Res 2016;118:970−83.

[80] Luger D, Lipinski MJ, Westman PC, Glover DK, Dimastromatteo J, Frias JC, et al. Intravenously delivered mesenchymal stem cells: systemic anti-inflammatory effects improve left ventricular dysfunction in acute myocardial infarction and ischemic cardiomyopathy. Circulation Res 2017;120:1598−613.

[81] Hare JM, Traverse JH, Henry TD, Dib N, Strumpf RK, Schulman SP, et al. A randomized, double-blind, placebo-controlled, dose-escalation study of intravenous adult human mesenchymal stem cells (prochymal) after acute myocardial infarction. J Am Coll Cardiol 2009;54:2277−86.

[82] de Oliveira HG, Cruz FF, Antunes MA, de Macedo Neto AV, Oliveira GA, Svartman FM, et al. Combined bone marrow-derived mesenchymal stromal cell therapy and one-way endobronchial valve placement in patients with pulmonary emphysema: a phase I clinical trial. Stem Cell Transl Med 2017;6:962−9.

[83] Chang YS, Ahn SY, Yoo HS, Sung SI, Choi SJ, Oh WI, et al. Mesenchymal stem cells for bronchopulmonary dysplasia: phase 1 dose-escalation clinical trial. J Pediat 2014;164 966-72e6.

[84] Wilson JG, Liu KD, Zhuo H, Caballero L, McMillan M, Fang X, et al. Mesenchymal stem (stromal) cells for treatment of ARDS: a phase 1 clinical trial. Lancet Respir Med 2015;3:24−32.

[85] Xu L, Wang S, Sui X, Wang Y, Su Y, Huang L, et al. Mesenchymal stem cell-seeded regenerated silk fibroin complex matrices for liver regeneration in an animal model of acute liver failure. ACS Appl Mat Interfaces 2017;9:14716−23.

[86] Jang YO, Kim YJ, Baik SK, Kim MY, Eom YW, Cho MY, et al. Histological improvement following administration of autologous bone marrow-derived mesenchymal stem cells for alcoholic cirrhosis: a pilot study. Liver Intern 2014;34:33−41.

[87] Moussa L, Pattappa G, Doix B, Mathieu N. A biomaterial-assisted mesenchymal stromal cell therapy alleviates colonic radiation-induced damage. Biomaterials 2017;115:40−52.
[88] Panés J, García-Olmo D, Van Assche G, Colombel JF, Reinisch W, Baumgart DC, et al. Expanded allogeneic adipose-derived mesenchymal stem cells (Cx601) for complex perianal fistulas in Crohn's disease: a phase 3 randomised, double-blind controlled trial. Lancet 2016;388:1281−90.

Polymeric nanoencapsulation for ameliorative application in rodent hepatic regeneration

14

Deba Brata Mondal[1], Jithin Mullakkalparambil Velayudhan[2], Aishwarya Lekshman[1], Ravi Shankar Kumar Mandal[3], Raguvaran Raja[1] and Naveen Kumar[4,]*

[1]Division of Medicine, ICAR-Indian Veterinary Research Institute, Izatnagar, Uttar Pradesh, India, [2]Department of Veterinary Medicine, College of Veterinary and Animal Sciences, Sardar Vallabhbhai Patel University of Agriculture and Technology, Meerut, Uttar Pradesh, India, [3]Department of Veterinary Medicine, College of Veterinary and Animal Sciences, Patna, Bihar, India, [4]Division of Surgery, ICAR-Indian Veterinary Research Institute, Izatnagar, Uttar Pradesh, India

14.1 Introduction

Design of nanosystem to deliver drugs with appropriate dosage and interval at the target site is the most striking areas of research at present. Nanocarrier systems are submicron particles containing entrapped drugs designed for enteral or parenteral administration, which may prevent or minimize the drug degradation, its metabolism as well as the cellular efflux [1,2]. Main objectives in the drug delivery are to design a system maintaining the chemical structure and activity of drug biomolecules and to release the therapeutic agent predictably over total course of time and defend it from degradation and/or degrade to nontoxic metabolites that are either absorbed or excreted.

Nanoparticles (NPs) include synthetic biodegradable polymers, natural biopolymers, lipids, and polysaccharides; become safe materials having long shelf-life; and have the potential for overcoming important mucosal barriers. Enormous deal of consideration has been directed to polymeric colloidal nanoparticulate formulations, which have been attained with polysaccharides, lipids, and specifically natural biopolymers. Interaction between biodegradable cationic and anionic biopolymers directing to the formation of polyionic hydrogels has demonstrated encouraging characteristics for drug entrapment and release [3]. Encapsulating drugs within nanoparticles can advance the solubility and pharmacokinetics of drugs, enabling further clinical improvement of disease. Major carrier materials of nanoparticles are synthetic biodegradable high molecular polymer and natural polymer. The former

*Present affiliation: Veterinary Clinical Complex, Apollo College of Veterinary Medicine, Jaipur, Rajasthan, India.

usually includes poly-α-cyanoacrylate alkyl esters, polyvinyl alcohol, polylactic acid, and polylacticcoglycolic acid. The latter is usually alienated into two classes: proteins (albumin, gelatin, and vegetable protein) and polysaccharides (cellulose, starch, and its derivatives, alginate, pectin, chitin, and chitosan).

Nanoparticles are colloidal systems with particles varying in size from 10 nm to 1000 nm [4,5]. Nanoparticle systems with mean particle size above the 100 nm have also been reported in literature [6–8]. In addition, nanoparticles could also be defined as being submicronic (<1 μm) colloidal systems [9]. Nanospheres have a matrix type structure in which active ingredient is dispersed throughout the particles, whereas the nanocapsules have a polymeric membrane and an active ingredient core. Nanoization possesses many advantages, such as increasing compound solubility, reducing medicinal doses, and improving the absorbency of medicines compared with the respective crude drugs preparations [9]. Use of polysaccharide polymers, namely, alginate, pectin, and chitosan have been accepted by European Pharmacopoeia as a formulation aid in controlled drug delivery systems and has become a significant area of research and development. Alginate has been used in many biomedical applications, including drug delivery systems, as it is biodegradable, biocompatible and mucoadhesive. Delivery systems are formed with monovalent, water-soluble alginate salts which under goes an aqueous sol-gel transformation to water-insoluble salts due to the addition of divalent ions such as calcium, strontium and barium [10]. Pectin is a high molecular weight, non toxic, anionic, biocompatible natural plant polysaccharide consisting of α-1−4-D-galacturonic acid residues forming a homo galacturon backbone. It has been used successfully for many years in the food and beverage industry as a thickening agent, a gelling agent and a colloidal stabilizer [11]. Ability of pectins to form gel depends on the molecular size and degree of esterification. Low pectin gelatinization in the presence of calcium salts are less water soluble than natural pectins; since calcium ions induce noncovalent associations of carbohydrate chains through the formation of the so-called egg box complexes [12]. Its biocompatible nontoxic effect, acid stability, mucoadhesive nature, and good biodegradability make it suitable carrier for drug delivery.

Chitosan (CS), the N-deacetylation form of chitin mostly found in the exoskeleton of crustacean, insects, and fungi, is a natural polysaccharide. CS is not only nontoxic, biodegradable, biocompatible, mucoadhesive with low immunogenicity but also possesses a high density of positive charge in an acid solution attributed to the glucosamine group on its backbone [13,14]. Because of these beneficial characteristics, increasing attention has been drawn to the applications of CS-based micro- and nanoparticles in the pharmaceutical and nutraceutical field [15].

Liver is the largest parenchymal organ to carry out essential biochemical functions [16]. It has tremendous ability to regenerate [17], but increasing oxidative stress can lead to further hepatic injury as a result, the treatment strategy has been more focused on the use of antioxidants [18−20]. Therapy of hepatobiliary dysfunction aimed for elimination of causative factors, reducing hepatic inflammation, minimizing fibrosis, controlling complications, and initiating hepatic regeneration. Drugs used for therapy against hepatic injury have limited therapeutic effects due to decreased bioavailability of drug and its therapeutic concentration on the target hepatic cell, which necessitates exploring novel and alternative approaches for hepatic regeneration (Fig. 14.1). Natural products with antioxidant property having free radical scavenging activity have attracted immense attention as potential functional ingredients for therapy of hepatopathy [19,20].

Catechins are predominant form of flavanols found in green and black tea, red wine, chocolate, fruits, etc. with various pharmacological effects including antioxidant, antidiabetic, antiinflammatory, antimutagenic, anticarcinogenic, and antimicrobial activities and have attracted particular attention due to their relatively high antioxidant capacity in biological systems [21]. However, the application of catechin is limited because of its unstable nature in solution with poor bioavailability in the body [18]. Stability of tea catechins is an important issue as they must be stable in finished products to deliver health benefits. Factors responsible for stability include pH, temperature, oxygen concentration, and humidity or metal ions. Autooxidation of catechin is pH dependent and it increases with increase in pH of solution. High temperature results in potency loss of catechin but the stability also is influenced by pH of the solution. Increasing pH causes degradation of catechins. Bioavailability of the catechin molecules is as low as 5% [22]. Lack of long-term stability [23] and sensitivity to light and temperature [24] make these molecules unpopular in pharmaceutical and other industries (Fig. 14.1).

Controlled drug delivery systems have shown their potential to protect sustained release and increase the action of different bioactive compounds [25,26]. Nanodrug delivery system needs a scientific approach to deliver the components in a sustained

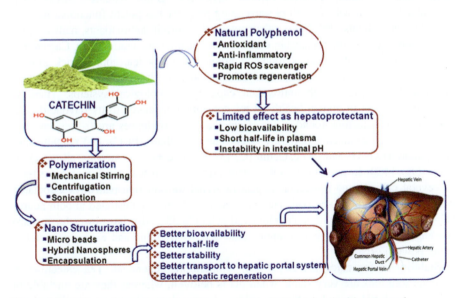

Figure 14.1 Schematic representation for nanostructurization of antioxidant toward nanodrug delivery system.

manner so as to increase patient compliance and avoid repeated administration. It could be achieved by designing novel drug delivery systems (NDDS), in addition to the drugs already available in the market. Novel drug delivery systems not only reduce the repeated administration due to its sustained release properties to overcome noncompliance but also help to increase the therapeutic value by reducing toxicity, increasing the bioavailability, stability, and target ability to the specific cell or organ due to its subcellular size. Importance of drug delivery approach with nanoparticles especially against chronic inflammatory diseases like hepatic injury, emphasis may be made for better understanding of nanoencapsulation of antioxidant with polymeric nanoparticles as an ameliorative application for rodent hepatic regeneration (Fig. 14.1). Any hepatoprotectant with reduced and prolonged dose interval may be further boon for better understanding of animal diseases and its therapeutic management.

14.2 Catechin: A potent antioxidant and hepatoprotectant

Catechins are phytochemical compounds found in high concentrations in variety of plant-based foods and beverages. Health benefits of green tea are mainly attributed for its antioxidant properties through the ability of its polyphenolic catechins to scavenge reactive oxygen species (ROS) [27]. Presence of the 3,4,5-trihydroxy B-ring has been shown to be important for antioxidant and radical scavenging activity [28,29]. Antioxidant potentially protects cellular structures against oxidative stress [20]. Natural antioxidant with free ROS scavenging property has potent functional hepatoprotectant ingredients [30]. Many other potent plant origin antioxidants, namely, piperine, quercitin, curcumin, catechin etc., help in hepatic regeneration [31]. Catechin is predominant form of flavanols found in green tea, black tea, apple, black grapes, blackberries, cherries, raspberries, fava beans, pears, etc. It attracted particular attention due to relatively high antioxidant capacity in biological systems. There is increasing interest in human health by tea consumption related to its flavonoid content. Principal catechins found in tea (*Camellia sinensis*) leaves are (−) epicatechin (EC), (−) epigallocatechin (EGC), (−) epicatechin gallate (ECG), (−) epigallocatechin gallate (EGCG) and (+) catechin (C) [32,33]; except (+) catechin, all are cytotoxic (Fig. 14.2). In addition to antioxidant effects, green tea catechins have effects on several cellular and molecular targets in signal transduction pathways associated with cell death and cell survival. These effects have been demonstrated in both neuronal cells and in tumor epithelial/endothelial cells [34,35]. Also has power for in vitro scavengers of ROS [36], in vivo antioxidant activity by rapid and extensive metabolizing power [29].

Stability of tea catechins is pH and temperature dependent. Tea catechins in aqueous solutions are very stable when pH is below 4, whereas they are unstable in solutions with pH > 6. In addition, storage temperature affects the stability of tea catechins significantly even at ambient conditions [37,38]. Catechin stability is influenced by oxygen concentration, the presence of free radicals as well as metal

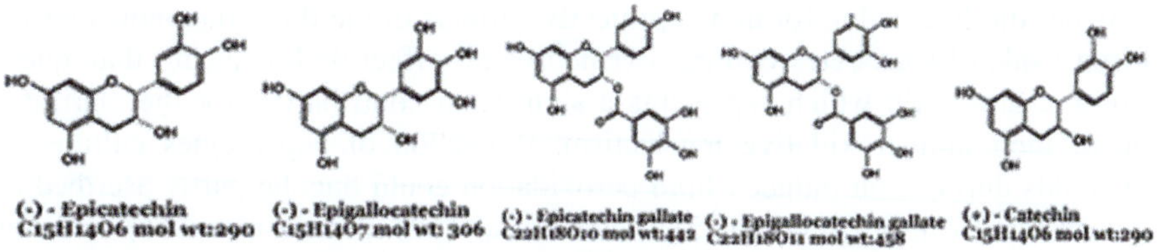

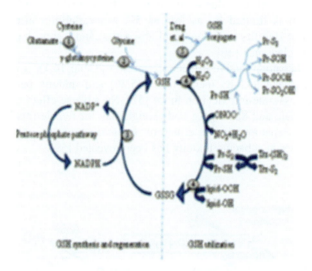

Figure 14.2 Chemical structure of different biological forms of catechin [33]. From Victoria KA, Amber S, Weibiao Z. Green tea catechins during food processing and storage: A review on stability and detection. Food Res Int 2013; 50: 469–79.

Figure 14.3 Pathways for GSH synthesis, regeneration, and utilization. (1) c-glutamycysteine synthetase, (2) GSH synthetase, (3) glutathione reductase, (4) glutathione peroxidase, (5) glutathione S-transferase.

ions. Sang et al. [39] reported that higher oxygen levels and low concentration of antioxidants increased catechin oxidation. Catechins chelate metal ions to form metal complexes [40]. Metal ions would affect antioxidative activity of catechins by their binding to catechins (Fig. 14.3).

Pharmacological property and health benefits of catechin hydrate are attributed with respect to its potential for preventing and treating cancer [41,42], cardiovascular diseases [43] and coronary heart disease [41,42]. It has antiinflammatory [43], antimutagenic, anticarcinogenic, antimicrobial [44], neuroprotective [45] and antiinflammatory [42] properties. Pharmacological effects as an antioxidant [46], nephroprotective [47] and neuroprotective [45] effect have been reported. Optimized dose of catechin (@ 5, 10, 20 and 50 mg/kg body wt) against neurodegenerative disease [46] and (@10 and 20 mg/kg body wt) orally for 21 days for antiinflammatory and antioxidant effects [46,48,49] have been emphasized. Active (−) OHgrs of (+) catechin [44] helps to treat diseases [50].

As per the IC_{50} value for hepatoprotective effects of the three flavonoids tested in iron-loaded hepatocytes culture revealed better effect with catechin than quercetin and diosmetin which represents a significant contribution for the clarification of their antiperoxidative mechanism. Protection of hepatocytes culture by flavonoids during iron induced lipid peroxidation could thus be partly ascribed to a mechanism of iron chelation [51]. Danni et al. [52]. demonstrated (+) catechin acts as a powerful free radical scavenger and an antioxidant (Fig. 14.4), protecting the rat liver endoplasmic reticulum against in vitro CCl_4 toxicity. Antioxidant property of catechin decreases when it is exposed to alkaline pH in intestine [53,54]. Catechins, especially those present in green tea, have received considerable attention in recent days due to its favorable biological properties though therapeutic potential is limited by as low as 5% bioavailability attributed by oral route due to poor stability with intestinal absorption and a short half-life owing to strong systemic clearance [30].

Among the hepatoprotective effects of catechin (Table 14.1), it protects against toxicity of halothane and its metabolites [55,56] and inhibits pathogenesis with CCl_4 and galactosamine-intoxication in rat [57]. No adverse effect of tea extract @ 764 mg/kg for male and 820 mg/kg body weight/day for female rats has been demonstrated [58]. Hepatoprotective effect of *Camellia sinensis* extract containing 41.6 mg catechin/g in subacute toxicity has been revealed [59].

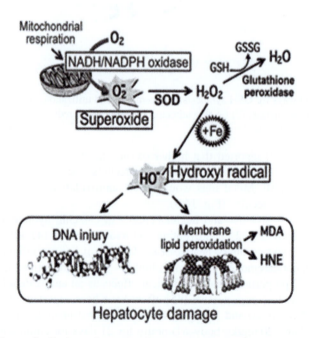

Figure 14.4 Free radical's formation and cell membrane damage [30].

Table 14.1 Biological, pharmacological and hepatoprotectant property of catechin: challenges and solutions.

Challenges	Probable solutions
Poor bioavailability in body [18]	Nanoparticles maintain structure, activity, and drug release [60]
Unstable in alkaline pH [61]	Formation of poly ionic hydrogels with biodegradable drug and anionic polymers is ideal DDS [26]
Degraded in gastric fluid [62]	Entrapment of drug with nanocarriers minimize drug degradation [1]
Lack of long-term stability [23]	Long shelf-life of nanoparticles overcomes mucosal barriers [2]
Inactive in hard water and milk [44]	Encapsulation improves solubility and drug pharmacokinetics [3]
Sensitivity to light and temp [24]	NDDS have potency to protect, control release, and increase drug action [25]

14.3 Excipients

Excipients were considered to be inert in that they should not exert any therapeutical and biological action. Stability and effectivity of drug depend on excipients [63]. It acts as adiluents, binders, disperser, glidants, lubricants, coatings, and coloring agents. Potential prospect of natural polymers, namely, alginate, pectin, chitosan, etc. [64–66] as excipients has been demonstrated [63,67]. It is now recognized that excipients can potentially influence the rate and/or extent of absorption of a drug (e.g., by complex formation). Dosage generally consists of one or more active principles together with a varying number of other substances (excipients) that have been added to the formulation in order to facilitate the preparation and administration, promote the consistent release and bioavailability of the drug, and protect it from degradation. Excipients strongly influence the physicochemical characteristics of the final products. Successful formulation of a stable and effective dosage depends on the careful selection of excipients.

14.4 Novel drug delivery systems

Novel drug delivery system (NDDS) designs the drug by maintaining structure and activity of biomolecules for delivering drug at the right place, dosage, and interval (Fig. 14.5).

NDDS increases therapeutic stability by reducing toxicity and increasing bioavailability [60]. It is helpful for sustained drug release to increase patient compliance and avoid repeated administration [68]. It increases targeting capacity to specific cell or organ due to subcellular size [31] and helps to treat chronic diseases

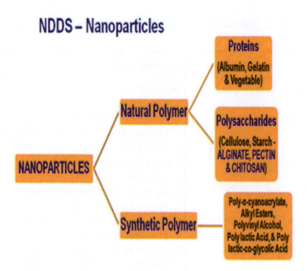

Figure 14.5 Schematic representation of NDDS.

like hepatopathy [68]. Effectiveness of drug delivery depends on the ability of carrier to deliver right amount of drug in the right form to the right place (pharmacokinetic) followed by pharmacological action of drug after it reaches its target (pharmacodynamc) site. Initiatives are underway globally to develop medical breakthrough based on nanotechnology. Nanotechnology is the new way of looking at how we manipulate and utilize matter on a very small scale—the nanoscale. Nanotechnology provides a means to address biological pathways with precisely targeted nanodrugs that are designed to fit the protein and nucleic acid sites associated with disease and disorders. Nanotechnology also provides the tools and techniques to deliver delicate organic macromolecules and peptides to their effective sites of action, protected from degradation, immune attack and shielded to pass through barriers that would normally block passage of large molecules.

Term nanomedicine or nanomedical particles are customarily applied to structures from 1 to 1000 nm. Particle sizes ranging from 10 to 100 nm are in a fortuitous zone for medical application. They are small enough not to become entrapped in the microstructure of the lungs capillaries, liver, and kidney and thereby cause blockage, while they are large enough that their clearance time in tissues and organs is long in relation to small molecules. Inflammation, fate in every disease in the surrounding tissues, and linings of the associated blood vessels lead to dilatation in the junction between the cells lining of vessels walls. This results in preferential delivery of nanosized particles through the lining of capillaries and small vessels into inflamed areas. Nanoparticles are especially adaptable for delivery of nonsoluble and hydrophobic drugs. Encapsulation strategies include polymers with absorbed drugs, dendrite molecules or coordination compounds with drugs bonded covalently or weakly attached, and artificial or natural micelles or liposomes vesicles containing nanodoses of insoluble or toxic drugs, which can be selectively released on targets. Release can be controlled by pH, antigen specific coating or functional groups on their surfaces then stimulating the release of drugs or heating up the particle by the application of electromagnetic radiation and light.

Major advantage of nanoparticle drug delivery includes control of pharmacokinetics due to its nanoscale size. Nanoparticles, surface property and function helps in the delivery of drug at the targeted site with prolong release time and reduce side effects. It separates pharmacokinetics from therapeutic activity, and multiple targets can be incorporated in a nanoparticle. Nanoparticles are about one to several hundred times larger than individual drug molecules, so it can aggregate a large number of drug molecules. Nanoparticles can deliver drugs at targeted site with desired concentration, evade undesirable effect on drugs, and penetrate biological barriers. Nanoparticulate pharmaceuticals agents can penetrate cell membranes effectively as well as being able to cross the blood brain barrier (BBB).

In recent years, the focus is on developing biodegradable polymeric nanoparticles for drug delivery (Fig. 14.6).

These particles apart from increasing the bioavailability provide sustained release of drug. The drug is dissolved, adsorbed, attached, or encapsulated in the polymeric matrix of nanometer size. Polymeric nanoparticles are also being explored for targeted drug delivery [69]. Depending upon the method of preparation, nanospheres or nanocapsules was obtained with different release and surface properties [70]. Polymeric nanoparticle delivery can prevent the degradation of antioxidants in the GI tract, which will help in improving the bioavailability of the antioxidants [4]. In case of the drugs where efflux mechanisms play major role, these nanoparticles, when taken orally, will help to improve their bioavailability by virtue of the unique absorption mechanism in lymphatic system.

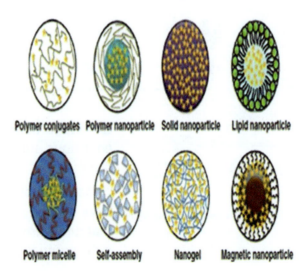

Figure 14.6 Different nanoformulations.

Considerable attention has been focused on the development of NDDS for phyto origin component. In phyto formulation research, nanodosage forms have a number of advantages over phyto drugs, including enhancement of stability, solubility, and bioavailability, protection from toxicity, enhancement of pharmacological activity, improving tissue macrophages distribution, sustained delivery, protection from physical and chemical degradation, etc. Thus, the nanosized novel drug delivery systems of phyto formulation have a potential future for enhancing the activity and overcoming problems associated with phyto medicines [31,60]. Nanotechnological process involving phyto medicine has gained much focus to researchers, who have developed several innovative delivery systems, including polymeric nanoparticles. Materials made from biodegradable and biocompatible polymers represent an option for controlled drug delivery. Polymeric nanoparticles are a promising formulation used for targeted drug delivery systems [8,71]. Polymeric nanoparticles can range from 10 to 1000 nm in diameter as nanocapsules (NCs) and nanospheres (NSs), while still protecting drugs efficiently though differ in their composition and structural organization (Fig. 14.7).

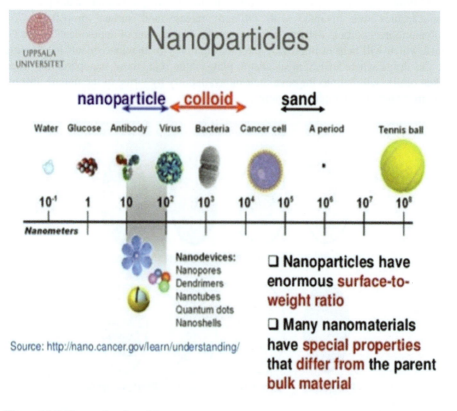

Figure 14.7 Nanoscale of particle.

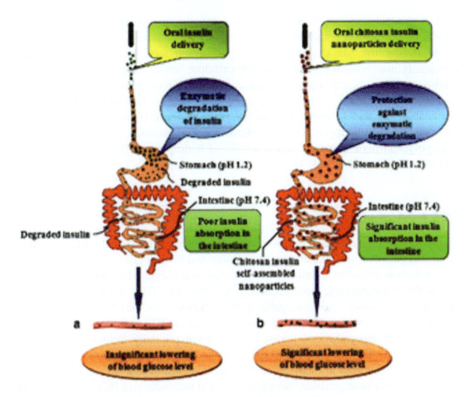

Figure 14.8 Oral insulin delivery by self-assembled chitosan NPs.

Nanocapsules contain an oily core surrounded by a polymeric membrane. Active constituent can be adsorbed to the polymeric membrane and/or dissolved in the oily core. Nanospheres are made only from a polymeric structure, where the active constituent is retained or adsorbed. Nanonization possesses many advantages, such as increasing compound solubility, reducing medicinal doses, and improving the absorbency of phyto medicines compared with the respective crude drugs preparations (Fig. 14.8) [9,31].

14.5 Polymers: biomedical excipients for nanostructurization

Polymeric nanoparticles (PNPs) are prepared from biocompatible and biodegradable polymers in size between 10 and 1000 nm where the drug is dissolved, entrapped, encapsulated, or attached to a nanoparticle matrix [72]. Depending upon the method of nanoparticles preparation, nanospheres or nanocapsules can be obtained. Nanocapsules are systems in which the drug is confined to a cavity surrounded by a unique polymer membrane, while nanospheres are matrix systems in which the drug is physically and uniformly dispersed [73]. Two types of polymers can be used in nanodelivery, which is natural and synthetic. Natural polymer matrix composites or biopolymers may be naturally occurring materials, which is formed in nature during the life cycles of green plants, animals, bacteria, and/or fungi. Commonly used natural polymers include cellulose, starch, chitosan, xantham gum, pectins, gellan gum, and alginate. Biodegradable polymers have an advantage over nondegradable systems in that they are nontoxic, biotolerable, biocompatible, biodegradable, and water soluble. Hybrid nanoparticles have been prepared mostly by the methods of ionotropic gelation, microemulsion, emulsification, solvent diffusion, polyelectrolyte complex, self-assembly of hydrophobically modified and dialysis (Figs. 14.9 and 14.10) [74].

14.5.1 Alginate

Carbohydrate polymers are extensively used in recent years in biomedical and pharmaceutical applications due to their biocompatibility and biodegradability. Polysaccharides represent most abundant industrial raw materials and subjected to intensive research due to their sustainability, biodegradability, and biosafety [76]. Alginates are random pH-dependent anionic and hydrophilic polysaccharide [10], linear long-term stabile [77], polymers consisting of varying ratio of glucuronic (G) and manuronic (M) acid unit. Abundant biosynthesized materials are derived primarily from brown seaweed [10]. Potential of alginate lies in development of alginate-controlled DDS [77]. Alginate has been used in many biomedical applications, including drug delivery systems, as they are biodegradable, biocompatible, and mucoadhesive [67]. These delivery

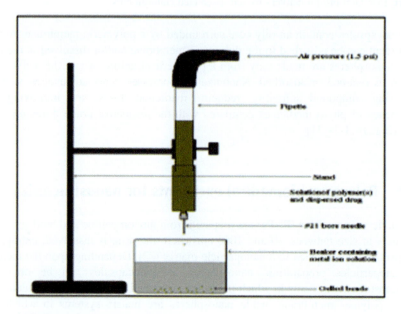

Figure 14.9 External ionotropic gelation.

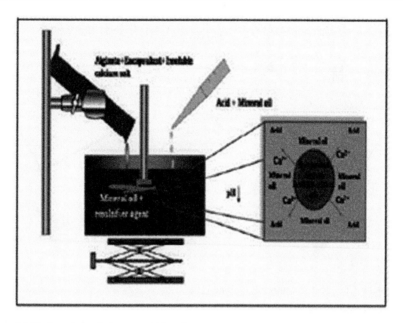

Figure 14.10 Internal gelation/emulsification [75].

systems worked when they are in monovalent, water soluble; alginate salts under goes an aqueous sol-gel transformation to water-insoluble salts due to the addition of divalent ions such as calcium, strontium and barium. Mainly calcium alginate matrix is used for drug delivery system including beads, gel, films, microparticles [10]. Calcium ions have unequal affinity for glucuronic and manuronic acid unit of alginate. As a result, calcium ions initially react with repeating unit of glucuronic acid unit to form egg box gel-shaped structures that stack upon each other. Additional calcium ions then interact with untreated glucuronic and manuronic acid unit to form a calcium alginate complex. Alginates with a high glucuronic acid contents form more rigid, porous gel due to their orientation within the egg box structure, and conversely gel with low glucuronic content is more randomly packed and less porous (Fig. 14.11).

Structurization of mean size of alginate lower than the critical diameter needed to be orally absorbed through the intestinal mucosa followed by their passage to systemic circulation and thus can be considered as a promising technology for insulin drug delivery (Fig. 14.8) system with 8% encapsulation efficiency [67]. Theophylline (TP)-loaded sodium alginate spherical microspheres as the hydrophilic carrier revealed extended in vitro drug release by decreased percent drug release with an increased alginate concentration thus potentially offers sustained release profile along with improved delivery of TP [78]. Encapsulation of curcumin in alginate-chitosan-pluronic composite nanoparticles for delivery to cancer cells has been reported [79]. Mucoadhesive chitosan (CS)-sodium alginate (ALG) nanoparticles as a new vehicle for prolonged topical ophthalmic delivery of gatifloxacin revealed a fast release during the first hour followed by a more gradual drug release

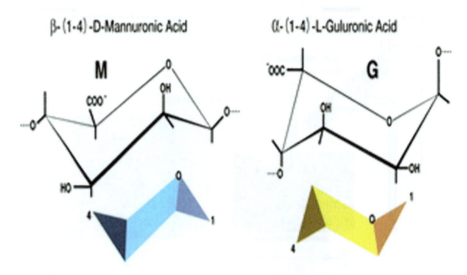

Figure 14.11 Chemical structure of alginate [10].

during next 24-hour period [3]. Alginate as delivery system for antituberculosis drugs in mice with induced tuberculosis has been reported with high drug concentration and bioavailability in the lungs, liver, and spleen where micro-organisms were in high concentration [80]. Potent and prolonged antioxidant effect of catechin incorporated into alginate can be used for increasing shelf-life of food [81].

14.5.2 Pectin

Pectin, natural sources in apple, apricot, cherry, orange, carrot, citrus peels etc., is a high molecular weight, non toxic, anionic, biocompatible natural plant polysaccharide consisting of α-1—4-D-galacturonic acid residues forming a homogalacturonan backbone (Fig. 14.12) [11].

It acts as hydrating agent and cementing material for the cellulosic network [82]. Two types of pectin are available based on degree of esterification, high methoxy pectin and low methoxy pectin, and it has been used successfully for many years in the food and beverage industry as a thickening agent, a gelling agent, and a colloidal stabilizer [83]. Ability of pectins to form gel depends on the molecular size and degree of esterification [84]. Low pectin gelatinization occurs in the presence of calcium salts, are less water soluble than natural pectins [85,86]. Calcium ions induce noncovalent associations of carbohydrate chains through the formation of so-called "egg box complexes" [12] but they maintain the selective biodegradation by pectinolytic enzymes of colonic bacteria microflora [87] resulting in the release of drug molecule without degrading [88,89]. Its biocompatibility, non toxic effect, acid stability, mucoadhesive nature, and good biodegradability make it suitable carrier for drug delivery.

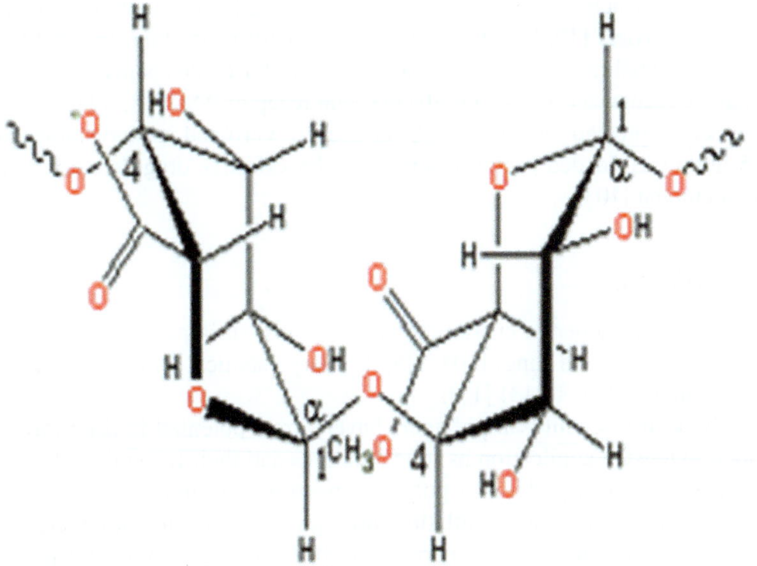

Figure 14.12 Chemical structure of pectin [11].

Pectin and its fragments possess immune-stimulant, antimetastasis, antiulcer hypoglycemic, and cholesterol lowering effect [90]. Natural pectin helps to prevent colon-specific cancer as dietary fiber [91]. It is used for the chelation of lead and mercury in toxicity conditions. Studies have shown that pectin-lecithin complex can have beneficial effect on the healing of gastric ulcers in horses [92]. It has been identified as protective agents against proteolysis, so selected as good candidate for protein drug delivery to prevent enzymatic proteolysis and acidic degradation of drugs [93]. Metronidazole-containing microparticles based on a pectin—4-aminothiophenol conjugate have been prepared for colon-specific drug delivery [94].

Pectin-based drug delivery system is developed for oral and parenteral administration. Pectin-coated tablets [95] and enteric-coated calcium pectinate microspheres of theophylline as nanodrug delivery system (NDDS) have been tested [96]. Modified pectin with various amine groups is tested as nonviral gene delivery carrier [97,98]. In order to improve intestinal absorption of calcitonin, self-assembling pectin—liposome nanocomplexes are prepared by a simple mixing of cationic liposomes with pectin solution [99]. Pectin-based nanoparticle has high ability to adhere to mucus layer and penetrate into GI epithelial cells, especially at duodenum and jejunum parts of small intestine, and thereby increases absorption. It is used for the ocular delivery of timolol maleate, which is widely used to decrease the intraocular pressure in glaucoma. Thiolation of pectin improved the mucoadhesive property and prolonged the release of timolol maleate in cul-de-sac [100]. Oxaliplatin-encapsulated magnetically functionalized spherical pectin nanocarriers are fabricated for targeted drug delivery to pancreatic cancer [101]. It exhibited 10-folds higher cytotoxicity effect than the free oxaliplatin, which has been attributed to sustained release of oxaliplatin from MP-OHP nanocarriers [102]. Pectin can be used for liver targeting drug delivery due to abundant terminal galactose residues available on RG-I side chains of pectin [103], which may be recognized by Asialo glycoprotein receptor ASGP-R, a lactose-binding pectin densely expressed on hepatic cell surfaces. In vitro and in vivo studies demonstrated 5-flurouracil-loaded pectin nanoparticles for effective drug delivery to hepatocellular carcinoma [103].

14.5.3 Chitosan

It is a natural biopolymer in crustacean shells (crabs, shrimps, and lobsters) and in some microbes, yeast and fungi [104,105]. Primary chemical unit is 2-deoxy-2-(acetyl amino) glucose (Fig. 14.13) [13].

Naturally occurring chitosan polymers have a great potential in drug formulation due to their extensive application as food additives and their recognized lack of toxicity. Naturally occurring chitosan polymers possesses a number of characteristics that render them useful as a formulation aid both as a conventional excipients and more specifically as a tool in polymeric-controlled drug delivery [63]. Chitosan has prospective applications in many fields such as food packaging, tissue regeneration, drug delivery, and pharmaceutical cosmetics, agricultural, and chemical industries [106]. Chitosan nanoparticles are very attractive carrier system for antibacterial, antiviral, antifungal, and antiparasitic drugs as they offer many advantages such as hydrophilic surface particles, nanosize of less than 100 nm [107]. Chitosan as antibacterial, antiviral, antifungal, antiinflammatory [108], antiparasitic, antineoplastic [109], immunomodulator [109], antiarteriosclerosis [110] property has been tested. Chitosan displays interesting properties such as biodegradability, biocompatibility, and its generated products are not toxic [111].

Chitosan has attracted attention of researchers working in the field of nanomedicine worldwide as effective drug delivery system [112] and has been reported to be very suitable for preparation of nanoparticles. Chitosan nanoparticles offer many advantages due to their better stability, low toxicity, simple and mild preparation methods, providing versatile routes of administration, and has gained more attention as a drug delivery carrier. Chitosan is widely used for nasal, vaginal, ophthalmic, transdermal and topical, buccal, parenteral, colon-specific and in implantable drug delivery, etc. due to its favorable biological properties [14]. Moreover, chitosan is a linear polyamine that contains a number of free amine groups that are readily available for crosslinking whereas its cationic nature allows for ionic crosslinking with multivalent anions [113].

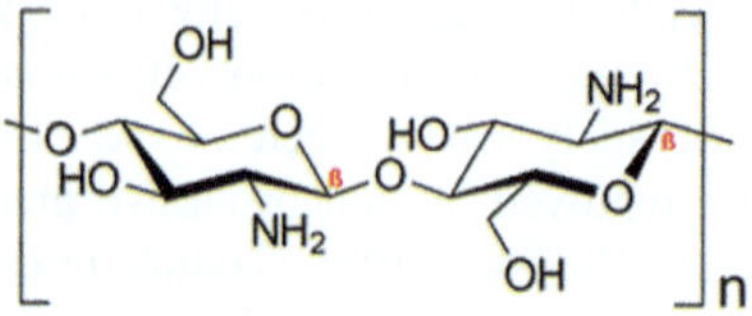

Figure 14.13 Chemical structure of chitosan [13].

14.6 Encapsulation of nanopolymer

Drug entrapped in polymeric beads offer novel drug delivery (NDD) approach for targeted locations and long-term delivery. Multiparticulate drug delivery systems have various well known advantages over single unit dosage forms, namely, masking of taste and odor, conversion of oils and other liquids to solids for the ease of handling, protection of drug against the environment for targeted and controlled sustained release. Microencapsulation is highly exploited techniques to formulate multiparticulate drug delivery systems (DDS). Different microencapsulation techniques are available, namely, solvent evaporation, solvent extraction, nanoemulsion, spray coating, etc. Microencapsulation in DDS poses serious drawbacks due to nonuniform coating, nonreproducible release kinetics and use of harsh chemicals during the process whereas principle of ionotropic gelation (polyelectrolyte complexation) involves neither the use of harsh chemicals nor elevated temperature (Figs. 14.12–14.15).

14.7 Ionotropic gelation (polyelectrolyte complexation)

Interaction of an ionic polymer with oppositely charged ion to initiate cross linking is ionotropic gelation. Three-dimensional structure and presence of other group influences the ability of cations/anions to conjugate with anionic/cationic with some kind of selectivity (Fig. 14.16).

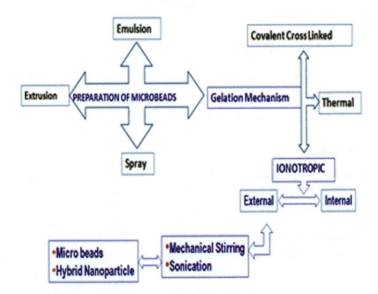

Figure 14.14 Nanoencapsulation of polymer for NDDS.

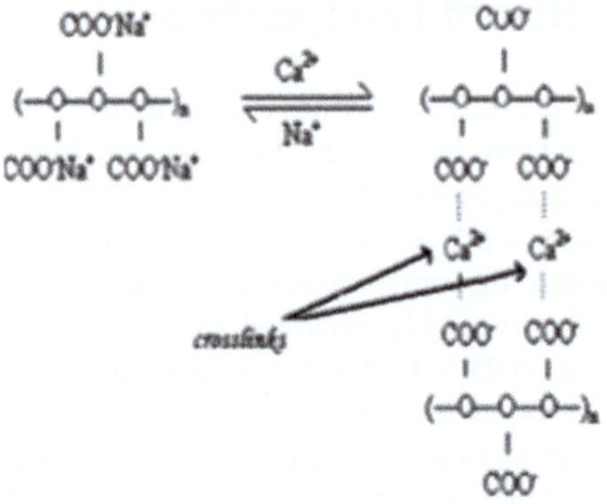

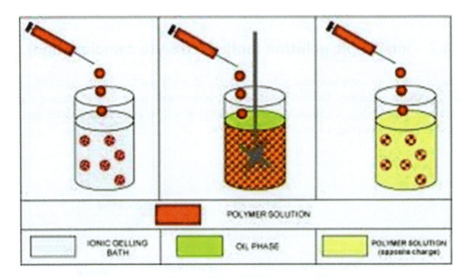

Figure 14.15 Electrostatic interaction between —COOgrs of alginate and cross linker Ca++ ions [75].

Figure 14.16 External ionotropic gelation technique.

14.7.1 Methods

Preparation of hydrogel beads can be done by two methods using ionotropic gelation technique, which differ from each other in the source of the crosslinking ion. In external ionotropic method [114], crosslinker ion is positioned externally (Fig. 14.10), and in internal ionotropic method [115] other crosslinker ion is incorporated within the polymer solution in inactive form (Fig. 14.17). External crosslinked produce thinner films with smoother surface, greater matrix strength, stiffness and permeability than internal crosslinked films. Externally crosslinked micropellets have greater drug encapsulation efficiency and slower drug release rate. Many natural and synthetic polymeric systems have been investigated for controlled drug release. Hydrophilic

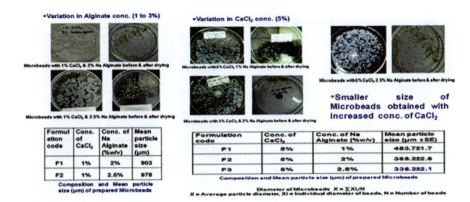

Figure 14.17 Structurization of CA-AL-NP by external ionotropic gelation technique.

polyionic carbohydrates such as alginate and chitosan have been paid much attention in recent years [116]. Since the preparation of beads by these materials involves use of aqueous solvents, environmental problems associated with organic solvents might be minimized. Variety of natural polymers and itsderivatives have been successfully employed in hydrogel system for various pharmaceutical applications. Uniform gel structure by sodium alginate and chitosan helps for stronger cross linked structure leading to more loaded entrapped material [75].

14.7.2 Difference in internal and external gelation on drug delivery system

Great extent of cross linkages could be produced by both the gelation techniques but distributions of crosslinkages were different. External crosslinkage produces thinner films with smoother surface, greater matrix strength, stiffness and permeability than internal crosslinked films. Externally crosslinked micropellets were also capable of greater drug encapsulation efficiency and slower drug release rate. Differences in the properties may be due to different gelation mechanisms involved and physical form of the matrix produced. External gelation is the preferred method in producing cross-linked alginate for coating and encapsulation purposes.

14.8 Catechin-loaded alginate polymeric nanoparticles (CA-AL-NP)

Facile coprecipitation method was followed for preparation of the catechin-alginate polymeric nanoparticles with minor modification [117]. Sodium alginate 0.5% prepared in deionized water at 30°C in which 10 mL of calcium chloride (0.02 M) added slowly and stirred under 1200 rpm for 30 minutes followed by 10 mL of sodium carbonate (0.02 M) added dropwise into the mixture and stirred under 1200 rpm for 3 hours.

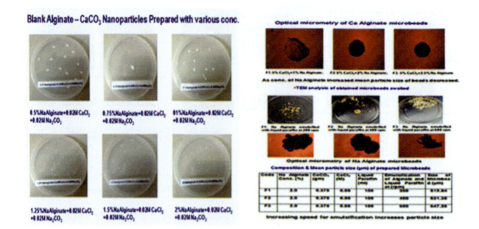

Figure 14.18 Structurization of CA-AL-NP by internal ionotropic gelation technique.

Mixture centrifuged at 18,000 rpm for 20 minutes, washed several times by deionized water, and freeze-dried to obtain alginate-calcium carbonate hybrid nanoparticles.

To prepare catechin hydrate loaded hybrid nanoparticle (CH-AL-NP), 50 mg catechin hydrate may be dissolved in 10 mL deionized water, and then 10 mg of freeze-dried nanoparticles added in solution and stirred for 12 hours at 600 rpm. Catechin hydrate loaded nanoparticles collected by centrifugation at 18,000 rpm for 20 minutes, washed several times with deionized water, and freeze-dried. Different concentration of alginate (0.5–1%) with different curing time (1 to 3 hours) and incorporation of (1–5 μm) amplitude of sonication standardized (Figs. 14.17 and 14.18).

14.9 Catechin-loaded pectin polymeric nanoparticles (CA-PC-NP)

Pectin-coated catechin nanospheres were prepared as per method described by Yu et al. [118] with minor modifications. To get the nanoparticle, 10 mL of 20% w/v solution of pectin was prepared under magnetic stirring at 55°C followed by dropwise addition of 2 mL of 0.01 M $Ca(OH)_2$ and kept for stirring under 800 rpm for 1 hour at 55°C. After that, 6 mL of 0.01 M $NaHCO_3$ added dropwise and kept for stirring under 800 rpm for 3 hours at 55°C. Obtained solution was sonicated to disperse the particles prior to drug loading. Catechin (10 mg/mL) was added dropwise to the solution and kept for stirring under 800 rpm for 24 hours. Pectin-coated catechin nanoparticles (PCNP) were collected after ultracentrifugation at 18,000 rpm for 1 hour at 4°C temperature of refrigerated condition. Sediments were washed several times using deionized water and lyophilized to obtain dried drug-loaded nanoparticles (Fig. 14.19).

Figure 14.19 Structurization of catechin-loaded pectin polymeric nanoparticles.

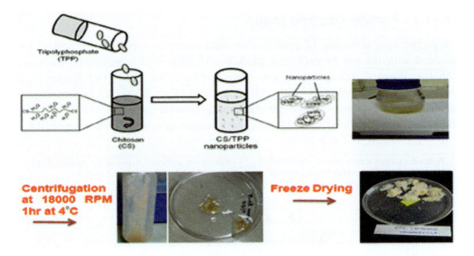

Figure 14.20 Structurization of catechin-loaded chitosan polymeric nanoparticles.

14.10 Catechin-loaded chitosan polymeric nanoparticles (CA-CH-NP)

Chitosan-coated catechin nanospheres were prepared as per method described by Calvo et al. [119] with minor modification. For chitosan blank nanoparticle, 0.2% chitosan (100 mg) was dissolved in distilled water containing 0.35% acetic acid and stirred at 900 rpm for 12 hours adjusting pH5.0 with 1 N NaOH. After that, 0.2% sodium tripolyphosphate (TPP) was added dropwise to 20 mL chitosan solution and kept for stirring at 900 rpm for 1 hour. Blank chitosan nanoparticle could be obtained by ultracentrifugation at 18,000 rpm with 4°C for 1 hour. Obtained sediment was washed several times with deionized water and freeze dried to obtain lyophilized powder(Fig. 14.20).

Forgetting catechin-encapsulated chitosan nanoparticle, chitosan was dissolved in distilled water containing 0.35% acetic acid and stirred at 900 rpm for 12 hours adjusting pH 5.0 with 1 N NaOH. Then 0.3% w/v catechin was dissolved in 20 mL distilled water containing 0.2% TPP and stirred at 900 rpm for 1 hour. Catechin solution should be added dropwise to chitosan solution on magnetic stirrer maintained at 37°C and cured for 1 hour at 900 rpm. Chitosan-coated catechin nanoparticles were collected after ultracentrifugation at 18,000 rpm for 1 hour at refrigerated condition of 4°C and the obtained sediments were washed several times using deionized water and finally lyophilization to obtain dried drug-loaded nanoparticles.

14.11 Characterization and evaluation of the polymeric nanoparticles

14.11.1 Particle size and shape

For preliminary screening of particle size, zeta potential polydispersity index was obtained with the aid of zeta sizer particle size (Fig. 14.21) analyzer (Microtrac Nanotrac Wave II, USA), which was confirmed by JOEL JEM 1011 transmission electron microscope (Fig. 14.22) with magnification of 80 kV to capture digital images [120]. A drop of freeze-dried desired polymeric (AL/PC/CH) nanoparticle (NP) solution (1 mg/mL) was placed on to a copper grid with a filter paper. After

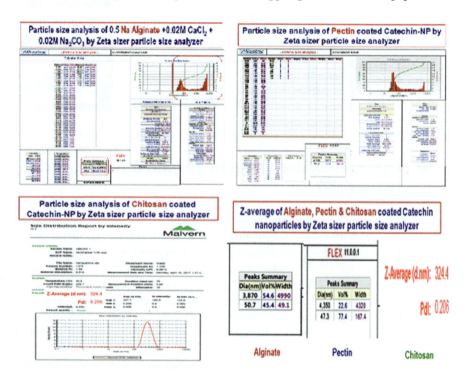

Figure 14.21 Particle size analysis and zeta potential of alginate, pectin, and chitosan-coated polymer.

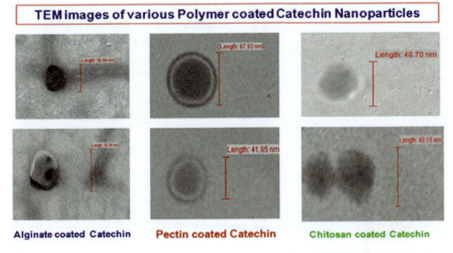

Figure 14.22 TEM images of various polymer-coated catechin nanoparticles.

15 minutes of nanoparticle deposition, the grid was tapped with filter paper for removal of excess water and stained using a solution of uranyl acetate (1%) for 10 minutes. Stained sample air was dried at room temperature following which the grid loaded in to the transmission electron microscope and size determined.

14.11.2 Encapsulation efficiency (%)

Encapsulation efficiency of the nanoparticles was determined by analyzing the supernatant after removal of the nanoparticles by centrifugation. For the estimation of unbound catechin hydrate in supernatant, the absorbance was measured spectrophotometrically at 425 nm and amount of catechin hydrate was calculated from calibration curve of known concentrations of catechin hydrate (Fig. 14.23). Amount of percent encapsulation was calculated as follows:

$$\text{Encapsulation Efficiency}(\%) = \{(\text{Total drug added} - \text{Unbound drug})/\text{Total drug}\} \times 100$$

14.11.3 Drug loading (%)

The drug loading was determined using the following formula:

$$\text{Drug loading}(\%) = \{(\text{Total drug added} - \text{Unbound drug})/\text{Total nanoparticle weight}\} \times 100$$

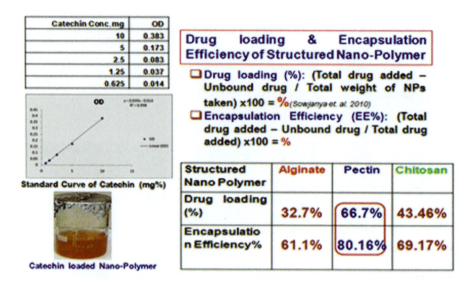

Figure 14.23 Drug loading and encapsulation efficiency of structured nanopolymer.

14.11.4 In vitro release of catechin

Two simulated digestive fluids phosphate buffer saline (PBS) solution adjusted to pH 1.5 with 1 M HCl as simulated gastric fluid (SGF) and phosphate buffer saline (PBS) solution adjusted to pH 7.4 with 1 M NaOH as simulated intestinal fluid (SIF) were utilized for release study of catechin-loaded polymeric nanoparticle as per the method of Dudhani and Kosaraju [121]. Accurately weighed amount of 10 mg nanoparticles placed in previously soaked dialysis bag (MWCO 12–14 kDa) in beaker containing 20 mL of enzyme free SGF and SIF separately and incubated at 37°C. Drug released in simulated fluids sampled predetermined time interval. Total concentration of catechin in simulated fluid determined spectrophotometrically.

14.12 Evaluation of in vitro antioxidant property of nanopolymer-coated catechin

14.12.1 Ferric-reducing antioxidant power assay

Accurately weighed amounts of the nanoparticle immersed in 20 mL of SGF and SIF separately, incubated at 37°C in the horizontal rotator, and mixtures sampled over time of incubation. Antioxidant activity in the supernatant determined by measuring ferric-reducing antioxidant power (FRAP) [122]. The FRAP reagent prepared freshly by mixing 2.5 mL of 20 mM $FeCl_3 6H_2O$, 2.5 mL of 10 mM 2,4,6-tripyridyl-s-triazine in 40 mM HCl, and 25 mL of 300 mM acetate buffer (pH 3.6) and warmed at 37°C prior to use. An aliquot of 900 µL of FRAP reagent mixed with 90 µL of distilled water and 30 µL of sample. Reaction mixture maintained at 37°C and its absorbance at 595 nm measured every 15 seconds up to 6 minutes using spectrophotometer. Solutions of $FeSO_4 7H_2O$ ranging from 50 to 1000 μM used as standard and values of FRAP expressed as μmol/L Fe equivalent.

14.12.2 Determination of free radical scavenging activity by 1,1-diphenyl-2-picryl-hydrazyl assay

Determination of free radical scavenging activity was done by 1,1-diphenyl-2-picrylhydrazyl (DPPH) radical scavenging assay (Fig. 14.24) [123]. Sodium acetate buffer of pH 5.5 prepared by adding 35.7 mL of glacial acetic acid solution (290 μL in 50 mL MilliQ) and 64.3 mL of sodium acetate solution (1.36 g in 100 mL MilliQ). To 40 mL of this sodium acetate buffer, 60 mL of methanol added to prepare buffered methanol. Working solution of 0.1 mM DPPH prepared by mixing 5 mL of 1 mM DPPH solution and 45 mL of buffered methanol. From this DPPH working solution, 1 mL was taken in a test tube. To this 0.5 mL supernatant was added and incubated at room temperature in a dark place for 30 minutes and after that analyzed spectrophotometrically at 517 nm.

14.12.3 Metal chelating activity assay

Chelating activity of Fe^{2+} could be measured adapting to previously established protocols of Afanas'ev et al. [124] and Ebrahimzadeh et al. [125]. Aliquots of sample from in vitro release study (500 μL) taken and centrifuged at 12,000 rpm for 5 minutes at

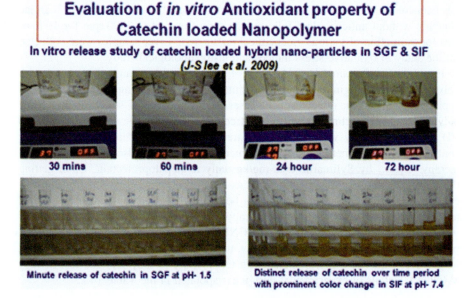

Figure 14.24 Evaluation of in vitro antioxidant property of catechin loaded polymer.

5°C using refrigerated high-performance centrifuge apparatus. Then 30 μL of supernatant and 150 μL of $FeCl_2$ added. This mixture was allowed to stand for 5 minutes at darkness, and then 60 μL of ferrozine (5 mM) added to the mixture, which then vigorously stirred and allowed to stand for 10 minutes in darkness. Absorbance of solution measured at 562 nm and percentage of iron chelating activity calculated.

14.13 Ameliorative application of nanopolymer-coated catechin for rodent hepatic regeneration

14.13.1 In vivo trial of nanopolymer-coated catechin in rat

Efficacy of the prepared polymeric nanoparticles assessed for its ameliorative effect as hepatoprotectant against CCl_4-induced hepatopathy in rats. Male Albino Wister rats 120–150 g were obtained from Laboratory Animal Resource (LAR) Section. Standard acclimatization period, standard feeding, and management condition were ensured before the onset of trial as per rules laid down by Institute Committee for the Purpose of Control and Supervision of Experiments on Animals (ICPCSEA) norms. Rats were randomly divided into desired number of groups having at least six rats ($n = 6$) in each group. Subacute hepatotoxicity was induced in rats by administering carbon tetrachloride (CCl_4) analar grade mixed with olive oil 1:1, (v/v) @ 4 mL/kg body weight by epigastric lavage twice a week for 4 weeks [126]. Following completion of trial, rats were humanized as per ICPCSEA norms and 2–3 mL of blood was collected by cardiac puncture into heparinized sterilized vial for hematological study. Serum was separated from blood collected by cardiac puncture without anticoagulant followed by centrifugation at 3000 rpm for 10 minutes for estimation of various biochemical parameters. Liver samples were collected for the study of oxidative stress indices and histopathological study.

In a pilot study, ameliorative effect of catechin antioxidant, nanocoated with polymers, namely, alginate, pectin, and chitosan, was studied against CCl_4-induced hepatopathy in rats in three distinct phases. In each phase of study, distinguished polymer (alginate/pectin/chitosan) coated catechin nanoparticle (test therapy I, Gp VI) @20 mg/kg body weight was compared with healthy control (Gp I), disease control (Gp II), treatment control I, treatment control II and also with conventional therapy (Gp V) by Sylimarin @20 mg/kg daily as hepatoprotectant. Upon getting the significant positive therapeutic effect as hepatoprotectant of polymer-coated catechin nanoparticle (PCCNP) @20 mg/kg body weight, the same PCCNP subjected to two reduced doserate, that is, test therapy II (Gp VII) PCCNP @10 mg/ kg body weight and test therapy III (Gp VIII) PCCNP @ 5 mg/kg body weight to find therapeutic effect with reduced dose. Similarly, to find therapeutic effect of PCCNP with longer dose interval, trial conducted with test therapy IV (Gp IX) and test therapy V (Gp X). Details of experimental design extrapolated in the Table 14.2.

Table 14.2 Experimental design for hepatoprotective effect of nanopolymer-coated catechin in rat.

Groups $n = 6$	Therapeutic group	Treatment	Duration
Gp I	Healthy control	Normal feeding	Induction of hepatopathy by giving CCl_4 epigastric lavage with olive oil (1:1 dilution) @ 4 mL/kg body wt. [126] twice a week. Study duration of 04 weeks with standard feeding and managemental condition
Gp II	Disease control	Normal feed + CCl_4	
Gp III	Treatment control I	Normal feed + CCl_4 + from day 15th onward Catechin AR @20 mg/kg body wt. daily [48]	
Gp IV	Treatment control II	Normal feed + CCl_4 + from day 15th onward Blank Polymer NP @ 20 mg/kg body wt daily	
Gp V	Conventional therapy	Normal feed + CCl_4 + from day 15th onward Sylimarin @20 mg/kg daily	
Gp VI	Test therapy I	Normal feed + CCl_4 + from day 15th onward CAT-NP @ 20 mg/kg body wt. daily	
Gp VII	Test therapy II	Normal feed + CCl_4 + from day 15th onward CAT-NP @ 10 mg/kg body wt. daily	
GpVIII	Test therapy III	Normal feed + CCl_4 + from day 15th onward CAT-NP @ 5 mg/kg body wt. daily	
Gp IX	Test therapy IV	Normal feed + CCl_4 + from day 15th onward CAT-NP @ 20 mg/kg body wt. alternate day	
Gp X	Test therapy V	Normal feed + CCl_4 + from day 15th onward CAT-NP @ 20 mg/kg body wt. every 4th day	

14.13.2 Clinical signs and body weight

Experimental rats to be observed closely for exhibition of any clinical and behavioral abnormality if any and body weight to be recorded at weekly interval during the course of study.

14.13.3 Hematological profile

Hemoglobin (Hb) and packed cell volume (PCV) in the whole blood estimated by using Sahli's hemoglobinometer with double prism standard. Packed cell volume (PCV) measured by Wintrobe hematocrit method. Total erythrocyte count (TEC) measured on counting chamber with improved Neubauer's ruling. Gower's RBC diluting fluid solution crystalline sodium sulfate 0.25 g, sodium chloride 0.5 g, and 100 mL distilled water used as diluting fluid for TEC and the results were expressed in million per microliter of blood. Total leukocyte count (TLC) measured on counting chamber with improved Neubauer's ruling. WBC diluting fluid solution used as diluting fluid for TLC and the results were expressed in thousand per microliter of blood.

14.13.4 Serum biochemical profile

Serum glutamate pyruvate transaminase (SGPT) activity and serum glutamic oxaloacetic transaminase (SGOT) activity were estimated as per method of Reitman and Frankel [127] using standard kit (Coral diagnostic Ltd.). Alkaline phosphatase was (ALP) estimated by PNPP method, and total protein was modified by Biuret end point assay and albumin by Dumas method using standard test kit (Coral Diagnostic Ltd.). Calculation of serum globulin by subtracting albumin from total protein and serum A/G ratio by dividing albumin value to globulin value may be done. Total bilirubin was estimated as per method described by Malloy and Eveluyun [128] using standard test kit (Coral Diagnostic Ltd.).

14.13.5 Assays of oxidative stress markers in the liver tissue sample

Estimation of various oxidative stress marker parameters in liver tissue executed. A double beam UV-VIS spectrophotometer used for recording the absorbance of the test samples.

14.13.6 Preparation of liver homogenates

For oxidative stress indices, 500 mg of liver tissue taken in 4.95 mL of ice-cold 1.15% KCl mixed with 25 µL PMSF. Another 200 mg liver tissue taken in 2 mL of 0.02 M EDTA in distilled water and used for estimation of reduced glutathione GSH. The liver homogenate of 10% concentration prepared with homogenizer under ice-cold conditions and centrifuged for 10 minutes at 3000 rpm. Final supernatant stored at $-80°C$ till completion of assay.

14.13.7 Assays lipid peroxidation (LPO) assay

Lipid peroxides level estimated in tissue homogenate following the method suggested by Rehman [129]. Briefly, tissue homogenate (1 mL) incubated at 37 ± 5°C for 2 hours. One milliliter of 10% (w/v) trichloro acetic acid (TCA) added to each 1 mL of homogenate sample and mixed thoroughly and centrifuged at 2000 rpm for 10 minutes. After centrifugation, 1 mL of supernatant taken into another test tube and equal amounts (1 mL) of 0.67% (w/v) thiobarbituric acid (TBA) mixed and kept in boiling water bath for 10 minutes. After boiling, cooled and 1 mL distilled water added in each sample. The absorbance could be measured at 532 nm against using extinction coefficient of MDA (1.56×10^{-5}) and results expressed in nmol MDA/g tissue. Concentration of LPO calculated as follows:

LPO in nmol MDA/gtissue = OD at 532 nm $\times$ 320.5128 using 1.56×10^{-5} as the extinction coefficient.

14.13.8 Glutathione assay (GSH)

GSH activity in liver homogenate determined by the method of Sedlak and Lindsay [130] using 5,5'-dithiobis(2-nitrobenzoic acid) (DTNB) with the supernatant of 10% tissue homogenates. In 1 mL of tissue, 0.8 mL water and 0.2 mL of 50% tricarboxylic acid (TCA) solutions added. It was then incubated at room temperature for 15 minutes and subjected to centrifugation @3000 rpm for 15 minutes. From the supernatant obtained after centrifugation, 0.4 mL taken and 0.8 mL of 1 M tris buffer and 0.2 mL of 0.01 M DTNB added. The absorbance was taken at 412 nm. Reagent blank should have no sample and sample blank devoid of DTNB was used.

μmol of GSH/mg of protein = OD of test/EC $\times$ Total volume of reaction mixture/volume of sample, where EC (extinction coefficient) taken as 13,100/M/cm.

14.13.9 Superoxide dismutase assay

The SOD activity in liver homogenate measured by method of Marklund et al. [131] modified by Masayasu and Hiroshi [132]. The tissue homogenates diluted suitably and results multiplied by dilution factor. In brief, 3 mL of Tris cocodylic acid used as reference blank at absorbance set to zero at 420 nm. 50 μL of pyrogallol added to it and OD taken at 30 seconds and 90 seconds. Subsequently, 2.9 mL Tris cocodylic acid, 50 μL of pyrogallol and 50 μL of diluted sample added and absorbance measured at 30 seconds and 90 seconds. The activity of SOD value expressed as unit/mg of protein.

14.13.10 Catalase assay

Catalase activity in 10% liver tissue lysate estimated using the method of Cohen et al. [133]. In brief, tissue samples diluted 20 times. The OD of phosphate H_2O_2 adjusted in such a way that it gave OD value in the range of 0.6 to 0.7 at 240 nm. The 50 μL of diluted sample mixed with 2.95 mL of phosphate buffer H_2O_2 solution and absorbance measured at 240 nm against reference cuvette containing 50 μL of phosphate buffer saline PBS and 2.95 mL of phosphate buffer H_2O_2. Initial OD taken at 30 seconds and time in second noted for decrease in initial absorbance by 0.05 (50 unit) and catalase activity values expressed in unit/mg of protein.

14.13.11 Gross pathology of liver

Liver of all the rats examined at the end of experiment for any gross pathological changes as per the method described by Runnels et al. [134].

14.13.12 Histopathological study

Liver tissues of rats fixed with 10% neutral buffer formalin for histopathological examination. Tissues after 24-hour fixation were further cut into 2−3 mm thickness and preserved for another 4−7 days. Prior to dehydration procedure, tissues washed with running tap water for 4−6 hours. Section dehydrated with ascending grades of ethanol, that is, 70% overnight, 80% 1 hour twice, 90% 1 hour twice and 100% 1 hour twice followed by three changes of xylene 1/2 hour each and finally tissues kept in paraffin wax overnight. Next day the tissues embedded in paraffin wax to prepare blocks, and kept at 0°C until cutting of sections and stained by hematoxylene and eosin staining protocol for microscopic evaluation of lesions. The lesions in different sites of liver graded by developing histopathological score system (HPS) [135]. The liver sections looked for pathological changes and graded on scale at 0 to 3 (0 = no change, 1 = mild change, 2 = moderate change, 3 = severe change). Maximum score was taken as most severe inflammatory changes.

14.13.13 May-Grünwald Giemsa staining of liver tissue homogenate

Method of Bharrhan [136] followed for May-Grünwald Giemsa staining of liver tissue homogenate. Liver tissues washed with chilled homogenizing buffer (24 mM sodium-EDTA buffer pH 7.5, containing 75 mM of NaCl). In a homogenizer, 200 mg of tissue homogenized in 5.0 mL of buffer at 500 rpm for 30 seconds. Homogenates then centrifuged at 3835 g for 10 minutes. Supernatant removed and fresh homogenizing buffer (0.4 mL) added to re-suspend the hepatocytes. Small drops of prepared suspension put at one end of precleaned, grease free microscopic slide. Drops were spread using another glass slide held at an angle of 45° into a smooth layer of smear. The air dried slides then stained according to the following sequence: fixed in methanol for 10 minutes; stained in undiluted May-Grünwald solution for 90 seconds; washed with water; stained with Giemsa (1:10 diluted) for 15−20 minutes; rinsed twice in distilled water; blot dried with filter paper; back side of slides cleaned with methanol; slides then cleared in xylene for 5 minutes. Minimum of 100 cells of hepatocytes counted per sample using light microscope and observed for any cellular changes and deformities. Parameters for hepatocytes evaluation included % normal hepatocytes, % hepatocytes with distinct or indistinct cell wall boundary, % hepatocytes with abnormal vacuolations in cytoplasm and % hepatocytes with pyknotic or necrosed nuclei. Result of May-Grünwald Giemsa staining study documented to compare with histopathological findings.

14.14 Conclusion

Nanopolymer-based encapsulation is an effective drug delivery system with diverse pharmaceutical applications for site-specific drug delivery, protection of enzymes, proteins and peptide from gastric environment, controlled drug delivery system, immunogenicity and bioadhesive properties. Varied crosslinking methods are available for the development of polymer-based nanoencapsulation in a simple, mild, and cost-effective manner. Antioxidants as a hepatoprotectant have limited therapeutic effects due to its unstable nature in solution with poor bioavailability in target hepatic cells or organ. Designing and nanofabrication will reduce repeated administration due to its sustained release properties and also will increase the therapeutic value by increasing the bioavailability and stability at the target cell or organ against chronic inflammatory diseases like hepatic injury. Nanofabrication of antioxidant by polymeric nanoencapsulation for ameliorative effect against hepatic regeneration in rodent may be an added boon for improved understanding of animal diseases and its therapeutic intervention.

14.15 Notes

14.15.1 Nanofabrication and assessment of catechin-loaded polymeric nanoparticles

14.15.1.1 Basis for selection of excipients

Different excipients namely alginate, chitosan, and pectin may be selected for the study on the basis of their chemical property (Table 14.3). Size of nanoparticle may vary with the variation in concentration of base polymer as well as $CaCl_2$ salt. Smaller size of microbeads may be obtained with increased concentration of $CaCl_2$ while increased concentration of polymer leads to decreased mean particle size. Variation in curing time by increased stirring and incorporation of sonication triggers better results for hybridization of nanoparticles.

14.15.2 Drug loading (%)

Drug loading of desired microbeads determined [141] as

$$DL(\%) = (\text{weight of drug in NPs}/\text{total weight of NPs taken}) \times 100.$$

Table 14.3 Chemical properties of different excipients namely alginate, chitosan, and pectin.

Character	Alginate	Chitosan	Pectin
Structure Sources Solubility Viscosity	Carbohydrate Brown algae Depend of deacetylation Low viscosity <240 mPas Medium viscosity 240–3500 mPas; high: > 3500 mPas (Rayid, 2009)	Cellulose analog Crab, shrimps Soluble in organic solvents Variable with acidic pH	Polysaccharide Citrus fruits, apple Depend on esterification Depend on sources and extraction methods
Molecular weight	30,000–400,000 g/mol [10]	100,000 to 1,200,000 g/mol [13]	Varies with fruit and esterification methods $1.5–2.3 \times 10^5$ g/mol [137]
Degradation	Depend on pH (mostly acid hydrolysis)	High temperature	No significant degradation
Serum biochemical profile			
Alanine amino transferase (ALT) and aspartate amino transferase (AST) Alkaline phosphatase (ALP) Total bilirubin Total protein Serum albumin			[127] PNPP method [128] Modified Biuret [138] Dumas method [138]
Assays of oxidative stress markers in liver tissue			
Lipid peroxidation (LPO) assay Glutathione peroxidase Superoxide dismutase (SOD) assay Catalase (CAT) assay			[139] [130] [140] [134]
Parameters			Group wise values in %
Normal hepatocytes cells Distinct hepatocytes cell wall boundary Indistinct hepatocytes cell wall boundary Abnormal vacuolations in hepatocytes cytoplasm Pyknotic hepatocytes nucleus Necrosed hepatocytes nucleus			
Parameters		Healthy group	Treated group
Gross pathology Histopathology score May-Grünwald Giemsa staining Mean score			

14.15.3 Encapsulation efficiency (%)

Encapsulation Efficiency (%) determined [141] as

$$EE(\%) = (\text{Practical drug loading}/\text{theoretical drug loading}) \times 100.$$

14.15.4 Parameters of study

1. Clinical signs and body weight: clinical and behavioral abnormality if any and body weight of experimental rats
2. Hematological profile: evaluation of Hb, TEC, PCV, WBC [142]
3. Serum biochemical profile
4. Assays of oxidative stress markers in liver tissue
5. Gross pathology of liver [134]
6. Histopathological study of liver tissues [135]
7. Percentage hepatocytes population by May-Grünwald Giemsa staining
8. Modified Histopathological Score (HPS) system between gross pathology, histopathology and May-Grünwald Giemsa staining

The lesions in different sites of liver may be graded by Modified Histopathological Score (HPS) System [135]. The liver sections looked for pathological changes and graded on scale at 0 to 3 (0 = no change, 1 = mild change, 2 = moderate change, 3 = severe change). The maximum score was taken as most severe inflammatory changes.

Statistical analysis was done as per reference [143].

References

[1] De S, Robinson D. Polymer relationships during preparation of chitosan-alginate and poly-L-lysine−alginate nanospheres. J Control Rel 2003;89:101−12.
[2] Gref R, Minamitake Y, Perracchia MT, Trubeskoy V, Torchilin V, Langer R. Biodegradable long-circulating polymeric nanospheres. Science 1994;263:1600−3.
[3] Motwani SK, Shruti C, Sushma T, Kanchan K, Farhan JA, Roop KK. Chitosan−sodium alginate nanoparticles as submicroscopic reservoirs for ocular delivery: formulation, optimization and in vitro characterization. Eur J Pharm Biopharma 2008;68:513−25.
[4] Ratnam V, Ankola DD, Bhardwaj V, Sahana DK, Ravi KMNV. Role of antioxidants in prophylaxis and therapy: a pharmaceutical perspective. J Control Rel 2006;113:189−207.
[5] Allémann E, Gurny R, Doelker E. Drug-loaded nanoparticles—preparation methods and drug targeting issues. Eur J Pharm Biopharm 1993;39:173−91.
[6] Tiyaboonchai W, Tungpradit W, Plianbangchang P. Formulation and characterization of curcuminoids loaded solid lipid nanoparticles. Int J Pharma 2007;337:299−306.
[7] Arica YB, Benoit JP, Lamprecht A. Paclitaxel-loaded lipid nanoparticles prepared by solvent injection or ultrasound emulsification. Drug Dev Ind Pharm 2006;32:1089−94.
[8] Mainardes RM, Evangelista RC. PLGA nanoparticles containing praziquantel: effect of formulation variables on size distribution. Int J Pharm 2005;290:137−44.

[9] Brigger I, Dubernet C, Couvreur P. Nanoparticles in cancer therapy and diagnosis. Adv Drug Deliv Rev 2002;54:631—51.
[10] Sun J, Tan H. Alginate-based biomaterials for regenerative medicine. Appl Mater 2013;6(4):1285—309. Available from: https://doi.org/10.3390/ma6041285.
[11] Wicker L, Kim Y, Kim MJ, Thirkield B, Lin Z, Jung J. Pectin as a bioactive polysaccharide: extracting tailored function from less. Food Hydrocoll 2014;42:251—9.
[12] Morris ER, Powell DA, Gidley MJ, Rees DA. Conformations and interactions of pectins: I. polymorphism between gel and solid states of calcium polygalacturonate. J Mol Biol 1982;155(4):507—16.
[13] Nagpal K, Singh SK, Mishra DN. Chitosan nanoparticles: a promising system in novel drug delivery. Chem Pharm Bull 2010;58(11):1423—30.
[14] Patel DP, Singh S. Chitosan: a multifacet polymer. Int J Curr Pharma Res 2015;7(2):21—8.
[15] Shah S, Pal A, Kaushik VK, Devi S. Preparation and characterization of venlaxine hydrochloride-loaded chitosan nanoparticles and in vitro release of drug. J Appl Poly Sci 2009;112:2876—87.
[16] Zakim D. Pathophysiology of liver disease. In: Smith LH, Their SO, editors. Pathophysiology: the biological principles of disease,. 2nd Edn. Philadelphia: W B Saunders and Co; 1985. p. 1253—98.
[17] Ettinger S.J., Feldman E.C. Liver and pancreatic diseases. Text book of veterinary internal medicine diseases of dogs and cats. 6th Edn. Philadelphia: Elsevier Saunders; 2005, 2—9, 1422—1492.
[18] Lui J, Lu JF, Wen XY, Kan J, Jin CH. Antioxidant and protective effect of inulin and catechin grafted inulin against CCl4-induced liver injury. Int J Biol Macromol 2005;72:1479—84.
[19] Wang R, Zhou W, Jiang X. Reaction kinetics of degradation and epimerization of epigallocatechin gallate (EGCG) in aqueous system over a wide temperature range. J Agric Food Chem 2008;56:2694—701.
[20] Jiang C, Jiao Y, Chen X, Li X, Yan W, Yu B, et al. Preliminary characterization and potential hepatoprotective effect of polysaccharides from *Cipangopaludina chinensis*. Food Chem Toxicol 2013;l59:18—25.
[21] Higdon JV, Frei B. Tea catechins and polyphenols: health effects, metabolism and antioxidant functions. Crit Rev Food Sci Nutr 2003;43(1):89—143.
[22] Catterall F, King LJ, Clifford MN, Ioannides C. Bioavailability of dietary doses of 3H-labelled tea antioxidants (+)-catechin and (-)-epicatechin in rat. Xenobiotica 2003;33:743—53.
[23] Volf I, Ignat I, Neamtu M, Popa VI. Thermal stability, antioxidant activity, and photooxidation of natural polyphenols. Chem Pap 2014;68:121—9.
[24] Munin A, Edwards-Levy F. Encapsulation of natural polyphenolic compounds: a review. Pharmaceutics 2011;3:793—829.
[25] Barras A, Mezzetti A, Richard A, Lazzaroni S, Roux S, Melnyk P, et al. Formulation and characterization of polyphenol-loaded lipid nanocapsules. Int J Pharm 2009;379:270—7.
[26] Weiss J, Decker EA, McClements DJ. Kristbergsson K, Helgason T, Awad T. Solid lipid nanoparticles as delivery systems for bioactive food components. Food Biophysics 2008;3:146. Available from: https://doi.org/10.1007/s11483-008-9065-8.
[27] Yang CS. Tea and health. Nutrition 1999;15(11—12):946—9.
[28] Nanjo F, Goto K, Seto R, Suzuki M, Sakai M, Hara Y. Scavenging effects of tea catechins and their derivatives on 1,1-diphenyl-2-picryhydrazyl radical. Free Radic Biol Med 1996;21(6):895—902.

[29] Valcic S, Annette M, Neil EJ, Daniel CL, Barbara NT. Antioxidant chemistry of green tea catechins. Identification of products of the reaction of (-)-epigallocatechin gallate with peroxyl radicals. Chem Res Toxico 1999;12:382–6.
[30] Zhang L, Kosaraju SL. Biopolymeric delivery system for controlled release of polyphenolic antioxidants. Eur Polym J 2007;43(7):2956–66.
[31] Ajazuddin SS. Applications of novel drug delivery system for herbal formulations. Fitoterapia 2010;81:680–9.
[32] Zaveri NT. Green tea and its polyphenolic catechins: medicinal uses in cancer and non-cancer applications. Life Sci 2006;78:2073–80.
[33] Victoria KA, Amber S, Weibiao Z. Green tea catechins during food processing and storage: a review on stability and detection. Food Res Int 2013;50:469–79.
[34] Mandel S, Weinreb O, Amit T, Youdim MB. Cell signaling pathways in the neuroprotective actions of the green tea polyphenol (-) epigallocatechin-3-gallate: Implications for neurodegenerative diseases. J Neurochem 2004;88(6):1555–69.
[35] Gouni-Berthold I, Sachinidis A. Molecular mechanisms explaining the preventive effects of catechins on the development of proliferative diseases. Curr Pharma Des 2004;10(11):1261–71.
[36] Hider RC, Liu ZD, Khodr HH. Metal chelation of polyphenols. Meth Enzymol 2001;335:190–203.
[37] Komatsu Y, Suematsu S, Hisanobu Y, Saigo H, Matsuda R, Hara K. Effects of pH and temperature on reaction kinetics of catechins in green tea infusion. Biosci Biotech Biochem 1993;57:907–10.
[38] Chen Z, Zhu QY, Tsang D, Huang Y. Degradation of green tea catechins in tea drinks. J Agric Food Chem 2001;49:477–82.
[39] Sang S, Lee MJ, Hou Z, Ho CT, Yang CS. Stability of tea polyphenol (-) epigallocatechin-3-gallate and formation of dimmers and epimers under common experimental conditions. J Agri Food Chem 2005;53:9478–84.
[40] Kumamoto M, Sonda T, Nagayama K, Tabata M. Effects of pH and metal ions on antioxidative activities of catechins. Biosci Biotech Biochem 2001;65(1):126–32.
[41] Ogata K, Mukae N, Suzuki Y. Effects of catechins on the mouse tumor cell adhesion to bronectin. Planta Med 1995;61:472–4.
[42] Midddleton E, Kandaswami C, Theoharides TC. The effect of plant flavonoids on mammalian cells implications for inflammation, heart disease and cancer. Pharmacol Rev 2000;52:673–715.
[43] Hollman PC, Feskens EJ, Katan MB. Tea flavonols in cardiovascular disease and cancer epidemiology. Proc Soc Exp Biol Med 1999;220:198–202.
[44] Rodrigues CF, Ascenção K, Silva FAM, Sarmento B. Oliveira MBPP, Andrade JC. Drug-delivery systems of green tea catechins for improved stability and bioavailability. Curr Med Chem 2013;20(37):4744–57.
[45] Ahmed ME, Khan MM, Javed H, Vaibhav K, Khan A, Tabassum R, et al. Amelioration of cognitive impairment and neurodegeneration by catechin hydrate in rat model of streptozotocin-induced experimental dementia of Alzheimer's type. Neuro Chem Intern 2013;62(4):492–550.
[46] Kalender Y, Kaya S, Durak D, Uzun FG, Demir F. Protective effects of catechin and quercetin on antioxidant status, lipid peroxidation and testis-histoarchitecture induced by chlorpyrifos in male rats. Env ToxicolPharmacol 2012;33(2):141–8.
[47] Sardana A, Kalra S, Khanna D, Balakumar P. Nephroprotective effect of catechin on gentamicin-induced experimental nephrotoxicity. Clin Exp Nephrol 2015;2:178–84.

[48] Daisy P, Balasubramanian K, Rajalakshmi M, Eliza J, Selvaraj J. Insulin mimetic impact of catechin isolated from *Cassia fistula* on the glucose oxidation and molecular mechanisms of glucose uptake on Streptozotocin-induced diabetic Wistar rats. Phytomedicine 2010;17:28−36.
[49] Ashafaq M, Raza SS, Khan MM, Ahmad A, Javed H, Ahmad ME, et al. Catechin hydrate ameliorates redox imbalance and limits inflammatory response in focal cerebral ischemia. Neuro Chem Res 2012;37:1747−60.
[50] Nagarajan S, Nagarajan R, Susan JB, Bruno F, Donna M, Lynne S, et al. Bio catalytically oligomerized epicatechin with potent and specific anti-proliferative activity for human breast cancer cells. Molecules 2008;13:2704−16.
[51] Morel I, Lescoat G, Cogrel P, Sergent O, Pasdeloup N, Brissot P, et al. Antioxidant and iron-chelating activities of the flavonoids catechin, quercetin and diosmetin on iron-loaded rat hepatocyte cultures. Biochem Pharmacol 1993;45(1):13−19.
[52] Danni O, Sawyer BC, Slater TF. Effects of (+)-catechin in vitro and in vivo on disturbances produced in rat liver endoplasmic reticulum by carbon tetrachloride. Biochem Soc Trans 1997;5(4):1029−32.
[53] Record IR, Lane JM. Simulated intestinal digestion of green and black teas. Food Chem 2001;73(4):481−6.
[54] Yoshino K, Suzuki M, Sasaki K, Miyase T, Sano M. Formation of antioxidants from-epigallocatechin gallate in mild alkaline fluids, such as authentic intestinal juice and mouse plasma. J Nutr Biochem 1999;10(4):223−9 1999.
[55] Karamanlioğlu B, Yüksel M, Temiz E, Salihoğlu YS, Ciftçi S. Hepatobiliary scintigraphy for evaluating the hepatotoxic effect of halothane and the protective effect of catechin in comparison with histo-chemical analysis of liver tissue. Nucl Med Commun 2002;23(1):53−9.
[56] Mehra P, Garg M, Ashwani K, Devi DB. Effect of (+) catechin hydrate on oxidative stress induced by high sucrose and high fat diet in male wister rats. Ind J Exp Biol 2013;53:823−7.
[57] Dong Yeon Y, Lee MY, Yun YP. Effect of green tea catechin on acute hepatotoxicity in rats. J Food Hyg Saf 2004;19(3):105−11.
[58] Takami S, Imai T, Hasumura M, Cho YM, Onose J, Hirose M. Evaluation of toxicity of green tea catechins with 90-day dietary administration to F344 rats. Food Chem Toxicol 2008;46(6):2224−9.
[59] Sharma R, Ahuja M, Kaur H. Thiolated pectin nanoparticles: Preparation, characterization and ex vivo corneal permeation study. Carbohydr Polym 2012;87(2):1606−10.
[60] Bonifácio BV, Patricia Bento da S, Matheus Aparecido dos SR, Kamila M, Silveira N, Taís MB, et al. Nanotechnology-based drug delivery systems and herbal medicines: a review. Int J Nanomed 2014;9:1−15.
[61] Dube A, Nicolazzo J, Ng K, Larson I. Effective use of reducing agents and nanoparticle encapsulation in stabilizing catechins in alkaline solution. Food Chem 2010;122:662−7.
[62] Mochizuki M, Yamazaki S, Kano K, Ikeda T. Kinetic analysis and mechanistic aspects of autoxidation of catechins. BiochemBiophys Acta 2002;1569:35−44.
[63] Tønnesen HH, Karlsen J. Alginate in drug delivery systems. Drug Devel Indus Pharm 2002;28(6):621−30.
[64] Pan JL, Bao ZM, Li JL, Zhang LG, Wu C, Yu YT. Chitosan-based scaffolds for hepatocyte culture. ASBM6: Adv Biomat 2005;VI:91−4.
[65] Park Y, Sugimoto M, Watrin A, Chiquet M, Hunziker EB. BMP-2 induces the expression of chondrocyte-specific genes in bovine synovium-derived progenitor cells cultured in three-dimensional alginate hydrogel. Osteoarthr Cartil 2005;13(6):527−36.

[66] Jana S, Gandhi A, Sen KK, Basu SK. Natural polymers and their application in drug delivery and biomedical field. J Pharm Sci Tech 2011;1:16−27.
[67] Reis CP, António JR, Ronald JN, Francisco V. Alginate microparticles as novel carrier for oral insulin delivery. Biotech Bioeng 2007;96(5):977−89.
[68] Ansari SH, Farha I, Sameem M. Influence of nanotechnology on herbal drugs: a review. J Adv Pharma Tech Res 2012;3(3):140−6.
[69] Peppas B, Blanchette JO. Nanoparticle and targeted systems for cancer therapy. Adv Drug Deliv Rev 2004;56:1649−59.
[70] Bhardwaj V, Hariharan S, Bala I, Lamprecht A, Kumar N, Panchagnula R, et al. Pharmaceutical aspects of polymeric nanoparticles for oral delivery. J Biomed Nanotech 2005;1:235−58.
[71] Khuda-Bukhsh AR, Bhattacharyya SS, Paul S, Boujedaini N. Polymeric nanoparticle encapsulation of a naturally occurring plant scopoletin and its effects on human melanoma cell A375. Zhong Xi Yi Jie He Xue Bao 2010;8(9):853−62.
[72] Tibbals HF. Medical nanotechnology and nanomedicine. CRC Press; 2011. p. 75−116.
[73] Rao JP, Geckeler KE. Polymer nanoparticles: preparation techniques and size-control parameters. Prog Poly Sci 2011;36(7):887−913.
[74] Sailaja AK, Amareshwar P, Chakravarty P. Formulation of solid lipid nanoparticles and their applications. Curr Pharma Res 2011;1(2):97−203.
[75] Ahirrao SP, Gide PS, Shrivastav B, Sharma P. Ionotropic gelation: a promising cross linking technique for hydrogels. Res Rev: J Pharm Nanotechnol 2014;1(2):1−6.
[76] Tabata Y, Ikada Y. Synthesis of gelatin microspheres containing interferon. Pharm Res 1989;6:422−7.
[77] Onsøyen E. Commercial applications of alginates. Carbohydr Eur 1996;14:26−31.
[78] Soni MK, Kumar M, Namdeo KP. Sodium alginate microspheres for extending drug release: formulation and in vitro evaluation. Int J Drug Deliv 2010;2:64−8.
[79] Das RK, Naresh K, Utpal B. Encapsulation of curcumin in alginate-chitosan-pluronic composite nanoparticles for delivery to cancer cells. Nanomed Nanotech Biol Med 2010;6:153−60.
[80] Zahoor A, Rajesh P, Sadhna S, Khuller GK. Alginate nanoparticles as antituberculosis drug carriers: formulation development, pharmacokinetics and therapeutic potential. Indian J Chest Dis Allied Sci 2006;48:171−6.
[81] Spizzirri UG, Ortensia IP, Francesca I, Giuseppe C, Francesco P, Manuela C, et al. Antioxidant-polysaccharide conjugates for food application by eco-friendly grafting procedure. Carbohy Poly 2010;79:333−40.
[82] Muralikrishna G, Tharanathan RN. Characterization of pectic polysaccharides from pulse husks. Food Chem 1994;50(1):87−9.
[83] Da Silva JL, Rao MA. Rheological behavior of food gels. Rheology of fluid and semi-solid foods. US: Springer; 2007. p. 339−401.
[84] Be Miller J.N. An introduction to pectin: structure and properties. In Fishman ML, Jen JJ, editors, Chemistry and Function of Pectins, ACS symposium Series 310, American Chemical Society, Washington DC; 1986, pp. 2−12.
[85] Oakenfull D, Scott A. Hydrophobic interaction in the gelation of high methoxylpectins. J Food Sci 1984;49(4):1093−8.
[86] Löfgren C, Hermansson AM. Synergistic rheological behaviour of mixed hm/lm pectin gels. Food Hydrocoll 2007;21(3):480−6.
[87] Sandberg AS, Ahderinne R, Andersson H, Hallgren B, Hultén L. The effect of citrus pectin on the absorption of nutrients in the small intestine. Hum Nutr Clin Nutr 1983;37(3):171−83.

[88] Sinha VR, Kumria R. Polysaccharides in colon-specific drug delivery. Int J Pharmacol 2001;224(1):19−38 2001.
[89] Vandamme TF, Lenourry A, Charrueau C, Chaumeil JC. The use of polysaccharides to target drugs to the colon. Carbohydr Polym 2002;48(3):219−31.
[90] Yamada H. Contribution of pectins on health care. Prog Biotechnol 1996;14:173−90.
[91] Bergman M, Djaldetti D, Salman H, Bessler H. Effect of citrus pectin on malignant cell proliferation. Biomed Pharmacother 2010;64(1):44−7.
[92] Venner M, Lauffs S, Deegen E. Treatment of gastric lesions in horses with pectin-lecithin complex. Equine Vet J 1999;31(S29):91−6.
[93] Liu L, Fishman ML, Hicks KB. Pectin in controlled drug delivery: a review. Cellulose 2007;14(1):15−24.
[94] Perera G, Barthelmes J, Bernkop-Schnürch A. Novel pectin-4-aminothiophenole conjugate microparticles for colon-specific drug delivery. J Control Rel 2010;145(3):240−6.
[95] Ugurlu T, Turkoglu M, Gurer US, Akarsu BG. Colonic delivery of compression coated nisin tablets using pectin/HPMC polymer mixture. Euro J Pharm Biopharm 2007;67:202−10.
[96] Maestrelli F, Cirri M, Corti G, Mennini N, Mura P. Development of enteric-coated calcium pectinate microspheres intended for colonic drug delivery. Euro J Pharm Biopharm 2008;69:508−18.
[97] Katav T, Liu L, Traitel T, Goldbart R, Wolfson M, Kost J. Modified pectin-based carrier for gene delivery: cellular barriers in gene delivery course. J Control Rel 2008;130:183−91.
[98] Opanasopit P, Apirakaramwong A, Ngawhirunpat T, Rojanarata T, Ruktanonchai U. Development and characterization of pectinate micro/nanoparticles for gene delivery. Aaps Pharm Sci Tech 2008;9(1):67−74.
[99] Thirawong N, Thongborisute J, Takeuchi H, Sriamornsak P. Improved intestinal absorption of calcitonin by mucoadhesive delivery of novel pectin−liposome nano-complexes. J Control Rel 2008;125(3):236−45.
[100] Sharma A, Gupta S, Sarethy IP, Dang S, Gabrani R. Green tea extract: possible mechanism and antibacterial activity on skin pathogens. Food Chem 2012;135:672−5.
[101] Verma A.K., Chanchal A., Kumar A. Potential of negatively charged pectin nanoparticles encapsulating paclitaxel: preparation and characterization. In Nanoscience Technology and Societal Implications NSTSI. International Conference; 2011. pp. 1−8.
[102] Dutta RK, Sahu S. Development of oxaliplatin encapsulated in magnetic nanocarriers of pectin as a potential targeted drug delivery for cancer therapy. Results Pharma Sci 2012;2:38−45.
[103] Yu CY, Wang YM, Li NM, Liu GS, Yang S, Tang GT, et al. In vitro and in vivo evaluation of pectin-based nanoparticles for hepatocellular carcinoma drug chemotherapy. Mol Pharma 2014;11(2):638−44.
[104] Shahidi F, Arachchi JKV, Jeon YJ. Food applications of chitin and chitosans. Trends Food Sci Technol 1999;10(2):37−51.
[105] Illum L. Chitosan and its use as a pharmaceutical excipients. Pharm Res 1998;15(9):1326−31.
[106] Li Q, Dunn ET, Grandmaison EW, Goosen MFA. Applications and properties of chitosan. J Bioact Compat Polym 1992;7:370−97.
[107] Janes KA, Calvo P, Alonso MJ. Polysaccharide colloidal particles as delivery systems for macromolecules. Adv Drug Delivery Rev 2001;47(1):83−97.
[108] Azuma K, Izumi R, Osaki T, Ifuku S, Morimoto M, Saimoto H, et al. Chitin, chitosan and its derivatives for wound healing: old and new materials. J Funct Biomat 2015;6:104−42.

[109] Cheng K, Lim LY. Insulin-loaded calcium pectinate nanoparticles: effects of pectin molecular weight and formulation ph. drug develop. Ind Pharma 2004;30(4):359–67.
[110] Zou Y, Yang Y, Li J, Li W, Wu Q. Prevention of hepatic injury by a traditional Chinese formulation, BJ-JN in mice treated with Bacille-Calmette-Guerin and lipopolysaccharide. J Ethnopharmacol 2006;107:442–8.
[111] Altiok D, Altiok E, Tihminlioglu F. Physical, antibacterial and antioxidant properties of chitosan films incorporated with thyme oil for potential wound healing applications. J Mater Sci: Mater Med 2010;21:2227–36.
[112] Davies NM, Farr SJ, Hadgraft J, Kellaway IW. Evaluation of mucoadhesive polymers in ocular drug delivery II. Polymer-coated vesicles. Pharm Res 1992;9(9):1137–44.
[113] Kotzé AF, Thanou MM, Luebetaen HL, De Boer AG, Verhoef JC, Junginger HE. Enhancement of paracellular drug transport with highly quaternized N-trimethyl chitosan chloride in neutral environments: in vitro evaluation in intestinal epithelial cells. J Pharma Sci 1999;88:253–7.
[114] Swarbrick J. Encyclopedia of pharmaceutical technology. Microsphere Technol Appl 2007;3(4):2329.
[115] Siddalingam P, Mishra B. Preparation and in vitro characterization of gellan based floating beads of acetohydroxamic acid for eradication of *H. pylori*. Acta Pharm 2007;57:413–27.
[116] Wayne RG, Siow FW. Protein release from alginate matrices. Adv Drug Delivery Rev 1998;1:267–85.
[117] Wu JL, Chao QW, Ren XZ, Si XC. Multi-drug delivery system based on alginate/calcium carbonatehybrid nanoparticles for combination chemotherapy. Coll Surf B: Biointerfaces 2014;123:498–505.
[118] Yu CY, Cao H, Zhang XC, Zhou FZ, Cheng SX, Zhang XZ, et al. Hybrid nanospheres and vesicles based on pectin as drug carriers. Langmuir 2009;25(19):11720–6.
[119] Calvo P, Remunan-Lopez C, Vilas-Jato JL, Alonso MJ. Chitosan and chitosan/ethylene oxide-propylene oxide block copolymer nanoparticles as novel carrier for protein and vaccines. Pharma Res 1997;14(10):1431–6.
[120] Liang HF, Yang TF, Huang CT, Chen MC, Sung HW. Preparation of nanoparticles composed of poly(gamma-glutamic acid)-poly(lactide) block copolymers and evaluation of their uptake by HepG2 cells. J Control Rel 2005;105:213–25.
[121] Dudhani AR, Kosaraju SL. Bioadhesive chitosan nanoparticles: preparation and characterization. Carbohydr Polym 2010;81:243.
[122] Benzie IF, Strain JJ. The ferric reducing ability of plasma (FRAP) as a measure of antioxidant power: the FRAP assay. Anal Biochem 1996;239(1):70–6.
[123] Basnet P, Matsumoo T, Neidlein R. Potent free radical scavenging activity of propolis isolated from Brazilian propolis. Z Naturforsh 1997;52(11–12):828–33.
[124] Afanas'ev IB, Dorozhko AI, Brodskii AV, Kostyuk VA, Potapovitch AI. Chelating and free radical scavenging mechanisms of inhibitory action of rutin and quercetin in lipid peroxidation. Biochem Pharma 1989;38(11):1763–9.
[125] Ebrahimzadeh MA, Nabavi SM, Nabavi SF. Correlation between the in vitro iron chelating activity and poly phenol and flavonoid contents of some medicinal plants. Pak J Biol Sci 2009;12(12):934–8.
[126] Doi K, Kurabe S, Shimazu N, Ingaki M. Systemic histopathology of rats with CCl_4 induced hepatic cirrhosis. Lab Ani 1991;25:21–5.
[127] Reitman S, Frankel S. A colorimetric method for the determination of serum glutamic oxalacetic and glutamic pyruvic transaminases. Am J Clin Pathol 1957;28(1):56–63.

[128] Malloy H.T., Eveluyun K.A. Test in liver and biliary tract disease. In: Varley H, editor, Practical clinical biochemistry, 4th edn. New Delhi: CBS Publications; 1988. pp. 353−355.

[129] Rehman SU. Lead-induced regional lipid peroxidation in brain. Toxicol Lett 1984;21(3):333−7.

[130] Sedlak J, Lindsay RH. Estimation of total, protein-bound, and nonprotein sulfhydryl groups in tissue with Ellman's reagent. Anal Biochem 1968;25:192−205.

[131] Marklund S, Marklund G. Improvement of superoxide anion radical in the auto oxidation of pyrogallol and a convenient assay for SOD. Eur J Biochem 1974;47:469−74.

[132] Masayasu M, Hiroshi Y. A simplified assay method of superoxide dismutase activity for clinical use. ClinicaChimica Acta 1979;92(3):337−42.

[133] Cohen G, Dembiec D, Marcus J. Measurement of catalase activity in tissue extracts. Anal Biochem 1970;34(1):30−8.

[134] Runnels RA, Monlux WS, Monlux AW. Principles of veterinary pathology. 7th Edn Calcutta: Scientific Book Agency; 1965.

[135] Culling CFA. Handbook of histopathological techniques (including museum technique). Washington DC: National Agricultural Library; 1963. p. 548−53.

[136] Bharrhan S, Ashwani K, Kanwaljit C, Praveen R. Catechin suppresses an array of signalling molecules and modulates alcohol-induced endotoxin mediated liver injury in a rat model. PLoS One 2011;6(6):1−9. Available from: https://doi.org/10.1371/journal.pone.0020635.

[137] Sayah MY, Chabir R, Benyahia H, Rodi Kandri Y, Ouazzani Chahdi F, Touzani H, et al. Yield, esterification degree and molecular weight evaluation of pectins isolated from orange and grape fruit peels under different conditions. PLoS One 2016;11(9): e0161751. Available from: https://doi.org/10.1371/journal.pone.0161751.

[138] Varley H., Grawlock A.H., Bell M. Practical Biochemistry. 5th Edn. London: William Heinmann Medical Book Ltd; 1980, pp 458-484.

[139] Ohkawa H, Ohishi N, Yagi K. Assay for lipid peroxides in animal tissues by thiobarbituric acid reaction. Anal Biochem 1979;95(2):351−8.

[140] Menami M, Yoshikawa H. Simplified assay method of superoxide dismutase activity of clinical use. Clin Chem Acta 1979;92:337−42.

[141] Sowjanya B, Prasanna RI, Devi KJ, Narayana S, Chetty CM, Purushothaman M, et al. Preparation and characterization of Cefadroxil loaded alginate microbeads. Int J Res Pharm Sci 2010;1(4):386−90.

[142] Benjamin MM. Outline of veterinary clinical pathology.. 3rd Edn Ludhiana: : Kalyani Publishers; 2007. p. 351−60.

[143] Snedecor CW, Cochran WG. Statistical methods. 6th Edn. USA: Anes Iowa State University Press; 1994.

Index

Note: Page numbers followed by "*f*" and "*t*" refer to figures and tables, respectively.

A

Abomasum, 75–77
Acellular antigen, 336
Acellular aortic matrix, 208, 224, 235–236
   graft, 221–223
   optimization of protocols for preparation of, 208
      group A, 208
      group B, 211
      protocol A1, 208–209
      protocol A2, 209
      protocol A3, 209–210
      protocol A4, 210
      protocol A5, 210
      protocol A6, 211
      protocol B1, 211–212
      protocol B2, 212
      protocol B3, 212
      protocol B4, 212
      protocol B5, 212–213
      protocol B6, 213–215
   in vivo biocompatibility determination of, 220–232
      immunological studies, 225
      lymphocyte proliferation assay, 225–231
      macroscopic observations, 220–222
      microscopic observations, 222–225
      molecular weight analysis, 231–232
      surgical procedure, 220
Acellular biological scaffold, 270
Acellular biomaterials of porcine, 34–35
Acellular bovine reticulum extracellular matrix
   preparation of, 107–113
      DNA contents analysis, 112–113
      gross and microscopic observations, 108–110
      scanning electron microscopic observations, 110–112
Acellular buffalo cholecyst-derived extracellular matrix
   preparation of, 163–167
      DNA contents analysis, 167
      macroscopic observations, 164
      microscopic observations, 164–166
      scanning electron microscopic observations, 166
Acellular buffalo omasum in wound healing in rat model
   experimental evaluation of, 141–154
      animals and ethics statement, 142
      cell-mediated immune response, 147–150
      gross observations/planimetry, 144–145
      histological observations, 150–154
      humoral response, 145–147
      immunological observations, 145–150
      open wound, 151–152
      skin wound creation and implantation, 142
      wound area and wound contraction, 142–144
      wound with acellular buffalo omasum laminae, 152–154
      wound with commercially available collagen sheet, 152
Acellular buffalo pericardium matrix
   crosslinking of native and, 245–256
   preparation of, 244–245
Acellular caprine reticulum extracellular matrix
   preparation of, 114–117
      calorimetric protein estimation, 116
      DNA contents analysis, 115

Acellular caprine reticulum extracellular matrix (*Continued*)
  microscopic observations, 114−115
  sodium dodecyl sulfate polyacrylamide gel electrophoresis analysis, 117
Acellular dermal matrix (ADM), 24, 34, 290
Acellular diaphragm matrix (ADiaM), 270−271, 289−290, 298−300
  matrices, 298
Acellular diaphragmatic scaffold of buffalo origin
  preparation of, 278−283
    agarose gel electrophoresis, 281−283, 282f
    (4,6-diamino-2-phenylindole) dihydrochloride staining, 280, 280f
    DNA quantification, 280−281
    gross observations, 278, 279f
    histological observations, 278−280, 279f
    materials and methods (II), 278
Acellular ECM, 135−137
Acellular fish swim bladder matrix, 313−314
Acellular goat cholecyst-derived extracellular matrix
  preparation of, 172−178
    DAPI staining, 176−177
    DNA contents analysis, 177−178
    microscopic observations, 175
    preparation of acellular goat gall bladder matrix, 173−175
    preparation of soapnut pericarp extract, 172−173
    scanning electron microscopic observations, 177
Acellular goat gall bladder matrix, preparation of, 173−175
Acellular goat pericardium, 243−244
  matrix
    crosslinking of, 243−244
    preparation of, 243
Acellular grafts, 322
Acellular group, 231
Acellular matrix (ACM), 94, 270
  caprine rumen, 95−99
    calorimetric protein estimation, 99
    DNA quantification, 97−98
    microscopic observations, 96
    scanning electron microscopic observations, 99
    SDS-PAGE analysis, 98−99
  bubaline rumen, 79−88
    cytocompatibility analysis, 87−88
    DNA quantification, 87
    macroscopic observations, 80
    microscopic observations, 80−87
    treatment with enzyme, 86−87
    treatment with ionic detergent, 84−85
    treatment with nonionic detergent, 83−86
    treatment with zwitter ionic detergent, 81−83
  buffalo omasum, 132−138
    DNA quantification, 137
    gross observations, 133
    microscopic observations, 133−137
    sodium dodecyl sulfate polyacrylamide gel electrophoresis, 137−138
  fish swim bladder, 309−312
  goat omasum, 138−141
    calorimetric protein estimation, 140
    DNA quantification, 139
    microscopic observations, 139
    sodium dodecyl sulfate polyacrylamide gel electrophoresis analysis, 140−141
  vascular allograft, 25
Acellular pericardium, 260
Acellular pig cholecyst-derived extracellular matrix, preparation of, 168−172
    DNA contents analysis, 171−172
    macroscopic observations, 169
    microscopic observations, 169−171
    scanning electron microscopic observations, 171
Acellular rumen matrices, 93
Acellular swim bladder matrix (ASBM), 328
    evaluation of ASBM for skin wound healing in rabbit model, 327−340
    preparation of ASBM and crosslinking with epoxy compounds, 328
Acellular tissue, 243
  matrices, 243−245, 283, 307−308, 310−311
Acid phosphatase (ACP), 286, 288−289
Acids (COOH), 241
ACM. *See* Acellular matrix (ACM)

Index

Adaptive immune response modulation, 375
Adherent cells, 312–313, 347
ADiaM. *See* Acellular diaphragm matrix (ADiaM)
Adipocytes, 371–372
Adipogenesis, 372
Adipogenesis:adipogenesis, 371–372
Adipogenic differentiation, 351
Adipose tissue–derived MSCs (ADSCs), 367, 371, 378
ADM. *See* Acellular dermal matrix (ADM)
Adult stem cells, 308–309
Adventitial progenitor cells (APCs), 371
Agar, 52
Agarose gel electrophoresis, 281–283, 282f
Alcian blue staining, 351
Alcoholic cirrhosis patients, 378
Alginate, 388, 398–400
Alizarin red staining, 351
Alkaline phosphatase (ALP), 286, 288–289, 414
　analysis, 288–289
Allogeneic stem cell, 323
Allogenic acellular dermal matrix, 24–25
Allogenic collagen-rich ECM, 339–340
ALP. *See* Alkaline phosphatase (ALP)
Alumina, 10
Amines (NH_2), 241
Amino propyl tri ethoxy silane-coated slides (APTES-coated slides), 176
Amplicons, 349
Animals, 142, 178–179, 286, 290–291, 316, 356
　clinical applications in different species of, 232–236
　and ethics statement, 142
Anionic biological detergent, 243
Anionic detergents, 214–215
Antibodies, 294, 320
　responses, 333–334
Antigen, 295, 322–323
　preparation of, 225
Antigen-presenting cells (APCs), 123, 228, 323
Antigenic epitopes, 327–328
Aorta, 214
　aorta-derived extracellular matrix
　　clinical applications in different species of animals, 232–236
　optimization of protocols for decellularization of buffalo aorta, 206–215
　preparation and characterization of buffalo aortic matrix, 215–220
　in vivo biocompatibility determination of acellular aortic matrix, 220–232
Aortic grafts, 221
Aortic tissue, 208
APCs. *See* Adventitial progenitor cells (APCs); Antigen-presenting cells (APCs)
Apoptosis, 28–29, 33
APTES-coated slides. *See* Amino propyl tri ethoxy silane-coated slides (APTES-coated slides)
Artificial extracellular matrix proteins, 66
Artificial heart valves, 5
ASBM. *See* Acellular swim bladder matrix (ASBM)
Aseptic preparation, 346
Aspirin, 64
Astrocytes, 345–346
Autoimmune diseases, role in, 377
Autologous grafts, 20–21
Autooxidation of catechin, 389
5-azacytidine, 374

B

B-cell response, 320–321
b-CEM. *See* Cholecystic-derived ECM of buffalo origin (b-CEM)
b-CS. *See* Bovine collagen sheath (b-CS)
b-FGF. *See* Basic fibroblast growth factor (b-FGF)
b-REM group, 125–126
BAM. *See* Buffalo aortic matrix (BAM)
BAMG. *See* Bladder acellular matrix graft (BAMG)
Basal medium, 367
Basic fibroblast growth factor (b-FGF), 204, 367
BBB. *See* Blood-brain barrier (BBB)
BDM. *See* Buffalo diaphragm matrix (BDM)
β cells, 374
β-mercaptoethanol (BME), 374
BG. *See* Buffalo gallbladder (BG)
Bioartificial organs, 10
Biochemical parameters, 359–360

Biocompatibility, 8–14
  assessment, 11–12
  evaluation of pig diaphragm, 283–285
    cell cytotoxicity, 283–285, 284f
    cell extraction, 283
    materials and methods (III), 283–285
  testing, 8–11
Biocompatible material, 8–9
Biodegradable polymers, 5, 397–398
Bioengineered fish swim bladder matrices for skin wound healing in rat model
  computerized planimetry, 319–320
  evaluation of, 316–327
    wound healing, 317
    gross observations, 317–319
      wound area, 317
      wound contracture, 317–319
    histopathological observations, 324–327
    immunological observations, 320–323
      indirect enzyme-linked immunosorbent assay, 320–321
      lymphocyte proliferation assay, 321–323
Bioengineered porcine small intestinal submucosal scaffolds, 38
Biological crosslinking, 53–55
  advantages of, 54
  disadvantages of, 54–55
Biological materials, 1, 126
Biological matrices, 204
Biological molecules, 53
Biological scaffolds, 322
Biomaterial(s), 1–2, 53, 246–247, 257, 270, 294
  basic features, 6–7
    characteristics of biomaterials, 6–7
    requirements for scaffolds used in tissue engineering, 7
  biocompatibility, 8–14
  classification of, 7
    class of materials used in body, 9t
    classification based on occurrence of biomaterials, 8t
  definitions of, 3–6
  engineering, 2
  science, 2
Biomedical excipients for nanostructurization, 397–402
Bioscaffolds, 277

Bladder acellular matrix graft (BAMG), 38–39
Blood, 225–226
  samples, 293–294
Blood-brain barrier (BBB), 345, 395
BME. See β-mercaptoethanol (BME)
BMP. See Bone morphogenetic proteins (BMP)
BMSCs. See Bone marrow–derived mesenchymal stem cells (BMSCs)
Bone, 2, 10
Bone marrow–derived mesenchymal stem cells (BMSCs), 313, 318–319, 367, 377
  collection and isolation of, 312–313
    from goat, 313
    from rat, 312–313
  isolation culture and expansion of, 346–348
Bone morphogenetic proteins (BMP), 373
Bovine, 241–242
Bovine collagen sheath (b-CS), 193–194
  group, 124–125
Bovine collagen-based material, 22–23
Bovine diaphragm, biochemical changes in rabbit organs after subcutaneous implantation of, 285–289
  acid phosphatase and alkaline phosphatase analysis, 288–289
  collection of organs, 286
  free amino group contents analysis, 286–287
  materials and methods (IV), 285
  protein contents analysis, 287–288
  subcutaneous implantation of buffalo diaphragm in rabbits, 286
Bovine pericardium, 25, 262
Bovine reticulum extracellular matrix in rat model
  evaluation of, 117–127
    gross observations, 119–121
    hematological observations, 121–122
    histopathological observations, 123–127
    immunological observations, 122–123
    wound area and percent contraction, 118–119
Bovine serum albumin (BSA), 145–146, 287, 320, 355

Bovine spongiform encephalopathy, 308
BSA. *See* Bovine serum albumin (BSA)
Bubaline aortic matrix, 235–236
Bubaline diaphragm, 298
Bubaline rumen
   development of 3-D bioengineered scaffolds from, 89
   evaluation of bubaline rumen matrix in clinical cases, 93–95
   preparation of acellular matrices from, 79–88
Bubalus bubalis. See Buffalo (*Bubalus bubalis*); Water buffalo (*Bubalus bubalis*)
Buffalo (*Bubalus bubalis*), 41, 271
   diaphragm, 278–280, 282, 298–300
      subcutaneous implantation of buffalo diaphragm in rabbits, 286
   preparation of acellular diaphragmatic scaffold of buffalo origin, 278–283
   preparation of acellular matrix from buffalo omasum, 132–138
Buffalo aorta, optimization of protocols for decellularization of, 206–215
   group A, 206–207, 207t
   group B, 207–208, 208t
   optimization of protocols for preparation of acellular aortic matrix, 208
Buffalo aortic matrix (BAM), 215
   preparation and characterization of, 215–220
      DNA extraction, quantification, and purity, 217–219
      FTIR spectroscopy, 219–220
      histological observations, 215–216
      SEM observations, 217
Buffalo cholecyst-derived extracellular matrix in rat model, experimental evaluation of, 178–189
   animals and ethics statement, 178
   gross observations/planimetry, 180
   histopathological observations, 184–189
   immunological observations, 180–184
      cell-mediated immune response, 182–184
      humoral response, 180–182
   skin wound creation and implantation, 178–179
   wound contraction, 179

Buffalo diaphragm matrix (BDM), 271–273, 275–276
   preparation and characterization of, 271–278
      DNA extraction, 274–275
      DNA quantification, 275–276
      Fourier transform infrared spectroscopy, 276–278, 276f
      histological observations, 271–272
      materials and methods (I), 271
      scanning electron microscopy, 272–273
Buffalo gallbladder (BG), 163
1,4-butanediol diglycidyl ether (BDDGE), 39, 41, 328
   crosslinked acellular aortic matrix graft, 223
   crosslinked grafts, 224–225
Butanol, 137–138

C

C-reactive protein (CRP), 359
CA-AL-NP. *See* Catechin-loaded alginate polymeric nanoparticles (CA-AL-NP)
CA-CH-NP. *See* Catechin-loaded chitosan polymeric nanoparticles (CA-CH-NP)
CA-PC-NP. *See* Catechin-loaded pectin polymeric nanoparticles (CA-PC-NP)
Calcification, 57, 63, 263
Calcium, 373
   ions, 398–400
Calorimetric protein estimation, 99, 116, 140
Camelids, 76
Camellia sinensis. See Tea (*Camellia sinensis*)
Caprine rumen
   evaluation of caprine rumen matrix in clinical cases, 100–102
   preparation of acellular matrices from, 95–99
Carbodiimides, 51, 64
Carbohydrates, 53, 398–399
Carbon nanotubes (CNT), 346
Carbon tetrachloride (CCl_4), 412
Carbon–carbon bonds (C–C), 62–63
Catalase
   activity, 415–416
   assay, 415–416

Catechin-loaded alginate polymeric
    nanoparticles (CA-AL-NP), 405–406
Catechin-loaded chitosan polymeric
    nanoparticles (CA-CH-NP), 407–408
Catechin-loaded pectin polymeric
    nanoparticles (CA-PC-NP), 406
Catechin(s), 389–392
  catechin-encapsulated chitosan
    nanoparticle, 407–408
  hydrate loaded hybrid nanoparticle, 406
  in vitro release of, 410
Catheters, 10
C–C. See Carbon–carbon bonds (C–C)
CD marker-based expression via RT PCR,
    349
CD34 marker, 371
CD73 marker, 370
CD90 marker, 370–371
CD105 marker, 371
CEF. See Chicken embryo fibroblasts (CEF)
CEF cell culture, 285
Cell-mediated immune response, 123,
    147–150, 182–184, 193–194,
    225–226, 228–229, 322–323, 336
Cell mediated immunity (CMI), 262, 294
Cell(s), 77–78, 244–245, 262, 308,
    313–314, 348
  adhesion, 33
  cell-surface marker expression, 370–371
    CD34 marker, 371
    CD73 marker, 370
    CD90 marker, 370–371
    CD105 marker, 371
    Stro-1 marker, 371
  culture, 373
  cytotoxicity, 283–285, 284f
  debris, 244–245
  differentiation, 33
  extraction, 283
    process, 297–298
  migration, 33
  pellets, 346–347
  percentage of cell seeding assessment, 354
  separation technique, 315
  systems, 285
  viability assay, 349–350
    MTT assay, 350
    trypan blue exclusion method, 349
Cellular antigens, 244–245, 310–311

Cellular goat pericardium, 243–244
Cellular matrix acellular matrix, 18–22
  decellularization methods, mode of action,
    and effects on ECM, 23t
CEM. See Cholecyst derived extracellular
    matrix (CEM)
Central nervous system (CNS), 345
Centrifugal seeding of MSC in scaffold,
    354
Ceramics, 5, 10, 13–14
Characterization methods
  of decellularized tissues, 26–34
    4′,6-diamidino-2-phenylindole, 28–29
    EM, 30–31
    hematoxylin and eosin, 27–28
    mechanical properties, 32–33
    MTT cell proliferation assay, 29
    SHG, 29–30
    zymography, 33–34
Chemical crosslinking, 50–51
  advantages of, 51
  disadvantages of, 51
Chemical decellularization methods, 18
Chemokine regulation, 375
Chicken embryo fibroblasts (CEF), 34–35,
    283–284
Chitosan (CS), 388, 399–400, 402
  chitosan-coated catechin nanospheres, 407
  nanoparticles, 402
3-[(3-cholamidopropyl)dimethylammonio]-
    1-propanesulfonate (CHAPS), 21–22
Cholecyst, 160
Cholecyst derived extracellular matrix
    (CEM), 159–163
Cholecystic-derived ECM of buffalo origin
    (b-CEM), 178–179
Chondrocytes, 373
Chondrogenesis, 372–373
Chondrogenic differentiation, 351
Clostridium histolyticum, 247
CMI. See Cell mediated immunity (CMI)
CNS. See Central nervous system (CNS)
CNT. See Carbon nanotubes (CNT)
Cobalt-chromium alloys, 5
Collagen, 10, 22, 32, 55, 241–242,
    269–270, 308, 322
  arrangement, 327
  collagen IV, 78
  collagen-based biomaterials, 249

collagen-based decellularized
    biomaterials, 35
collagen-rich matrices, 327
    fibers, 110–112, 166, 171, 310, 336–339
    sheet, 183
        wound with commercially available,
            152
Collagenous tissues, 242
Colony forming assays, 350
    colony forming units-fibroblasts assay,
        350
    colony forming units-osteoblasts assay,
        350
Color digital image processing, 193,
    319–320
Color photographs, 291
Composite materials, 5
Computerized planimetry, 92, 319–320
Con A. *See* Concanavalin A (Con A)
Concanavalin A (Con A), 321
Contact lenses, 5
Controlled drug delivery systems, 389–390
Coprecipitation method, 405–406
Crohn's disease, 378
Crosslinked acellular aortic matrix grafts,
    225
Crosslinked acellular swim bladder matrix
    for skin wound healing in rabbit
    model, evaluation of, 327–340
Crosslinked cellulose, 47, 50
Crosslinked collagen, 47, 50
Crosslinked hyaluronic acid, 47, 50
Crosslinked hydrogels, 47, 50
Crosslinking, 48, 242, 261
    agents, 55–67
        1-ethyl-3-(3-dimethylaminopropyl)
            carbodiimide hydrochloride, 64–65
        diphenyl phosphoryl azide, 60–61
        enzymatic crosslinking, 67
        ethylene glycol diglycidyl ether, 61–63
        formaldehyde, 58–59
        glutaraldehyde, 56–58
        glyoxal, 59–60
        hexamethylene diisocyanate, 65–66
        PEG, 63–64
        physical, 66–67
    of biomaterials
        advantages of crosslinking biomaterials,
            48

disadvantages of crosslinking
    biomaterials, 49
of native and acellular buffalo
    pericardium matrix, 245–256
    concentration of solution, 246
    duration of treatment, 246–247
    free amino group contents
        determination, 249–250
    gross observations, 246–247
    moisture content analysis, 250–252
    molecular weight analysis, 252–254
    temperature, 247
    in vitro cell cytotoxicity, 255–256
    in vitro enzymatic degradation,
        247–249
of native and acellular goat pericardium
    matrix, 243–244
types of, 49–55
    biological crosslinking, 53–55
    chemical crosslinking, 50–51
    physical crosslinking, 51–53
CRP. *See* C-reactive protein (CRP)
Crystallinity, 52
CS. *See* Chitosan (CS)
Culture
    of mesenchymal stem cells,
        367–370
        basal medium, 367
        growth factors and supplements,
            367–369
        incubation conditions, 370
        passaging, 370
        serum supplementation, 367
    of p-MEF, 290
Curcumin, 399–400
Cyanamide, 64
Cytocompatibility analysis, 87–88
Cytokines, 323
    regulation, 375
Cytotoxicity, 49
    testing, 11

D

Damage associated molecular pattern
    (DAMP), 308
DAMP. *See* Damage associated molecular
    pattern (DAMP)
DDS. *See* Drug delivery systems (DDS)

Decellularization process, 18–22, 41, 88, 137, 205–206, 243–245, 262, 310–311
  cellular matrix acellular matrix, 18–22
  of forestomach matrix, 78–79
  optimization of protocols for decellularization of buffalo aorta, 206–215
  protocol, 21, 79, 214
Decellularized aortic samples, 217
Decellularized biomaterials, 3, 22–25
Decellularized ECMs (dECM), 20, 26
Decellularized goat rumen matrices, 100
Decellularized naturally derived biomaterials
  development of, 34–38
  in vitro evaluation of, 38–41
Decellularized scaffolds, 270
Decellularized tissues, 19–20
  characterization methods of, 26–34
dECM. See Decellularized ECMs (dECM)
Delaminated bovine omasum, 135–137
Delaminated bubaline rumen, 88
Delaminated gallbladder, 165, 169
Delaminated reticulum, 109
Delaminated rumen matrix, 84
Delivery systems, 398–399
Dendritic cells, 376
  modulation of, 376
Dense fibrous connective tissue, 221
Dental implants, 5
Dermal regeneration, 339–340
Dermis, 25
Detergents, 282–283
Diabetes mellitus, 89–91
  in rats, 91
Diabetic rat model, testing efficacy of 3-D bioengineered scaffolds in, 89–93
4′,6-diamidino-2-phenylindole (DAPI), 26, 28–29, 353–355
  staining, 176–177, 354–355
4,6-diamino-2-phenylindole dihydrochloride staining, 280, 280f
Diaphragm, 25
Diaphragm-derived extracellular matrix scaffolds
  biochemical changes in rabbit organs after subcutaneous implantation of bovine diaphragm, 285–289
  biocompatibility evaluation of pig diaphragm, 283–285
  clinical applications, 298–300, 299t
  dermal wound healing using primary mouse embryonic fibroblasts seeded buffalo acellular diaphragm matrix in rat model, 289–298
  preparation
    of acellular diaphragmatic scaffold of buffalo origin, 278–283
    and characterization of buffalo diaphragm matrix, 271–278
Diaphragmatic tissue, 278, 280–281
Differential leukocyte count (DLC), 121–122
Dimethyl sulfoxide, 350
3-(4,5-dimethylthiazol-2-yl)-2,5-diphenyltetrazolium bromide (MTT), 29, 122, 147, 182–183, 193–194, 334–335, 350
  assay, 350
  cell proliferation assay, 29
Diphenyl phosphoryl azide (DPPA), 41, 60–61, 243–244
  DPPA-crosslinked collagen membranes, 61
1,1-diphenyl-2-picrylhydrazyl (DPPH), 411
Disease modeling, 19
5,5′-dithiobis(2-nitrobenzoic acid) (DTNB), 415
DLC. See Differential leukocyte count (DLC)
DNA, 274–275
  band, 277–278
  contents analysis
    acellular buffalo cholecyst-derived extracellular matrix, 167
    acellular caprine reticulum extracellular matrix, 115
  extraction, 274–275
  quantification, and purity, 217–219
  isolation, 281
  quantification
    acellular diaphragmatic scaffold of buffalo origin, 280–281
    acellular matrices from bubaline rumen, 87
    acellular matrices from caprine rumen, 97–98

acellular matrix from buffalo omasum, 137
acellular matrix from goat omasum, 139
buffalo diaphragm matrix, 275–276
DNase, 26
Donor
  organs, 307
  tissue, 308
Double acellular matrix, 188–189
Double stranded DNA (dsDNA), 173
DPBS. *See* Dulbecco's phosphate buffer saline (DPBS)
DPPA. *See* Diphenyl phosphoryl azide (DPPA)
Drug
  discovery, 19
  loading, 409
Drug delivery systems (DDS), 5, 10, 48, 403
  difference in internal and external gelation on, 405
dsDNA. *See* Double stranded DNA (dsDNA)
Dulbecco's phosphate buffer saline (DPBS), 346–347

E
EC. *See* Epicatechin (EC)
ECG. *See* Epicatechin gallate (ECG)
ECM. *See* Extracellular matrix (ECM)
Ecto-5'-nucleotidase, 370
Ectodermal derivatives, 374
EDC. *See* 1-ethyl-3-(3-dimethylaminopropyl)-carbodiimide (EDC)
EEO. *See* Electroendosmosis (EEO)
EGC. *See* Epigallocatechin (EGC)
EGCG. *See* Epigallocatechin gallate (EGCG)
EGDGE. *See* Ethylene glycol diglycidyl ether (EGDGE)
EGF. *See* Epidermal growth factor (EGF)
"Egg box complexes", 400
Elastase enzymatic degradation, 247–249
Elastin, 32, 55, 241
Elastomers, 53
Electroendosmosis (EEO), 274–275
Electromagnetic wave, 30
Electron microscopy (EM), 30–31
  SEMs, 31
  TEMs, 31
Electrophoresis, 33–34, 281–282

Electrospinning process, 351
ELISA. *See* Enzyme-linked immunosorbent assay (ELISA)
EM. *See* Electron microscopy (EM)
Embryonic stem cells, 308–309
Encapsulation
  efficiency, 409
  of nanopolymer, 403
Endodermal derivatives, 374
Endoglin, 371
Endonucleases, 26
Enrofloxacin, 329–330
Enzymatic crosslinkers, 67
Enzymatic crosslinking, 67
Enzymatic decellularization methods, 19
Enzymatic degradation, 66
Enzyme-linked immunosorbent assay (ELISA), 149, 229–231, 320, 332–334
  procedure, 122, 180–182
Enzymes, 289
  treatment with, 86–87
Eosinophil worm (*Haliclystus octoradiatus*), 27
Epicatechin (EC), 390
Epicatechin gallate (ECG), 390
Epidermal growth factor (EGF), 367
Epigallocatechin (EGC), 390
Epigallocatechin gallate (EGCG), 390
Epithelialization, 153–154, 326, 336
Epoxides, 51
Epoxy compounds, 61–62
  preparation of acellular swim bladder matrix and crosslinking with, 328
Ethanol-dried acellular bovine omasum matrix, 37
1-ethyl-3-(3-dimethylaminopropyl) carbodiimide hydrochloride, 64–65, 330
1-ethyl-3-(3-dimethylaminopropyl)-carbodiimide (EDC), 38–39, 55–57, 246
  crosslinked acellular aortic matrix graft, 223–224
Ethylene glycol diethyl ether, 61–62
Ethylene glycol diglycidyl ether (EGDGE), 41, 61–63, 243–244, 252
Excipients, 393
Exogenous collagen, 326

Exogenous collagen (*Continued*)
 supplementation, 188
Experimental animal models, 355
External gelation on drug delivery system,
 difference in, 405
External ionotropic method, 404–405
Extracellular matrix (ECM), 2–3, 17–18,
 26, 32, 77–78, 89–91, 131,
 159–160, 204, 241, 307–308, 373
 proteins, 269–270
 scaffolds, 87–88, 105–106, 203–204
Extracts, 275–276

F
FA. *See* Formaldehyde (FA)
FBS. *See* Fetal bovine serum (FBS); Fish
 swim bladder scaffolds (FBS)
FDA. *See* United States Food and Drug
 Administration (FDA)
Ferric-reducing antioxidant power (FRAP),
 410–411
 assay, 410–411
Fetal bovine serum (FBS), 350, 367
Fibrin, 55
Fibroblast proliferation, 324–325
Fibroblastic cells, 285
Fibroblasts, 224–225, 297–298
Fibronectin attachment sites, creation of, 353
Fibroplasia, 258–259, 336
Fibrous connective tissue, 221
Fish processing wastes, 327
Fish swim bladder scaffolds (FBS), 313
Fish swim bladder–derived tissue scaffolds
 attachment and growth of mesenchymal
  stem cells, 314–316
 collection and isolation of bone
  marrow–derived mesenchymal stem
  cells, 312–313
 evaluation of acellular and crosslinked
  acellular swim bladder matrix,
  327–340
 evaluation of bioengineered fish swim
  bladder matrices, 316–327
 preparation of acellular matrix from fish
  swim bladder, 309–312
 seeding of stem cells on acellular fish
  swim bladder matrix, 313–314
Flavonoids, 392
Flushing out method, 315

FM. *See* Forestomach Matrix (FM)
Food production, 19
Forestomach, 79
 ECM, 106
Forestomach Matrix (FM), 105–106, 132
 decellularization of, 78–79
 scaffolds, 105–106
Formaldehyde (FA), 38–39, 55–59, 246
 formaldehyde-treated native pericardium
  graft, 259
Fourier transform infrared (FTIR), 276–277
 spectroscopy, 219–220, 276–278, 276*f*
FRAP. *See* Ferric-reducing antioxidant
 power (FRAP)
Free amino group contents analysis, 250,
 286–287
Free amino group contents determination,
 249–250
Free radical scavenging activity by 1,1-
 diphenyl-2-picrylhydrazyl assay,
 determination of, 411
Frequency doubling, 29–30
Fresh water fish (*Labeo rohita*), 309
FTIR. *See* Fourier transform infrared (FTIR)
Full thickness skin wounds, 153, 188,
 197–198
 healing, 153–154

G
GA. *See* Glutaraldehyde (GA)
GAGs. *See* Glycosaminoglycans (GAGs)
Gall bladder-derived extracellular matrix
  scaffolds
 anatomy of gallbladder, 160
 cholecyst-derived extracellular matrix,
  160–163
 experimental evaluation
  of buffalo cholecyst-derived
   extracellular matrix in rat model,
   178–189
  of pig cholecyst-derived extracellular
   matrix in rat model, 190–198
 preparation of
  acellular buffalo cholecyst-derived
   extracellular matrix, 163–167
  acellular goat cholecyst-derived
   extracellular matrix, 172–178
  acellular pig cholecyst-derived
   extracellular matrix, 168–172

Index

Gallbladder, anatomy of, 160
GAP-43. *See* Growth Associated Protein-43 (GAP-43)
Gastrointestinal diseases, 378
Gelatin, 52
Genes by real time PCR, relative expression of different, 360–361
Genipin, 252
Genotoxicity testing, 11
Glucuronic acid, 398–399
Glutaraldehyde (GA), 38–39, 41, 51, 55–58, 61–62, 243–244, 246, 252, 285
Glutathione assay, 415
Glycosaminoglycans (GAGs), 26, 121, 145, 159–160, 214, 282–283
Glyoxal (GO), 41, 59–60, 243–244
GO. *See* Glyoxal (GO)
Goat
   bone marrow–derived mesenchymal stem cells from, 313
   gall bladder, 172–175
   preparation of acellular matrix from goat omasum, 138–141
   rumen, 102
Goat bone marrow, derived mesenchymal stem cells (G-BMSC), 314
Government agencies, 11
Graft
   graft-assisted healing, 121, 145, 180
   rejection, 229
Graft-versus-host disease (GvHD), 376
Granulation tissue, 184–186, 194, 295–296
Growth Associated Protein-43 (GAP-43), 360
Guinea pigs, 227–228
GvHD. *See* Graft-versus-host disease (GvHD)

H

H&E. *See* Hematoxylin and eosin (H&E)
Haematoxylum campechianum.
   See Logwood tree (*Haematoxylum campechianum*)
Haliclystus octoradiatus. See Eosinophil worm (*Haliclystus octoradiatus*)
Hb. *See* Hemoglobin (Hb)
Heart, 287–288
   valves, 10

HEC. *See* Hydroxyethyl cellulose (HEC)
Hematological profile, 414
Hematoxylin and eosin (H&E), 26–28, 215, 271–272
   staining, 27–28, 79, 108, 133–134, 316
Hemocompatibility testing, 11
Hemoglobin (Hb), 414
Hemorrhage, 142, 178–179
Heparan sulfate, 106
Heparin, 64
Hepatocyte growth factor (HGF), 374
Hernial contents, 235
Heterologous collagen, 24–25
Hexamethyldisilazane (HMDS), 177
Hexamethylene diisocyanate (HMDC), 38–39, 55–57, 65–66, 285
   HMDC-crosslinked groups, 288
HGF. *See* Hepatocyte growth factor (HGF)
Hip replacements, 10
Histopathological score system (HPS), 416
Histopathological study, 416
HMDC. *See* Hexamethylene diisocyanate (HMDC)
HMDS. *See* Hexamethyldisilazane (HMDS)
Holstein–Friesian calves, 234–235
Homogenate, 249–250, 252
Host
   cells, 243
   inflammatory reaction, 224–225
HPS. *See* Histopathological score system (HPS)
Human bone marrow, 366
Humoral immune response, 148–149, 228–229, 322–323
Humoral response, 145–147, 180–182, 193, 320
Hybrid nanoparticles, 397–398
Hydrogel, 53, 404–405
   bonding, 52
Hydrolysable bonds, 270
Hydrophilic polyionic carbohydrates, 404–405
Hydroxyethyl cellulose (HEC), 59
Hydroxyls (OH), 241
Hypoalbuminemia, 288–289

I

IGF. *See* Insulin-like growth factor (IGF)
IgG. *See* Immunoglobulin G (IgG)

Immune cell functions, inhibition of, 376
Immune response to xenogenic transplantation, 230–231
Immune tolerance, induction of, 376
Immunogenicity, 24
Immunoglobulin G (IgG), 332–333
Implantation, skin wound creation and, 142
Implanted tissues, 257
In vitro biocompatibility
  of acellular aorta, 41
  testing of BAMG, 38–39
In vitro calcification, 66
In vitro cells, 283–284
  cytotoxicity, 255–256
In vitro cytotoxicity testing, 353
In vitro enzymatic degradation, 247–249
  in vitro collagenase enzymatic degradation, 247
  in vitro elastase enzymatic degradation, 247–249
  in vitro trypsin enzymatic degradation, 249
In vitro evaluation
  of bubaline acellular small intestinal matrix, 41
  of decellularized naturally derived biomaterials, 38–41
In vitro release of catechin, 410
In vivo biocompatibility determination of acellular aortic matrix, 220–232
In vivo evaluation in rabbit model, 256–263
In vivo trial of nanopolymer-coated catechin in rat, 412–413
Incubation conditions, 370
Indian Institute of Technology (IIT), 351
Indirect enzyme-linked immunosorbent assay, 293–294, 320–321
Indoleamine 2,3-dioxygenase (IDO), 376
Induced pluripotent stem cells (iPSCs), 377
Inflammation, 336
Inflammatory cells, 295–296
Inflammatory environment, modulation of, 376
Inflammatory reaction, 262–263
Infrared light (IR light), 30
Inhibition of immune cell functions, 376
Injured mammalian skin, 339–340
Innate immune response interaction, 375
Institute Animal Ethics Committee, 328
Institute Animal Ethics Committee of Indian Veterinary Research Institute, 142
Institute Committee for Purpose of Control and Supervision of Experiments on Animals (ICPCSEA), 412
Insulin-like growth factor (IGF), 346
  IGF-1, 355, 374
  preparation of, 355
Internal gelation on drug delivery system, difference in, 405
Internal ionotropic method, 404–405
International Society for Cellular Therapy (ISCT), 366
Ionic detergents, 135–137, 311–312
  treatment with, 84–85
Ionic interactions, 52
Ionotropic gelation, 403–405
  difference in internal and external gelation on drug delivery system, 405
  methods, 404–405
iPSCs. *See* Induced pluripotent stem cells (iPSCs)
IR light. *See* Infrared light (IR light)
Irritation testing, 11
ISCT. *See* International Society for Cellular Therapy (ISCT)
Isocyanates, 51
Isolation
  of bone marrow–derived mesenchymal stem cells, 312–313
  culture and expansion of bone marrow-derived mesenchymal stem cell, 346–348
    stem cell collection, 346
    stem cell culture, 347
    stem cell expansion, 347–348
    stem cell isolation, 346–347
  of mesenchymal stem cells, 366–367
    adipose tissue–derived mesenchymal stem cells, 367
    bone marrow–derived mesenchymal stem cells, 367
    tissue source, 367
    umbilical cord blood–derived mesenchymal stem cells, 367
  of p-MEF, 290

J

Joint replacements, 5

K

Karyotyping, 29
KDP. *See* Potassium dihydrogen phosphate (KDP)
Kidney, 286–287
Knee Injury and Osteoarthritis Outcome Scores (KOOS), 377
KOOS. *See* Knee Injury and Osteoarthritis Outcome Scores (KOOS)

L

Labeo rohita. *See* Fresh water fish (*Labeo rohita*); Rohu fish (*Labeo rohita*)
Laboratory Animal Resources (LAR), 355 section, 412
Laboratory Animals Resource, 328
Lamina propria, 80, 105–106, 108–109, 133–135
LAR. *See* Laboratory Animal Resources (LAR)
Lipid peroxidation assay (LPO assay), 415
Lipid peroxides level, 415
Liquid nitrogen, 281
Liver, 287, 388–389
  assays of oxidative stress markers in liver tissue sample, 414
  diseases, 378
  gross pathology of, 416
  May-Grünwald Giemsa staining of liver tissue homogenate, 416–417
  preparation of liver homogenates, 414
  tissues, 416–417
Logwood tree (*Haematoxylum campechianum*), 27
LPA. *See* Lymphocyte proliferation assay (LPA)
LPO assay. *See* Lipid peroxidation assay (LPO assay)
Lung, 287
Lymphocyte proliferation assay (LPA), 123, 149, 225–231, 321–323, 335–336
  ELISA, 229–231
  peripheral blood lymphocytes, 225–227
  preparation of antigen, 225
  splenocytes culture, 227–229
Lymphocyte proliferation evaluation, 294–295, 334–336
Lymphocytes, 259

M

Macro-scale measurement, 33
Macrophages, 224–225, 259, 376
Macroscopic observations, 257–258
Major histocompatibility complex (MHC), 123, 228, 270–271, 323, 335–336
Manuronic acid, 398–399
Masson's trichrome stain (MTS), 168–169, 215, 271–272
Matrix metalloproteinases (MMPs), 34, 40
Mature neurons, 374–375
May-Grünwald Giemsa staining of liver tissue homogenate, 416–417
Medical devices, 10, 19
Medical implant, 10, 48
Mesenchymal stem cells (MSCs), 318–319, 323, 346, 365–366, 370–375, 377–378. *See also* Bone marrow–derived mesenchymal stem cells (BMSCs)
  attachment and growth of, 314–316
  centrifugal seeding of MSC in scaffold, 354
  characterization of, 370–377
    cell-surface marker expression, 370–371
    differentiation potential, 371–375
  in disease treatment, 377–378
  history of, 366
  immunomodulatory properties of, 375–377
    adaptive immune response modulation, 375
    cytokine and chemokine regulation, 375
    immunosuppressive factors, 376
    induction of immune tolerance, 376
    inhibition of immune cell functions, 376
    innate immune response interaction, 375
    microenvironment modulation, 376
    modulation of dendritic cells, 376
    modulation of inflammatory environment, 376
    role in autoimmune diseases, 377
  isolation and culture of, 366–370
    culture conditions, 367–370
    isolation methods, 367
  MSC-derived exosomes, 376

Mesenchymal stem cells (MSCs) (*Continued*)
  MSC-loaded hydrogels, 378
Mesengenesis, 374
Metals, 5, 10
  chelating activity assay, 411–412
  ions, 390–391
MHC. *See* Major histocompatibility complex (MHC)
Micro-scale measurement, 33
Microbial TGase (MTGase), 67
Microencapsulation techniques, 403
Microenvironment modulation, 376
Microglia, 345–346
Microscopic observations, 258–263
Mitochondrial dehydrogenases, 350
MMPs. *See* Matrix metalloproteinases (MMPs)
Moisture content analysis, 250–252
Molecular weight analysis, 231–232, 252–254
Mononuclear cells, 346–347, 367
Mouse embryonic fibroblast cells, 35–36
MSCs. *See* Mesenchymal stem cells (MSCs)
MTGase. *See* Microbial TGase (MTGase)
MTS. *See* Masson's trichrome stain (MTS)
Mucopolysaccharides, 241
Multiwalled carbon nanotubes (MWCNT), 351
MWCNT. *See* Multiwalled carbon nanotubes (MWCNT)
Myoblasts, 374
Myocardial injury, 378
Myogenesis, 374
Myotubes, 374

N
Nanobiomaterial construct
  stem cell loading on, 353–354
    centrifugal seeding of MSC in scaffold, 354
    creation of fibronectin attachment sites, 353
    percentage of cell seeding assessment, 354
Nanocapsules (NCs), 396
Nanocarrier systems, 387
Nanocomposite scaffold, 353

Nanodrug delivery system (NDDS), 389–390, 401–402
Nanoparticles (NPs), 5, 387–388, 394–395, 407–408
  drug delivery, 395
Nanopolymer, encapsulation of, 403
Nanopolymer-coated catechin
  ameliorative application, for rodent hepatic regeneration, 412–417
    assays lipid peroxidation assay, 415
    assays of oxidative stress markers in liver tissue sample, 414
    catalase assay, 415–416
    clinical signs and body weight, 414
    glutathione assay, 415
    gross pathology of liver, 416
    hematological profile, 414
    histopathological study, 416
    May-Grünwald Giemsa staining of liver tissue homogenate, 416–417
    preparation of liver homogenates, 414
    serum biochemical profile, 414
    superoxide dismutase assay, 415
    in vivo trial of nanopolymer-coated catechin in rat, 412–413
  evaluation of in vitro antioxidant property of, 410–412
    determination of free radical scavenging activity by 1,1-diphenyl-2-picrylhydrazyl assay, 411
    ferric-reducing antioxidant power assay, 410–411
    metal chelating activity assay, 411–412
Nanospheres (NSs), 396
Nanostructurization, biomedical excipients for, 397–402
Nanosystem, 387
Nanotechnological process, 396
Nanotechnology, 393–394
National Institutes of Health, 328
Native aortic matrix graft, 222–223
Native aortic samples, 217
Native bovine reticulum, 108–109
Native bovine rumen, 80
Native buffalo gallbladder, 164
Native buffalo omasum, 135
Native buffalo pericardium, 249
  crosslinking of native buffalo pericardium matrix, 245–256

Native diaphragm (ND), 271−272, 275−276
Native ECMs, 17−18
Native gall bladder tissue, 177
Native goat pericardium matrix, crosslinking of, 243−244
Native pericardium, 254, 258
Natural antioxidant, 390
Natural biomaterials, 7, 269−270, 308
Natural bioscaffolds, 270
Natural killer cells (NK cells), 376
  activity, 376
Natural materials, 5, 10
Natural polymers, 392
Natural rubber, 52
Naturally derived biomaterials, 3
NBF. See Neutral buffered formalin (NBF)
NCs. See Nanocapsules (NCs)
ND. See Native diaphragm (ND)
NDD. See Novel drug delivery (NDD)
NDDS. See Nanodrug delivery system (NDDS); Novel drug delivery system (NDDS)
Neovascularisation, 325−326, 336
Nerve, scanning electron microscopic evaluation of, 361−362
Neural scaffold
  degradation test, 353
  electrical property of, 352
  surface roughness of, 352
Neurological evaluation, 356−359
  sciatic function index, 356−358
  toe out angle, 358−359
Neuropillin-1 (NRP-1), 360
Neuropillin-2 (NRP-2), 360
Neutral buffered formalin (NBF), 96, 114−115, 138−139
  solution, 114
New Zealand white rabbits (*Oryctolagus cuniculus*), 328
Ninhydrin, 286
  assay, 249−250
NK cells. See Natural killer cells (NK cells)
Nonionic detergent, treatment with, 83−86
Nonionic surfactants, 214
Normal saline solution (NSS), 38−39, 246, 258
Normal tissue, 269−270
Novel drug delivery (NDD), 403

Novel drug delivery system (NDDS), 389−390, 393−397
NPs. See Nanoparticles (NPs)
NRP-1. See Neuropillin-1 (NRP-1)
NRP-2. See Neuropillin-2 (NRP-2)
NSs. See Nanospheres (NSs)
NSS. See Normal saline solution (NSS)
Nuclear transfer, 308−309
Nuclear transplantation, 308−309
Nucleated cells, 347
Nucleic acids, 53

O
OA. See Osteoarthritis (OA)
OFM. See Ovine forestomach matrix (OFM)
Oil red O staining, 351
Oligodendrocytes, 345−346
Omasum, 77
  omasum-derived extracellular matrix scaffolds
    experimental evaluation of acellular buffalo omasum, 141−154
    preparation of acellular matrix from buffalo omasum, 132−138
    preparation of acellular matrix from goat omasum, 138−141
Open wound, 151−152
Organs, 19−20, 286
  collection of, 286
Oryctolagus cuniculus. See New Zealand white rabbits (*Oryctolagus cuniculus*)
Osmotic process, 78
Osteoarthritis (OA), 377
Osteoblasts, 373
Osteogenesis, 373
Osteogenic differentiation, 351
Ovine forestomach matrix (OFM), 78
Oxidative stress, 388−389
  assays of oxidative stress markers in liver tissue sample, 414

P
p-MEF. See Primary mouse embryo fibroblasts (p-MEF)
p-MEF cells, 290
  seeding of, 290
PAA. See Peracetic acid (PAA)
Packed cell volume (PCV), 414
Pancreatic islets, 374

Panniculus carnosus, 117−118, 142, 178−179
Papillae, 77−78
Parietal layer, 241
Parkinson's disease, 374−375
Passaging, 370
PBMC. *See* Peripheral blood mononuclear cell (PBMC)
PBS. *See* Phosphate buffer saline (PBS)
PCCNP. *See* Polymer-coated catechin nanoparticle (PCCNP)
PCL. *See* Polycaprolactone (PCL)
PCNP. *See* Pectin-coated catechin nanoparticles (PCNP)
PCV. *See* Packed cell volume (PCV)
PDGF. *See* Platelet-derived growth factor (PDGF)
Pectin, 388, 400−402
　pectin-based drug delivery system, 401−402
　pectin-based nanoparticle, 401−402
　pectin-coated catechin nanospheres, 406
Pectin-coated catechin nanoparticles (PCNP), 406
PEG. *See* Polyethylene glycol (PEG)
Penicillin-streptomycin (PS), 346
Peracetic acid (PAA), 21−22
Pericardium, 241, 246
　pericardium-derived extracellular matrix scaffolds
　　crosslinking of native and acellular buffalo pericardium matrix, 245−256
　　crosslinking of native and acellular goat pericardium matrix, 243−244
　　preparation of acellular buffalo pericardium matrix, 244−245
　　preparation of acellular goat pericardium matrix, 243
　　in vivo evaluation in rabbit model, 256−263
Perioperative antibiotic prophylaxis, 329−330
Peripheral blood lymphocytes, 225−227
Peripheral blood mononuclear cell (PBMC), 255−256
Peripheral nerve injury (PNI), 345
Peripheral nervous system (PNS), 345
PG. *See* Pig gallbladder (PG)
PGE2. *See* Prostaglandin E2 (PGE2)

PHA. *See* Phytohemagglutinin (PHA)
Phosphate, 373
Phosphate buffer saline (PBS), 66, 79, 132, 162−163, 172−175, 215, 217, 243−244, 271, 278, 309, 313, 328, 354−355, 410
　solution, 271
Physical crosslinking, 51−53, 66−67
　advantages of, 52−53
　disadvantages of, 53
Physical decellularization methods, 19
Physical methods of decellularization, 21−22
Phytohemagglutinin (PHA), 122, 147, 182−183, 193−194, 294, 321
Pig cholecyst-derived extracellular matrix in rat model
　experimental evaluation of, 190−198
　　cell-mediated immune response, 193−194
　　gross observations/planimetry, 191−193
　　histological observations, 194−198
　　humoral response, 193
　　immunological observations, 193−194
　　skin wound creation and implantation, 190
　　wound contraction, 190−191
Pig diaphragm, biocompatibility evaluation of, 283−285
Pig gallbladder (PG), 168
　tissue, 169
Planimetry, 144−145, 180
Platelet-derived growth factor (PDGF), 367
PMA. *See* Premarket approval (PMA)
PMMA. *See* Polymethyl methacrylate (PMMA)
PNI. *See* Peripheral nerve injury (PNI)
PNPs. *See* Polymeric nanoparticles (PNPs)
PNS. *See* Peripheral nervous system (PNS)
Polarization, 30
Poly amino acids, 241
Polyacrylamide gel, 252
Polycaprolactone (PCL), 351
　scaffolds, 346
Polyelectrolyte complexation, 403−405
Polyethylene glycol (PEG), 38−39, 55−57, 63−64
　grafting of bovine pericardium, 63

PEG-grafted tissues, 63–64
Polymer-coated catechin nanoparticle (PCCNP), 412
Polymeric materials, 13–14
Polymeric nanoencapsulation
  ameliorative application of nanopolymer-coated catechin for rodent hepatic regeneration, 412–417
  catechin, 390–392
  catechin-loaded alginate polymeric nanoparticles, 405–406
  catechin-loaded chitosan polymeric nanoparticles, 407–408
  catechin-loaded pectin polymeric nanoparticles, 406
  characterization and evaluation of polymeric nanoparticles, 408–410
  encapsulation of nanopolymer, 403
  evaluation of in vitro antioxidant property of nanopolymer-coated catechin, 410–412
  excipients, 393
  ionotropic gelation, 403–405
  novel drug delivery systems, 393–397
  polymers, 397–402
Polymeric nanoparticles (PNPs), 395, 397–398
  characterization and evaluation of, 408–410
    drug loading, 409
    encapsulation efficiency, 409
    particle size and shape, 408–409
    in vitro release of catechin, 410
Polymers, 5, 10, 397–402
  alginate, 398–400
  chitosan, 402
  pectin, 400–402
Polymethyl methacrylate (PMMA), 5
Polymorphonuclear leukocytes, 224–225
Polypeptides, 241
Polysaccharides, 387–388, 398–399
  polymers, 388
Polyurethane, 10
Polyvinyl alcohol (PVA), 52
Porcine dermal collagen, 66
Porcine diaphragm, 283
Porcine small intestinal submucosa, 38
Porous matrices, 65

Potassium dihydrogen phosphate (KDP), 29–30
Powdered tissue, 280–281
Preadipocytes, 372
Premarket approval (PMA), 7
Preosteoblasts, 373
Primary mouse embryo fibroblasts (p-MEF), 289–290
  dermal wound healing using p-MEF seeded buffalo acellular diaphragm matrix in rat model, 289–298
    histologic observations, 295–298
    immunologic observations, 293–295
    indirect ELISA, 293–294
    isolation, culture, and seeding of p-MEF over ADM, 290
    lymphocyte proliferation evaluation, 294–295
    materials and methods (V), 289–290
    wound creation and implantation, 290–291
    wound healing evaluation, 291–292
Procaine, 241–242
Propria-submucosa, 105–106
Prostaglandin E2 (PGE2), 376
Protein(s), 26–27, 53, 387–388
  concentration, 288–289
  content analysis, 287–288
  protein–protein interaction mapping, 54
Proteoglycans, 241
Pseudo ruminants, 76
Pulmonary diseases, 378
PVA. See Polyvinyl alcohol (PVA)
Pyrogenicity testing, 11

R

R-BMSC. See Rat bone marrow–derived mesenchymal stem cells (R-BMSC)
r-MSC. See Rabbit mesenchymal stem cells (r-MSC)
Rabbit mesenchymal stem cells (r-MSC), 38
Rabbits, 327–328
  biochemical changes in rabbit organs after subcutaneous implantation of bovine diaphragm, 285–289
  model
    evaluation of acellular and crosslinked acellular swim bladder matrix for skin wound healing in, 327–340

Rabbits (*Continued*)
　macroscopic observations, 257–258
　microscopic observations, 258–263
　in vivo evaluation in, 256–263
　subcutaneous implantation of buffalo diaphragm in, 286
Rat
　bone marrow–derived mesenchymal stem cells from, 312–313
　model
　　dermal wound healing using p-MEF seeded buffalo acellular diaphragm matrix in, 289–298
　　evaluation of bioengineered fish swim bladder matrices for skin wound healing in, 316–327
　　evaluation of bovine reticulum extracellular matrix in, 117–127
　　experimental evaluation of acellular buffalo omasum in wound healing in, 141–154
　　experimental evaluation of buffalo cholecyst-derived extracellular matrix in, 178–189
　　experimental evaluation of pig cholecyst-derived extracellular matrix in, 190–198
　in vivo trial of nanopolymer-coated catechin in, 412–413
Rat bone marrow-derived mesenchymal stem cells (rMSCs), 87–89
Rat bone marrow–derived mesenchymal stem cells (R-BMSC), 314
Reactive oxygen species (ROS), 390
Real Time PCR
　CD marker-based expression via, 349
　relative expression of different genes by, 360–361
Reconstructive surgery, 101–102, 203–204, 255–256
Regenerative medicine, 6, 19, 48
Regulatory T cells (Tregs), 375–376
Reticulum, 77
　of buffalo, 107
　of caprine, 114
　reticulum-derived extracellular matrix scaffolds
　　evaluation of bovine reticulum extracellular matrix in rat model, 117–127
　　preparation of acellular bovine reticulum extracellular matrix, 107–113
　　preparation of acellular caprine reticulum extracellular matrix, 114–117
Reversible crosslinks, 52
Ritha fruits (*Sapindus mukorossi*), 172–173
rMSCs. *See* Rat bone marrow-derived mesenchymal stem cells (rMSCs)
RNase, 26, 285
Rodent hepatic regeneration, ameliorative application of nanopolymer-coated catechin for, 412–417
Rohu fish (*Labeo rohita*), 328
　collagen, 333–334
ROS. *See* Reactive oxygen species (ROS)
Rosewell Park Memorial Institute (RPMI), 334–335
RPMI. *See* Rosewell Park Memorial Institute (RPMI)
Rumen, 77
Rumen-derived extracellular matrix scaffolds and clinical application
　anatomy of ruminant stomach, 77–78
　decellularization of forestomach matrix, 78–79
　development of 3-D bioengineered scaffolds from bubaline rumen, 89
　evaluation of bubaline rumen matrix in clinical cases, 93–95
　gross and clinical observations, 94–95
　surgical technique, 94
　evaluation of caprine rumen matrix in clinical cases, 100–102
　preparation of acellular matrices from bubaline rumen, 79–88
　preparation of acellular matrices from caprine rumen, 95–99
　testing efficacy of 3-D bioengineered scaffolds in diabetic rat model, 89–93
Rumeno-reticulum, 77
Ruminant, 75
　anatomy of ruminant stomach, 77–78

S

SB-16. *See* Sulfobetaine-16 (SB-16)
SB10. *See* Sulfobetaine-10 (SB10)
Scaffold
   centrifugal seeding of MSC in, 354
   preparation and characterization, 351–353
      electrical property of neural scaffold, 352
      neural scaffold degradation test, 353
      surface roughness of neural scaffold, 352
   requirements for scaffolds used in tissue engineering, 7
   scaffold-specific antibodies, 193
Scanning electron microscope (SEM), 26, 31, 173–175, 177, 272–273
   evaluation of nerve, 361–362
   examination, 290, 310, 314, 316
   observations, 99, 110–112, 217
Scanning electron microscopy, 314, 361–362
Sciatic function index, 356–358
Sciatic nerve, 360
   preparation of sciatic nerve injury model, 355–362
      biochemical parameters, 359–360
      neurological evaluation, 356–359
      relative expression of different genes by real time PCR, 360–361
      scanning electron microscopic evaluation of nerve, 361–362
SDC. *See* Sodium deoxylate (SDC)
SDS. *See* Sodium dodecyl sulfate (SDS)
SDS-PAGE. *See* Sodium dodecyl sulfate polyacrylamide gel electrophoresis (SDS-PAGE)
Second harmonic generation (SHG), 26, 29–30
   imaging, 30
Seeding of primary mouse embryo fibroblasts, 290
Seeding of stem cells on acellular fish swim bladder matrix, 313–314
SEM. *See* Scanning electron microscope (SEM)
Sensitization testing, 11
Serum
   serum-free media, 367
   supplementation, 367

Serum biochemical profile, 414
Serum glutamate pyruvate transaminase activity (SGPT activity), 414
Serum glutamic oxaloacetic transaminase activity (SGOT activity), 414
SFB. *See* Society for Biomaterials (SFB)
SFI, 358
SGF. *See* Simulated gastric fluid (SGF)
SGOT activity. *See* Serum glutamic oxaloacetic transaminase activity (SGOT activity)
SGPT activity. *See* Serum glutamate pyruvate transaminase activity (SGPT activity)
SHG. *See* Second harmonic generation (SHG)
SI. *See* Stimulation index (SI)
SIF. *See* Simulated intestinal fluid (SIF)
Silicone, 5, 10
   hydrogel, 5
Simulated gastric fluid (SGF), 410
Simulated intestinal fluid (SIF), 410
SIS. *See* Small intestinal submucosa (SIS)
Skin
   defects, 270–271
   evaluation of acellular and crosslinked acellular swim bladder matrix for skin wound healing in rabbit model, 327–340
      enzyme-linked immunosorbent assay, 332–334
      gross observations and planimetry, 330–331
      histological observations, 336–340
      immunological observations, 332–336
      lymphocyte proliferation evaluation, 334–336
      materials and methods, 328
      preparation of acellular swim bladder matrix and crosslinking with epoxy compounds, 328
      wound area and contraction, 332
      wound creation and implantation, 329–330
   evaluation of bioengineered fish swim bladder matrices for skin wound healing in rat model, 316–327
   flaps, 270–271

Skin (*Continued*)
  wound creation and implantation, 142, 178–179, 190
Small intestinal submucosa (SIS), 161–162
SMCs. *See* Smooth muscle cells (SMCs)
Smooth muscle cells (SMCs), 216
Soapnut pericarp extract, preparation of, 172–173
Society for Biomaterials (SFB), 4
Sodium acetate, 411
Sodium alginate, 405–406
Sodium deoxylate (SDC), 24, 95–96, 107–108, 138–139, 289–290, 311–312
Sodium dodecyl sulfate (SDS), 25, 79, 84–85, 95–96, 114, 138–139, 164, 211, 215, 278
  detergents, 206
  SDS-treated diaphragm, 271–272
    tissues, 271–272
  solution, 271
  treatment, 25, 214–215
Sodium dodecyl sulfate polyacrylamide gel electrophoresis (SDS-PAGE), 34, 98, 137–138, 231, 252
  analysis, 117, 140–141
    for native buffalo omasum laminae, 138
Sodium tripolyphosphate, 407
Sonication, 19
Spleenocytes culture, 227–229, 294
Squamous epithelium, 76
Stainless steel, 10
Stem cells, 308–309
  collection, 346
  culture, 347
  expansion, 347–348
  isolation, 346–347
  loading MWCNT-based bioactive scaffold CD marker-based expression via RT PCR, 349
    cell viability assay, 349–350
    colony forming assays, 350
    diamidino-2-phenylindole staining, 354–355
    isolation culture and expansion of bone marrow-derived mesenchymal stem cell, 346–348
    preparation of IGF-I, 355
    preparation of sciatic nerve injury model, 355–362
    preparation of stem cell loaded MWCNT-based bioactive nanoneural construct and assessment, 353–355
    scaffold preparation and characterization, 351–353
    stem cell characterization, 349
    stem cell loading on nanobiomaterial construct, 353–354
    tri-lineage staining characterization, 350–351
  in vitro cytotoxicity testing, 353
  research, 365
  seeding of stem cells on acellular fish swim bladder matrix, 313–314
  stem cell-seeded composite random polymeric neural scaffold, 354–355
Stents, 10
Sterile plastic template, 117–118
Stimulation index (SI), 122, 147, 183, 193–194, 225–227, 255, 321, 334–335
Stomach submucosa compositions, 77
Streptomyces mobaraensis, 67
Stro-1 marker, 371
Stromal vascular fraction (SVF), 367
Subcutaneous implant studies in rats, 62
Sulfobetaine-10 (SB10), 21–22
Sulfobetaine-16 (SB-16), 21–22
Superoxide dismutase assay, 415
Surface epithelialization, 152–153
Surgical instruments, 10
Surgical technique, 94
SVF. *See* Stromal vascular fraction (SVF)
Swim bladder, 327–328
  tissues, 327–328
Synthetic biomaterials, 3, 7
Systemic toxicity testing, 11

T
T lymphocytes, 57–58
T-cell
  responses, 321
  subsets, 229, 262, 335–336
TBA. *See* Thiobarbituric acid (TBA)
TBST. *See* Tris buffered saline Tween-20 (TBST)

Index

TCA. *See* Tricarboxylic acid (TCA);
    Trichloro acetic acid (TCA)
TE. *See* Tissue engineering (TE)
Tea (*Camellia sinensis*), 390, 392
TEC. *See* Total erythrocyte count (TEC)
TEMs. *See* Transmission electron
    microscopes (TEMs)
TGase. *See* Transglutaminase (TGase)
TGF-β. *See* Transforming growth factor beta
    (TGF-β)
Th2 lymphocytes, 323
Theophylline (TP), 399–400
Therapeutic cells, 366
Therapeutic cloning, 308–309
Thiobarbituric acid (TBA), 415
Three dimension (3D)
    development of 3-D bioengineered
        scaffolds from bubaline rumen, 89
    testing efficacy of 3-D bioengineered
        scaffolds in diabetic rat model,
        89–93
Thy1. *See* CD90 marker
Timolol maleate, 401–402
Tissue, 243, 278, 283
    biopsies, 336
    engineered blood vessel, 205
    homogenates, 415
    homogenization, 99, 116, 140
    regeneration, 257–258
    source, 367
Tissue engineering (TE), 6, 19, 48, 89,
    270–271, 307–309
    requirements for scaffolds used in, 7
    scaffolds, 10
Titanium, 5, 10
TLC. *See* Total leukocyte count (TLC)
TNBP. *See* Tri-n-butyl phosphate (TNBP)
Toe out angle, 358–359
Total erythrocyte count (TEC), 414
Total leukocyte count (TLC), 121–122, 414
TP. *See* Theophylline (TP)
TPP. *See* Tripolyphosphate (TPP)
Tracheal segment replacement, 234
Traditional lineages, 371–375
    adipogenesis:adipogenesis, 371–372
    chondrogenesis, 372–373
    ectodermal derivatives, 374
    endodermal derivatives, 374
    mature neurons, 374–375

mesengenesis, 374
myogenesis, 374
osteogenesis, 373
Transdermal patches, 10
Transforming growth factor beta (TGF-β),
    204
Transglutaminase (TGase), 67
Transmission electron microscopes (TEMs),
    31
Tregs. *See* Regulatory T cells (Tregs)
Tri (n-butyl) phosphate, 107–108
Tri-lineage staining characterization,
    350–351
    alcian blue staining, 351
    alizarin red staining, 351
    oil red o staining, 351
Tri-n-butyl phosphate (TNBP), 79, 81–83
Tricarboxylic acid (TCA), 415
Trichloro acetic acid (TCA), 415
Triglycerides, 372
Tripolyphosphate (TPP), 407
Tris buffered saline Tween-20 (TBST), 176
Tris cocodylic acid, 415
Triton X-100, 24, 79, 83–84, 107–109, 134,
    206
Trypan blue exclusion method, 349
Trypsin, 24, 86–87, 107–108
Tumor necrosis factor-inducible gene 6
    protein (TSG-6 protein), 376
Tunica muscularis, 77
Tween-20, 85–86
Two-way ANOVA, 317

U

Ultraviolet light (UV light), 30
Umbilical cord–derived mesenchymal stem
    cells (UC-MSCs), 367, 377
Umbilical hernia, 94–95, 232–234
    repair, 234–235
United States Food and Drug Administration
    (FDA), 7, 11
Unmodified natural materials, 205–206
Urinary bladder, 25
UV light. *See* Ultraviolet light (UV light)

V

Van der Waals forces, 52
Vascular endothelial cell growth factor
    (VEGF), 204

Vascular stem cells (VSCs), 371
VEGF. *See* Vascular endothelial cell growth factor (VEGF)
Vimentin, 88, 135–137
Visceral layer, 241
VSCs. *See* Vascular stem cells (VSCs)

W

Water buffalo (*Bubalus bubalis*), 79
Water molecules, 241
Water vapor transmission rate (WVTR), 161
Weigert's resorcin fuchsin (WRF), 215
Wharton's Jelly–derived MSCs (WJ-MSCs), 377
White connective tissue, 222
WJ-MSCs. *See* Wharton's Jelly–derived MSCs (WJ-MSCs)
Wound
  with acellular buffalo omasum laminae, 152–154
  area, 142–144, 317, 332
    and percent contraction, 118–119
  with commercially available collagen sheet, 152
  contraction, 119, 142–144, 179, 190–191, 332
  contracture, 317–319
  creation and implantation, 290–291, 329–330
  edges, 320
  experimental evaluation of acellular buffalo omasum in wound healing, 141–154
  healing, 48, 131
    evaluation of, 317
    rate, 119, 144, 179, 318–319
  healing evaluation, 291–292
    macroscopic observation and planimetry, 291–292
    wound area and contraction, 292
WRF. *See* Weigert's resorcin fuchsin (WRF)
WVTR. *See* Water vapor transmission rate (WVTR)

X

Xenogeneic collagen-rich ECM, 339–340
Xenogenic acellular scaffolds, 22–23
Xenogenic biomaterials, 282
Xenogenic bubaline aortic matrix, 235–236
Xenogenic tissue, 224, 277–278
Xenografts, 20–21
Xylazine, 220

Z

Zirconia, 5, 10
Zwitter ionic detergent, treatment with, 81–83
Zymography, 33–34

Printed and bound by CPI Group (UK) Ltd, Croydon, CR0 4YY
31/08/2024
01030803-0001